Policy-Based Network Management

The Morgan Kaufmann Series in Networking
Series Editor, David Clark, M.I.T.

Policy-Based Network Management: Solutions for the Next Generation
John Strassner

Computer Networks: A Systems Approach, 3e
Larry L. Peterson and Bruce S. Davie

Network Architecture, Analysis, and Design, 2e
James D. McCabe

MPLS Network Management: MIBs, Tools, and Techniques
Thomas D. Nadeau

Developing IP-Based Services: Solutions for Service Providers and Vendors
Monique Morrow and Kateel Vijayananda

Telecommunications Law in the Internet Age
Sharon K. Black

Optical Networks: A Practical Perspective, 2e
Rajiv Ramaswami and Kumar N. Sivarajan

Internet QoS: Architectures and Mechanisms
Zheng Wang

TCP/IP Sockets in Java: Practical Guide for Programmers
Michael J. Donahoo and Kenneth L. Calvert

TCP/IP Sockets in C: Practical Guide for Programmers
Kenneth L. Calvert and Michael J. Donahoo

Multicast Communication: Protocols, Programming, and Applications
Ralph Wittmann and Martina Zitterbart

MPLS: Technology and Applications
Bruce Davie and Yakov Rekhter

High-Performance Communication Networks, 2e
Jean Walrand and Pravin Varaiya

Internetworking Multimedia
Jon Crowcroft, Mark Handley, and Ian Wakeman

Understanding Networked Applications: A First Course
David G. Messerschmitt

Integrated Management of Networked Systems: Concepts, Architectures, and their Operational Application
Heinz-Gerd Hegering, Sebastian Abeck, and Bernhard Neumair

Virtual Private Networks: Making the Right Connection
Dennis Fowler

Networked Applications: A Guide to the New Computing Infrastructure
David G. Messerschmitt

Modern Cable Television Technology: Video, Voice, and Data Communications
Walter Ciciora, James Farmer, and David Large

Switching in IP Networks: IP Switching, Tag Switching, and Related Technologies
Bruce S. Davie, Paul Doolan, and Yakov Rekhter

Wide Area Network Design: Concepts and Tools for Optimization
Robert S. Cahn

Frame Relay Applications: Business and Technology Case Studies
James P. Cavanagh

For further information on these books and for a list of forthcoming titles, please visit our website at http://www.mkp.com

POLICY-BASED NETWORK MANAGEMENT

Solutions for the Next Generation

John Strassner
Intelliden Corporation
Colorado Springs, Colorado

AMSTERDAM • BOSTON • HEIDELBERG • LONDON
NEW YORK • OXFORD • PARIS • SAN DIEGO
SAN FRANCISCO • SINGAPORE • SYDNEY • TOKYO
Morgan Kaufmann Publishers is an imprint of Elsevier

Senior Editor: *Rick Adams*
Publishing Services Manager: *Simon Crump*
Production Editor: *Julio Esperas/Troy Lilly*
Developmental Editor: *Karyn Johnson*
Cover Design: *Eric DeCicco*
Cover Image: *Getty Images*
Full-Service Provider: *Kolam USA*
Printer: *The Maple-Vail Book Manufacturing Group*

Morgan Kaufmann Publishers
An imprint of Elsevier Science
500 Sansome St., Suite 400
San Francisco, CA 94111
www.mkp.com

Printed in the United States of America

08 07 06 05 04 5 4 3 2 1
Library of Congress Control Number: 2003107477
ISBN: 1-55860-859-1

This book is printed on acid-free paper.

Contents

Preface

The world has always held out great hope for policy-based network management (PBNM). When PBNM first hit the mainstream network arena, the expectations, as well as the hype, were overwhelming. Companies rushed to build products that used policy management mechanisms, standard bodies considered its impact on the Internet in general and in managing networks in particular, and conferences sprung up devoted to PBNM.

Then reality set in. People realized that PBNM was *hard*. There was no simple, easy method to solve complex management problems. More people, using more sophisticated applications, were using the network. These applications often had conflicting requirements of the network. This led to the "*policy is a means of applying QoS*" mantra. In other words, despite the various meanings that QoS has to different people, PBNM was going to be the way to implement QoS.

Unfortunately, this meant that the capabilities of PBNM were not fully realized. Many people concluded that PBNM was an academic research project and couldn't be used in "real world" network management situations. Yet, policies surround us, not just in networking, but in other disciplines as well. In networking, the simple act of requiring someone of a specific skill set to approve a configuration change to a device is, of course, a policy. The desire of Service Providers to offer differentiated, value-added services, where different users could pay for different service levels, is another form of policy. Even implementing a script that backs up information to a server is a form of policy.

The answer to realizing the true potential of PBNM lies in the realization that PBNM is not "just" a way to manage QoS, nor is it "just" a way to determine who can perform what function. Rather, PBNM is a way to *define business needs and ensure that the network provides the services that its clients require.*

This is indeed a powerful statement. Imagine a world in which the network is no longer a "fat dumb pipe" but is instead an "intelligent provider of services." The recent flurry of attention to on-demand computing, which espouses the simple but compelling theory that an on-demand business is one that can instantly respond to changing market needs and conditions in real time, cannot be realized without policy management techniques.

This book explains exactly what PBNM is. It debunks common misconceptions, explains how policies are used in a PBNM system, discusses the architectural principles governing PBNM systems, and describes several policy models. One of these models, the DEN-ng policy model, is used as the cornerstone for the discussion of PBNM, because it was explicitly constructed to bind information and data models to management architectures.

DEN-ng is a set of information models, of varying levels of abstraction, that are each oriented towards business, system, and implementation viewpoints. The use of different viewpoints is important, because it enables PBNM to be understood by different people having different needs. A business analyst does not care about queuing algorithms—he or she just wants to define which applications get prioritized access to network resources. Similarly, a network technician wants these business requirements to be expressed in networking terminology.

Thus, the notion of a *Policy Continuum* was defined, which enabled a policy to be translated into a form that different constituencies could utilize. The key point is that this enables people from multiple organizations, roles, and responsibilities to finally be united and work together in defining and implementing a PBNM solution. It is this simple fact—enabling different people of different disciplines to work together—that sets the approach defined in this book apart from previous approaches. The result is an end-to-end prescription of how policy can be used to manage and control network services, and how architectures can be built today to realize this new PBNM approach.

Following this mantra, every attempt has been made to make this book accessible to a wide audience. Clearly, some of the chapters are oriented towards people who need to understand the technology at a low level. However, even those chapters have been sprinkled with business use cases and examples. It has been assumed that readers have a certain amount of knowledge in information modeling and networking, though background material has been provided where appropriate.

Objectives

There are several reasons that this book was written. First, it is important that PBNM is better understood, so that its benefits can finally be realized. Hence, this book provides a matter-of-fact assessment of current policy management theory, and proposes a new approach that achieves the objective of tying the business, system, and implementation worlds together. This is very important, as it enables existing and potential network operators to better evaluate vendor offerings, and hopefully will guide them in building a more efficient and scalable implementation.

Second, this book is based on an open standard: DEN-ng. This work is being done in the TeleManagement Forum (www.tmforum.org), and ties in closely with the eTOM (for business process automation) and NGOSS (defining an overall architecture). No other forum provides this combination of tools and work, which is required if the goal of weaving together the business, system and implementation viewpoints is to be achieved.

Third, there are precious few examples at any level of how policy works. It is hoped that the various use cases and examples given throughout this book give readers of all backgrounds a better, more thorough understanding of PBNM.

Intended Audience

This book was written so that it would be accessible to as wide an audience as possible. Specifically, it is oriented to four types of people:

- CxOs and other decision-makers who are trying to figure out what PBM is and if it can help them
- Application and network architects in large enterprises, telcos, and service providers who want to better understand PBM and/or learn how to customize PBNM solutions for their own application-specific needs

- Vendor strategists who want to better understand PBM, as well as to better evaluate policy servers and policy-compliant products from different vendors
- Network administrators and application developers who want to develop network-aware applications

There is an important secondary audience for this book. These are the developers of new network architectures and applications that want to take advantage of the services that the network provides seamlessly, without requiring complicated and painful application and network configuration. Such solutions are too complex to build without policy-based management.

Organization of This Book

This book is organized into chapters that fall into three basic sections: Chapters 1-4 introduce PBNM, Chapters 5-8 examine PBNM in detail, and Chapters 9-11 provide examples of PBNM.

Chapter 1 provides a brief retrospective of PBNM. This chapter identifies two fundamental problems of previous solutions—the lack of use of an information model, and the inability to integrate the business, system, and implementation viewpoints of a system. A path forward is described, which is explained in detail in the second of this book.

Chapter 2 supplies the technical foundation for discussing policy throughout the rest of this book, including a robust terminology and a simplified conceptual model that will be elaborated on as more of the DEN-ng model is introduced. The rest of this chapter focuses on using business rules to drive the construction and deployment of device and network configuration.

Chapter 3 describes a novel approach to define different, yet related, information views of the managed environment. This is built around the notion of sets of information models to describe policy and the entities that are managed by policy. More importantly, it enables *behavior* to be defined, modeled, and implemented.

Chapter 4 explains how policies are used in a PBNM system. It describes issues covering definition, organization, coordination, and distribution of policies, as well as the management of policies. This focuses on understanding how policies are applied, and the effect of applying policies to a system.

Chapter 5 introduces the Ponder system and then defines extensions of this system that are designed to enhance the structure of high-level policies and their application in a PBNM system. These policies are aimed at the business and system levels of the Policy Continuum. This chapter serves as a bridge to connect the introductory concepts presented in Chapters 2 to 4 with the advanced concepts in Chapters 6 to 8.

Chapters 6 and 7 first describe the motivation for creating a new policy model and second the DEN-ng policy model in detail. This model strives to reuse established concepts in other forums while enhancing their semantics. These chapters work together to explain how the business, system, and networking worlds can be related to each other.

Chapter 8 takes the previous concepts and maps them to a new, more robust, PBNM architecture. The proposed architectural principles are then explained, including a brief glimpse into the planned "cookbook" that explains how the novel use of information modeling and a language, coupled by using a data dictionary, support a distributed implementation of a PBNM system.

Chapter 9 focuses on how the business, system, and implementation views can be coordinated using the DEN-ng information model. The Policy Continuum serves as the basis for a set of examples focused on managing different aspects of a Multi-Protocol Label Switching–Virtual Private Network (MPLS–VPN).

Finally, Chapters 10 and 11 provide several different examples of how policy is used in a system and new directions for PBNM, respectively. These two chapters concentrate on describing the current state-of-the-art, commercial implementations, and finally provide an example of a new PBNM product from Intelliden.

An important benefit of this book is its extensive listing of resources and references. Every chapter has a brief introduction to focus the reader on the content of the chapter, and a summary of the contents of the chapter at the end. This summary is followed by a comprehensive list of references.

The Web Site

A comprehensive web site has been created to accompany this text and is available at http://www.mkp.com. Questions about the content of the book are invited to be sent to Morgan Kaufmann Publishers, which will forward them to the author.

Acknowledgments

The writing of this book would not have been possible without the help and support of many people. First, I'd like to thank all of the wonderful people of Morgan Kaufmann, especially Rick Adams and Karyn Johnson, who provided excellent support and just the right blend of patience and urgency to help complete this book.

I am greatly indebted to the technical input of my reviewers: Jeff Wheeler, Jane Butler, Nigel Warren, Kevin D'Souza, and Lisa Phifer. I would like to thank all of them for their encouragement, insight, and recommendations.

I would also like to specifically thank some of my colleagues in the modeling and policy worlds: Cliff Faurer of the TeleManagement Forum for his meticulous comments on the DEN-ng models and putting up with my eccentricities; John Reilly of MetaSolv and Dave Raymer of Motorola for their detailed comments, unique humor, and close friendship; the (in)famous Red Team of the TeleManagement Forum (myself, Cliff, Dave, Joel Fleck of HP, Jerry McCollum of Agilent, Michael Ogg of Valaran, and Giuseppe Covino of Telecom Italia, the latter of whom I dearly ask for the negatives to a certain two pictures . . .) during the time that this book was written for their early review and helpful comments on the DEN-ng Policy Model; and Mike McLoughlin, Dean Rayne, Steve Fisher, Martin Iddon, Eoin McDonnell, and Jabir Shaikh (all of British Telecom) for having the courage to try and implement my models. I would also like to thank the other core members of the TeleManagement Forum's Shared Information and Data (SID) team: John Reilly, Dave Raymer, Ram Garg of MetaSolv, Chris Hartley and Wayne Sigley of Telstra, Helen Hepburn of British Telecom, Matt Izzo of Agilent, Roger Cutts of Vodafone, and Josh Salomon of Amdocs. Your dedication to reviewing the DEN-ng inputs to the SID helped more than I can express.

Of course, people at Intelliden who supported this effort are too numerous to name. However, I would especially like to thank Mike Courtney, Jeff Schenk, Andrew Hewlett, Brendan Kelly, and Bret Taylor for translating my vision to reality, as well as Cheryl London, Andrea Wescott, Mark Vaske, and Bob Kollar Ravi for believing in the vision and, more importantly, selling it and providing a hearty cheerleading section. Mark Scherbring, Dave Thompson, John Boyd, and Steve Shuman will be supporting these efforts before this book is published and before they know it, so thanks in advance.

Finally, I'd like to thank the powers-that-be of the TeleManagement Forum (Martin Creaner, CTO; Jim Warner, President; and Keith Willets, Chairman and Founder) for having faith in the SID and Red Teams and providing unquestioning support in pursuit of NGOSS, Macallan, and Bonanza (not necessarily in that order).

Dedication

This book is dedicated to three special people. First, it is dedicated in loving memory to my Father. It is also dedicated to my Mother and, last but certainly not least, to my loving wife, Kim. Her constant support and willingness to stay home so I could think and write made finishing this book possible.

Feedback and Comments

Thank you for taking the time to read this book. If you have already purchased it, it is my sincere hope that you find it to be a useful resource wherever you deploy, manage or study PBNM. If you have questions or comments about this book, please feel free to contact me through Morgan Kaufmann at: mkp@mkp.com.

John Strassner

The Foundation of Policy Management

This chapter will first provide a brief retrospective of how PBNM has been conceived in the past. This will be used to point out two fundamental problems of previous solutions—the lack of use of an information model, and the inability to use business rules to drive configuration of devices, services, and networks. A path forward is described, and benefits resulting from this improved approach are described.

1.1 Introduction—A Retrospective

Policy management means many things to many people. As Michael Jude[1] writes "When first conceived in the late 1990s, policy-based network management (PBNM) promised enterprise information technology shops the ability to control the quality of service (QoS) experienced by networked applications and users." Jude goes on to write: "In fact, the hype went further than that: Vendors promised that CIOs or CEOs would soon be able to control policies through a simple graphical interface on their desk. Behind the scenes, those instructions would translate into specific traffic management adjustments, bypassing traditional network operations."

QoS means many things to many people. Contrary to popular belief, QoS does not mean "just" an increase or decrease in bandwidth speed. Rather, it means differentiated treatment of one or more metrics. These metrics are completely dependent on the type of application(s) that the QoS is being designed for. Thus, QoS for a voice application is usually different than QoS for a mission-critical data application, because the characteristics of each application are different. This causes the specific QoS mechanisms to be made different.

My favorite definition of QoS is "managed unfairness." This describes the differences in how network elements are programmed to provide different QoS mechanisms to treat various application traffic streams differently. Clearly, this is complex to perform for the same type of devices; the complexity of this configuration increases dramatically if different devices with different capabilities and commands are used in the same network.

Differentiated QoS, which is the ability to provide different configurations of QoS for different types of applications, is the key to opening up new avenues of revenue. Because providing QoS is currently very difficult, the application of policy to provide differentiated QoS is one of the primary drivers for implementing PBNM solutions and will be repeatedly examined throughout this book.

The emphasis on managing and implementing QoS describes some of the buildup and excitement that followed the dawn of PBNM. The reason, of course, is because networks are complex, and running different services, each of which have different requirements on the same network, is very difficult. People who were looking for a "quick fix" to their network problems were disappointed; PBNM was found to be a time intensive, complex, and expensive. There were several reasons for this:

- Most early PBNM solutions were single-vendor approaches and could only manage some of the devices on the network. As a result, multiple incompatible PBNM solutions had to be introduced to manage the entire network, which caused hard-to-solve integration problems.
- PBNM solutions were focused on particular technologies and devices. For example, a QoS policy server might be able to control most (but probably not all) of the QoS functions of a particular vendor's device or device families. However, it probably could not control other types of technologies, such as security and internet protocol (IP) address management.
- PBNM solutions focused on the IP world. This caused disruption in organizations that have different technologies present in their networks.

- PBNM solutions were misunderstood.
- PBNM solutions rushed forth without a solid set of standards in place.

Although the first three problems are important, the last two are fundamental problems that prevented the first wave of PBNM solutions from realizing their goals.

In addition, two other problems prevented wide adoption. First, the solutions initially available were not very scalable, and hence could not easily be used in large service provider networks despite the fact that they provided some attractive technology (e.g., configuring QoS functions).

Second, network monitoring technology lagged behind the new provisioning technology promoted by PBNM solutions to control the network. As a result, there was no easy way to monitor whether the PBNM solutions were actually working.

1.1.1 Early PBNM Solutions Missed the Point

In its early days, PBNM was characterized (and unfortunately, this characterization continues somewhat today) as a sophisticated way to manipulate different types of QoS. The motivation for this was to avoid overprovisioning the network, (i.e., enough resources are present for the network to respond to any anticipated need). The problem with this approach is that it is static and cannot adjust to the changing environment. Thus, if the network is provisioned according to the maximum expected load, resources will be wasted most of the time. Furthermore, if that load is exceeded for some reason (e.g., a heavy day of stock trading), then the network will still be unable to perform.

PBNM was used to set the QoS levels based on inspecting different fields in the header of traffic that was being sent. People then reasoned that PBNM could also be used for other applications (such as ensuring that high-priority traffic was delivered ahead of less important traffic and that different services received the level of service that they were contracted for) and for different types of security applications (such as refusing traffic from an unknown source to enter the network or starting an accounting application when a connection was completed).

The common theme to each of these (and other) applications is the desire to link the way the business runs to the services that the network provides. Regardless of application, PBNM was defined as reacting to a particular condition and then taking an appropriate action. The missing point is that some centralized authority has to decide which users and applications get priority over other users and applications.

Business rules are defined as the set of rules, regulations, and practices for operating a business. They often define and sometimes constrain business processes. Business processes are defined as the means by which one or more activities are accomplished in operating business practices. They take the form of an interconnected set of business functions (perhaps constrained by various business rules) to obtain a specific set of business goals.

Recently, the focus has turned to integrating business rules and processes with PBNM solutions. This focus makes intuitive sense, as it is certainly natural to want the network to provide services according to business contracts. However, the relationship can be, and should be, deeper than that. Business rules and processes govern how a system is run. They are responsible for the many decisions that must be made for every action performed by the system.

If policies are the reasons for doing something and business rules and processes are the means for doing it, why not connect them together? Although this seems obvious in retrospect, precious few information models have been constructed with this direction and capability.

An important corollary of this decision is as follows:

PBNM solutions require information models that contain business and system entities that can be easily implemented.

This book describes a unique object-oriented information model, called DEN-ng (Directory Enabled Networks-new generation). It is being developed in the TM Forum. The development is led by this author, and many different companies are involved. The author's company, Intelliden, is also actively involved in implementing DEN-ng and has incorporated it into the latest release of its product. Other companies, such as British Telecom, Telecom Italia, Telstra, MetaSolv, Hewlett Packard, and Agilent, have participated in reviews of DEN-ng.

An object-oriented information model is a means to represent various entities in a managed environment. An entity can be a person, a computer, a router, or even a protocol message—anything that needs a uniform and consistent representation for configuration and management is a possibility for definition and representation in DEN-ng.

An object-oriented information model provides a common language in which different types of management entities can be represented. This common language is of the utmost importance. Operational support systems (OSSs) are large, complex sets of applications that are composed of best-of-breed applications. This tendency to use best-of-breed applications encourages the use of "stovepipe" applications, which are applications that maintain their own definition of data. Much of the data used by each stovepipe application should be shared with other stovepipe applications. Unfortunately, this simply cannot be accomplished unless a common language exists to represent these common data.

One difficulty in building an OSS lies in the large variety of different management objects that must be harmonized and shared among the different management applications being used. Further exacerbating this problem is the fact that different types of management data have different characteristics. For example, very volatile data, such as statistical interface measurements, changes much too fast to be placed in a directory. Other data are very appropriate to put into a directory. Thus, an OSS needs to use multiple repositories to accommodate the different characteristics and uses of different management information.

An object-oriented information model, such as DEN-ng, is independent of any specific type of repository, software usage, or access protocol. Therefore, DEN-ng can be used as a single authoritative means for describing how different management information are related to each other.

To put this into perspective, Figure 1-1 shows five exemplary management applications that comprise an OSS. Notice that for two of these applications, the

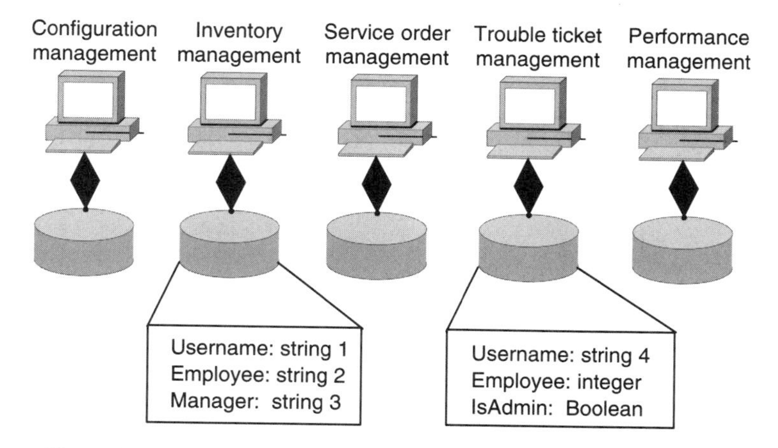

Figure 1-1 Problems in not using a single information model.

same data appears. For the username attribute, two different names are given. This makes it very difficult for applications to realize that these two different names actually refer to the same attribute of the same object. Furthermore, both applications define the same Employee attribute. However, the data types are different. This can cause problems in trying to write a single query to gather data based on this and other attributes across these two repositories.

Thus, unless there is a way to relate different information that are implemented using different data models to each other, it will be impossible to share and reuse management information. This raises the cost of the OSS and increases the probability that errors (resulting from the inability to share and reuse management data) will be embedded in the system. Furthermore, it means that entire processes will be repeated to derive and/or retrieve the same data (because the data cannot be shared). Instead, what is desired is a single, unified information model that relates the differences in data model implementations to each other.

DEN-ng is unique because it contains business and system entities that can be used to build management representations and solutions. In fact, in the Intelliden implementation, the DEN-ng information models are translated to two types of data models (Java and directory models). Specifically, business and system entities are represented in generic form in the information model and are then translated to platform-specific implementations. The Intelliden product uses these models to define business rules to activate network services. Other companies, such as MetaSolv (in their case, primarily a database), are using different repositories to implement DEN-ng and the SID.

1.1.2 Early PBNM Solutions Were Ahead of the Standards

The IETF took the DEN policy model and, in August of 1998, formed a working group to start modeling policy. This working group was originally co-chaired by myself and was based on using the DEN policy model.[17] This model concentrated on the generic representation of policy and chose QoS as a representative application that would be modeled as a separate set of extensions of the generic representation of policy. This is shown in Figure 1-2 below.

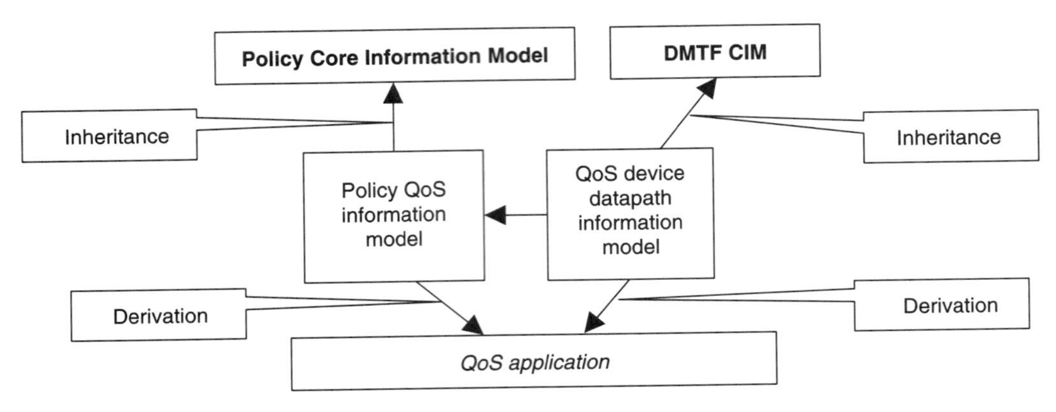

Figure 1-2 The structure of the IETF information models.

The policy core information model[13] defined a framework of classes and relationships that could represent the *structure* of policy of any discipline. This is an important point. The use case in 1998 is still the same as it is now—to build a single PBNM solution that can be used to manage different types of policies required by different applications. For example, QoS for voice applications is fundamentally different than QoS for data applications. As such, the target of the PCIM was to be able to represent how a policy was defined—it was not targeted at defining the *content* of the policy.

The policy QoS information model[15] refined this framework and added semantics to represent policies that could be used to control QoS mechanisms. The QoS device datapath information model was derived from the DMTF's common information model and represented much of the original DEN network model. The QoS device datapath information model[16] was used to represent the various mechanisms that the policy QoS information model would be used to manage. Both information models were designed to at provide *content* within a common overall *representational structure*.

The DMTF CIM was proposed as a way to provide a high-level representation of network elements. Thus, the policies could be "grounded" and applied to a network device. For example, a policy could describe a change in a function of a device; the content of this change could be represented by the policy QoS information model, and the structure of the policy could be represented in PCIM.

Unfortunately, the CIM model was too high-level and confused many people in how policy would be applied. For example, the CIM had no representation of either a physical port or a logical device interface (and this is true even today). This made it very difficult for people to picture how policies were going to be applied and built. Furthermore, the DMTF CIM was not really an information model—it was more of a data model. An information model is supposed to be independent of platform and technology. The DMTF CIM is based on the use of "keys"—special attributes that are used to name and uniquely identify a particular object. Keys are really a database construct, and their use must be carefully considered or else mapping to other types of data models that do not use keys (or have different keys than those of a database) will be much harder. This is why specific constructs used in one type of data model should not be part of a more general information model. In contrast, DEN-ng is a true information model in that it does not contain keys or other technology-specific concepts and terms. It instead concentrates on defining managed objects and their interrelationships. This is also true of the TMF SID, of which the DEN-ng information model is one component.

The approach shown in Figure 1-2 was good. It took a very long time, however, to get the participants in the IETF to agree to these models. The PCIM was not published as an RFC until February 2001. Although the policy QoS information model is ready, it still has not been published as an RFC as of June 2002. The QoS device datapath information model is further behind.

There are many reasons for the holdup. This was the first time that the IETF was working with information models. Second, policy models of this depth had not been done before in the industry. The main holdup was the fact that the IETF is composed of many different people; each of whom are there primarily to represent the companies that they work for. Each network vendor had by then launched its own set of policy applications. No one wanted a standard to come out that would brand their products as noncompliant! Thus, the standards were worked on, and watered down, and watered down some more, until they represented something that everyone could agree on. The delay in issuing standards is due to these reasons plus the delay in getting different companies (through their IETF members) to announce consensus. Members are always fearful that a last-minute change in the standard will adversely impact their companies' products, and so consensus building is a relatively long process.

However, there was another, more serious, problem. The above models focused "just" on network devices. Although the PCIM was generic in nature, it was also limited (this will be discussed throughout this book). For example, there was no model of how a policy rule would be evaluated. More importantly, there were no business entities in these models and very few non-network entities. Thus, there was no formal way to define how business rules could use policy to control network services. The primary motivation for building the DEN-ng model was to address these problems.

1.2 Where We Are Today

Today, work has proceeded in various standards bodies and fora to rectify these problems. Prominent among these is the work of the TeleManagement Forum (TMF). Two examples of this work are in the new generation operational systems and software (NGOSS) architecture[18] and the TMF's shared information and data (SID) model.[23]

1.2.1 The NGOSS Architecture

The NGOSS Architecture is characterized by the separation of the expression and execution of business processes and services from the software that implements these business processes and services. Fundamentally, NGOSS is concerned with defining an architecture that automates business processes.

For example, policies can be used to choose which set of processes are used to perform a function. Feedback from executing processes can then be used to change which policies are in force (or even applicable) at any given time. Thus, although either policy management or process management can be used by itself to manage an NGOSS system, to do so is to fail to realize the greater potential afforded by using both to manage the same system.[16]

The NGOSS behavior and control specification defines in high-level terms the architectural ramifications of using policy management.[2] The NGOSS policy specification, defines in high-level terms the definition of a policy model that includes business, system and implementation viewpoints.[3,24] This is based on work from ISO[4] and UML.[5]

Although these are evolving specifications, credit should be given to the TMF to have the vision to try and specify these important concepts and also to develop them for all to use. A good example of this is the Catalyst programs of the TMF. These team demonstrations are usually led by a Service Provider or independent software vendor (ISV) and are designed to demonstrate one or more concepts of the NGOSS architecture. This work is important because it defines architectural and implementation ramifications of using PBNM solutions. This is one of the few forums in the world where this is being studied in depth by commercial, academic, and industrial players.

One of the prominent differences between the design of DEN-ng and the design of other information models is that DEN-ng was built to support the needs of the NGOSS architecture. All other information models that the author is familiar with were *not* built to support any particular architecture.

The TMF approach is inherently better suited to produce useful standards faster. First, it is centered on real-world work, that is proven to be implementable through its Catalyst programs. Second, the TMF has as one of its goals the production of a shared information model. While the IETF emphasizes protocol development, the TMF emphasizes architecture and information modeling. Finally, because the different vendors are all united in achieving common goals (architecture and information modeling), it is easier for them to come to agreement than in the IETF.

1.2.2 The TMF Shared Information and Data Model

The TMF's shared information and data (SID) model is a federated model, which means that it is composed of different "sub-models," which have either been contributed by companies, mined from other standards, or developed within the TMF.

The communications industry is seeking technological advances that will improve time to market for new products and services. Service providers and enterprises like to use best-of-breed software. However, this software is hard to integrate with other software products constructed in a similar manner. Furthermore, each software product that is produced in a "vacuum" more than likely redefines concepts that are used by other applications.

To achieve true interoperability (where data from different components can be shared and reused), a common language needs to be developed and agreed on.

This goal is even more important in an NGOSS system, because one of its key architectural principles is to use component-based software, interacting through contracts. Therefore, the TMF embarked on building a shared information model that could be used as a single source for defining common data.

The SID consists of inputs from Intelliden, MetaSolv, British Telecom, Telstra, Vodaphone, Motorola, Agilent, AT&T, and others. Material donated includes DEN-ng and several models and model snippets from many of these companies. The objective of the SID is to provide the industry with a common language, defined using UML, for common shared data. By agreeing on a common set of information/data definitions and relationships, the team sets forth a common language used in the definition of NGOSS architectures.

Another important feature of the SID is that it contains multiple models that concentrate on different disciplines. Most other information models concentrate on a single subject, such as networking. In contrast, the charter of the SID is to define business and system concepts for a variety of different domains. These domains characterize how network elements and services are represented, used, and managed in business and system environments.

1.2.3 The Ingredients to a Compelling PBNM Solution

The industry is now starting to appreciate the complexity of PBNM solutions. PBNM is more than writing a policy rule and more than building elaborate UML models; it is about a paradigm shift.

Historically, network management has focused on setting parameters of individual interfaces of a device one at a time. Recent innovations of policy management, ranging from new protocols to the use of information models to represent policy rules, have helped simplify this daunting task. However, in and of themselves these are insufficient to develop PBNM solutions that will solve network configuration problems and help make network services profitable once again.

We need a more extensible, more robust solution. The key to implementing this solution is to think more holistically about policy management. Most people consider policy to be a set of rules that express a set of conditions to be monitored and, if those conditions are met, one or more actions will be executed. This definition fails to take into account two key issues: users and process.

First, different types of people use policy. Business people do not want to express their policies in networking terminology, and networking people do not want policies written using business concepts. However, business and network personnel must work together to ensure that network services are managed according to the business goals of the organization. A new form of policy is needed that can translate business needs into device configuration.

However, this by itself is not enough. The second missing feature is process. No matter how simple or how sophisticated, every configuration change has an underlying set of business rules that govern its deployment. Business procedures will define who checks the change for correctness (sometimes from a technical and a business point of view). They identify who must approve the change and who must implement the change. They also describe how to verify that the change has been properly implemented and what to do if a problem is discovered.

Policies define how the shared resources of the organization are accessed and allocated. Different users and services have different needs, and policy is the tool that enables the appropriate process to be applied as a function of business priority. This enables network services to be adjusted in response to the changing environment (e.g., new users logging on, different application usage, and so forth) by providing dynamic and automatic (re)configuration of the appropriate network devices according to the business rules of the organization.

The realization that business rules and processes, device configuration, and service activation are all tightly bound together provides the clue to our answer. We need a robust, extensible information model that can represent the managed environment as a set of entities. If policies are also entities that exist in this information model, then we can be assured that policies are represented using the same tools, and therefore can be applied to users, applications, device interfaces, services, and other managed objects. The information model provides a set of formalisms through which we can build a robust system.

1.3 Definition of Policy Management

Policy is typically defined as a set of rules. Each policy rule consists of a condition clause and an action clause. If the condition clause is TRUE, then the actions in the action clause are allowed to execute.

Therefore, our first definition of policy management is:

Policy management is the usage of rules to accomplish decisions.

Policy is usually represented as a set of classes and relationships that define the semantics of the building blocks of representing policy. These building blocks usually consist of a minimum of a policy rule, a policy condition, and a policy action and are represented as follows:

Figure 1-3 A simplistic policy model.

Figure 1-3 represents these three building blocks as classes. This simple Universal Modeling Language (UML),[5-7] model shows the relationships between these three classes. Attributes and methods have not been shown to keep the discussion simple.

Figure 1-3 shows that a PolicyRule contains a set of conditions and a set of actions. These are represented by the *hasConditions* and *hasActions* aggregations, respectively (an aggregation is a special type of relationship that is used to represent whole-part dependencies). In this book, UML is used to represent and define entities of interest. Appendix A is a brief overview of the essential concepts and notation of UML.

PBNM solutions usually use an information model to represent policy. Some of the better ones also use an information model to represent the subject and target of the policy. DEN-ng is unique, in that it does this for business, system, and implementation viewpoints. By representing what you want the policy to do and how you want it to act, you can use the power of an information model to represent how different entities relate to each other. For example, two different users can be logged on to the same system but receive different classes of service, which dictate how the applications that each operate are handled in the network.

An information model is a means for defining common representation of information. This enables management data to be shared, reused, and altered by multiple applications. The DEN-ng policy model will be used as the policy model in this book. The DEN-ng policy model is different than other policy models in the industry and will be discussed in detail in Chapter 7. However, three differences are important to discuss now. The first is the use of an event model to trigger the evaluation of the policy condition clause. This changes Figure 1-3 to the following:

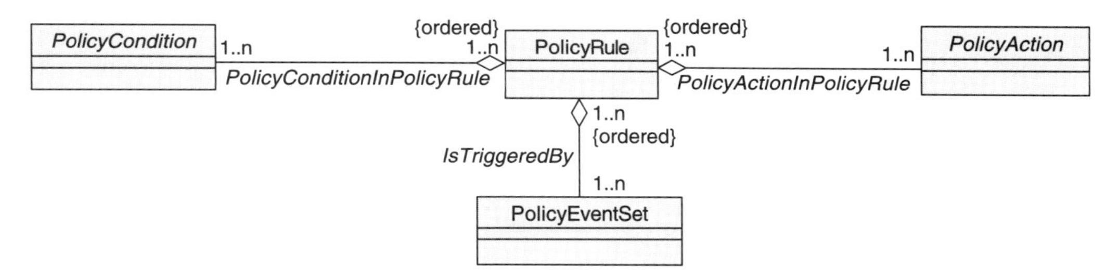

Figure 1-4 A simplistic view of the DEN-ng policy model.

Figure 1-4 can be read as follows:

On receipt of an Event, evaluate the policyCondition of a policyRule. If it evaluates to TRUE, then execute the set of policyActions that are associated with this policyRule.

The second difference is the use of constraints to better define (through restriction and more granular specification) what the model represents. For example, it makes no sense to define a PolicyRule that does not have any conditions. This is allowed in the simplistic model of Figure 1-3, because the cardinality on each end of the *hasConditions* aggregation is 0. However, this is not the case in Figure 1-4, as the cardinality is 1..n on each side of the *policyConditionInPolicyRule* aggregation. Another example is the Object Constraint Language (OCL)[5] expression "{ordered}." This expression requires that the policyEvents, policyConditions, and policyActions are each ordered when aggregated in the policyRule.

The third difference is that DEN-ng uses a finite state machine to represent the state of a managed entity. Most current information models, such as those from the DMTF, the IETF, and the ITU, are *current-state* models, (i.e., they define a managed entity to represent a state of an object). Although important, that does not make a closed-loop system. In particular, it does not enable the life cycle of the managed object to be represented.

Therefore, DEN-ng defines a finite state machine and instantiates multiple current state models to represent the different states that a managed object can take. This enables behavior of an individual or a group of managed objects to be represented. More importantly, the behavior of an object or set of objects can be related to the value of one or more attributes that are used to represent the current state of the attribute. This helps simplify the design of closed-loop PBNM solutions. For example, suppose that a particular state transition sets the attribute of an entity to a particular value and that this represents a bad or failed state. The changing of this attribute value is in fact an *event*, which can be used to trigger the evaluation of a PolicyRule. The PolicyRule can cause a state transition back to a valid state, which is checked by ensuring that the value of the attribute is changed to an acceptable value.

Without events or a state machine, such closed-loop control is not possible. More importantly, policy is represented as a means to control when a managed object transitions to a new state.

This notion is simple, yet powerful. It succinctly captures the connotation of "control" that policy has and shows how policy can be used to govern the behavior of a managed object throughout its life cycle. Furthermore, it provides a means to control the behavior of a managed system in a predictable and consistent fashion. Events represent external stimuli that correspond to changes in state. If policies are used to control state transitions, then policies can be defined that govern each state of the managed object—from creation to deployment to destruction. This guarantees that the correct state of the managed object is achieved in response to a given event, in a simple and consistent manner. This will be discussed more in Chapters 3 and 5.

1.4 Introduction and Motivation for Policy Management

The promises of policy management are varied, powerful, and are often conceptualized as a single, simple means to control the network, as illustrated in Figure 1-5.

The simplicity of the components shown in Figure 1-5 is part of the appeal of policy management. In particular, the ability to hide vendor-specific interfaces behind a uniform information model is very important. Without this ability, a common interface to programming the same function in different network devices cannot be accomplished. This is one of the toughest problems a network manager needs to deal with—how to string a network of multivendor equipment together to provide a seamless set of customer-facing services. Furthermore, the growth of large ISP networks that seek to provide multiple specialized services exacerbates this problem.

This drive for simplicity has led to six commonly heard value propositions for policy management that position policy management as a means of:

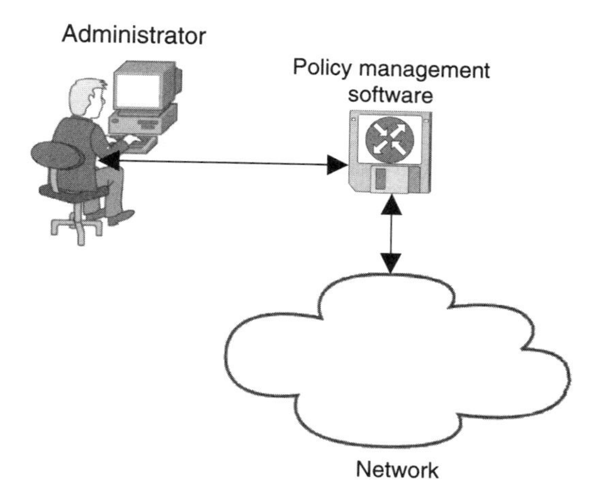

Figure 1-5 The promise of policy management.

- Providing better-than-best-effort service to certain users
- Simplifying device, network, and service management
- Requiring less engineers to configure the network
- Defining the behavior of a network or distributed system
- Managing the increasing complexity of programming devices
- Using business requirements and procedures to drive the configuration of the network

These six points are discussed in more detail below.

1.4.1 Providing Different Services to Different Users

The Internet was built to handle traffic on a best-effort basis. Clearly, people will not be satisfied with best-effort service. People want *predictable* services—services that they can rely on for providing information and functionality that they desire (whether the Internet is being used or not). This is the fundamental motivation for QoS.

When I worked at Cisco, we used to describe QoS as "managed unfairness." This complements the above desire for information and functionality that meet a specific set of needs. QoS is not just about providing faster downloads or more bandwidth. Rather, it is about providing the right set of functionality to provide a user with the service(s) that the user is requesting. Although this may mean faster downloads or more bandwidth, the point is that such metrics in and of themselves are not a good definition of QoS.

QoS is harder to provision and manage than it may first appear because of two main factors:

- Complexity of implementing QoS
- Variety of services that can use QoS

The complexity of implementing QoS is caused by two main factors: (1) network vendors continue to add additional types of mechanisms that can be used (by themselves or with other mechanisms) to implement QoS and (2) different devices have different QoS mechanisms. This makes it hard to ensure that the same relative levels of service are implemented by different devices that use different mechanisms.

Another problem is the lack of specificity in standards. For example, the IETF has completed a set of RFCs that specify different approaches for implementing differentiated services (e.g., the Differentiated Services RFCs[8]). However, these RFCs by themselves are not sufficient to build an interoperable network because they concentrate on specifying behavior without specifying how to implement that behavior. For example, none of the RFCs specify what type of queuing and drop algorithms to use to implement a particular type of behavior. This is in recognition of the IETF—this is in fact in recognition of the fact that network vendors have designed a vast arsenal of different mechanisms to condition traffic as well as recognizing that different services uses different QoS mechanisms.

Thus, we have the first motivation for policy management—the promise of using a set of standard declarations for managing the different QoS mechanisms required to implement a service. This desire is amplified by the fact that multiple users want different services. Clearly, a service provider or enterprise cannot provide tens or hundreds of different services because of the complexity of managing these different services coupled with the fact that most approaches (such as DiffServ) define far less than these. DiffServ, for example, provides a set of 64 total code points, but these are divided into 32 standard and 32 experimental code points. Most service providers offer between three and ten different services. This provides the second motivation for policy management—the promise of providing a small set of standard rules that can be used to manage the set of services provided to multiple customers.

1.4.2 Simplifying Device, Network, and Service Management

PBNM was conceptualized as a set of mechanisms that can be used to "fine-tune" different network services. Similarly to how a stereo equalizer gives the user control over the response of the stereo to different frequencies, a PBNM-based system provides a set of mechanisms that can be used to condition traffic flowing through the network. PBNM systems also have the ability to define a complex set of mechanisms that can be used to implement a predefined service. This is a particularly attractive characteristic—choosing a single command to implement what previously consisted of a set of commands.

In addition, the real power of PBNM systems is through abstraction. Imagine a network where a switch feeds a router. The switch uses the IEEE 802.1q specification for delivering QoS, while the router uses DiffServ. This causes a problem, because there is not a defined set of standards for relating an 802.1q marking to a DiffServ code point (DSCP). Now, assume that the switch is programmed using simple network managment protocol (SNMP) set commands, while the router is programmed using Command Line Interface (CLI) commands. The network administrator is now forced to learn two different ways to program a single network connection.

The motivation for PBNM is one of simplification through abstraction. By providing an intermediate layer of policy rules, PBNM users can concentrate on the task at hand, rather than the myriad of programming models and traffic conditioning mechanisms used to program a device.

However, an equally powerful motivation exists—recovery from changes and failures. Networks present an ever-changing infrastructure for providing services. The day-to-day management of this infrastructure includes making subtle changes to how different components are configured. Sometimes, these changes can adversely affect network services. These changes are hard to find, because most of the time, the change being made is not obviously related to the service that was being changed. In addition, networks can develop faults. That impair the ability for the network to provide services that people and applications depend on. When this happens, administrators tend to fix the fault by changing the configuration of the device.

These and other factors culminate in a set of changes that, over time, impact the ability of the device to support one or more of its services. When this happens, PBNM systems can be used to restore the configurations of devices to their original state. Thus, PBNM provides a means to fix the fault and to also keep track of the state of various network devices. This requirement for tracking state is one of the reasons why DEN as well as DEN-ng both use finite state machine models.

1.4.3 Requiring Less Engineers to Configure the Network

There is an acute shortage of engineers that understand new technologies and mechanisms implemented by network vendors. There are even less engineers that understand these technologies and are able to deploy and manage them on a network. In addition, the cost of using an emerging technology is very high, interactions with other legacy technologies is not completely known, and management costs associated with initially deploying the technology often outweigh the advantage provided by that technology. For example, many network operators choose to over-engineer their networks to address any performance concerns rather than deploy QoS techniques. This is because the cost associated with learning the new technologies (and the tools used to deploy them) and managing them is much higher than the savings in bandwidth-related costs that would result from deploying these technologies. Another factor is the previous lack of specificity mentioned—if different technologies are being used, then they can only interoperate if their functionality is specified at a sufficiently detailed level. For example, there is no standard that defines how to map ATM's concept of QoS to the different DSCP values that are present in an IP network.

The theory behind being able to use fewer personnel to run a network is based on distributing intelligence to managed devices and applications that manage devices so that dynamically changing environments can be more easily managed and controlled. Although the number of skilled individuals may be reduced, it is wrong to think that PBNM applications will eliminate the need for specialized network engineers. Skilled personnel will always be needed to build and operate systems.

However, PBNM systems provide two important benefits. First, the majority of network configuration tasks are simple in nature and do not require a specialist. Many of these are also repetitive. If the PBNM system can be programmed to

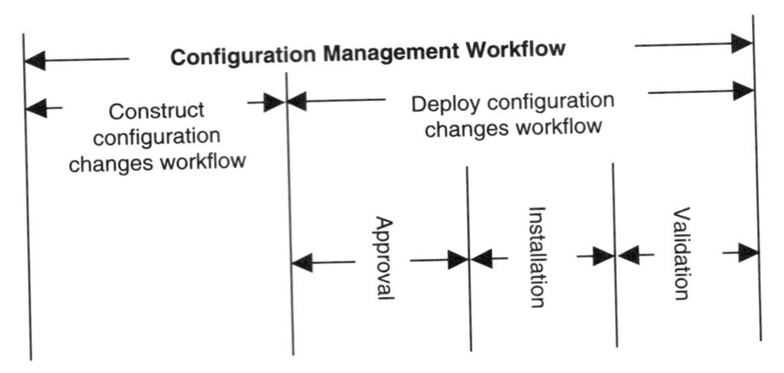

Figure 1-6 Processes used in configuration management.

deal with these mundane changes, then they enable more highly skilled engineers to be used on other, more strategic, problems. Second, PBNM systems enforce process. Figure 1-6 illustrates this.

PBNM can be used to define processes, such as:

- Which personnel are qualified to build a configuration change
- Which personnel must approve a configuration change
- Which personnel must install a configuration change
- Which personnel must validate a configuration change

These four processes[10] are meant to be exemplary in nature and should not be construed as being the "only" processes that are involved in device configuration.

The strength of PBNM is that these four processes (and others) can be enforced by a PBNM system independent of whether the PBNM system is used to implement a configuration change or not. For some reason, this message has not been emphasized by most vendors. Even some researchers tend to ignore it, concentrating instead on the representation of policy. Two counterexamples to this trend are Intelliden and Metasolv, both of whom are building software to help in this area.

PBNM systems also offer the ability to ensure that the same approved processes are used to consistently implement specific types of configuration changes. Chapter 10 describes the Intelliden product, which is a good example of offering both of these benefits.

1.4.4 Defining the Behavior of a Network or Distributed System

Networks are growing in complexity because of several factors, including an increasing number of people using networks, a growing number of different applications used, and an increase in the number of different services required by network users.

These factors all help to create an ever-growing overhead of operating and administrating networks. As a result, it is very difficult to build management systems that can cope with growing network size, complexity, and multi-service operation requirements. There is also a need to be able to dynamically change the behavior of the system to support modified or additional functionality after it has been deployed.

A single network device can have thousands of interfaces or subinterfaces. Clearly, if an administrator has to manually configure each of these, the network cannot scale. For example, assume each device interface takes 10 minutes to configure and that there are 10,000 total interfaces. This works out to requiring 69.44 days, or 9.92 weeks, to program this set of interfaces. Without software, this is simply not possible. In addition, the chances of making an error without automation software are enormous.

PBNM software can help in several ways. First, it can be used to define policy rules once and mass deploy them. For example, the Intelliden product has a concept called "command sets," that enable sets of configuration changes (which are controlled by policy) to be deployed to multiple devices concurrently. Second, as shown in Chapter 6, policy rules can be used in either an ad hoc or reusable fashion. Although ad hoc policy rules are intended to be used once, reusable policy rules (or even policy components) are designed to be used multiple times by different applications. This concept can be used to help simplify the arduous process of configuring different interfaces. For example, an access control list can be defined that filters on certain fields in the IP header and then performs a set of actions if those fields matched or not. This is a fundamental building block that can be used for many different types of policies. Third, large systems will execute many different policies. PBNM systems should enable different sets of policies to be analyzed to ensure that they do not result in conflicting actions. This will be covered more in Chapter 4.

However, most importantly, PBNM software can be used to capture business logic that is associated with certain conditions that occur in the network. Although centralizing the development and management of this business logic is important, coordinating its proper application is mandatory for large systems. This last point raises four important questions that the reader should ask when evaluating PBNM systems:

- How many physical devices is the PBNM software capable of managing?
- How many logical components (e.g., subinterfaces) is the PBNM software capable of managing?
- How many changes per time period (e.g., minute or hour) can the PBNM software execute?
- How does the PBNM solution handle errors?

Each of these points is important. The third point is especially important, because most organizations operate using a "time window" in which changes must occur. The point, then, is how many changes can your PBNM software physically perform during that time window? The reader will find that this is often the limiting factor in choosing a PBNM system. The fourth point is also critical, because one of the reasons for deploying a PBNM solution is to automate complex tasks. The form of this question is different than a simple "can it scale" question. Vendors will all claim that their solutions scale. Thus, a much easier way to be sure of what you are buying is if it can provide a common error handling methodology for large deployments. This is a simpler and better test of what you are buying.

1.4.5 Managing the Increasing Complexity of Programming Devices

Present-day IP networks are large, complex systems that consist of many different types of devices. Different devices are chosen for cost and functionality. However, from the end-user's point of view, it is imperative that the end-user not have to be explicitly aware of these differences. In other words, the network should appear as a single homogenous entity that provides services for the end-user.

Therefore, when most services are defined, they are characterized as having a set of properties that exist from one end of the network to the other. For example, think of a service level agreement that specifies availability (which in this example is defined as remote access accessibility without busy signals). While the service provider will likely specify different times for different networks (e.g., a connection to a U.S. network versus a connection to a European network), it certainly will not specify availability between different parts of the network. Not only is this too hard to do (and very costly for the service provider), it doesn't really matter, because the service is specified as an end-to-end service. The end-user does not care what devices or QoS mechanisms are used or what the latency or drop rate is along an intermediate path in the network as long as the service that was contracted for is successfully delivered.

Network engineers do not have this luxury. In fact, ensuring that all of the different devices that comprise a network interoperate smoothly is far from a trivial task. This is because different devices have different functionality, represented as different commands that are available to the network developer. The problem is that these different network devices are each responsible for doing their best in providing *consistent* treatment of the traffic. Clearly, if the two devices have different commands, then this is harder to achieve, because a mapping needs to be defined to map the different commands to each other.

For example, consider two Cisco devices, one running a pre-12.x release of IOS (a common operating system used by Cisco routers and other types of devices) and another running a 12.x release of IOS. Suppose that the task is to provide QoS for traffic that contains voice, video, and data. Both devices can do this. However, the actual commands that are used are very different. As an example, the design for the 12.x device is likely to use low latency queuing, which is not available in pre-12.x IOS releases.[9] Thus, someone (or something) has to provide a mapping between the set of commands used in each version of an operating system. Clearly, if different devices are using different operating systems, this mapping becomes both harder and more important. Mapping the commands is a good start, but even that is not sufficient. Other factors must also be taken into account. Two important ones are side effects and dependencies in executing each command.

Sometimes, when a command is executed, effects occur that cause other entities to be affected besides the ones that are targeted by the command. These are called side effects, because though these changes were not intended, they nevertheless happened. If these commands have any side effects, then they must be noted, and if the side effects affect the traffic, then they must be emulated for each device.

Exacerbating this situation is the notion of hardware and software dependencies. For example, a device that uses an older processor may be unable to perform the same functions as a device that uses a newer processor past a certain line rate. This is a hardware dependency and must be accounted for to ensure that each device performs traffic conditioning in a consistent manner. Similarly, software dependencies exist; if they affect the flow of the traffic, then their effect must be emulated in devices that do not have these same software dependencies.

If that is not bad enough, new technologies have emerged or will continue to emerge to either address current limitations or to perform a task better. Thus, the PBNM system must be capable of addressing new commands and features of the devices that it supports. This is best done using an information model to abstract the different functionality that is present in multiple devices. For example, Figure 1-7 shows a simplified approximation of the DEN-ng QoS model, which is an information model designed to represent QoS.

(Remember that a line with an arrowhead denotes inheritance in UML. Thus, GoldPackage is a subclass of ServiceBundle, which is a subclass of Customer FacingService, which is a subclass of Service.)

In DEN-ng, there are two types of services: CustomerFacingServices and ResourceFacingServices. This is modeled as two separate subclasses that inherit from the Service superclass.

CustomerFacingServices are services that a customer is directly aware of. For example, a VPN is a service that a customer can purchase. ResourceFacingServices are network services that are required to support the functionality of a CustomerFacingService, but which the customer cannot (and should not!) know about. For example, a service provider doesn't sell Border Gateway Protocol (BGP, a means of advertising routes between networks) services to a customer. Yet, BGP is required for different types of CustomerFacingServices to operate correctly. Thus, BGP is an example of a ResourceFacingService.

A ServicePackage is an abstraction that enables different CustomerFacing Services to be packaged together as a group. Thus, a GoldService user may access high-quality voice, video, and data, whereas a SilverService user may be unable to use voice.

Several types of ResourceFacingServices are shown in Figure 1-7. QoSService is an abstraction that relates the particular networking architecture to its ability to provide QoS. For example, ToSService uses the three-bit Type of Service bits in IPv4 to define the QoS that can be given, whereas DiffServService uses the six-bit code point to define much more granular QoS for IPv4. Because a given network may have both DiffServ-compliant and DiffServ-unaware devices, the information model provides a formal way to synchronize their configurations, so that a given ToS setting provides the same QoS as a particular DiffServ setting.

Finally, NetworkForwardingService defines how traffic is conditioned. This consists of two types of "subservices": the ability to identify traffic and the ability to affect the order in which packets are transmitted from the device. Again, because these are two distinct concepts, two distinct subclasses (TrafficIdentificationService

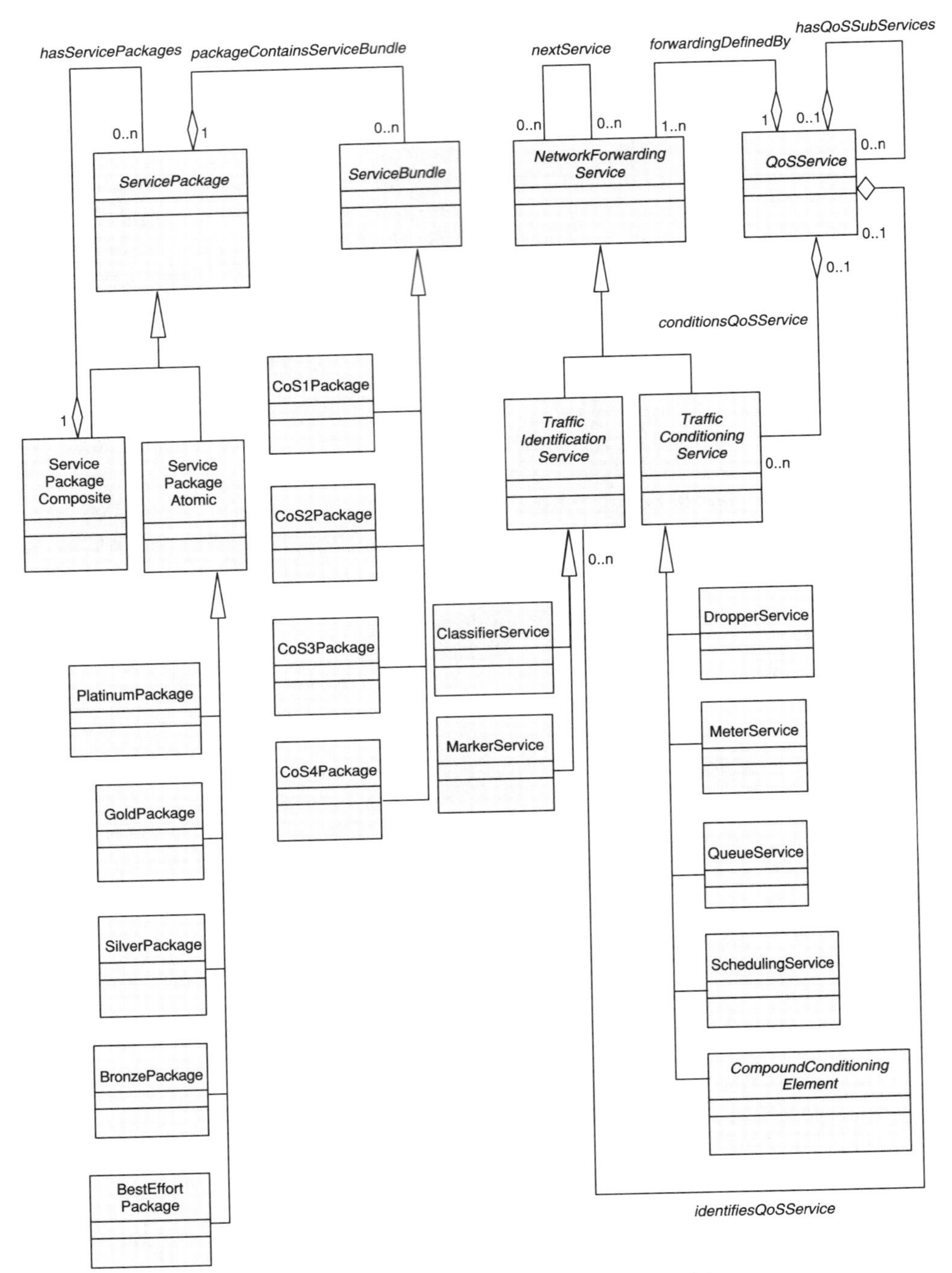

Figure 1-7 Simplified DEN-ng QoS model.

and TrafficConditioningService, respectively) are used to represent these concepts. With respect to TrafficIdentificationServices, ClassifierService performs the separation of traffic into distinct flows that each receive their own QoS, whereas MarkerService represents the ability of a device to mark or remark the ToS or DiffServ bits. This marking tells the other devices what type of QoS that flow should receive. With respect to TrafficConditioningServices:

- DropperService drops packets according to a particular algorithm, which has the effect of telling certain types of sending applications to slow their transmission
- MeterService limits the transmission of packets
- QueueService delays the transmission of packets
- SchedulingService defines which queue (of multiple output queues) should send packets
- CompoundConditioningService models advanced features, which are combinations of the above basic services

The objective in such a model is to describe a particular feature (such as metering) and how that feature relates to other features (e.g., classification and dropping) in a particular function (e.g., traffic conditioning) using classes and relationships. The idea is that if the abstractions are defined properly they can be used to model the types of functions that are present in different vendor devices and accommodate new functionality. This model will be discussed more in Chapters 9 and 10.

Put another way, the model can be used as a design template for constructing commands that are to be applied to a device or set of devices. The advantage of such a model is that the model can be used to represent the functionality desired and can hide the intricacies of translating to different implementations from the user. In fact, this is one of the principles on which the Intelliden R-Series was founded.[10] This will be discussed more in Chapter 11.

Sometimes, such models are all that are needed, and enable vendor-specific programs that are derived directly from these models to be used. Often, however, additional information is required. In the DEN-ng information model, this will take the form of subclasses that are used to model vendor-specific differences from the model. These concepts will be defined in more detail in Chapters 9 and 10.

1.4.6 Using Business Rules to Drive Network Configuration

The thesis of *A New Paradigm for Network Management* is that existing network management architectures prevent business processes from being used to drive the configuration and management of the network.[11] In essence, this paper states that businesses must define and implement network services according to their own business processes and policies. Although this is true for all businesses, it even more true for the so-called "next generation network" initiatives and corporations that are striving to become more profitable by changing the network services that they provide.

Business driven device management (BDDM) is one example of using business rules to drive network configuration. As defined by Strassner,[11] BDDM is a new paradigm that enables business rules to be used to manage the construction and deployment of network configuration changes. The difference is that BDDM controls both the construction and the deployment of configuration changes using a combination of policies and processes.

Most of the current research in PBNM systems revolves around the definition of policy class hierarchies that can be used to represent functionality of a network device. BDDM leverages this work, but combines it with policies and processes that define how configuration changes are created, deployed, and modified in a scalable and consistent manner. Part of the desire to use business rules to drive the configuration of a device is because business rules provide a higher-level view of what needs to be accomplished. This is necessary to ensure that those changes will not disrupt the operation of the device or the network. This, in turn, requires other entities besides devices and services (such as users and their various different roles) to be modeled.

Although abstractions that are used to represent business entities can still be modeled in UML, their content and detail is significantly different than that used for device and service entities. The administrator does not have to understand the low-level details of the technology used to support a particular business need to direct its usage. For example, suppose that a network operator needs to define three levels (gold, silver, and bronze) of customers. An administrator can easily assign each customer to a particular level based on their contract. A variety of techniques can be used to implement these three services in the network; one such example is to use DiffServ. However, there is a difference between the business person (whose job is to assign a particular service level to a customer) and a network administrator, (who is responsible for implementing commands that will enable the network to recognize and enforce these three network service levels). Both the business person and the network administrator can use policies. For example, a business person may need to write policies for handling service outages or interruptions, whereas a network administrator will be more interested in writing policies that control how the configuration of a device is changed. This difference is fundamental to how policies for each are used and expressed and mandates that different representations of policy should be used for the business person and the network administrator.

If business rules and processes are not used to manage changes made to the configuration of network devices, the device's configuration is reduced to changing lines in a file. This doesn't reflect how the business operates! Even worse, this means that the network is probably not reflecting the proper set of services that the organization needs to run its business. The semantics of what to change, when to change it, and who can change it are all captured using business rules and processes. These semantics must be used to drive how the configuration is constructed and deployed

However, the problem is more complex than "just" modeling QoS commands or defining which user can make a configuration change. Fundamentally, different types of people—having different responsibilities and different functions in the organization—use policy for different reasons. Business people do not want to express their policies in networking terminology, because network terminology is not needed to express their requirements. Similarly, networking people do not want policies written using business concepts, because these concepts are usually not precise enough to enable them to program the devices that they are managing. However, both business and network personnel must work together to ensure that network services are managed according to the business goals of the organization. A set of policies that supports the translation between one type of policy and another is therefore needed.

This translation between different types of policies is called the Policy Continuum. Each level in the Policy Continuum addresses a specific type of user that has a very specific understanding of the managed entities operating at that particular level of abstraction. The Policy Continuum is shown in Figure 1-8.

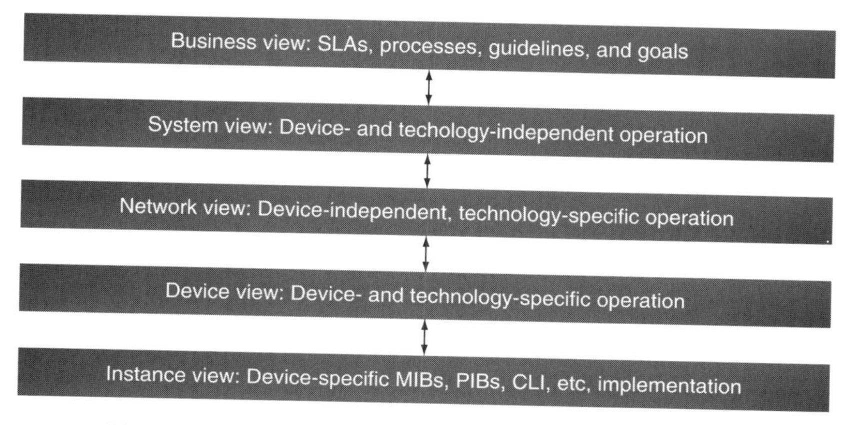

Figure 1-8 The policy continuum as defined in DEN-ng.

The PBNM system must translate these entities and concepts between layers of the Policy Continuum. This will be discussed more in Chapters 2, 3, and 7–9. The DEN-ng model is the only information model that uses the concept of a Policy Continuum.

This book will use the new DEN-ng information model to represent managed entities, people, and applications that use those managed entities and policy entities. The advantage of using a single information model that has multiple domains is that it is easier to relate different elements in each domain to other elements in other domains.

1.4.7 Summary of the Benefits of PBNM

The traditional promise of PBNM is that people will be able to deploy more complex services across a wider array of devices with fewer highly skilled individuals. This will in turn simplify network and service management. This is augmented by newer promises, such as those envisioned by BDDM, which use business requirements to drive the configuration of the network. This forms a tight closed-loop system, in which decisions governing the behavior of the network and the services that it provides are driven by business rules.

The results of these promises are compelling: increased revenue, faster time to activate services, and decreased expenses.

The next two sections will focus on two key themes:

- The need for and use of a shared information model
- The benefits of using PBNM

1.5 The Need for a New Shared Information Model

The two big issues that face us today concerning network devices and network management are:

- Lack of a consistent product model prevents predictable behavior
- No standard for shared data

1.5.1 Lack of a Consistent Product Model

The lack of a consistent product model means that, despite all the standards that you hear about, different vendors build devices with different hardware and software. One router can have vastly different characteristics and functionality than another router. This situation is exacerbated when mergers, acquisitions, and divestitures occur, as the customer ends up buying completely different devices that happen to have the same logo and vendor name on them. Therefore, when different devices are used in the same network, predictable behavior cannot be obtained. Standards help define invariant parts of the programming model. However, they are usually not explicit enough to guarantee interoperability. For example, Nichols et al.[12] define the structure of a differentiated services code point (DSCP), which is used to indicate how to condition traffic. The invariant portion of this RFC includes the fact that a DSCP is six bits long, and certain bit patterns are already defined. However, this RFC does not define which dropping and queuing algorithms to use for different bit patterns. Thus, multiple vendors can be compliant with the differentiated service standard (of which this RFC is one element) without being able to interoperate.

This is also true, but to a lesser degree, of the emerging policy information model standards. Moore et al.[13,14] define a class hierarchy and relationships for representing generic policy elements, while Snir et al.[15] and Moore et al.[16] extend these to use provide QoS models. There is even an LDAP mapping,[25] and the beginnings of one for policy core extension LDAP schema.[26] These classes and relationships help define how policy is used to control various QoS mechanisms. However, these models have very limited semantics and are subject to interpretation by different applications. For example, these networking concepts are not linked closely enough to network device and service entities to specify how policy could be used to program device features (let alone commands). As a simple example, because these models do not specify the concept of a device interface they cannot be used to specify how to program a device interface.

More importantly, these models do not contain any associations to business entities, such as Product and Customer. Thus, they cannot be used to define which Services from which Products are assigned to which Customers. This also contributes to the complexity of building a management system, because now additional components must be used if business rules and processes are used to drive the configuration of the network.

A networking model that is associated with other models that represent users and targets of networking services and a policy model that controls how networking services are implemented and provided to users are needed. This requires a layered, integrated information model. This book will use the DEN-ng model to meet these needs.

1.5.2 Lack of a Standard for Representing Shared Data

Until the TMF launched its SID model effort, no standard existed for sharing and reusing data for network devices and services. The Common Information Model (CIM) of the DMTF is rooted in instrumentation of desktop applications. Although the model has reached out over the last few years to encompass additional concepts, it still lacks many telecommunications concepts that enterprise and service provider networks need. For example, its physical device model has no physical port, and its logical model has no device interface. Without these, the model cannot be used in telecommunications applications. The CIM is not a bad model; it is simply not a self-contained model that can be used for telecommunications applications.

This is precisely why the DEN-ng and the SID efforts were started. The DEN-ng effort was designed to extend and enhance the original DEN[17] effort to tie it more closely to the NGOSS effort of the TMF.[18] The design of the DEN-ng model is unique, because one of its use cases is to support the NGOSS architecture specification.

The DEN-ng effort focuses on modeling network elements and services. However, it provides a business, system, and implementation viewpoint of these models. The focus of the SID is on the entire NGOSS environment. The SID uses many models, including DEN-ng, to provide comprehensive coverage of entities and concepts present in an NGOSS environment.

DEN-ng is being developed in the TMF because the TMF Catalyst programs can be used to validate and spread the model across different service providers, vendors and ISVs. This distribution vehicle (which also provides detailed feedback) is lacking in other standards bodies and fora and is one of the main reasons why DEN-ng was developed in the TMF.

The DEN-ng policy model was developed using an iterative top-down, bottom-up approach. Business concerns were first considered, which provided a high-level structure for and helped define key concepts of the overall policy information model. This model was then augmented by adding detail necessary to build a system. This is currently where the public part of the DEN-ng set of specifications exists.

Intelliden's vision is to take this a step further in its product line. Once these business and system views were defined, a set of tools will be produced that will focus on translating the information model to two different data models: a directory data model and a Java model. This will enable the information model to be implemented in software. A second set of tools will be developed, which will focus on ease of implementation (Figure 1-9).

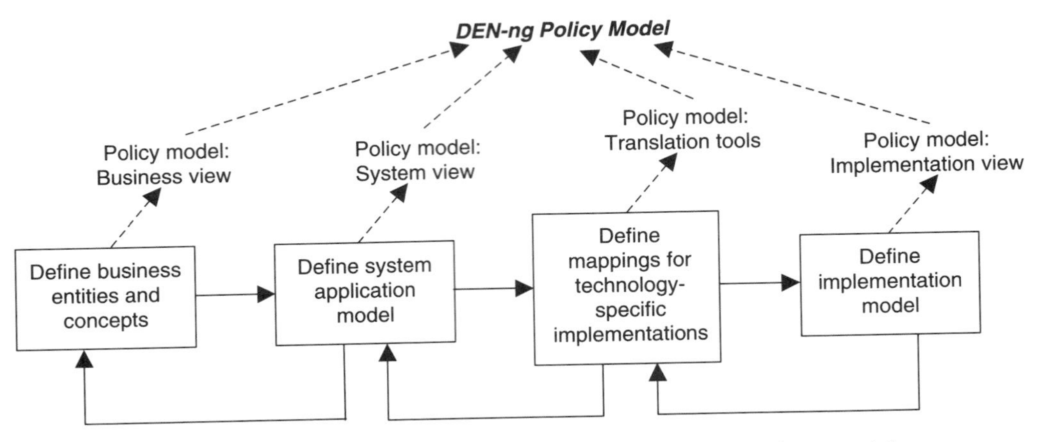

Figure 1-9 Design approach for building the DEN-ng policy model.

This brief description is meant to show the reader that information models can and should be used to drive software implementations. By embedding the information model in a product, that product is better able to adjust to changing features and functions. For example, in the Intelliden R-Series, the different functions of IOS are modeled using extensions of the DEN-ng logical model. When a new IOS train is released, Intelliden only has to update the model. Code is then generated that understands these features, and these features are updated as part of that release's product catalog. These features are assigned as capabilities to that particular IOS version.

This is a great example of building to accommodate the future. The information model provides a formal structure to represent different device capabilities. Software can then be written that uses this formal structure to represent these capabilities in the R-Series product. This enables the structure of the R-Series to be fixed; when new IOS releases are produced by Cisco, Intelliden updates the information model, new Java code is generated, and the *rest of the interface and APIs of the product stay the same.*

As another example, in the Intelliden implementation, the information model is used for the following tasks:

- Represent different functions that can be programmed using the IOS software
  - Routing and forwarding functions
  - Peering with other hosts and devices
  - Traffic classification and conditioning functions
- Represent different commands and their structure as a function of a particular software release
- Represent different hardware and software capabilities of a given device
- Define business policies that control
  - Who can perform what changes on which devices
  - Who must approve a certain change (or category of changes)
  - When a change is deployed

- Define system policies that control when certain actions happen (e.g., when a device is examined for changes to its configuration and/or to its physical composition)
- Define implementation policies that control how changes are made and how services are activated

The information model serves as the centralized authority that links different parts of the managed environment to each other. As shown in the preceding example, the information model is used to define different types of policies that are used to control various types of behavior. Business, system, and implementation parts of the product are all seamlessly integrated using the DEN-ng information model.

1.5.3 Why An Information Model is Important

An information model is more than just a representation of a set of objects. The most important feature of an information model is its ability to describe relationships between managed objects. From this, other types of models and diagrams, such as defining how data flows within the system, can be defined.

The information model serves as a normalization layer. By concentrating on invariant aspects of an object (e.g., a device has physical ports over which information flows), a framework can be defined that can represent the different features and functions of heterogeneous devices. Device-specific differences can then be modeled by extending the common framework to accommodate the features and functions of these different devices. Without a common framework, different device features and functions cannot be easily accommodated because there is no common set of objects that can be used to build them from. In other words, to accommodate ten new features, a system that does not have a common information modeling framework must define ten new sets of objects (and more if interactions between these objects are to be modeled). If it is desired to interoperate between these ten new features, then in the worst case, all of the permutations of each new object operating with not just the other objects, but existing objects, must be defined.

Compare this to a system that uses a common framework. Adding ten new features means that the framework itself will be expanded to accommodate as many of these as extensions (i.e., subclasses) as possible. Furthermore, by developing these new features as extensions, interoperability with existing concepts and information is guaranteed.

It is not feasible to have a single information model that can represent the full diversity of management information that is needed. This is because the characteristics of managed data are very different and require many different subject matter experts. DEN-ng solved this problem by defining a layered information model that used patterns and roles.

A layered information model is one in which a common framework is built that supports different domain models. A simplified view of the DEN-ng layered information model is shown in Figure 1-10.

The DEN-ng common framework model consists of a set of classes and relationships that enable the different lower level models to be associated with each other. Because DEN-ng and SID are complementary, the DEN-ng model takes the work of the SID team and either uses it in an unaltered state (as shown by

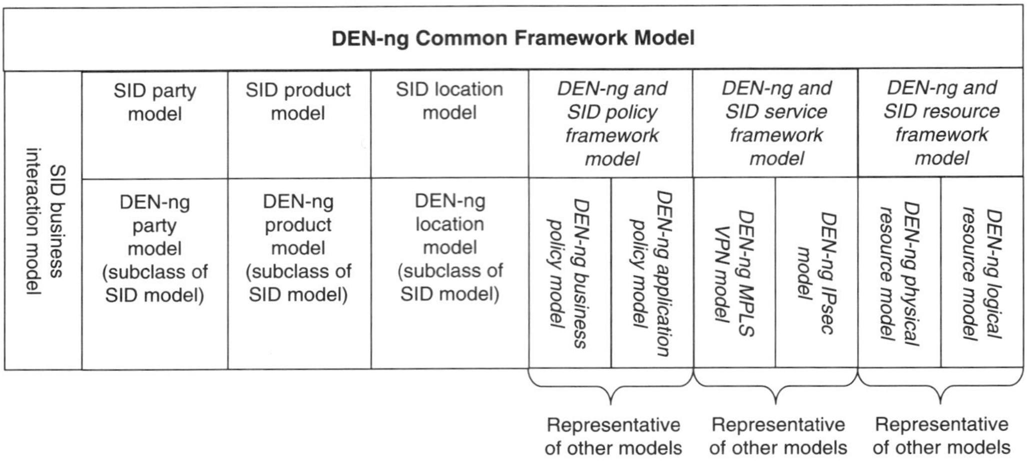

DEN-ng Common Framework Model

SID business interaction model

SID party model

SID product model

SID location model

DEN-ng and SID policy framework model

DEN-ng and SID service framework model

DEN-ng and SID resource framework model

DEN-ng party model (subclass of SID model)

DEN-ng product model (subclass of SID model)

DEN-ng location model (subclass of SID model)

DEN-ng business policy model

DEN-ng application policy model

DEN-ng MPLS VPN model

DEN-ng IPsec model

DEN-ng physical resource model

DEN-ng logical resource model

Representative of other models

Representative of other models

Representative of other models

Figure 1-10 A simplified view of the DEN-ng layered information model.

the business interaction model) or makes minor modifications to it (as is done in the party, product, and location models). Note that for the Party, Product, and Location models, DEN-ng takes the SIG models and defines new subclasses wherever possible. This means that the DEN-ng versions are more granular versions of the SID models. If DEN-ng needs to change something in the SID, then it is submitted as a change for review by the SID team. In addition, many parts of the DEN-ng model are in the process of being contributed to the SID team, as is shown in the policy, service and resource models. Each of these is in reality another framework model, into which additional submodels "plug into." For example, the DEN-ng policy model provides a generalized framework into which business policy, application use of policy, and other policy models can each plug into.

To provide as extensible a framework as possible, DEN-ng uses patterns and roles to model common concepts in as generic a way as possible. This differentiates DEN-ng from most other models (e.g., DMTF, IETF, and ITU), as they do not use roles and patterns.

Modeling objects describes entities in a system, their inter-relationships and behavior, and how data flows within the system. This provides the ability to represent and understand the programming model of the device. Three examples are CLI, SNMP, and Transaction Language One (TL1). TL1 is a set of ASCII instructions that an OSS uses to manage a network element—usually an optical device. More importantly, it provides the ability to understand dependencies between hardware and software. For example, a router may have a line card that has a main CPU and memory that are dedicated to performing traffic conditioning functions. This may work fine at low speeds (e.g., a fractionalized T1). However, at high speeds, such as OC-48, suppose that this particular type of CPU cannot keep up. Or even if it could, suppose that there was not enough memory.

This is an example of a dependency that most current PBNM systems will not catch. That is, the card has the correct operating system version, and the operating system says that it can perform this type of function. However, the physical media is simply too fast for this card to perform this type of function. The reason that most PBNM systems will not catch this dependency is because there is no convenient way to represent it. In contrast, any PBNM system that uses an information model, such as DEN-ng, will be able to model this and other dependencies naturally.

Information modeling provides a common language to represent the features and functions of different devices. DEN-ng uses the concepts of *capabilities* to represent functions of an entity and *constraints* as restrictions on those functions. Think of the information model as defining a common language that enables the different capabilities of each device to be represented in a common way. This enables them to be programmed together to deliver a common service. But sometimes, a particular environment might restrict the use of certain commands. For example, export control laws might restrict different encryption or other features from being used. These are modeled as constraints.

The combination of capabilities and constraints form a set of powerful abstractions that can be used to model current and future devices and services.

1.5.4 Linking Business, System, and Implementation Views

Most information models have focused on policy as a domain that is isolated from the rest of the managed environment. Here, domain is used to signify a set of related information and concepts. In contrast, the main use case for the DEN-ng policy model is to define a policy model that is closely integrated with the rest of the managed environment. The DEN-ng policy model is best thought of as an information model that defines how policy interacts with the rest of the managed environment (which is also represented as an information model). This has three important consequences.

First, it was apparent that building a policy information model in isolation of other information models was not going to work. The original DEN specification, as well as CIM, each had many different domains in addition to policy. However, little effort was made to associate policy in detail with these other domains. In addition, the original DEN and CIM models did not specify in enough detail how policy could be applied to a managed object. The DEN-ng model takes a different approach. It builds out the policy model as one of the last domain models and then concentrates on associating appropriate parts of the policy model with appropriate parts of other domain models.

Second, the existing models concentrated on representing policy. They either did not address or addressed in a very superficial manner how policy affected other managed entities. The difference here is subtle but important. Current policy models concentrate on defining the structure of a policy rule, what its condition terms are, and so forth. Although there was a lot of talk about policy changing a value in a device configuration file, the details of *how* that was accomplished were left unspecified. For example, the IETF and DMTF models do not specify the concept of a device interface or physical port. If the device uses CLI to change its configuration, how then can policy be used if these fundamental concepts are not modeled? The DEN-ng policy model fixes this unfortunate situation by developing other domain models alongside the policy model and ensuring that appropriate elements in the policy model can be associated with appropriate elements in other models.

The goal of DEN-ng is the ability to translate policy expressions *directly* to running code—something that cannot be done with existing models.

Third, the original models (and almost all current additions to those models) are still thinking of policy in a very static way (i.e., they use policies to express the static configuration of target devices). Most models concentrate solely on the network layer and do not provide an information model for representing business entities and how they affect target devices. In fact, there is very little literature on detailed information models that are designed with business views in mind, and even less literature describing how business information models can be linked to information models of other domains. For example, how does a changed SLA affect device configuration files? Clearly, the SLA defines how traffic should be treated, but when it is changed, the policy of treating that traffic is changed—how is that accomplished? Or how does a customer, who has just bought a new product with a higher class of service, get that service installed and running? These are two simple examples of linking the business world, with its set of entities and concepts, to the system and networking worlds, which have different expressions for those concepts. Although policy is required, unless the representations are equated, the business, system, and networking domains will always remain disconnected. This adversely affects service activation and deployment.

Both the IETF and the DMTF approaches do not attempt to represent business entities and objectives. Although a few other approaches[19] do, none have addressed building a set of models that are designed to support business, system, implementation, and run-time views (as espoused by Faurer et al.)[20] that are closely tied to an overall architecture. This concept is shown in Figure 1-11.[21]

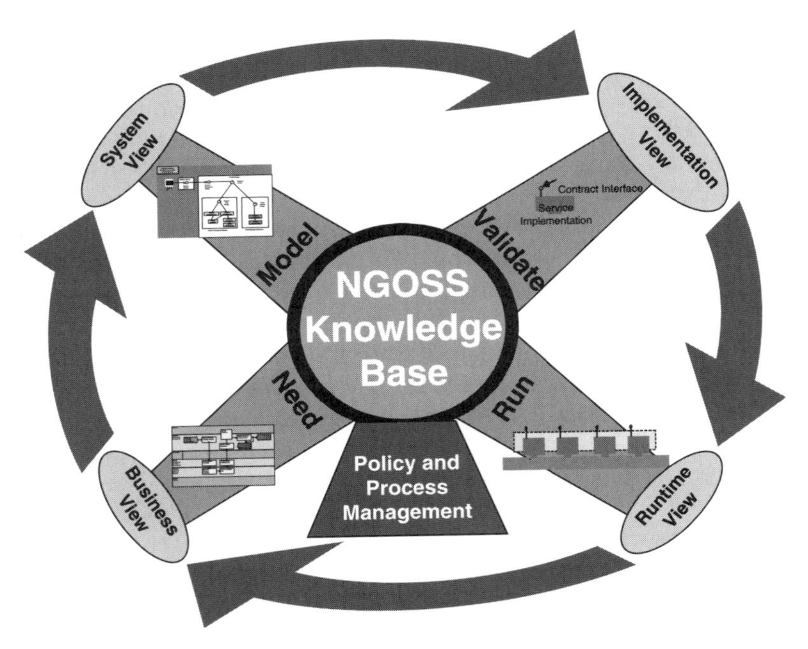

Figure 1-11 The TMF NGOSS architecture.

Figure 1-11 shows a conceptual view of the NGOSS architecture. The NGOSS knowledge base is a collection of information and data models, specifications, contracts, code, and supporting documentation that collectively and cooperatively describe how to build an NGOSS system. The four quadrants represent the business, system, implementation, and run-time views. The overall behavior of the system is driven by the holistic combination of policy and process management functions.

The TMF has developed a set of principles and procedures to coordinate each of these four processes. This takes the form of the various architectural specifications (e.g., TMF053 series), the TMF documents,[22,23] the contract work, and other elements, which together form the NGOSS knowledge base. Each of the DEN-ng domain models were built to fit into this approach.

A key objective of the NGOSS methodology is the development of models that focus on particular characteristics and procedures in an NGOSS system. These are characterized by the four viewpoints shown in Figure 1-11. The viewpoints are in turn tied together through the use of common shared information and a common means to exchange that information—contracts. The combination of the SID and contracts allow interoperability to be realized.

The SID (as well as DEN-ng) was built to provide a set of entities that model business, system, implementation, and run-time concepts. Put another way, the SID (and DEN-ng) were built to help realize the NGOSS architecture shown in Figure 1-11 by providing a common language to represent the transition of a concept from the business through the run-time views.

One main goal of the DEN-ng policy model was to accommodate the NGOSS architecture as shown in Figure 1-11. The DEN-ng policy model accomplishes this goal in two ways. First, it uses the different layers of the policy continuum to define different abstractions that must be modeled. This enables the different users of policy to work with and express concepts in their own terminology, rather than having the terminology and concepts of a static model given to them, never to be changed. Second, the different viewpoints are each supported by different views of the DEN-ng information model. This is realized by focusing on different entities for each of the DEN-ng domain models.

1.6 The Benefits of PBNM

There are many benefits to PBNM solutions. Some of the original drivers were listed at the beginning of this chapter. This section describes some of the more popular current benefits.

1.6.1 An Intelligent Alternative to Overprovisioning the Network

The economic downturn has forced companies to stop overprovisioning their networks and instead look to more intelligent means of delivering needed network services.

Overprovisioning may be done for several reasons. An organization may be running several mission-critical applications that must run in a timely, noninterrupted fashion. Or, it may use overprovisioning to achieve the desired levels of application delivery, such as quality of service, which its users require. However, the real reason that most networks have been overprovisioned is that it is supposedly easier and cheaper than its alternative—classifying, prioritizing, and conditioning the different types of traffic that exist in the network.

In truth, although overprovisioning can deliver on some of its promises, it cannot really solve QoS, and it is very expensive. With respect to QoS, overprovisioning attempts to solve the problem by making it go away. That is, its approach is to provide more resources than will be needed. However, QoS is all about *levels*. For example, although the following mechanisms all take a different approach to QoS, they all use a particular level on which to act:

- Congestion management methods, which essentially are different ways to sort and schedule traffic
- Congestion avoidance methods, which use various techniques to avoid congestion from occurring
- Policing and shaping enable the input and output rates of traffic to be controlled

Part of managing different types of traffic is planning on which types of traffic need which resources and trying to ensure that those resources exist. The problem with overprovisioning is that it never establishes a minimum level of performance. In addition, you must be careful what you overprovision. For example, providing extra bandwidth for certain types of applications, such as SNA and voice, does nothing; these applications need strict requirements on jitter, latency, and delay.

Of course, other problems exist with overprovisioning. The most important of these is that your network is being severely underutilized most of the time. Overprovisioning means that you will provision the network for a particular capacity. The longer you run at less than that capacity, the less your valuable (and expensive!) equipment is used.

PBNM solutions can be used to intelligently allocate resources. There is no free lunch here, however. A lot of work must be done, and the amount of work is arguably more than simply throwing equipment at the problem, as is done in over-provisioning. This is because what is important is ensuring that different applications having different needs of the network can peacefully coexist. This is more difficult than simply "throwing bandwidth" at the problem. However, the benefits are a more efficient, cost-effective, streamlined operation. Plus, as PBNM is implemented to classify traffic, it can also be used for a variety of other tasks (such as providing better security) at the same time.

1.6.2 Providing Better Security

As the number of users and applications proliferate, networks get more complex, and with complexity, comes risk. One important form of risk is resource abuse.

The benign form of resource abuse is when authorized users misuse their network privileges (e.g., downloading large music or video files when the network is congested, playing network games, and other acts). Users often do not realize what an adverse effect such acts can have on a network. PBNM solutions can help by simplifying the enforcement of policies that clamp down on these abuses and prevent them from happening.

The worrisome form of resource abuse is when unauthorized users attempt to gain access to corporate information. A variant of this is when malicious users attempt to disrupt the operation of the network by either a denial of service attack or by sending a worm or virus into the network. PBNM can help categorize traffic into expected and unexpected types and assign rules to deal with each. For example, if a web-based virus is detected, a PBNM product can easily shut down the ability for routers to forward web traffic. This helps contain the problem while it is being diagnosed.

The dangerous form of resource abuse is when an employee or similarly trusted user decides to willfully misuse his or her privileges and violate a company's intellectual property rights. Studies show that the greatest threats to intellectual property come from within a company. PBNM lets administrators restrict users to only those applications and information sources that they need during their current session.

Any one of these forms can stop unauthorized applications from using shared resources that they should not have access to. For example, if the goal is to meet a particular SLA that has availability levels specified, the seemingly innocent use of the network to download research information may cause periods of congestion that cause the SLA to fail. An SLA is a business concept. Therefore, it makes sense to let the business and IT personnel define which users can use which shared resources. This allows the company to define its network utilization based on the real requirements of the business contract. PBNM solutions are a good match for business policies that seek to optimize the performance of the network—the PBNM tools can be used to catch such unwanted occurrences and help ensure that the SLA is met. PBNM solutions can also be used to reveal traffic usage patterns, so that policies can be finetuned on an ongoing basis.

The common thread in all of these examples is that PBNM tools operate by first classifying traffic. Just as classification is used to decide what type of traffic conditioning to give to a particular flow, it can also be used to determine whether a particular user can access a resource or not. Depending on the capabilities of the PBNM tool, it may be able to do even more. For example, some PBNM tools can perform "deep packet inspection" and examine the contents of URLs. Security improvements can be done if the PBNM tool enables administrators to write policies to perform these checks and actions.

1.6.3 Managing Device Complexity

Network devices can be classified along several different dimensions. Some of the more important ways of classifying network devices are:

- What is the role of this device? For example, will it be on the edge or in the core? Is it a border router?
- What is the physical capacity of this device? For example, how much of a particular resource (e.g., number of ports) does a device have?

- What is the logical capacity of this device? For example, how many VPNs can a particular device support?
- What is the programming model (e.g., CLI, SNMP, TL1, etc.) used to program the device?
- What is the programming model used to monitor the device?
- What version of the operating system is this device going to use?
- What are the critical features (i.e., commands) that this device must support?
- What types of cards are available for this device?
- Is the configuration small enough to fit in flash memory, or does it require RAM?
- What types of services are planned to be activated on this device?

This is a very short list of many of the different factors that need to be considered. An information model is well-suited for managing this complexity, as it is able to represent these different device characteristics, and relate them to each other. For example, the simplified DEN-ng model shown in Figure 1-10 provides separate physical and logical resource models. Associations and constraints can be defined that relate different logical features to different physical features, thereby building up a more complete picture of the device. Similarly, policy can be applied to control which combinations of features can be used in a given situation. Separating the different domain models (e.g., physical resource from logical resource in the preceding example) enables each domain model to change without adversely impacting the other domain models. All that needs to be updated are the relationships between the different domain models. Furthermore, the ability to work on each domain model in parallel enables the information model to be more rapidly updated to accommodate new devices.

The benefit of using an information model to model device features and functionality is that this method is robust enough to justify the investment in understanding the capabilities of the information model. It provides a robust starting point for managing device and service complexity and offers an extensible and scalable platform to accommodate the future requirements of new devices and services.

1.6.4 Managing Complex Traffic and Services

The world has changed. Today, more types of applications are available that generate more types of traffic than ever before. Some sophisticated applications generate several types of traffic of different types (e.g., H.323 traffic, which generates both UDP and TCP flows). Other applications provide unpredictable behavior (e.g., they open random ports for communication).

In addition, networks have increased in complexity. Security more important than ever, because a network can carry many types of different traffic. Many of the individual flows representing this traffic load have different requirements. In the typical converged network (i.e., a network that carries data, voice, and video application traffic), some of the flows are sensitive to delay and jitter, whereas others are not. Thus, different flows require different types of traffic conditioning. For example, using any of the weighted fair queuing approaches will adversely affect voice traffic. Instead, voice traffic demands priority queuing so that jitter, latency, and delay can be controlled. However, if priority queuing is used for data traffic, relatively unimportant flows can swamp the priority queue and effectively starve other types of traffic. As another example, some traffic is classified as mission critical. If this traffic is to share the same network resources, then it demands completely different treatment to avoid compromising its usage.

Therefore, simply throwing bandwidth at network traffic is no longer the answer (not that it ever was for certain types of flows, such as SNA traffic, but people keep stubbornly associating PBNM with bandwidth). The real problem that network administrators face today is how to enable multiple applications that each demand slightly different resources from the network to not just peacefully coexist, but to work and consume shared resources according to their importance.

PBNM solutions are natural choices for these types of applications. PBNM solutions are predicated on analyzing traffic and classifying it into one of several predefined categories. Each category will correspond to preprovisioned traffic conditioning that is suited to the type of traffic that is being carried by that application. Advanced network technologies, such as MPLS or DiffServ (or even both), can be used to mark this traffic so that appropriate traffic conditioning is applied.

1.6.5 Handling Traffic More Intelligently

Because PBNM solutions rely on classification, they provide the opportunity to make other more intelligent decisions regarding how to handle all types of traffic. In addition to deciding how the flow is to be conditioned, the classification decision itself can be used to help direct different types of traffic. For example:

- Nonauthorized users, as well as other forms of unwanted traffic, can be denied access to network resources. This is not to say that firewalls or VPNs are no longer needed; rather, it means that an additional measure of security is present and available
- Business-critical applications can be identified immediately and transported using special mechanisms, such as policy-based routing (i.e., based on a classification decision, traffic can be instructed to use a special path that normal traffic is not allowed to use)

Many more examples could be given. PBNM solutions provide the inherent intelligence to be used to accomplish more tasks than those that were originally intended.

1.6.6 Performing Time-Critical Functions

PBNM solutions can simplify and better implement two basic types of time-critical network functions.

- Change device configurations within a specific time-window
- Perform scheduled provisioning functions

The first point reflects the need to address common maintenance functions. Most organizations perform maintenance operations on their network at night or during other non-business hours to avoid any inadvertent adverse effects on the operation of network services. The second point addresses small, simple changes for a specific customer or set of customers. This is the "network equivalent" of setting up a conference call.

Part of the allure of PBNM solutions is that they can address both of these functions.

1.7 Summary

This chapter provided a quick retrospective on how PBNM was designed. Despite many of its early limitations, such as being single-vendor approaches and being focused on a particular technology, great promise was envisioned for PBNM solutions. Accordingly, vendors poured resources into making various types of policy solutions, and the press hyped these new solutions.

Unfortunately, these early solutions were misunderstood and were quickly developed without supporting technology and, most importantly, standards. Interoperability was destroyed, and PBNM started to get a bad reputation.

Fortunately, the TMF rejuvenated this effort. It brought a completely different approach—one predicated on tying policy to an architecture that used a shared information model—to the forefront. The TMF's NGOSS architecture emphasized the importance of business rules and processes, which was something that was lacking in previous efforts. Furthermore, it resurrected early work done using viewpoints to help provide an integrated, multifaceted approach for defining policy. This was picked up by the TMF's SID effort. The SID is a federated approach that incorporates DEN-ng and other models and information definitions.

The result is that policy has re-emerged as a new approach that is tightly integrated with other domain models.

The DEN-ng effort was based on this premise. It added additional insight, such as the use of a policy continuum and a finite state machine, to transform it to a *collected set of models, each of which represented a state of a managed object. Policy, then, was redefined as the means to control when a managed object transitioned to a new state.*

With this introduction in place, the motivation for PBNM was examined in more detail. Part of the allure of PBNM was its simplicity. Other benefits were also its ability to provide different services to different users, its promise of simplifying device, network, and service management, and its promise of requiring less engineers to do the work. Newer promises, such as helping to define the behavior of a system and managing the ever-increasing complexity of devices and services, were added.

However, the true breakthrough was when PBNM was defined as a means for business rules to drive the configuration of the network. This brought forth the promise of changing the network from a cost center to a profit center. Although the other benefits are very important, they only incrementally affect profitability. Transforming the network into a profit center is very compelling, as it affects the bottom line of the entire organization.

To complete this transformation, two key ingredients were needed. The first was the establishment of a shared information model. This was needed for many reasons, but one of the most important ones was interoperability. Modern-day

OSSs are not purchased from a single vendor, as they are too complex. Instead, they are built from best-of-breed applications. For these applications to scale, they should be constructed as components. For the components to share and reuse data, they need to use the same data, defined in a "universal language" that any OSS component that needs to share data can use. This universal language takes the form of a layered information model. DEN-ng and the SID are part of that solution.

The second ingredient is a revolution in how management applications are built. Management applications should be constructed using models to define their data and architecture. This revolutionary idea is epitomized by the NGOSS architecture. Its design process uses four viewpoints—business, system, implementation, and run-time—to define the functionality and processes of the architecture. Interoperability is achieved using the SID and contracts, which define how data are communicated using XML.

Finally, five new benefits of PBNM solutions were provided. Two of these focused on providing more intelligence to routing and managing traffic. Instead of overprovisioning the network and wasting valuable resources, PBNM can be used to intelligently assign different traffic to preprovisioned paths that already have the appropriate traffic conditioning in place. In addition, managing complex traffic and services, where different types of traffic having different needs compete for the same shared resources, can be efficiently managed using PBNM solutions.

Additional benefits were provided by realizing that the classification portion of PBNM solutions can be used for providing better security, accommodating the needs of confidential and mission-critical traffic, and others.

Finally, PBNM can be used to manage device complexity. Combined with an information model, a system can be built that can accommodate new types of devices that have new types of functionality by changing the information model and ensuring that software can be used to translate changes in the information model to code. In other words, the structure, GUI, and APIs of the application remain constant; only the internals (which are governed by the information model) change. An example of this new avant-garde application is the Intelliden R-Series, which will be discussed in Chapter 10.

1.8 Recommended Further Reading and References

1. Jude, M., *Policy-Based Management: Beyond the Hype*, Business Communications Review, March 2001, p. 52–56.
2. The TeleManagement Forum, *TMF 053: The NGOSS™ Technology Neutral Architecture Specification, Annex C: Behavior and Control Specification*, version 0.4, November 2002.
3. The TeleManagement Forum, *TMF 053: The NGOSS™ Technology Neutral Architecture Specification, Annex P: Policy Specification*, version 0.3, work in progress.
4. ISO, *RM-ODP Part 1. Overview and Rationale*, ISO/IEC 10746-1:1998(E).
5. The UML 1.4 specification is downloadable from the following URL: *http://www.rational.com/uml/resources/documentation/*.
6. Alhir, S., *UML in a Nutshell—A Desktop Quick Reference*, O'Reilly, 1998 Sebastopol, California.

7. Rumbaugh, J., Jacobson, I., Booch, G., *The Unified Modeling Language Reference Manual*, Addison-Wesley, 1999, New York.
8. The home page of the Differentiated Services working group of the IETF is *www.ietf.org/html.charters/diffserv-charter.html*
9. Low latency queuing combines strict priority queuing with class-based weighted fair queuing. Please see: *http://www.cisco.com/en/US/products/sw/iosswrel/ps1830/products_feature_guide09186a0080087b13.html*
10. An innovation policy-driven configuration management and activation product. Please see *www.intelliden.com*
11. Strassner, J., *A New Paradigm for Network Management: Business Driven Network Management*, Presented at the SSGRR Summer Conference, L'Aquila, Italy, July 2002.
12. Nichols, K., Blake, S., Baker, F., Black, D., *Definition of the Differentiated Services Field (DS Field) in the IPv4 and IPv6 Headers*, RFC2474, December 1998.
13. Moore, B., Ellesson, E., Strassner, J., Westerinen, A., *Policy Core Information Model—Version 1 Specification*, RFC 3060, February 2001.
14. Moore, B., Rafalow, L., Ramberg, Y., Snir, Y., Westerinen, A., Chadha, R., Brunner, M., Cohen, R., Strassner, J., *Policy Core Information Model Extensions*, draft-ietf-policy-pcim-ext-06.txt, November 2001.
15. Snir, Y., Ramberg, Y., Strassner, J., Cohen, R., Moore, B., *Policy QoS Information Model*, draft-ietf-policy-qos-info-model-04.txt, November 2001.
16. Moore, B., Durham, D., Strassner, J., Westerinen, A., Weiss, W., *Information Model for Describing Network Device QoS Datapath Mechanisms*, draft-ietf-policy-qos-device-info-model-08.txt, May 2002.
17. Strassner, J., *Directory Enabled Networks*, Chapter 10, Macmillan Technical Publishing, 1999, New York.
18. TeleManagement Forum, *TMF 053: The NGOSS™ Technology Neutral Architecture Specification*, version 3.0, April 2003.
19. *www.ebxml.org*
20. Faurer, C., Fleck, J., Raymer, D., Reilly, J., Smith, A., Strassner, J.: *NGOSS: Reducing the Interoperability Tax*, TMW University Presentation, TMW, October 2002.
21. Strassner, J., *NGOSS Technology Overview*, TMW Asia-Pacific Conference, August 2002.
22. The TeleManagement Forum, *GB921: eTOM—the Business Process Framework, version 2.6*, March 2002 (TMF member document).
23. The TeleManagement Forum, *GB922: Shared Information/Data (SID) Model: Concepts, Principles, and Business Entities and Model Addenda v1.5*, May 2002 (TMF member document).
24. The TeleManagement Forum, GB922: *Common Business Entity Definitions Addenda 1P*, May 2002 (TMF member document).
25. Strassner, J., Moore, B., Moats, R., Ellesson, E., *Policy Core LDAP Schema*, draft-ietf-policy-core-ldap-schema16.txt, October 2002.
26. Reyes, A., Barba, A., Moron, D., Brunner, M., Pana, M., *Policy Core Extension LDAP Schema*, draft-reyes-policy-core-ext-schema-02.txt, June 2003.

Chapter 2

Policy Management Fundamentals

This chapter will set the foundation for discussing policy throughout the rest of this book. It will introduce basic terms and definitions that are used in the study of policy management, as well as a simplified conceptual model of policy. This will be followed by describing the high-level system requirements of a policy-based network management system. Key among these is the notion of business rules driving the construction and deployment of device and network configuration. This new approach enables the network to be operated as profit center instead of a cost center.

2.1 Introduction

This chapter describes where policy-based management (PBM) systems fit in the overall scheme of management systems and provides an introduction to their operating context. As such, it will answer three fundamental questions:

- *What is policy-based management?*
- *Why is it important?*
- *How is it used?*

Policy-based management was briefly defined in Chapter 1. This is a very active research area, with entire conferences[1] and parts of prestigious conferences[2,3] covering various aspects of policy. However, to understand policy, we need to define some common terms. Therefore, this chapter begins by defining those terms. Next, a simple conceptual model of policy will be built to provide brief answers to the three questions (what, why and how) above. Subsequent sections of this chapter will then examine each question in detail.

Terminology is very important. The third section of this chapter reviews and summarizes current accepted terminology and provides some additional terms to help explain how PBM systems are implemented and used. This is important because words, such as "policy" and "goal," have many levels of meaning and understanding. Next, common requirements of PBM systems will be discussed. This section builds on the "Benefits of Policy-Based Management" section of Chapter 1 by defining the common features needed to realize these benefits. Finally, the notion of a workflow-based process for implementing policy is introduced. This process is essential to enabling PBNM systems to integrate the needs of the business world with the services that the network can provide.

2.2 The Need for Object-Oriented Analysis, Design, and Modeling in PBNM Systems

This section will describe the need for object-oriented analysis (OOA), object-oriented design (OOD), and object-oriented information modeling (OOIM) for PBNM systems. First, important concepts of object orientation will be defined. Second, OOA and OOD will be briefly explained, along with a description of their essential benefits. Finally, a brief description of why OOIM is important will be provided. This book assumes familiarity with object-oriented concepts.[4-6]

2.2.1 Guide to Object-Oriented Concepts for PBNM Systems

Object-oriented technology models the real world in terms of *objects*. This important point is often overlooked. Many people are intimidated by the set of formalisms used by object-oriented technology. However, this intimidation is unfounded. The basic object-oriented paradigm uses human description of the environment to create software models. In other words, object-oriented tools seek to represent what already exists in the real world in a consistent format.

An *object* is an abstraction of something that consists of a collection of characteristics and behavior. This collection is treated by the system as a named entity that has state and a unique identity. This book defines objects to represent system entities that can be managed, configured, and reported on, as well as objects to represent "things," places, and concepts. Objects promote understanding of the system components, the interaction between these components, and the overall system. They are, in effect, a set of "Lego building blocks" that can be used to describe the entity or concept in a reusable manner. In addition, they provide an extensible basis for implementation of a system.

For our purposes, an object has four fundamental parts: attributes, methods, relationships, and constraints.

An *attribute* (also called a *property*) represents data that defines fundamental characteristics of a particular object. The data can be defined using a simple data type (e.g., integer), a complex data type that is composed of simple data types (e.g., a TimePeriod, which is made up of two Date attributes), or as a reference to another object whose semantics are used to define the attribute (e.g., AuthenticationMethod). Not all attributes that are defined in a class have to be created; some are mandatory, and some are optional. Two criteria are used to decide whether an attribute should be created or not. The first criterion is imposed by the model. This criterion defines whether an attribute is required or not as a function of the integrity of the model. For example, a salary attribute of an employee will probably be designated as a required attribute because it represents common information that is present in all employees and because it also models a key aspect of what makes an employee an employee. Conversely, including the middle name of the employee might be nice, but not required, because some people do not have a middle name, whereas others have several. The second criterion is common sense: if that attribute describes a key characteristic of what is being modeled, then it should be created; otherwise, it is extraneous and does not need to be created (as long as the model does not declare it to be required, of course).

A *method* is a function or procedure that represents some behavior that is characteristic of the class. This behavior can be simple and fundamental to using the object or can be more complex. For example, the accessing and storing of its attributes (e.g., via "getter" and "setter" methods) enable the object's data to be *encapsulated* and not directly manipulateable. This fundamental tenet of object-oriented systems ensures that the *specification* of what an object can do is kept separate from its *implementation*. As another example, methods could be defined that enable the transformation and/or sharing of that attribute with other objects, perhaps by performing a computation using one attribute and writing the result into another attribute. Complex operations may also have side effects, which may also be modeled. This book assumes that all attributes of a class are read, written, and edited using the methods of the class (direct manipulation of attributes without using methods violates the principles of object orientation, which are stated later in this section). Methods may therefore be viewed as performing a transformation, or mapping, of the attribute(s).

A *relationship* is a construct that defines how two or more objects are related to each other. The nature of the relationship defines how the objects are semantically related to each other. Four types of relationships will be used in this book: generalization, associations, aggregations, and compositions. The following definitions are based on the Universal Modeling Language (UML; the *de facto* standard for defining and implementing information models) definitions of these terms.[7]

- A generalization defines how more generic objects (i.e., "superclasses" or parent classes) are used to define more specific objects (i.e., "subclasses" or child classes). Generalization enables separate concepts that are common to many objects to be used by each of those objects through *inheritance* (i.e., if an attribute is defined once in a superclass, that same attribute is inherited by all of its subclasses).
- An *association* is a semantic relationship (e.g., "depends on," "requires") between two or more classifiers. A *classifier* is a mechanism that describes behavioral and structural features of interest, such as classes and datatypes. An association can be thought of as a way to "connect" two or more objects and is therefore the "glue" that is used in an object model to connect different objects together to describe a concept. *Without associations, the object model is nothing more than a set of disparate classes that are unrelated to each other.*
- An *aggregation* is a special form of association that specifies a whole–part relationship between the aggregate (whole) and a component part. Thus, if an object "depends on" or "requires" a set of other objects to exist, then it is an aggregation.
- A *composition* is a strong form of aggregation that requires that a part instance be included in at most one composite at a time and that the composite object is responsible for the creation and destruction of the parts. It is this propagation of operations from the composite to the constituent parts of the composite that differentiates a composition from an aggregation. Composition may be recursive. Compositions are used for many things, but one of their essential uses is to represent containment.

Finally, a *constraint* is a semantic condition (often in the form of a *restriction*) that in UML is represented as a Boolean expression using the Object Constraint Language (OCL).[7] Constraints are used to represent global properties, conditions, or restrictions that apply to an object. For example, a constraint may be defined to say that an object can be created only after some operation to one or more other objects has been successfully completed. Appendix A introduces the notation used in UML to define these concepts.

Two additional constructs deserve mention. The first is the notion of a class, and the second is an introduction to abstraction. The structure and behavior of similar objects is defined by their common object *class*. A *class* is a description of one or more objects that have a uniform set of data (called *attributes*) and functions that manipulate and operate on the data (called *methods*). *Concrete* classes can be instantiated, whereas *abstract* classes cannot be instantiated. Abstract and concrete classes are used to organize and classify information in a class *hierarchy*, which is a set of classes that is used to organize information about a management domain. For example, Figure 2-1 defines the high-level construct of a *NetworkDevice*. Although *Routers*, *Switches*, and *Firewalls* are

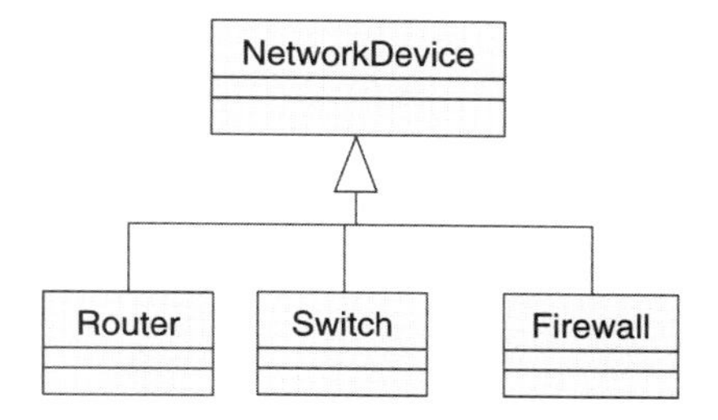

Figure 2-1 Different types of *NetworkDevices.*

each different types of *NetworkDevices*, all have a set of attributes and concepts that are in common. Thus, these different objects are grouped together in a hierarchy of related classes.

2.2.2 Introduction to Object-Oriented Modeling by Way of Example

As a brief introduction to object-oriented modeling, we will take the preceding model and improve and enhance it.

The model shown in Figure 2-1 has several inherent problems. The most basic problem is that it defines each of the three subclasses to be a "fixed" concept. To illustrate this point, consider a switch. A switch is traditionally classified as a "layer 2" device, meaning that it operates at the data link layer (below the network layer and the above physical layer).

Layer 3 and Layer 4 are two of the seven layers defined in the International Standards Organization (ISO) Open Systems Interconnect (OSI) model. This model is used to describe defined layers in a network operating system. The layers provide clearly defined functions that can better enable different devices from different manufacturers to interoperate. Each layer has a standard defined input and a standard defined output.

Many manufacturers advertise traditional switches and so-called "layer 3 switches"[8] and "layer 4 switches."[9] One possible solution that accommodates these additional types of switches is shown in Figure 2-2.

The problem with the approach in Figure 2-2 is that it misses the point of why these devices are called *Layer3Switches* and *Layer4Switches*. A *Layer2Switch* forwards packets based on the unique address of each device connected to the network. Layer 3 is traditionally used to partition subnetworks. Layer 4 is known as the transport layer and is the communications path between end-user devices and the network infrastructure. Thus, layer 4 is associated with a "higher intelligence."

Layer3Switches and *Layer4Switches* add intelligence to the switching function by switching traffic at higher network layers. Thus, subclassing a *Layer3Switch* or a *Layer3Switch* from a *Layer2Switch* is wrong, because the behavior defined for switching functions in each of these different layers is at different levels of abstraction.

A similar problem arises if a manufacturer makes a *router* that also has some *firewall* capabilities; this is shown in Figure 2-3.

This example trys to define a new class by using multiple inheritance (i.e., the new *RouterWithFirewall* class is derived from both the *Router* class and the

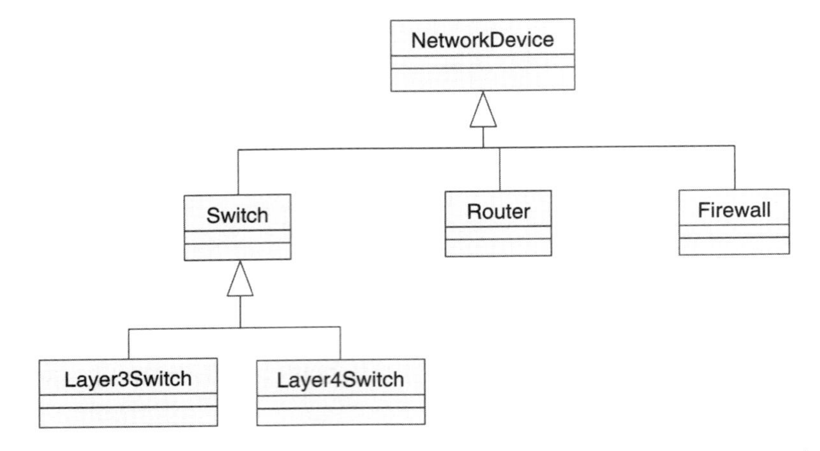

Figure 2-2 Trying to model layer 3 and layer 4 switches—a bad approach.

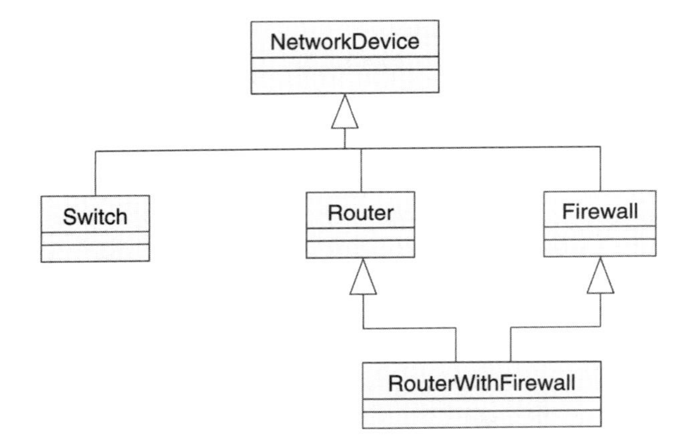

Figure 2-3 Trying to model a router with firewall capabilities—a bad approach.

Firewall class). Although this makes intuitive sense, the problem is that many systems cannot support multiple inheritance. The model is useless if it cannot be implemented!

A better approach is to use the concept of *roles*. *Roles* define different concepts an object can play with in the context of its related objects. The key to unraveling this puzzle is to think differently. Instead of conceptualizing a *Router*, a *Switch*, and a *Firewall* as an atomic device, separate the device from its functionality. Then, abstract the different functionality defined and associate each bit of functionality back to a device. Thus, instead of embedding functionality into a device, the role approach used by DEN-ng defines *roles* to represent different functions and then enables a device to aggregate one or more *roles*. This provides a much more extensible approach to defining functionality.

In the approaches illustrated in Figures 2-2 and 2-3, if a new function is added, an entire new subclass must be built. In the DEN-ng approach, a new *role* is created and then it can be decided if it should be associated with a particular device or not.

The DEN-ng role-based approach for solving this problem is shown in Figure 2-4.

The simplified DEN-ng role model below enables a *LogicalDevice* to take on zero or more roles. (In DEN-ng, a device such as a router is divided into a

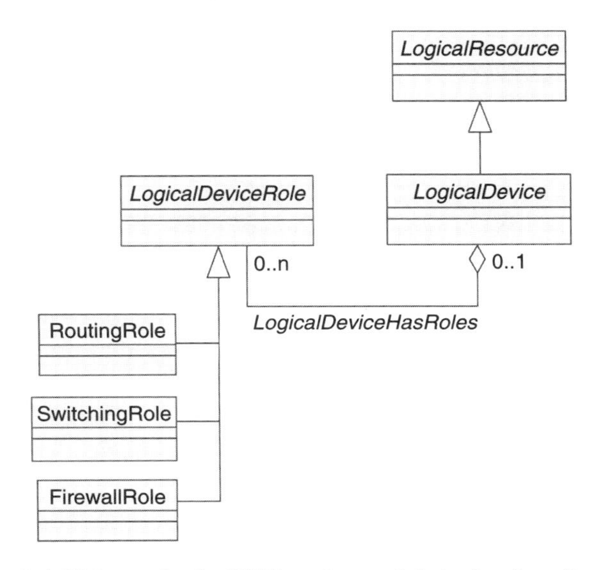

Figure 2-4 Using roles in DEN-ng to model device functionality.

PhysicalDevice and a *LogicalDevice*; the *PhysicalDevice* models physical aspects of the device, such as cards and power supplies, whereas the *LogicalDevice* models logical aspects, such as routing and switching traffic.) Different functionality is now abstracted as a *LogicalDeviceRole*, and new roles, such as *Layer3SwitchingRole*, can be added without affecting the definition of the *LogicalDevice*.

DEN-ng also defines various physical roles. Roles will be discussed later in this chapter and also in much more detail in Chapters 6, 8, and 9.

2.2.3 What is OOA and OOD?

OOA is the process of understanding a problem domain, determining what the responsibilities of the various components of the system(s) and participants using the system are, and developing a specification of how the objects of that domain function are related to each other. Implicit in this definition is the use of abstraction (i.e., the principle of ignoring certain aspects of a subject that are not relevant to the current purpose, so that the aspects of a subject that are relevant can be focused on exclusively). This is in effect a "divide-and-conquer" strategy, where a difficult problem is divided into smaller parts to better understand each part. Thus, by understanding each constituent part, a better and *more thorough* understanding of the whole problem can be achieved.

Object-oriented design is the process of taking a specification produced by OOA and adding enough detail to enable it to be implemented. The implementation is developed using the object-oriented principles defined earlier.

Traditionally, OOA and OOD have been defined and used as separate disciplines. However, most methodologies suggest or implement a "blurring" between the hard lines of OOA and OOD. This important principle deserves further elaboration. In traditional approaches, the strict definitions of OOA and OOD defined a set of differences that required different languages and notations. There are two obvious problems with this approach. First, if different notations and languages are used, how can all of the appropriate concepts, principles, and objects from OOA be represented in OOD? A good analogy is in the problems encountered when a literal translation between dissimilar human languages is attempted. Literal word-for-word translation and replacement of these words often forms gibberish or meaningless phrases. The second problem is that if the representation cannot be guaranteed, how can it be implemented correctly?

Instead, it should be realized that, fundamentally, both OOA and OOD are based on object-oriented principles. Therefore, the opportunity is to bridge these traditionally separate disciplines and concentrate on the unified representation of objects and concepts. After all, the basic object-oriented paradigm uses fundamental human linguistic and cognitive mechanisms to create software models. Why, then, cannot this same paradigm be used to analyze, design, and implement PBNM systems? This book argues that it can and should be used.

2.2.4 The Benefits of OOA and OOD

When blended together, OOA and OOD provide several compelling benefits over alternative methods that do not use object-oriented concepts. The most important of these are as follows:

- Provide a consistent underlying representation for representing the information being analyzed and modeled
- Improve the ability for analyst, subject matter experts (i.e., theorists), and implementation personnel to interact
- Improve the ability to cope with the complexity of distributed system
- Improve the ability to better recognize common concepts and behavior within a system and to realize them as objects that can be shared and/or reused by other system components
- Build specifications that are robust and resilient to change (a key feature of the DEN-ng and SID object-oriented information models)

Providing a Consistent Frame of Reference

OOA is concerned with understanding a given problem domain. It aids this process by bringing a set of formal methods and procedures to help understand the problem. One key advantage of this approach is that it is resistant to redefining the same concepts. Rather, an initial concept will instead become more detailed as additional analysis and design is performed.

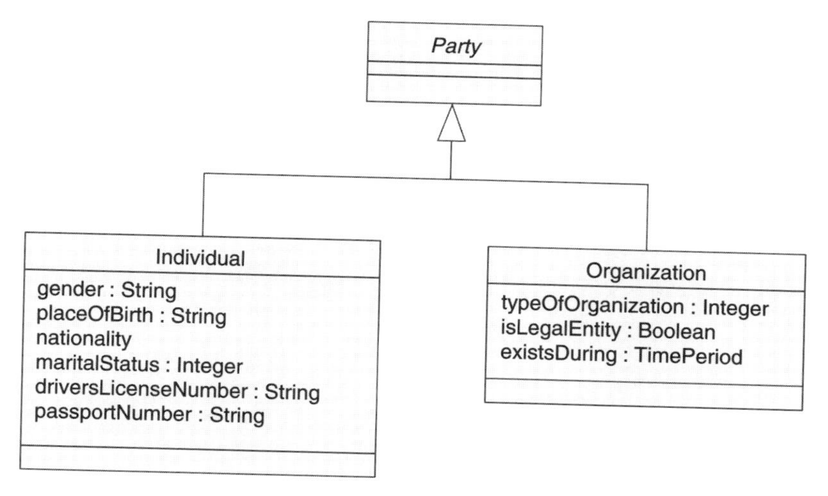

Figure 2-5 Simple business model of a party.

For example, Figure 2-5 shows a simple UML model of an *Individual* suitable for business analysts. This figure is taken from the DEN-ng specification, which is a system view of the higher-level (as in more abstract) business view of the TeleManagement Forum's Shared Information and Data (SID) Model Common Business Entity Definition Party specification.[10] This figure shows that two common concepts—*Individual* and *Organization*—are *generalized* into a new concept called *Party*. This abstraction makes it easier to represent behavior where an *Individual*, a *Group of Individuals*, or an entire *Organization* can be the subject and/or target of *Policy* (policy subject and target are defined later in this chapter).

Although useful for a variety of uses, Figure 2-5 is not very extensible. Consider, for example, the simple case where an employee can play several *roles* in an organization. If the approach of Figure 2-5 is used, many different subclasses of *Individual* and *Organization* will have to be created, and a complex set of associations linking them will also have to be defined. This is because *Individuals* and *Organizations* exhibit complex behavior that can change over time. Subclasses can only represent a single particular *fixed* concept. Thus, if a *Party* has to play more than one *role* at any given point in time, multiple subclasses with a complicated set of associations must be used.

Figure 2-6 shows an improved model,[10] where the concept of a *PartyRole* is introduced (attributes have been suppressed to make the figure simpler).

Figure 2-6 uses the role object pattern[11] to simplify this problem. By separating the information held about *Individuals* and *Organizations* from the *roles* that they perform, this model enables any *Party* to aggregate one or more *PartyRoles*. Note that the Party model has been significantly enhanced by building on existing simpler concepts. In fact, the DEN-ng specification, and each-Addendum of the SID, is written to lead the reader through a progressively more complex representation of an entity. This approach enables different abstractions of the same concepts to be used by different types of users.

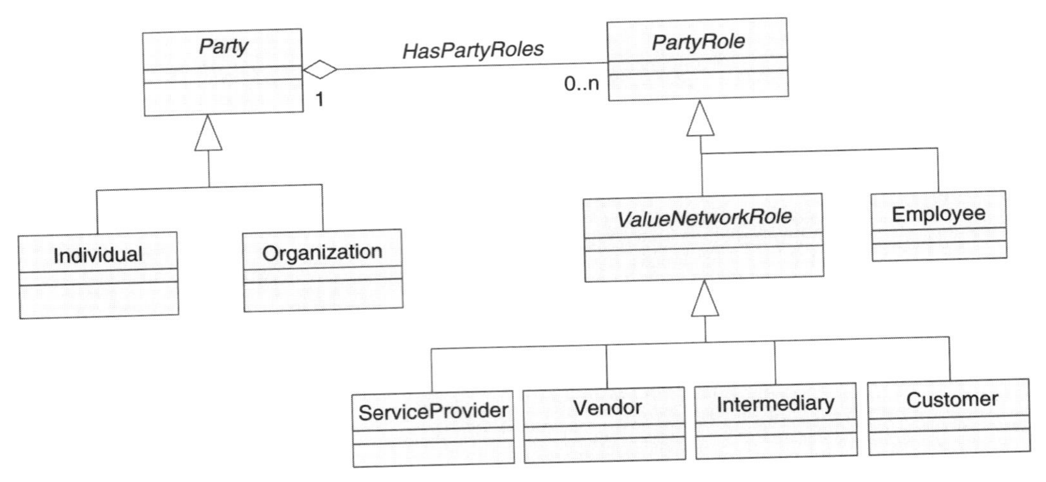

Figure 2-6 Improved business model of a party, using the role object pattern.

This use of roles is very similar to the previous use of roles to abstract device functionality. In fact, both DEN-ng and SID use roles throughout the information model to make each part of the information model inherently extensible.

Another important point in Figure 2-6 is how easily varying amounts of detail can be accommodated using OOIMs. The amount of detail shown on any given OOIM is determined by the use of that model and the needs of the users of that model. This enables a single "infrastructure" to be built, whose details are shown and hidden to suit the needs of the users.

Details of each class (e.g., attributes and methods) and constraints and relationships can be shown or hidden to suit the needs of the users and/or applications that are using the OOIM. Additional subclasses and components of those classes can be defined where needed and can be used to define additional functionality. For example, the full SID model has a flexible and robust set of mechanisms for dealing with how different *Parties* and *PartyRoles* are used. For example, different *Individuals* and *Organizations* can be referred to by different names. Consequently, the full SID model has a robust naming model that satisfies this need. This advanced functionality is rarely needed and was hidden (along with other features of the *Party* model) in Figure 2-6 because it was not needed.

Enable Different Types of Users to Interact Better With Each Other

Complex systems require a variety of different users, ranging from business analysts to system designers to implementation specialists, to work together to build a working system. OOA and OOD both rely heavily on classification, which is a formal way to organize the knowledge represented by an information model. This enables different people with different responsibilities to better add their knowledge to the collective whole without having to understand every detail of every part of the design. By merging the OOA and OOD disciplines, specific knowledge can be applied to a consistent underlying representation. In effect, a continuum of knowledge is built—some of the continuum is applicable at the analysis stage, some at the design stage, and other at the implementation stage.

Complexity in PBNM Systems

One reason that policy is needed in network management is to cope with the inherent complexity of networks. OOA and OOD provide formal, yet intuitive, methods for tackling the complexity of network management systems. Various forms of abstraction, which are represented naturally using OOA and OOD methods, can be used to reduce system complexity. Three examples are procedural, data, and entity abstraction.

Procedural abstraction enables any operation that provides a well-defined result to be treated by its users as a single process, even though the operation may in fact have been realized through a sequence of lower-level operations. This also enables the various entities that participate in this set of operations to be similarly abstracted.

Data abstraction enables a data type to be defined in terms of the operations that are applied to objects of that type. This enables complex data types to be designed and used by many different objects. For example, DEN-ng defines a TimePeriod complex data type that can be used to define the starting and ending points of a period of time. It is an object because it contains many different features that enable the starting and ending points of the time period to be specified. Because it is an object, it can be used by any other DEN-ng object that supports the concept of a time period. As we will see, this is an important concept in representing policy rules.

The third and final example of abstraction is the principle of entity abstraction. For example, consider a router. The *wrong* way of modeling a router is as a single class with a large number of attributes because such a model is not reusable. When one aspect of the router changes, the entire router model has to change.

Abstraction enables us to separate distinct aspects and functions of the router, as shown in Figure 2-7. (This is a simplified and abstracted view of six types of different aspects of a managed device, as modeled in DEN-ng and the SID.)

This fundamental abstraction is the DEN-ng definition of a managed device, which is also used in the SID. This abstraction enables each of these different aspects of a device to be defined in its own submodel. Development of each submodel can then proceed in parallel. More importantly, each of these different models can be reused by other types of devices. For example, the concepts of *Statistics* and *Alarms* are applicable not only to *Routers*, but also to many other types of *NetworkDevices*. In Figure 2-3, a *Router* is defined as a type of *NetworkDevice*. Similarly, in Figure 2-4, the function of routing is defined as a type of *LogicalDeviceRole*. Therefore, both *Statistics* and *Alarms* can be applied to various types of *NetworkDevices* and to *roles* that *NetworkDevices* play.

This design also provides inherent flexibility. For example, if the *Statistics* of one type of device differ from those of a different type of *NetworkDevice*, then this difference can be accommodated by defining different subclasses of the *Statistics* class. Because the overall model still relates *Statistics* to *NetworkDevices*, the new subclasses of *Statistics* will also be "auto-magically" related to *NetworkDevices*.

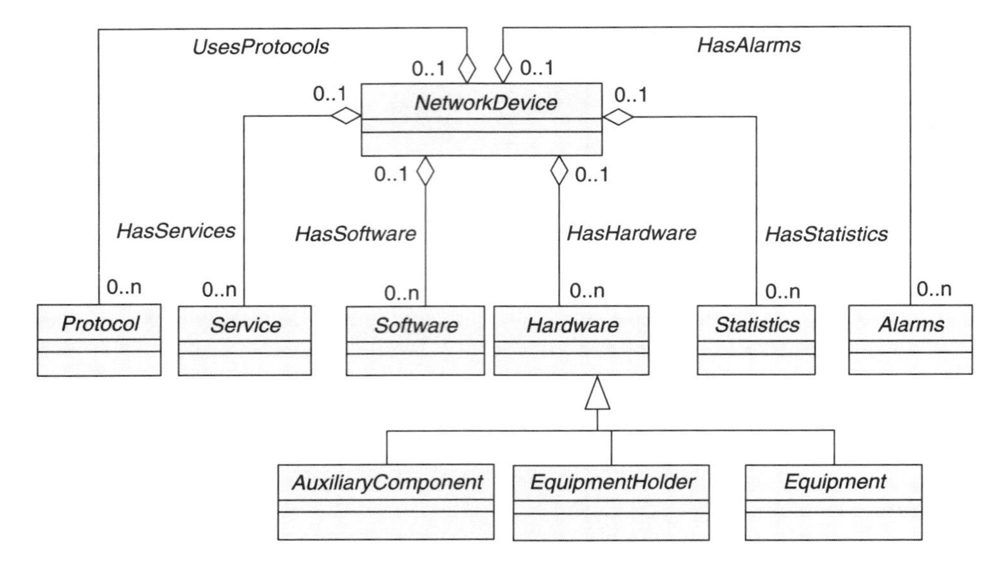

Figure 2-7 Using abstraction to define different aspects of a device.

Improvements in Reusing and Sharing Objects

OOA uses inheritance to identify common objects and concepts that can be shared and reused by different components. When combined with OOIM, a set of powerful class hierarchies—one for each different concept—can be defined. Furthermore, if both OOA and OOD are merged, then a common language and representation enables the best of OOA (representing what is to be built) and OOD (describing how to build the components identified in the OOA stage) to be holistically merged. This combination establishes a continuum for the objects being defined and specifies how knowledge can be systematically expanded and applied to different applications.

The above example of defining new types of *Statistics* classes is a good example of this point. The model shown in Figure 2-7 captures the fundamental relationship of a device aggregating a set of *Statistics*. Because a subclass inherits the relationships defined by its superclasses, new types of *Statistics* classes can be defined whenever needed. This provides an inherent robust extensibility to the model. Furthermore, because *NetworkDevice* can be subclassed as well, different *NetworkDevices* can use different *Statistics* classes. Thus, by taking care and building a rich infrastructure, different *NetworkDevice* having different statistical capabilities can be easily accommodated, and new classes can be reused by other classes.

Building Robust, Resilient Object Specifications

OOA and OOD (as well as OOIM) mechanisms use classification to organize information. This has the wonderful side effect of limiting changes to objects

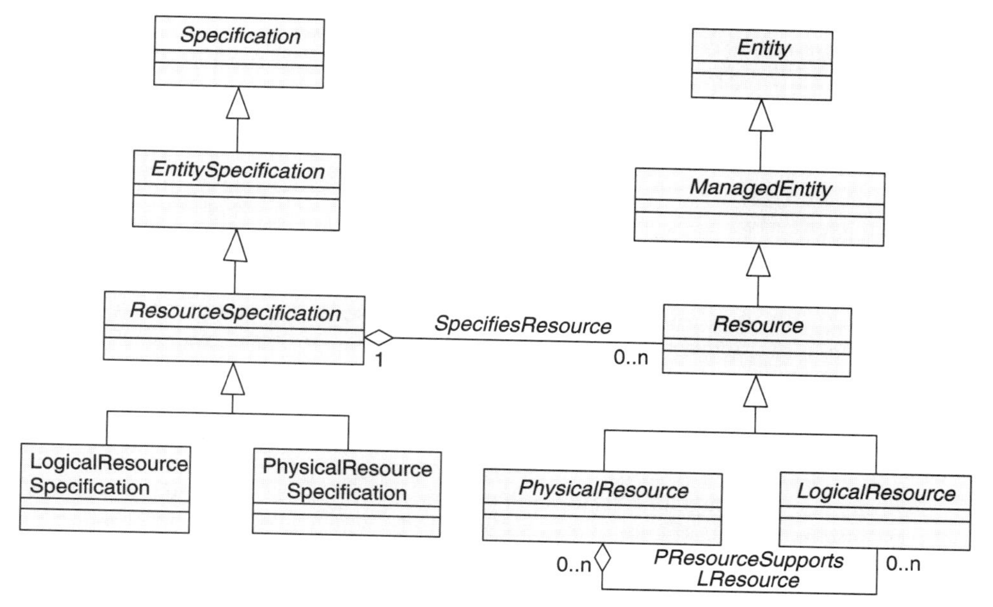

Figure 2-8 The use of *Specifications* in the DEN-ng model.

to a particular portion of the model. DEN-ng and the SID take this concept a step further by defining the concept of a *Specification* class (Figure 2-8).

DEN-ng and the SID represent different types of managed objects using different types of *patterns*. *Patterns* can be thought of as specifying a repeatable way to represent and/or implement a concept. This particular pattern is called the Entity-EntitySpecification pattern and features prominently in DEN-ng, SID, and OSS/J (the OSS for Java effort[12]). It is used to define how a managed object is represented. The idea is for the invariant (i.e., nonchanging) characteristics and behavior of a *ManagedEntity* to be represented by an *EntitySpecification* (or an appropriate subclass). This enables different *ManagedEntities* to specify the changeable characteristics and behavior of the managed object being modeled. Put another way, the *EntitySpecification* defines characteristics and behavior that different *Entities* have, and each of the *Entities* can then define its own unique characteristics and behavior that can be used to differentiate its instance from other instances. For example, suppose we want to model a phone. The *EntitySpecification* for the phone is used to define common things that all phones share. Differences in individual phone capabilities, such as whether it is a single, dual-band, or tri-band phone, can be captured in the specific phone *Entity*. Continuing the example, if there are multiple types of tri-band phones, then a subclass of the more general *EntitySpecification* can be created to model this capability.

The *SpecifiesResource* aggregation defines which *ManagedEntities* are related to which particular *ResourceSpecification*. For example, a vendor could make a *Layer2Switch* and a *Layer3Switch*; the difference is that the *Layer3Switch* also has some simple routing capabilities. If the only difference is additional software capabilities, then a *ResourceSpecification* could be defined for both the *Layer2Switch* and the *Layer3Switch*, and the differences between the two switches captured with two different *ResourceRole* classes.

In Figure 2-8, a *Specification* is an object that represents the invariant (i.e., nonchanging) characteristics and behavior of an object that it is associated with. The *EntitySpecification* class is the particular class (along with the *ManagedEntity* class) that establishes the Entity-EntitySpecification pattern. The *ResourceSpecification* class is one of several subclasses of the more generic *EntitySpecification* class, just as *Resource* is one of the more specific subclasses of *ManagedEntity*. *ResourceSpecification* is an abstract base class used to define the invariant characteristics (attributes, methods, relationships, and constraints) of a *Resource*. Other examples include *ProductSpecification* and *ServiceSpecification*.

An *Entity* represents the base class of all discrete devices in the managed environment. Its subclass, *ManagedEntity*, represents all devices that can be managed. This is an important distinction. An Entity is not manageable, However, even if the Entity is not manageable, it may still provide an important function that needs to be represented. A good example is a legacy hub, which provides connectivity. Thus, DEN-ng and the SID are capable of representing all entities in the environment. A *Resource* is the abstract base class for all entities that are inherently manageable and comprise a *Product* (note that *ManagedEntities* are not necessarily tied to *Products*). Because *ResourceSpecification* is related to *Resource* through the *SpecifiesResource* association, more specific subclasses of *Resource* and *ResourceSpecification* are also related to each other. Thus, the *PhysicalResource* and *LogicalResource* subclasses, which represent the physical and logical aspects of a *Resource* (such as a *NetworkDevice*), are related to the *PhysicalResourceSpecification* and *LogicalResourceSpecification* subclasses, respectively.

This pattern is a robust and extensible way to model managed objects of different types. Although Figure 2-8 shows this pattern applied to *Resources*, it is also used for *Products, Services*, and many other DEN-ng objects.

2.2.5 Why Object-Oriented Information Modeling is So Important to PBNM Systems

Object-oriented modeling is a design methodology that applies object-oriented analysis and design techniques to describe a system. A complete object-oriented model will describe both the physical and the logical aspects of the system. To do this, object-oriented terminology is introduced.

2.2.6 Basic Terminology

Objects that are grouped into the same class can also share similar relationships between other objects as well as share common semantics. Through *inheritance*, a subclass (child) of a superclass (parent) automatically has the same attributes, methods, constraints, and relationships defined as its superclass does. Therefore, a class can be thought of as a template that defines attributes, methods, constraints, and relationships that describe the class in a uniform way.

An *instance* refers to an actual object that belongs to a particular class. Each instance of a class can potentially have the same attributes and methods.

However, it can contain different values for its attributes. It also does not need to instantiate all of the possible attributes that are defined by its class. The class defines the attributes, methods, and relationships that the instance can possess, and the instance defines and differentiates objects that belong to the same class.

2.2.7 Object-Oriented Thinking

Object oriented implies a particular way of organizing and using information to build software and systems. Specifically, it means that a system is thought of as composed of a set of objects; each of which encapsulates data and behavior. Six fundamental axioms are required by this type of thinking:

- Identity—the ability to distinguish each object within a system; this implies that two instances of the same class can be identified even if they have the exact same attribute values.
- Abstraction—the process of focusing on a subset of essential characteristics of an object, even if other characteristics of that object must be ignored.
- Classification—the process of grouping objects that have the same characteristics and behavior (e.g., attributes, methods, and relationships) into the same class. Classification further implies that there is a structured hierarchy for organizing all classes in a system.
- Encapsulation—the process of separating the external characteristics and behavior of an object from its internal implementation. Other objects in the system can then depend on those characteristics and behavior without having to know how the object is implemented. Encapsulation conceals the structure and implementation of an object, and instead focuses on its externally visible characteristics and behavior. This enables the implementation of an object to change without drastically affecting other objects in the system.
- Inheritance—a mechanism for expressing that two classes are related to each other. Specifically, a *subclass* inherits all of the attributes, methods, and relationships of its *superclass*. Thus, common characteristics and behavior are expressed once in a superclass, enabling subclasses to represent specific refined behavior of the superclass.
- Polymorphism—a mechanism that enables several classes to share the same basic type of functionality, but implement the same operation differently in each of the classes. For example, a draw function that is used by two different shapes (a line and a circle) renders each shape in slightly different ways. Furthermore, each function may have different prerequisites and side effects.

2.3 Conceptual Policy Model

It will be easier to understand what PBM is all about if we first start by understanding one of its fundamental building blocks: the policy rule.

The underlying model of a policy rule is deceptively simple and is illustrated in Figure 2-9 using UML notation. The lines with the diamonds at the end are aggregations, with the diamond denoting the aggregate part of the relationship.

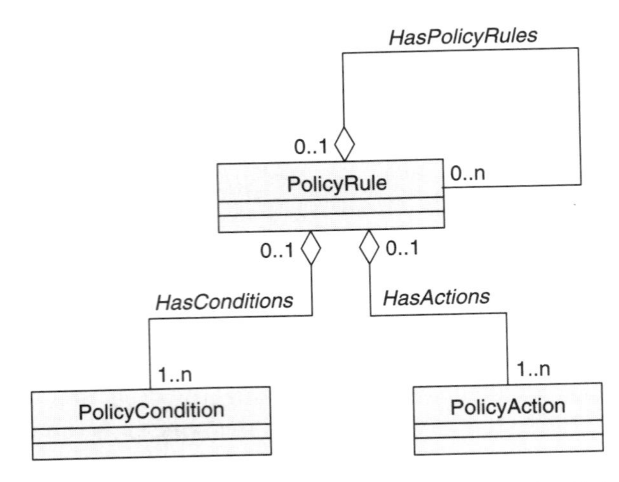

Figure 2-9 Simplified conceptual model of a PolicyRule.

In this model, a *PolicyRule* is a container that aggregates a set of *PolicyConditions* and a set of *PolicyActions*. The cardinalities of these two aggregations are defined so that at least one condition and at least one action must be defined as part of the *PolicyRule*.

Note that the Internet engineering task force (IETF) and distributed management task force (DMTF) made these two cardinalities zero-or-more to zero-or-more. This was to allow for policies that are in an incomplete state. However, this is wrong, because this mixes the state of constructing a *PolicyRule* with the state of an already constructed *PolicyRule*. If the PBM system is supposed to manage the construction of a *PolicyRule*, then as a minimum, the process of constructing a *PolicyRule*, including its life-cycle aspects, must be modeled. This results in a set of inter-related models that work together to model various aspects of the life cycle of a *PolicyRule*. The simplified model shown in Figure 2-9 is just one portion of this set of models and is used to model the current state of a *PolicyRule*. We will discuss this more in later chapters.

Directory enabled networks (DEN) and DEN-ng both model the process of constructing and deploying a *PolicyRule* using a finite state machine. Figure 2-10 shows the other major processes needed to model the life cycle of a *PolicyRule*. (Note that conflict detection and resolution, as well as the basic notion of editing a *PolicyRule*, are not specifically included in this figure. This is because they need more description, as each are potentially complicated operations. They will be addressed in Chapters 5 and 6.)

The semantics of this simple *PolicyRule* are as follows. The *PolicyRule* is itself a container that consists of three things:

- Metadata, which defines the overall behavior and function of the policy rule
- Boolean condition clause
- action clause

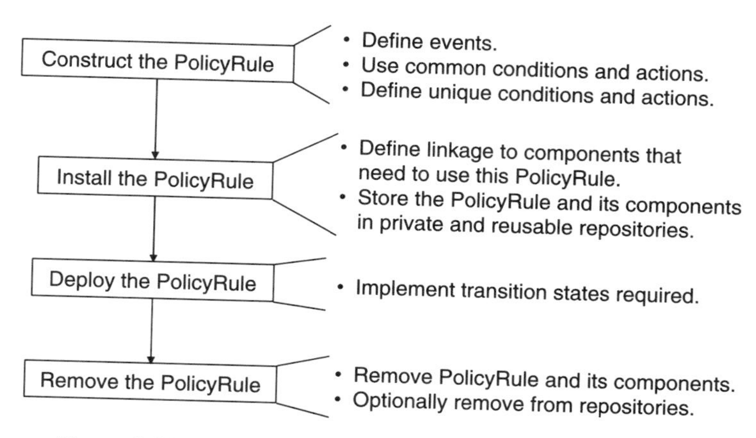

Figure 2-10 Simplified life cycle model of a PolicyRule.

Although the Boolean *PolicyCondition* clause may be composed of a set of *PolicyCondition* clauses, the end result is a single result that says whether or not the *PolicyCondition* is satisfied. If the *PolicyCondition* is satisfied, then the actions aggregated by this *PolicyRule* may be evaluated. In theory, one could define action clauses that define what should happen if the *PolicyCondition* clause is not satisfied. Practice has shown, however, that the (dramatic) increase in complexity of doing this results in little practical return. The increase in complexity results from many factors, such as the interaction of a *PolicyRule* with other *PolicyRule*, as well as the side effects that can result from the execution of a *PolicyRule*. In addition, this greatly complicates the policy conflict detection and resolution process.

Similarly, the *PolicyAction* clause is composed of one or more *PolicyAction*. *PolicyAction* can be optionally prioritized. This prioritization, along with what we will later refer to in Chapter 6 as an "execution strategy," enables a set of common semantics to be applied that govern how these *PolicyActions* execute with respect to each other. For example, one *PolicyAction* could have an execution strategy of execute and exit (i.e., prevent further processing of the actions), whereas another could have the semantics of execute and continue (i.e., try and execute the next action).

Chapters 5 and 6 will examine this structure and more advanced policy structures in more detail. For now, it is sufficient to note that this simple definition will be expanded in three important ways:

- Concept of nested (as in hierarchical) *PolicyRules* and *PolicySubRules* will be added (true of the IETF, DMTF, and DEN-ng approaches)
- Concept of rule-specific versus reusable policy components will be added to enhance the (simplistic) definitions of conditions and actions in the *PolicyRule* (true of the IETF, DMTF, and DEN-ng approaches, although the DEN-ng approach enhances this beyond the IETF and DMTF approaches)
- the concept of events that trigger the evaluation of the conditions in the *PolicyRule* will be added (true *only* of the DEN-ng approach)

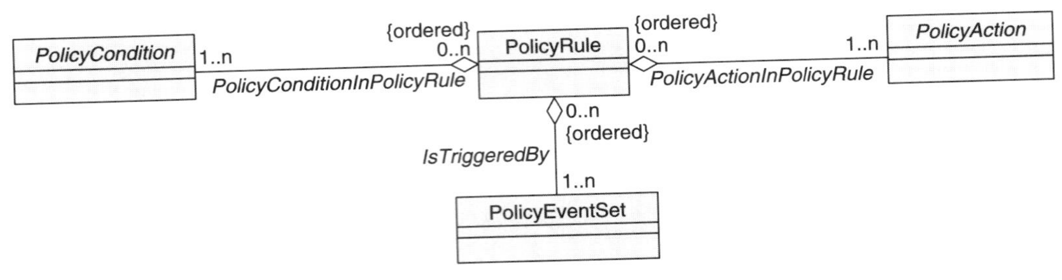

Figure 2-11 Simplified DEN-ng model of a *PolicyRule*.

The IETF (and DMTF) have defined the first and second of these additional concepts, but not the third. This is important, because events can be used as a mechanism to explicitly indicate how a *PolicyCondition* is evaluated. Without this triggering mechanism, there is no way to indicate when a *PolicyCondition* is going to be evaluated. This is why DEN-ng expands this model to an event-condition-action triplet (as shown in Figure 2-11).

The preceding model of a DEN-ng *PolicyRule* has three important differences (even at this simplified level) compared with the IETF and DMTF approaches (Figure 2-12). They are:

Functionality	IETF/DMTF Approach	DEN-ng Approach
Triggering evaluation of a policy condition	Not specified	Specified using a set of events
Cardinality of aggregations	0..n to 0..n for all	0..n to 1..n for all
Use of OCL	None	Specified in each aggregation

Figure 2-12 High-level comparison between the IETF/DMTF and DEN-ng *PolicyRule* models.

These differences exist for the following three reasons. First, if the triggering mechanism for evaluating a *PolicyCondition* is not specified, then it is impossible to guarantee interoperability between different implementations of PBNM systems. Second, it is important for the model to be precise. Specifying a cardinality of 1..n prevents degenerate structures (e.g., a *PolicyRule* with no *PolicyActions*) and enables a higher degree of interoperability. Finally, the use of OCL enables rule-specific semantics to be specified in a simple way.

2.4 Definition of a PBM System

We can now provide a more formal definition of a PBM system:

Policy-based management is defined as the usage of policy rules to manage the configuration and behavior of one or more entities.

In this book, the use of an object model that expresses the *PolicyRules* as instances of a class hierarchy, along with the use of a management method (i.e., a finite state machine), is implied in this definition. This last point is critical. Most models are "open-loop" systems (i.e., they represent just the state of a set of managed entities). DEN and DEN-ng, as well as the emerging policy work in the TMF, all use a finite state machine to model the various states that a managed entity can have in its life cycle and the operations required to transition a managed entity to a new state (or to keep it in a current state). This enables models to be built to represent the entire life cycle of the managed system. Given these definitions, we can now answer the three questions posed at the beginning of this chapter. Note that the IETF and DMTF approaches do not specify any type of management method and hence are current state models.

Simply stated, PBM is a methodology for managing systems. It does this by modeling the different entities in the environment to be managed as a set of objects. However, it does not stop here—it also models the various relationships between objects and constraints placed on those objects. For example, two different users can be logged on to the same system but receive different classes of service, which dictate how the applications that each operate are handled in the network.

The models used by PBM provide a common representation of information that different system components and applications can use to build more intelligent, easier to manage systems. A common representation of information enables management data to be shared, reused, and altered by multiple applications. This common representation has as one of its foundations the principle of abstraction, which enables us to construct class and relationship hierarchies that model different aspects of managed objects. For example, instead of having one class to model a router, with many (as in unmanageable!) attributes, we can instead focus on different aspects of the router (e.g., its physical composition, the protocols it runs, the management information bases [MIBs; a body of knowledge that defines characteristics about an aspect or function of a network device] it supports, the traffic it filters, and so forth) and develop classes to represent each of these aspects. This enables us to reuse valuable concepts (e.g., a card that contains memory is not specific to a router and can be used to add memory to a wide variety of objects) and apply them to different objects. This in turn enables consistent representation of these same concepts.

PBM is a methodology that describes one or more applications that manage one or more systems according to a set of rules. These rules take the form of policies that are applied to components of the system to better and more efficiently manage those components. The application of the rules is governed by the finite state machine that describes how to manage the system. In this way, we can achieve true end-to-end control, as opposed to having "just" device- or element-level control without PBM control because the behavior of each component is captured by the states defined in the finite state machine.

PBM differs from other approaches in its use of policies to control the behavior of managed entities. As stated previously, implicit in this book's definition of a PBM system is the use of a management methodology—in our case, a finite state machine—to manage the life-cycle aspects of entities. This concept will be discussed more in Chapter 5.

PBM uses policies to control the behavior of a managed system in a predictable and consistent fashion. To do this, the characteristics of the system that is being managed must be represented in as much detail as required. Then, policies can be defined that govern each state of the managed object—from creation to deployment to destruction. Without policies, there is no way to coordinate the behavior (e.g., the state and state transitions) of the objects being managed, and there is no way to guarantee consistent behavior and reaction to events. This will be discussed more in Chapter 3.

How PBM uses policies is critical to the implementation of a PBM system. Many current PBM systems are focused on a particular component in a system, or a set of features, that must be controlled. For example, many quality of service (QoS) PBM systems are designed to control a small subset of the features of a device, such as a router. The worry, of course, is the interaction between the QoS features and other features of the router: what if the QoS PBM system makes an adjustment that adversely affects the delivery of some other service or feature that the router is supporting? The answer, of course, is for a PBM system to holistically manage the different components in a system and the different services that each device supports. This will be discussed in more detail in Chapter 3.

2.5 Policy Terminology—An Approach

Now that we understand what a PBM system is, we can now identify the essential terminology needed to study PBM. These terms provide a basic overview of the key components comprising a policy system and give insight into how PBM systems are implemented and used.

The motivation for a consistent set of policy terminology is that without common terminology, interoperable PBM systems will be impossible to describe, let alone implement and use. However, until relatively recently, no document defining a policy terminology existed.

In November 2001, RFC3198, an informational RFC that took the first step to formally define terminology for PBM systems, was released.[13] This delay wasn't for lack of effort. Rather, the problem was one of getting agreement from the major vendors who were implementing PBM systems. To understand this, realize that virtually all of the major network vendors have at least one, and usually multiple, PBM systems. Therein lies the problem—no vendor wants their *deployed* product branded as "nonstandard." Furthermore, because the IETF operates by consensus, until there is a majority consensus, a specification of the IETF will not be advanced regardless of how seminal the work is. Even then, the process of advancing an Internet draft can be delayed based on the review of one or more respected people. Given these considerations, the fact that this RFC was produced is indeed a positive step forward.

One may ask why such specifications are required, given that vendors are already producing products? The answer, is to promote interoperability. Sometimes, one can buy a policy server to control all of the functions of a device. However, the usual case is that a policy server is built to have a purposely narrow scope. For example, vendors will build a "QoS Policy Server" that controls *some* of the QoS functions for a set of different devices. Most policy servers that control a particular function, such as QoS, do not control *all* of the different commands of a given device corresponding to that particular function; let alone control all of the commands for other functions. Furthermore, if a network vendor builds a policy server, then that policy server will almost certainly be limited to supporting devices from that vendor. Thus, people need to integrate different policy servers, so that either the appropriate functionality and/or necessary vendor devices are managed.

RFC3198 is important because it is the first specification that seeks to formally define policy terminology from the IETF. However, because the information model that this RFC was describing was not completely defined and because this was arguably the first information model that the IETF had dealt with, there are several errors in this document. The author's name is on the document because more good would come out of a partially correct document than not having a document be released at all and because the author was at that time still actively involved in the IETF. Because that is no longer the case (the author having shifted most of his "standards" time to other fora), the policy terminology section for this book will be organized as follows:

- The remainder of this section will contain *just* the essential policy terminology, with due accreditation, that the author uses
- Appendix 2 will contain comments on other (not all!) terms found in RFC3198, so that the advanced practitioner can use them to further policy research

2.6 Essential Terminology for PBM Systems

This section will summarize only those terms that are relevant to generic PBM system design. There are in general two sources for these terms:

- RFC3198
- From the author, which are part of the ongoing DEN-ng specification work

RFC3198 is focused on policy-based network management, and in particular is strongly influenced by current IETF work in the areas of differentiated services,[14] policy representation,[15] and security.[16] As such, it is not generic to PBM, but rather is specific to policy-based network management.

Some of these definitions will be changed, based on implementation experience of the author. In other cases, the definitions of RFC3198 will be modified to make these terms more generic to PBM systems (i.e., when a single definition is overloaded with conflicting meanings resulting from different uses that are forced into a single definition). In every case, the RFC definition will first be analyzed to explain why it needs modification and then a new definition will be provided.

This RFC is a perfect example of how the standards process works in real-life. The reader will notice that I am a co-author of this RFC, yet I do not agree with all of its definitions. This is because the IETF standards process is driven by consensus. Co-authors can influence the content of the draft, but at the end of the day, the working group chairs need to go for consensus. Plus, vendors will often become more involved as the Internet draft nears completion, to avoid the stigma of having their product labeled as "nonconforming" with the standard.

The pertinent definitions from RFC3198 now follow in alphabetical order, except when terms are dependent on each other. Each definition is given in its own section, and comments (where appropriate) are provided immediately below each definition. The new definition, if needed, follows in *italics*.

2.6.1 Terms Relating to the Object-Oriented Foundations of PBM

This initial set of terms are formal definitions for terms used by PBM. They are either conceptual or architectural in nature, and although they don't define directly what a PBM term is, they do define needed concepts for PBM.

Data Model (defined in RFC3198; modified by DEN-ng)

RFC3198 defines a data model as: "A mapping of the contents of an information model into a form that is specific to a particular type of data store or repository. A 'data model' is basically the rendering of an information model according to a specific set of mechanisms for representing, organizing, storing and handling data."

The problem with the above definition is that it is not so much a *mapping* as it is the definition of an *implementation* (which requires not just a mapping, but other factors, such as implementing for ease of querying or performance). A better definition is as follows:

A data model is a concrete implementation of an information model in terms appropriate to a specific type of repository that uses a specific access protocol or protocols. It includes data structures, operations, and rules that define how the data is stored, accessed, and manipulated.

Information Model (defined in RFC3198; modified by DEN-ng)

RFC3198 defines an information model as: "An abstraction and representation of the entities in a managed environment. This includes definition of their properties, attributes and operations, and the way that they relate to each other. It is independent of any specific repository, software usage, protocol, or platform."

This definition is workable, although there are some small problems with it. These include: (1) what's the difference between a property and an attribute (they are actually synonyms!), (2) "the way they relate to each other" lacks the formalism to mandate the use of object-oriented relationships, such as associations, aggregations, and compositions, and (3) platform doesn't have any effect on the structure or implementation of an information model. Therefore, this definition is modified as follows:

An information model is an abstraction and representation of the entities in a managed environment. This includes definition of their attributes, operations, constraints, and relationships. It is independent of any specific type of repository, software usage, or access protocol.

An information model can be thought of as the defining document that is used to model all of the different managed objects in a managed environment.

Figure 2-13 illustrates the relationship between an information model and a data model in DEN-ng:

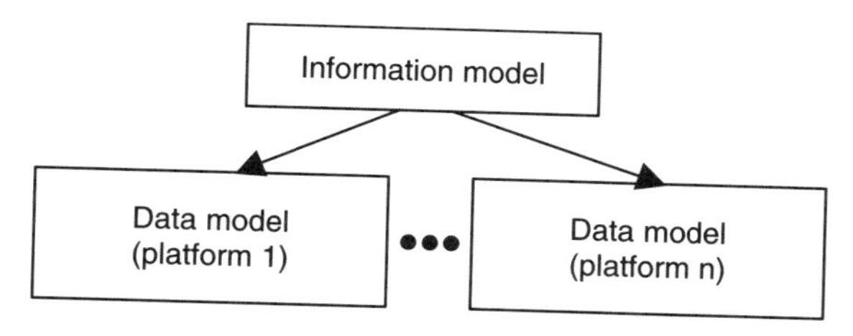

Figure 2-13 Relationship between an information and its data models.

DEN-ng specifies that a single information model is to be used. Multiple data models must be used, because management information is diverse and requires different types of repositories to facilitate the storage, querying, and editing of these data. This single information model will thus serve as the basis for all data models that are used. This helps ensure that different data models that represent parts of the same object will be able to interoperate.

Model Mapping (from DEN-ng)

RFC3198 does not define the term *model mapping*. It defines the term "policy translation"; however, this isn't quite the same thing, as this relates just to policies, not to entire models. Policy translation is covered later in this section.

Model mapping is used to enable different types of models to be related to each other. It is defined here so that other definitions in this section can use the term.

A model mapping is a translation from one type of model to another type of model. Model mapping changes the representation and/or level of abstraction used in one model to another representation and/or level of abstraction in another model.

The most common form of model mapping is from an information model to a data model; another important form is from a vendor-neutral data model to a vendor-specific data model. Another important form of model mapping is being done in the SID modeling working group of the TMF,[17] where different types of models (e.g., business domain models and system analysis models) are being integrated to form a common continuum of shared data. (see Chapters 5 and 6).

Figure 2-14 shows one form of model mapping. In this figure, a particular Data Model (e.g., a Directory) is mapped from an information model. This mapping produces a directory implementation that conforms with the appropriate standards (e.g., LDAP or X.500). The second tier of mapping accounts for the fact that different vendors provide varying degrees of compliance with the standard. In addition, some vendors provide features that are not yet standardized. This second tier of mapping enables these differences to be normalized, so that

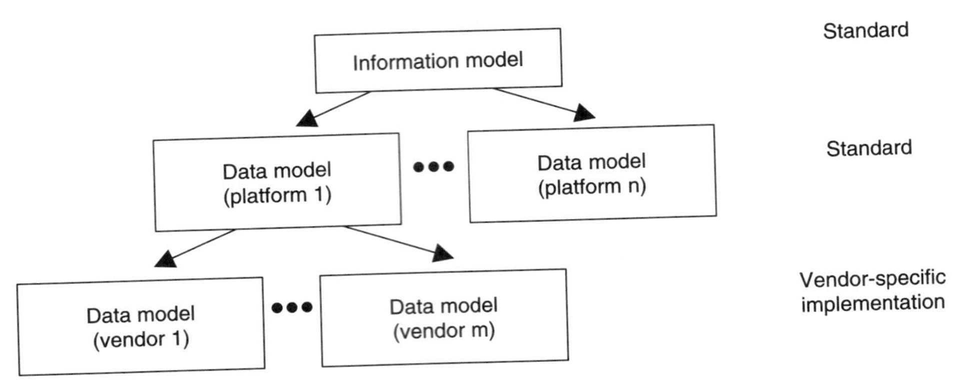

Figure 2-14 The concept of model mapping.

different implementations of the same data model can better interoperate. Thus, our hierarchy shows an information model standard being mapped to the appropriate data model standard, from which various vendor-specific implementations are built.

DEN (defined in RFC3198; modified by DEN-ng)

RFC3198 defines DEN as: "A data model that is the LDAP mapping of CIM (the Common Information Model). Its goals are to enable the deployment and use of policy by starting with common service and user concepts (defined in the information model), specifying their mapping/storage in an LDAP-based repository, and using these concepts in vendor/device-independent policy rules." This definition is wrong and misleading; the DEN specification was defined as two things:

- An information model (as defined earlier)
- A mapping to a specific data model that used the directory access protocol (DAP; part of the ITU X.500 suite) or lightweight directory access protocol (LDAP; defined by the IETF) access protocols to access data stored in a directory.

Thus, we see that DEN is not "just" an LDAP mapping, and it is not solely a mapping of CIM. CIM is not a pure information model by the above definition because it contains elements of data models (e.g., keys and weak references, which are both database concepts) that cause problems when mapping to various types of data models. The correct definition of DEN is as follows:

DEN is two things. First, it is a specification of an object-oriented information model describing the elements and entities in a managed environment, and how they are related to each other. Second, it also specifies a model mapping to a format that can be stored in a directory that uses (L)DAP as its access protocol.

DEN-ng (defined in DEN-ng)

DEN-ng is the next version of the DEN standard. It is being constructed in the TeleManagement Forum, not the DMTF, because it is very strongly tied to the NGOSS architecture effort. In addition, DEN-ng, like DEN, is UML-compliant. The DMTF CIM is not UML compliant.[30] Thus, it is easier to develop DEN-ng outside of the DMTF DEN effort.

The definition of DEN-ng is as follows:

DEN-ng is an object-oriented information model that describes the business and system views of managed entities and their relationships. This definition is done using UML and is strongly tied to the definition of NGOSS.

Because there was confusion regarding DEN being both a specification for building an information model and a data model, these efforts have been split apart in DEN-ng. Thus, there is a DEN-ng information model and a set of DEN-ng data models.

2.6.2 The Main Worker Terms of Policy

This section contains a set of formal definitions for the policy-specific terms *Policy*, *PolicyRule*, *PolicyGroup*, *PolicyCondition*, and *PolicyAction*.

Policy (defined in RFC3198; modified by DEN-ng)

RFC3198 defines policy in a very generic fashion (too generic, in the author's opinion!) as follows: "Policy can be defined from two perspectives:

- A definite goal, course or method of action to guide and determine present and future decisions. 'Policies' are implemented or executed within a particular context (such as policies defined within a business unit).
- Policies as a set of rules to administer, manage, and control access to network resources [RFC3060]."

Both perspectives of this definition are too generic to be useful. This definition fails to account for the differences among business, system, and network policies. For example, no difference is offered between a low-level queuing policy (to control how network traffic is conditioned) versus a high-level policy governing what resources can be accessed as a function of how the user is logged on (e.g., intranet versus public Internet), time of day, and other factors.

There are more problems with this definition. Looking at the "first perspective," it specifically avoids the use of the word "rule." This means that no definitive mechanism exists to define, implement, and use policy. More importantly, the "context" referred to in the first perspective is not present in moore et al.[18] Therefore, it conflicts with the RFC that defines the policy information model in the first place. (Note that Policy Core Information Model Extensions[19] make mention of defining a "context" by using a role. I think that this is a misuse of roles as defined in Moore et al.[18] This will be discussed more in Chapter 5.)

Sloman et al.[31] define a policy as follows: "a policy is a rule that can be used to change the behavior of a system." The only thing lacking from this definition is that it is specific to changing behavior. Let's instead use the following definition:

Policy is a set of rules that are used to manage and control the changing and/or maintaining of the state of one or more managed objects.

In this definition, "behavior" is replaced with "changing and/or maintaining state." This emphasizes the relation of policy to a management methodology; in our case, a finite state machine. The changing and/or maintaining of state could indeed denote a behavioral change, but it does not have to. Hence, the above definition is more flexible.

The preceding definition is influenced by the design of DEN and DEN-ng, which both use a finite state machine model. As discussed in Chapter 5, this is fundamental to the design of DEN-ng, which consists of three sets of classes: to model state, the changing of state, and policies to control when state is being changed.

PolicyRule (defined in RFC3198; modified by DEN-ng)

RFC3198 defines a policy rule as: "a basic building block of a policy-based system. It is the binding of a set of actions to a set of conditions—where the conditions are evaluated to determine whether the actions are performed."

This definition is somewhat obtuse. Furthermore, this definition is not accurate enough for our purposes, because the actions are not bound to the conditions (as implemented in moore et al.[18], which is quoted in the definition). Rather, the actions and the conditions are both aggregated by the policy rule, and the actions are enabled by the condition being satisfied. This was shown in Figure 2-11. Furthermore, metadata contained in the *PolicyRule* itself defines how the different actions will be executed (e.g., in which specific order) and whether execution should continue if a problem is encountered. In other words, the *PolicyRule* is an intelligent container that plays a vital role in determining how the events, conditions, and actions all work together.

This leads to the following revised definition for a *PolicyRule*:

A PolicyRule is an intelligent container. It contains data that define how the PolicyRule is used in a managed environment as well as a specification of behavior that dictates how the managed entities that it applies to will interact. The contained data is of four types: (1) data and metadata that define the semantics and behavior of the policy rule and the behavior that it imposes on the rest of the system, (2) a set of events that can be used to trigger the evaluation of the condition clause of a policy rule, (3) an aggregated set of policy conditions, and (4) an aggregated set of policy actions.

In our usage, the conditions and actions each form clauses, and the action clause is only executed if the condition clause is satisfied. Although many conditions can exist in a condition clause, the end result of all such policy conditions is to determine whether this policy rule is applicable. If applicable, then additional logic residing in the *PolicyRule* container is used to determine which policy actions are executed and how they are executed.

One or more events, or a combination of events, can be used to trigger the evaluation of the *PolicyCondition*. Similarly, many actions may exist in a condition clause, and one or more of them will execute as a function of the logic contained in the *PolicyRule*. This will be covered in more detail in Chapters 5 and 6.

A simplified picture of the DEN-ng PolicyRule was already shown in Figure 2-11. As with all other classes in this section, this class will be examined in much more detail in Chapters 6 and 7.

Policy Group (defined in RFC3198; modified by DEN-ng)

RFC3198 defines a policy group as: "an abstraction in the Policy Core Information Model.[18] It is a class representing a container, aggregating either policy rules or other policy groups. It allows the grouping of rules into a Policy, and the refinement of high-level Policies to lower-level or different (i.e., converted or translated) peer groups."

This definition should not be restricted to Moore et al.[18] Otherwise, interoperability is impaired. In addition, the last part of the last sentence is wrong. Policy Group classes do not cause *refinement* or *translation* of policy abstraction levels; they are used simply to coordinate the actions of separate policy rules.

This results in the following modified definition of a *PolicyGroup*:

A PolicyGroup is a container that can aggregate Policy Rule and/or PolicyGroup objects.

In DEN-ng and Policy Core Information Model Extensions[19] (PCIMe) a superclass of *PolicyGroup*, called *PolicySet*, is defined. By defining a recursive aggregation on *PolicySet*, both *PolicyRule* and *PolicyGroup* can inherit this relationship. This also enables us to define compound *PolicyRules*, as we will see in Chapter 6. Thus, a simplified picture of the DEN-ng *PolicyGroup* is shown in Figure 2-15.

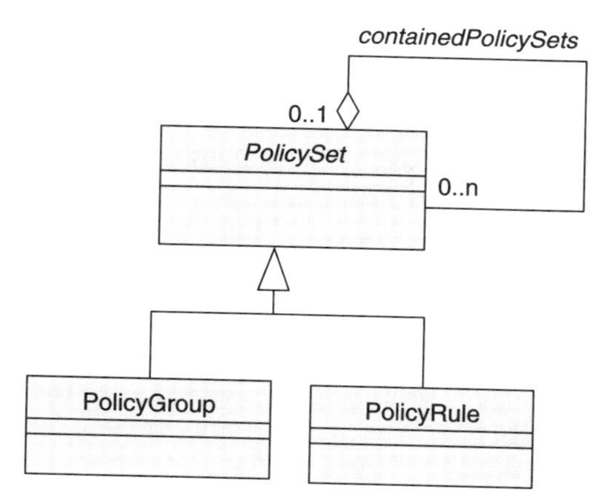

Figure 2-15 The DEN-ng simplified view of *PolicyRules* and *PolicyGroups*.

PolicyCondition (defined in RFC3198; modified by DEN-ng)

RFC3198 defines a policy condition as: "a representation of the necessary state and/or prerequisites that define whether a policy rule's actions should be performed. This representation need not be completely specified, but may be implicitly provided in an implementation or protocol. When the policy condition(s) associated with a policy rule evaluate to TRUE, then (subject to other considerations such as rule priorities and decision strategies) the rule should be enforced."

The problem with the preceding definition is its insistence that "this representation need not be completely specified." The original thinking in the RFC was that part of the representation would serve as a guide for implementation. However, this makes interoperability impossible, as there is no single standard to tie together different implementations. Another problem in the preceding definition is its lack of specificity.

The final problem with this definition is that it does not say what happens when multiple *PolicyConditions* are present. Although a simple *PolicyRule* may only require a single *PolicyCondition*, most *PolicyRules* need a set of *PolicyConditions* that must be evaluated together. In DEN-ng, this is called a *PolicyCondition* **clause**.

Experience has shown that implementing the *PolicyCondition* clause as a Boolean expression clause is simple and flexible enough to handle most PBM implementations. In this context, the task is to evaluate a Boolean expression to see if the *PolicyActions* of the *PolicyRule* should be executed or not. As with the nested rules above, there can be complex conditions, which are conditions that are composed of many individual condition clauses that are bound together. In such cases, one *PolicyCondition* in a *PolicyCondition* clause may need evaluation first for efficiency reasons (i.e., if it fails, then there is no need to evaluate the other condition clauses). The formal definition of a *PolicyCondition clause* follows.

A PolicyCondition clause is an aggregation of individual PolicyConditions and is treated as an atomic object that is aggregated by a PolicyRule. It is represented as a Boolean expression and defines the necessary state and/or prerequisites that define whether the actions aggregated by that same PolicyRule should be performed. This is signified when the PolicyCondition clause associated with a PolicyRule evaluates to TRUE.

A *PolicyCondition* is typically associated with the occurrence of an event (i.e., something significant that has happened). The PBM system itself will define what events are of interest. Common examples include a user logging onto the system, a link failing, and someone logging on to a router to change its configuration.

A *PolicyCondition* is usually represented by an expression that typically consists of three elements: a variable, an operator, and another variable or constant.[19,20]

A simplified view of the DEN-ng *PolicyCondition* class is shown in Figure 2-16. OCL expressions (such as the ordering of the *PolicyConditions*) have been omitted for simplicity and so as to emphasize the three fundamental relationships that affect *PolicyConditions*.

The recursive aggregation *ContainedPolicyConditions* will be used to build compound *PolicyConditions*. This will be discussed in Chapter 7. The aggregation *IsTriggeredBy* defines the set of *PolicyEvents* that can trigger the evaluation of the *PolicyCondition*. This is associated to the *PolicyRule* using the *PolicyConditionInPolicyRule* aggregation.

Policy Action (defined in RFC3198; modified by DEN-ng)

RFC3198 defines a policy action as: "definition of what is to be done to enforce a policy rule, when the conditions of the rule are met. Policy actions may result

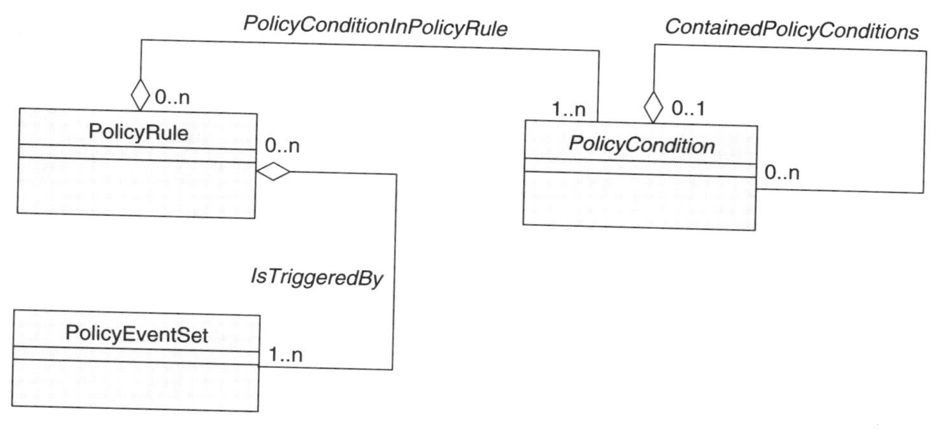

Figure 2-16 The DEN-ng simplified view of a *PolicyCondition*.

in the execution of one or more operations to affect and/or configure network traffic and network resources."

The problem with the preceding definition is its use of the word "enforce." Because a *PolicyRule* is an aggregation of *PolicyConditions* and *PolicyActions*, what exactly is being enforced? Although some representations implicitly define this, it is incorrect to abstract *all* definitions of *PolicyActions* to have this form. For example, consider the simple *PolicyRule*: "If traffic originated from this range of IP addresses, mark it with this DSCP."[21] It's easy to see that "enforcing" this *PolicyRule* means that anytime packets are detected with a particular range of source IP addresses, they should be marked a special way (e.g., certain bits in the IP header should be changed to a special value). Consider a policy that says: "Only relegate 30% of my core bandwidth to streaming video applications." How do you enforce this policy as it is currently written? First, you have to define what network elements are in the "core" of your network, and then you have to define which traffic corresponds to "video streaming" traffic. But what happens when there is no video traffic—does 30% of your network remain idle?

The other problem with this definition is that the representation of policy, in Moore et al.[18] lacks the semantics and metadata required to enforce anything because there is no specification of what is to be enforced. (As a side note, we could not get agreement in the working group about how to include a specification of what to enforce, which is why it is lacking. Vendor implementations played a large part in removing semantics and metadata from the definition.)

The final problem with this definition is that it does not say what happens when multiple *PolicyActions* are present. Although a simple *PolicyRule* may only define a single *PolicyAction*, most *PolicyRules* need a set of *PolicyActions* that must be executed together. In DEN-ng, this is called a *PolicyAction* **clause.**

The formal definition of a *PolicyAction* clause is as follows:

A PolicyAction clause is an aggregation of individual PolicyActions and is treated as an atomic object that is aggregated by a PolicyRule. It represents the necessary actions that should be performed if the PolicyCondition clause evaluates to TRUE.

These actions are applied to a set of managed objects and have the effect of either maintaining an existing state, or transitioning to a new state, of those managed objects. Note that DEN-ng differentiates between executing a PolicyAction and enforcing the results of that PolicyAction. This concept is missing in the IETF and DMTF approaches. It will be covered in more detail in Chapter 7.

A simplified view of the DEN-ng *PolicyAction* class is shown in Figure 2-17. OCL expressions (such as the ordering of the *PolicyActions*) have been omitted

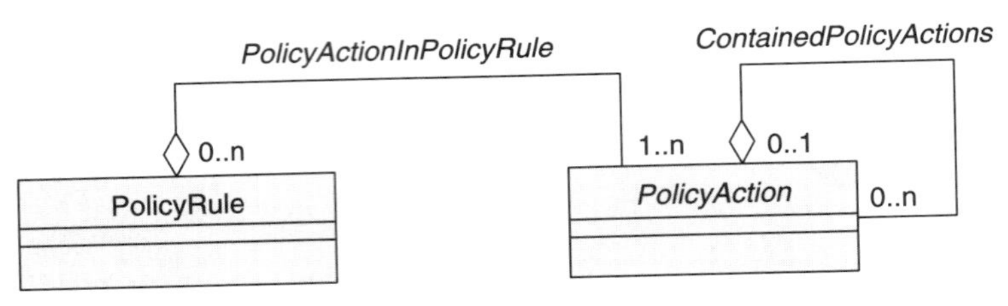

Figure 2-17 The DEN-ng simplified view of a *PolicyAction.*

for simplicity and to emphasize the two fundamental relationships that affect *PolicyActions*.

The *ContainedPolicyActions* aggregation is used for defining nested *PolicyActions* and is discussed in Chapter 7. The *PolicyActionInPolicyRule* aggregation is used to associate a set of *PolicyActions* with a particular *PolicyRule*.

2.6.3 Terms For Controlling Policy

This section contains a set of formal definitions for the policy-specific terms *PolicyConflict, PolicyDecision, PolicyServer, PDP*, and *PEP*.

Policy Conflict (defined in RFC3198; modified by DEN-ng)

RFC3198 states that a policy conflict, "occurs when the actions of two rules (that are both satisfied simultaneously) contradict each other."

Although correct, this definition is terse. The following expands on this definition, making its purpose clearer:

A policy conflict occurs when the conditions of two or more PolicyRules that apply to the same set of managed objects are simultaneously satisfied, but the actions of two or more of these PolicyRules conflict with each other.

Actions can conflict with each other in several ways. For example, two actions may generate two different values for an object. In general, conflicting actions will cause conflicting states to be specified for the same managed object.

Policy Decision (defined in RFC3198; modified by DEN-ng)

RFC3198 defines a policy decision as: "two perspectives of 'policy decision' exist:

- A 'process' perspective that deals with the evaluation of a policy rule's conditions
- A 'result' perspective that deals with the actions for enforcement, when the conditions of a policy rule are TRUE."

The problem with the preceding definition is that it is overloading a single definition with two different meanings that arise from two different uses. Each of these different meanings should be given its own definition, as follows:

- *A Policy Evaluation is the set of computations necessary to determine if the PolicyCondition clause is satisfied.*
- *A Policy Decision is the determination that one or more PolicyActions that are aggregated by a PolicyRule should be applied to a set of managed objects. These PolicyActions correspond to either maintaining the current state, or transitioning to a new state, of each of the managed objects that it is affecting.*

Policy Server (defined in RFC3198; modified by DEN-ng)

RFC3198 defines a policy server as: "a marketing term whose definition is imprecise. Originally, Yavatkar et al.[22] referenced a 'policy server.' As the RFC evolved, this term became more precise and known as the policy decision point (PDP). Today, the term is used in marketing and other literature to refer specifically to a PDP or for any entity that uses/services policy."

This definition needs further clarification. Although it is an imprecise marketing term, it should not be perceived as "just" a PDP. A "server" connotes more than just requesting and providing decisions; it also implies a broader interaction with the rest of the system. As a bare minimum, the definition should be enhanced to include a PDP and a PEP. (By the way, this was not agreed to in the IETF because of differing vendor implementations, as some products were "just" PDPs and others combined the notion of a PDP with a PEP, and *both* wanted to call their products a policy server.)

This thinking is predicated on the IETF decision to define a simple policy system, as illustrated in Figure 2-18. The DMTF also uses this definition; however, the TMF and DEN-ng do not for reasons that will soon become apparent.

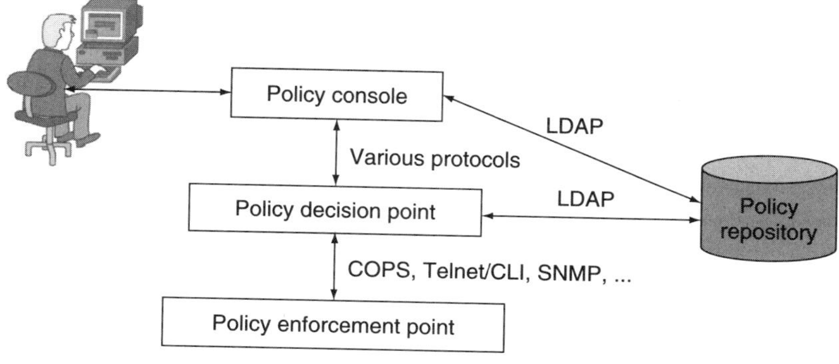

Figure 2-18 The IETF/DMTF conceptual model of a policy-based system.

There is more to a functioning PBM system than just these simple components. Two examples include the task of detecting (let alone resolving!) policy conflicts and translating policies to legacy entities that are not aware of policy requires functionality that is not present in a PDP or a PEP. This is because, in the most general case, conflict detection between different technologies, such as QoS and security, requires the PDP or PEP to be knowledgeable in each of those domains. When you add to this the needs of different devices having different programming models, a monolithic solution to conflict detection and resolution quickly becomes nonimplementable.

Therefore, a policy server needs to include these additional entities as a minimum. This is clearly a much longer discussion and will be covered in detail in chapters 4, 6, and 7. However, it is instructive to see the differences between what I think is a minimal specification of a realistic PBM system and the conceptual model shown in Figure 2-18. Therefore, Figure 2-19 shows my definition of a realistic PBM system.

Several important differences exist between Figures 2-18 and 2-19. The main points of this architecture are as follows. First, this architecture defines PBM systems as distributed systems, which means that all of the components of a PBM system must communicate with each other. This can be accomplished in a variety of ways. The figure uses a message bus; however, other methods, such as distributed communication using Jini, are also possible. Second, this architecture provides for three different types of conflict detection: global, neighborhood (inter-PDP), and local (intra-PDP). Global detection catches conflicts that are technology independent; inter-PDP detection catches conflicts between different tech-

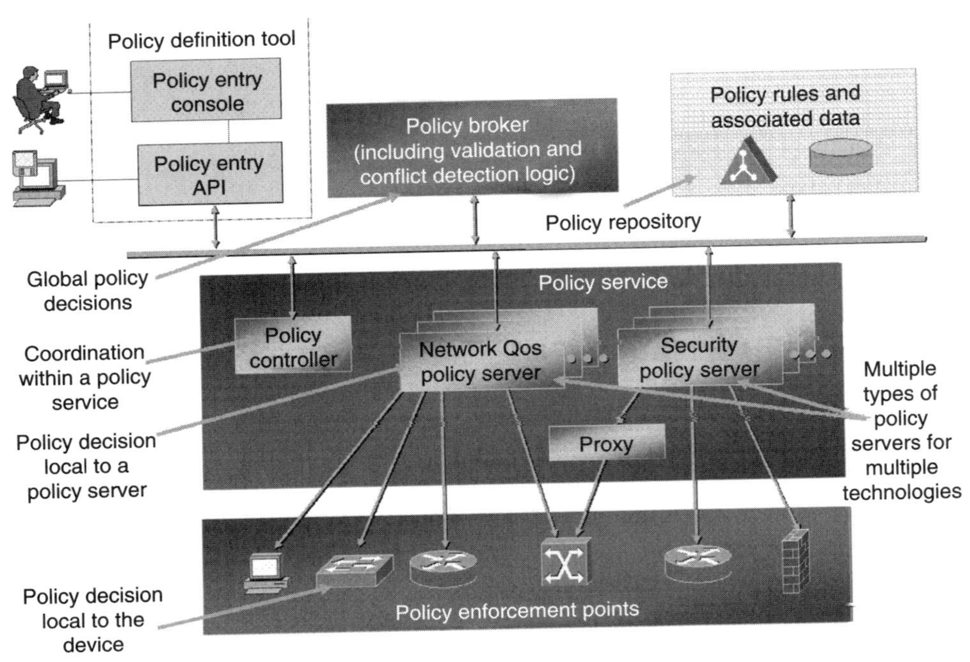

Figure 2-19 The IETF/DMTF conceptual model of a policy-based system.

nologies and/or vendors; and intra-PDP detection catches conflicts within a particular device or family of devices (or, depending on the complexity of the implementation, conflicts within particular functions of a device or device family). Furthermore, legacy devices may be unable to communicate with the PBM system; they may use different protocols and/or object models and/or programming mechanisms and may not have the ability to translate policy rules into their own configuration commands. (By the way, this gives rise to the definition of a *policy-unaware entity* in the following section.) Therefore, a policy proxy is used to translate to PBM policy rules into legacy configuration commands, such as CLI.

Policy Decision Point (PDP) (defined in RFC3198; modified by DEN-ng)

RFC3198 defines a PDP as: "a logical entity that makes policy decisions for itself or for other network elements that request such decisions."[22]

The obvious problem with this definition is the explicit reference to a network element. In addition, it does not define what a "decision" is. This latter problem is solved by the above definition of a policy decision. Our definition of a PDP is:

An entity that makes Policy Decisions for itself or for other entities that request such decisions.

Policy Execution Point (PEP) (defined in RFC3198; modified by DEN-ng)

RFC3198 defines PEP as: "A logical entity that enforces policy decisions."[22] This definition is terse and does not define what is meant by "enforcement." Our definition of a PEP is:

An entity that is used to verify that a prescribed set of PolicyActions have been successfully executed on a set of PolicyTargets. Note *that DEN-ng differentiates between enforcement and execution (see Chapter 7).*

2.6.4 Policy Container Terms

This section contains a set of formal definitions for the policy-specific terms *PolicyDomain* and *PolicyRepository*.

Policy Domain (defined in RFC3198; modified by DEN-ng)

RFC3198 defines a policy domain as: "a collection of elements and services, and/or a portion of an Internet over which a common and consistent set of policies are administered in a coordinated fashion.[21] This definition of a policy domain does not preclude multiple sources of policy creation within an organization, but does require that the resultant policies be coordinated."

To be honest, I've never understood this definition. Nichols et al.[21] doesn't even contain the term "policy domain" in it. Second, a domain is used to contain things, so that common operations can be executed against a group of the contained entities. The preceding IETF definition mentions "coordinated fashion," but this isn't necessarily true—domains are more about *who* than *how*. Third, elements and services are limiting—what about managing the allocation of common IP addresses, which are neither elements nor services?

The key to fixing this definition is as follows:

- A "managed entity" is what we want to collect in a Domain—a *ManagedEntity* can be a network device, a network service, an IP address, a route, or anything else that we need to manage in a common way.
- The purpose of defining a *Domain* is to define a set of managed entities that are all operated on in the same way. Although administration is important, it is only one of a set of operations that are targeted on entities in a domain.

Thus, the following definition of a policy domain is used instead:

A PolicyDomain is a collection of managed entities that are operated on using a set of policies. The policies are used to administer and control the set of characteristics and behavior of these managed entities.

A simplified diagram of the DEN-ng *PolicyDomain* is shown in Figure 2-20.

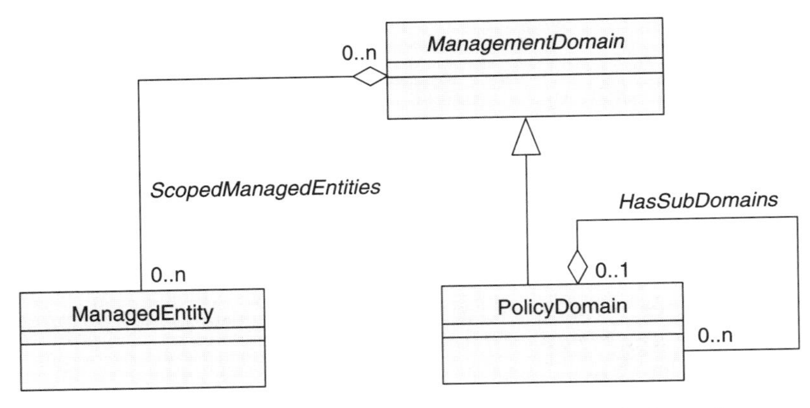

Figure 2-20 The DEN-ng simplified view of a *PolicyDomain*.

Policy Repository (defined in RFC3198; modified by DEN-ng)

RFC3198 defines a policy repository from three perspectives:

- A specific data store that holds policy rules, their conditions and actions, and related policy data. A database or directory would be an example of a store.
- A logical container representing the administrative scope and naming of policy rules, their conditions and actions, and related policy data. A QoS policy domain would be an example of a container.
- In Moore et al.,[18] a more restrictive definition than the prior one exists. A PolicyRepository is a model abstraction representing an administratively defined, logical container for reusable policy elements.

Again, the preceding definitions suffer from overloading different meanings into a single definition. The original definition of a Policy Repository was the third definition (the term "model abstraction" should really be replaced by object class). The text stating that it was "a more restrictive definition" was added to signify that the original purpose of the Policy Repository was "just" for reusable elements. Vendors with existing implementations, however, were using repositories for rule-specific elements, and hence the attempt to make a "better" (as in more flexible) first definition. (Basically, a rule-specific component is one that conceptually is "attached" to a particular policy and is not usable by other policy rules; a reusable component is one that can be used by multiple rules; this will be explained in more detail in Chapter 6.) The problem with the first definition is that it loses the essential semantic of being an administratively defined container (as opposed to an entire data store, as the RFC3198 definition states). The second definition arose from some vendors who believed that their implementations were providing extra semantics. That is, a policy repository was more than "just" a container; it was a container that had a specific purpose.

In real-life implementations, there is no reason that the second and the third definitions cannot be combined. There is also no reason that a Policy Repository be restricted to a particular type of object (reusable or rule-specific). This allows us to ignore the first definition, because it now represents a subset of the second and third definitions. This leads to the following definition:

A policy repository is an administratively defined logical container that is used to hold policy information. For the purposes of this definition:

- Administratively defined means that it resides in a single policy domain.
- Logical container means that it may be implemented as either a separate data store or a special area of a data store that is used expressly to contain policy information.
- Policy information means policy rules and groups, their constituent elements, and related data that may be used in the evaluation and/or execution of policy conditions and actions.

A simplified view of a DEN-ng *PolicyRepository* is shown in Figure 2-21.

A *PolicyRepository* is subclassed from *Collection*, because a *Collection* provides the necessary semantics to iterate over, group, and select sets of entities

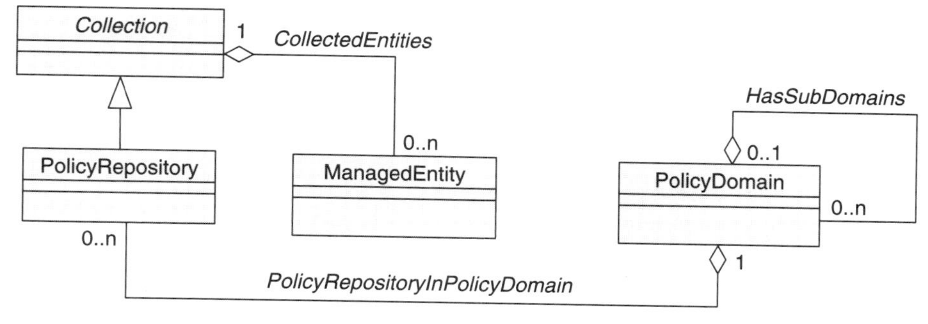

Figure 2-21 The DEN-ng simplified view of a *PolicyRepository*.

residing in the *PolicyRepository*. In the preceding figure, the *CollectedEntities* aggregation enables different types of ManagedEntities to be placed in a *PolicyRepository*. Sets of *PolicyRepositories* can be defined as being contained in a particular *PolicyDomain* using the *PolicyRepositoryInPolicyDomain* aggregation. This enables different policies to be applied to a particular *PolicyRepository*.

2.6.5 Terms Defining Roles, Policy Subjects, and Policy Targets

This section contains a set of formal definitions for the terms *Role* and *RoleCombination*, as well as Policy Subject and Policy Target.

Role (defined in RFC3198; modified by DEN-ng)

This is a difficult concept. The Policy Framework Working Group wanted to constrain the concept of a role to an attribute, per the following definition from RFC3198:

"Role is defined from three perspectives:

- A business position or function, to which people and logical entities are assigned[23]
- The labeled endpoints of a UML (Unified Modeling Language) association . . .
- An administratively specified characteristic of a managed element (e.g., an interface). It is a selector for policy rules and Provisioning Classes (PRCs) to determine the applicability of the rule/PRC to a particular managed element.[18]

Only the third definition (roles as selectors of policy) is directly related to the management of network policy. However, the first definition (roles as business positions and functions) may be referenced in policy conditions and actions."

The preceding definition is confusing, because it overloads different meanings and uses into one word. Furthermore, it completely ignores the existence of the role object pattern.[11] This design pattern is used in DEN-ng and in the TMF SID modeling group. Thus, this definition needs clarification.

Looking at the third definition from RFC3198, we see that a *role* as far as the IETF and DMTF are concerned is nothing more than an attribute. Thus, we need to differentiate between the use of the role object *pattern* and the role *attribute*. The role object pattern is used when different abstractions of the same entity (e.g., two different types of people) are required. Rather than having one large "bloated" class that contains different attributes, methods, and relationships corresponding to the different abstractions, the role object pattern defines an aggregate object that consists of a "core" object to which separate "role" objects are dynamically attached to and removed from as needed (Figure 2-22).

An important benefit of this pattern is that different roles can be defined to suit different clients needs. Object interfaces are kept pure and simple because each interface can be optimized to serve the needs of particular clients. The reader is encouraged to read Baumer et al.[11] for a basic introduction to roles, as

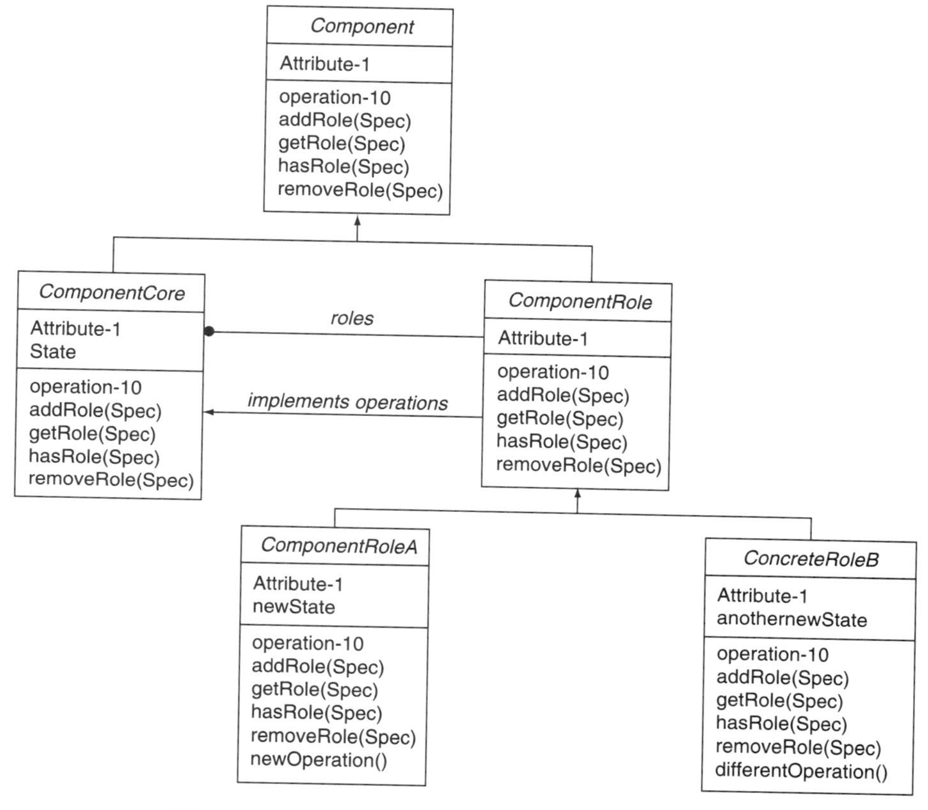

Figure 2-22 The structure of the role-object pattern.

well as Fowler.[24,25] The use of the role object pattern in DEN-ng and in the TMF SID group will be explained in more detail in Chapters 5, 6, 8, and 9.

One obvious way to differentiate between the role object pattern and the role attribute is to qualify their usage with different names and to avoid using the term "role" unless its meaning is unambiguous. Therefore, the following terms and definitions are proposed:

- *A role attribute is a fundamental characteristic of an object that is used to define the purpose or function of that object.*
- *A role object pattern is as defined in Baumer et al.*[11]*—an aggregate set of objects that enable a component object to be adapted to different needs through transparently attached role objects.*
- *A role object is an object that is not meant to stand on its own; rather, it is meant to supply a combination of common and unique functionality that can augment the basic definition of another object. The unique functionality may be supplied in the form of additional attributes, methods, and/or relationships. (Note that this latter is not specifically identified in Baumer et al.,*[11] *but is a characteristic of DEN-ng and the work in the SID.)*

In terms of quantifying the semantics of roles, the following definitions are used:

- *A role-selector is a means of grouping together a set of objects, so that a set of policies can be applied to them.*
- *A role-behavior is a means of explicitly defining behavior that is expected of one or more objects.*

Finally we have teased apart the essence of using a role as an attribute—sometimes it is used to group objects together for operating on them, and sometimes it is used for specifying behavior. These essential points will be used through the remainder of this book. Note that DEN-ng uses "role" in this "object pattern" sense by default, since this provides the greatest flexibility and extensibility.

Role-Combination (defined in RFC3198; modified by DEN-ng)

RFC3198 defines a role-combination as: "a lexicographically ordered set of roles that characterize managed elements and indicate the applicability of policy rules . . . A policy system uses the set of roles reported by the managed element to determine the correct rules . . . to be sent for enforcement. That determination may examine all applicable policy rules identified by the role combination, its sub-combinations and the individual roles in the combination. The final set of rules . . . for enforcement are defined by the policy system, as appropriate for the specified role combination of the managed element."

There is little use in addressing this definition in detail, as we have already decided not to use "role" without a qualifier and because we have changed the semantics of roles as defined in RFC3198. However, the essence of this definition is the ability to combine roles in some meaningful way. For example, we might have the roles "edge interface," "core interface," "Ethernet interface," and "OC-48 interface." It would be beneficial to be able to differentiate between ports on a router in the core of the network that carry Ethernet traffic and ports on that same router that carry OC-48 traffic. By allowing a role-selector to contain more than one attribute value, we achieve this. (The same beneficial effect can be achieved for allowing a role-behavior to specify more than one type of behavior, but that gets more complicated.)

However, the "devil is in the details." In the preceding example, what objects get selected by the combination of the "core interface" and "OC-48" roles—only those objects that contain both roles or all objects that contain either one (or both) of the roles? The IETF decided that (quoting from RFC3060): ". . . the selection process for a role combination chooses policies associated with the combination itself, policies associated with each of its sub-combinations, and policies associated with each of the individual roles in the role-combination." In other words, the kitchen sink.

This is simply too general to be useful. If the purpose of the role-selector is to indeed select objects, then it stands to reason that fewer is better. This yields the following slight modification to the definition of role-selector:

A role-selector is a means of grouping together a set of objects, so that a set of policies can be applied to them. Multiple role-selectors can be combined to select a set of objects, in which case only those objects that contain all attributes specified by the role-selector will be selected.

However, the above changes should *not* be applied to the definition of role-behavior, because the crispness of the role-behavior specification will then be lost. Practice has shown that combining multiple behavioral specifications into a single specification makes implementation very difficult, if not impossible.

The definitions of role-selector and role-behavior give rise to the definitions of policy subject and policy target.

Policy Subject (defined in RFC3198; modified by DEN-ng)

Subject is defined in RFC3198 is defined as: "an entity, or collection of entities, which originates a request, and is verified as authorized/not authorized to perform that request."

There are several problems with this definition. Although originating a policy information (or decision) request is certainly interesting, the subject of a policy is the identification of the theme, or focus, of the policy. Therefore, the subject of a policy will do more than simply request information. Furthermore, whether or not a subject is authorized to perform an operation (such as requesting information) is completely separate from the act of performing the operation. Therefore, we'll use the following definition instead:

A policy subject is a set of entities that is the focus of the policy. The subject can make policy decision and information requests, and it can direct policies to be enforced at a set of policy targets.

Policy Target (defined in RFC3198; modified by DEN-ng)

Target is defined in RFC3198 is defined as: "an entity, or collection of entities, which is affected by a policy. For example, the 'targets' of a policy to reconfigure a network device are the individual services that are updated and configured."

There are also several problems with this definition. First, the term "target" implies an object that a set of operations is being directed at, which is lacking in the preceding definition. Second, the example is confusing at best. To reconfigure a network device, its configuration must change, which means that one or more interfaces on the device will be changed. Services may be changed, but this is a second-order effect that is a direct function of the changing of the configuration of the device interfaces. In other words, in the preceding example, the "target" should be the configuration that is applied to the set of interfaces over which the services run. Therefore, we will use the following definition instead:

A policy target is a set of entities that a set of policies will be applied to. The objective of applying policy is to either maintain the current state of the policy target or to transition the policy target to a new state.

With respect to the preceding example on reconfiguration, a policy target could be a device (e.g., power it on), a device interface (e.g., check if it is up or down), or a device configuration (as in the preceding example). The new definition is able to link in the notion of using a finite state machine to control the behavior of the policy target—this is one of the foundations of PBM.

2.7 New Terminology Not Covered in RFC3198

This section will define additional terminology that is not covered in RFC3198, but which is important for a clear understanding of the design and implementation of PBM systems.

Capabilities

One difficulty in providing an end-to-end service is that the path that traffic will take usually traverses different devices. These devices often use different means to provide a common function, such as the mechanisms used to condition traffic (e.g., classification, dropping, queuing, and so forth). Unless these mechanisms are abstracted into a common layer, they cannot be controlled in a unified manner. DEN-ng uses the term "capabilities" to describe this abstraction:

The capabilities of a device represent the set of features that the device supports that can be harnessed to perform a service. This set of capabilities is independent of any particular protocol, repository or programming mechanism, and enables different devices having different implementation mechanisms to coordinate their features to apply the equivalent function.

An example will help clarify this concept. Imagine a scenario that defines three different classes of service (CoS). Traffic is flowing through two routers made by two different vendors. Each router uses different mechanisms to implement various traffic conditioning functions, such as dropping and queuing, which are required by each CoS. The notion of capabilities enables the different high-level features of each device to be abstracted from their low-level implementations. This in turn allows the high-level abstraction of CoS to be understood by each router. Furthermore, each router is free to use different low-level mechanisms to do its part in conditioning the traffic in accordance with the high-level CoS. The end result is consistent traffic conditioning according to the capabilities of each device, even though different commands and mechanisms are being used. This powerful concept is crucial to providing end-to-end service in heterogeneous environments.

Constraints

Another important concept is constraints, which are defined in DEN-ng as follows:

Constraints represent invariant conditions that must hold for the system being modeled. These conditions do not have side effects and can not alter the state of the system that they are applied to. Rather, they represent limitations and/or restrictions on using certain aspects of the system.

Although constraints are *not* policies, there are many similarities between them. Figure 2-23 illustrates these differences.

The difference is subtle, yet important. Constraints represent predefined restrictions or limitations on the system that *do not change over time.* Constraints are expressions (usually Boolean) that restrict or limit the operation or behavior of the managed entity that they apply to. Because constraints are

Functionality	Policy	Constraint
Pre-defined	Almost always is	Can be
Triggering mechanism	PolicyEvent	PolicyEvent or statically defined
Invariance	Different policies apply at different times	Constraint applies once
Object	Yes	No

Figure 2-23 High-level comparison between DEN-ng Policy and Constraint functionality.

expressions, they can be triggered statically or dynamically. In contrast, policies are objects that are evaluated, which can give different results at different times in response to different conditions. Policies can use constraints, but constraints normally do not use policies. (see later chapters).

Policy-Aware Entity

It is important to be able to distinguish between entities that can operate using policies versus those that do not. That represents one of the first design decisions in a PBM system—will policy be used to control all, or just some, of the objects in the system? The definition of a policy-aware entity is:

A policy-aware entity is one that can understand and use policies to make present and future decisions. These decisions are used to manage and control the changing and/or maintaining of the state of one or more managed objects that are the targets of the policy.

Policy-Unaware Entity

The definition of a policy-unaware entity is:

A policy-unaware entity is one that can neither understand nor use policies to make present and future decisions. A policy-unaware entity cannot use policies to manage and control the changing and/or maintaining of the state of one or more managed objects.

Policy-Enabled System

Similarly, it is important to be able to distinguish between systems that can operate using policies versus those that do not. The definition of a policy-enabled system is:

A policy-enabled system is one that can operate using policies to make present and future decisions. These decisions are used to manage and control the changing and/or maintaining of the state of one or more managed objects that are the targets of the policy.

A policy-enabled system can be explicitly disabled from processing policies.

Reusable and Rule-Specific Policy Components

As we will see in Chapter 3, policy rules may consist of several components, not just conditions and actions. Reusable policy components are defined as follows:

A reusable policy component is one that can be associated with multiple policy rules. This implies that reusable policy components are stored in a different location than the policy rules that are using them.

Similarly, the components of a policy rule that are designed to be used by just a single policy rule are called rule-specific policy components and are defined as follows:

A rule-specific policy component is one that is only associated with a single policy rule.

The difference between a rule-specific policy component and a reusable one is based solely on the intent of the policy *administrator*, not on how many policy rules are using a policy component. This means two things:

- There is no limitation in functionality between what a reusable policy component and a rule-specific policy component can be used to represent
- A policy component that a policy administrator has created to be reusable may at some point in time be associated with exactly one policy rule, without thereby becoming a rule-specific policy component

Figure 2-24 provides a simple example of the difference between reusable and rule-specific *PolicyRules*.

In DEN-ng, many different types of *ManagedEntities* can be placed in a *PolicyRepository*. This enables these entities to be used by multiple containing entities. In the preceding example, both *PolicyRuleA* and *PolicyRuleB* are reusable *PolicyRules*, because they both reuse at least one reusable entity from

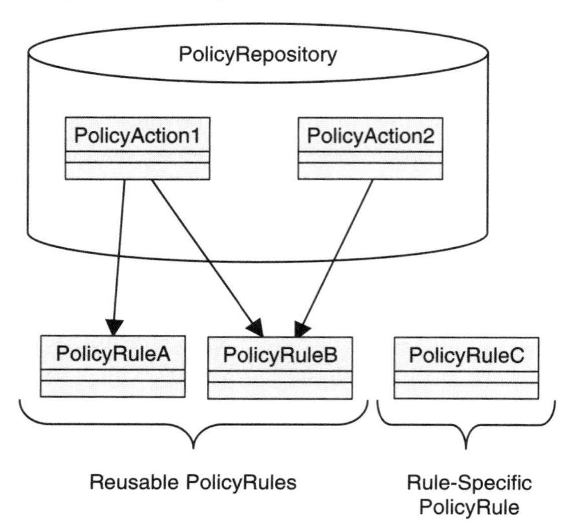

Figure 2-24 Conceptual difference between reusable and rule-specific policies.

the *PolicyRepository*. *PolicyRuleC* is a rule-specific policy because it does not use any reusable entities.

A rule-specific *PolicyRule* can be thought of as an object that has embedded in it all of the information (e.g., the *PolicyCondition* and *PolicyAction* clauses) that it needs to be used. Although this clearly makes it nonreusable, it just as clearly enables it to be accessed more efficiently (because it is in effect a single object). In contrast, a reusable *PolicyRule* requires a set of access operations (one for the *PolicyRule* itself and one for each reusable component). However, the components are all able to be used by other entities.

2.8 Definition of Policy-Based Management

Policy-based management (PBM) controls the state of the system and objects within the system using policies. Control is implemented using a management model, such as a finite state machine. It includes installing and deleting policy rules as well as monitoring system performance to ensure that the installed policies are working correctly. PBM is concerned with the overall behavior of the system and adjusts the policies that are in effect based on how well the system is achieving its policy goals.

2.9 Definition of Policy-Based Network Management

Policy-based network management (PBNM) is an area of network management that treats the system being managed as a policy-enabled system. As in a PBM system, policies are used to control the state of objects within a network (such as the ports on a router). This includes installing and deleting policy rules in network devices and monitoring network performance related to the installed policy. PBNM is specifically concerned with the overall behavior of the network (e.g., the end-to-end or edge-to-edge services provided by the network) and uses policy to provide consistent and predictable network services across the entire network, not just on a device-by-device basis. As such, PBNM treats the network as a provider of intelligent services and assigns these services based on the needs of clients using the network.

2.10 High-Level Requirements of a PBNM System

IT administrators are searching for tools to better manage their service offerings. PBNM systems offer the promise of being able to better manage large and dynamic environments. The basic approaches to building such tools form the basic requirements for a PBNM system and are discussed in this section.

2.10.1 Controlling Access to Shared Resources

Data traffic continues to grow, and the number of users keeps on increasing. Furthermore, applications are also increasing in complexity, and they demand different services from the network. The issue is no longer bandwidth; rather, the real issue is which applications get priority usage of shared system resources. For example, imagine three classes of service (CoS) that govern, at a high level, which applications get preferred access to network resources. Different applications are assigned to each CoS, as shown in Figure 2-25.

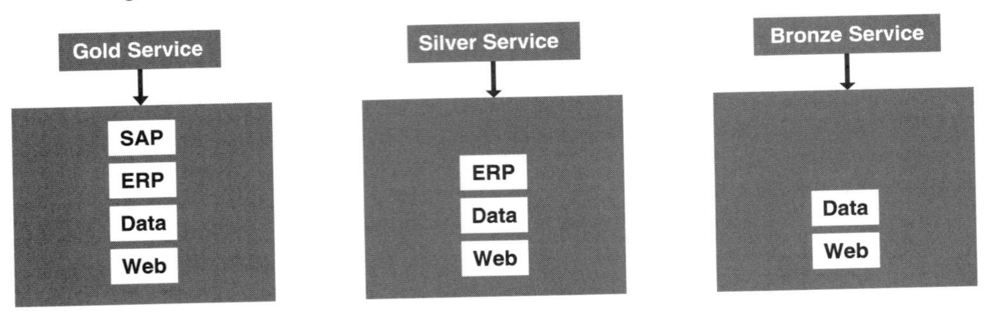

There are three differences between Gold, Silver, and Bronze Service:

- Functionality
- Quality
- Gold Services are treated better collectively than Silver Services, which are treated better collectively than Bronze Services

Figure 2-25 Different classes of service for prioritizing application traffic.

Each CoS (Gold, Silver, and Bronze) has its own set of services. Some of the services represent new functionality that is only available at that (and higher) levels of service (e.g., ERP and SAP applications). Other services, such as Data and Web, are available at all levels—the difference is quality. For example, in a TCP-based environment, the lowest CoS would have the highest probability of dropping packets, which signals some types of sending applications to slow down their transmission rates; similarly, the highest CoS would have the lowest drop probability. These drop probabilities work together to tell the router to drop traffic belonging to a lower CoS when there is congestion.

Sharing the definition of CoS and the specifications that govern router behavior (e.g., through DiffServ[14]), enables these relationships to be more easily implemented in a consistent fashion. Abstracting the capabilities of each router enables these definitions to be built in a vendor- and technology-independent fashion. The set of model mappings that transform these vendor- and technology-independent specifications to (ultimately) a set of device-specific configuration commands can then be done in a structured fashion using different model mappings (as will be shown in Figure 2-30).

Another advantage in this grouping is that all of the applications in Gold Service will receive "better" service than similar applications in Silver and Bronze Service. Better takes many forms, such as more bandwidth, less jitter, and optimized costs are three examples. Each metric requires preferential access to different types of shared resources in the network. Without the ability to abstract different services into different CoSs, this becomes much more difficult to implement consistently.

Thus, the requirement is for a set of abstractions, structured as a set of models, to represent the translation of business goals to device configurations.

2.10.2 Integrating the Business and Networking Worlds

One current problem with network management is that it is not linked to the business processes that run the network. For example, people should not be allowed to telnet into a router and start changing its configuration! This violates fundamental business processes and makes it very difficult for the overall state of the network to be tracked and updated. For every configuration, regardless of how large or how small, defined processes govern how a configuration file is built, who must approve it, when it can be scheduled for installation, and what to do if something goes wrong. ***Ultimately, the business and operational policies that govern the construction and deployment of configuration changes are more important than the configuration changes themselves!***

Process is everything. The network is not a "fat dumb pipe" that is composed of individual interfaces; businesses do not operate or sell interfaces! Businesses operate and manage services according to the priority and contractual obligations that the business enters into. This mandates intelligent processes that can manage the rich functionality of your network and ensure that changes to your network devices follow approved processes.

This philosophy can be recursively applied to different types of services. Specifically, many processes are associated with the management of a service. In the DEN-ng model, there is a difference between a *customer-facing* service and a *resource-facing* service. A good example of a *customer-facing* service is a 2547bis VPN[26] which is a type of virtual private network (VPN). A VPN is a private network (e.g., a network that ensures confidential communications) that is constructed within a larger public network, where "virtual" means that the private network has no physical counterpart and is in reality a virtual connection. This service is termed a *customer-facing* service in DEN-ng because *the customer is explicitly aware of and can purchase such a service*. In contrast, a *resource-facing* service is one that the customer is *not* explicitly aware of (and hence cannot purchase), but nevertheless is required for the proper operation of the *customer-facing* service. *Resource-facing* services include protocols, such as BGP or OSPF. Customers buy VPN services and are not interested in the *sub-services* that are used to build up the customer-facing service. This informal model shown in Figure 2-26 reflects this.

The "simple things" that we often take for granted, such as corporate intranet connectivity using a VPN, can be quite complicated to manage. Furthermore, if business processes are used to manage complicated networking concepts, it is imperative that scalable, extensible, structured management approaches, such as PBNM, be used. The requirement—integrating the business and networking worlds—is achieved by using an information model to represent the entities to be managed and by using policy to control their management. Specifically, PBNM systems that use an information model with policy *enable the different business rules that govern how a configuration change is constructed, approved, installed, and verified to drive network and configuration management*. This enables

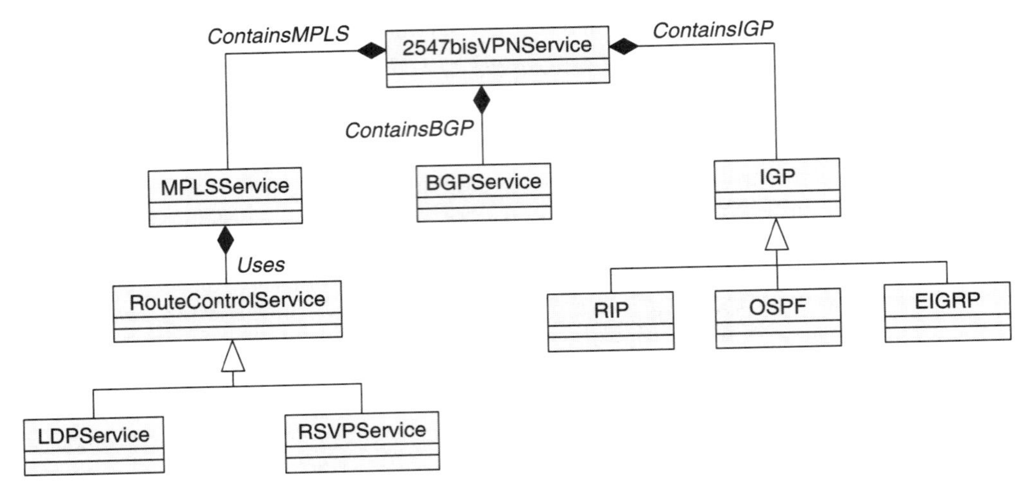

Figure 2-26 Different services in a 2547bis VPN Service Hierarchy.

business rules to drive service activation, making the network more responsive to business needs.

2.11 Using Modeling to Solve Information Overload

Information overload comes in many forms. Two of them are the sheer increase in the number of entities to be managed, and the increase in the complexity of the solution to be managed. The solution to both of these problems is to use an information model to provide a set of abstractions to simplify the differences in the entities being managed.

2.11.1 Managing an Increased Number of Devices

The soaring demand for advanced IP services is leading service providers and enterprises to build IP backbones as fast as they can. Recent studies indicate that enterprise IP traffic (e.g., text, images, video, and audio) is doubling every year. In fact, at the time of this writing, more than 50,000 terabytes of IP traffic are created around the world every day, driving the demand for responsive data networks.

However, it is not just this sheer demand in moving data that is increasing the number of devices. Different data require very different types of traffic conditioning (i.e., behavior) in the network to provide appropriate quality, responsiveness, and other metrics that guarantee a good end-user experience. For example, Figure 2-27 shows that different types of traffic require different types of conditioning. This is because they are sensitive to different characteristics in how data are transmitted.

Most companies built out separate networks to support these diverse needs. People then realized that all of the different costs in running separate networks outweighed the benefits of having separate networks. The lack of control means

	Voice	FTP	ERP and mission-critical
Bandwidth	Low to moderate	Moderate to high	Low
Random drop sensitive	Low	High	Moderate To high
Delay sensitive	High	Low	Low to moderate
Jitter sensitive	High	Low	Moderate

Figure 2-27 Different traffic characteristics are not the same.

increased costs, because different networks require different protocols and management tools to configure devices. Reliability and availability suffered, because the overall reliability (and availability) of a system is the product of the reliability (and availability) of each of its individual components. Furthermore, it became increasingly problematic to find qualified people to run the different networks. Fundamentally, therefore, lack of control means lack of service.

Convergence is the coalescing of separate voice, data, and video networks onto a single (IP) network. Although the number of devices does not necessarily decrease, the number of management tools and protocols should decrease. This is in spite of the proliferation of different device-specific functions. For convergence to work, we must be able to abstract the differences among these different tools, protocols, and device functions and instead provide a unified method of provisioning and service activation. Thus, the requirement is for a *set of information models to abstract the differences in heterogeneous devices, so that their capabilities can be more easily harnessed to support different services.*

2.11.2 Managing the Proliferation of Device-Specific Functions

The combination of more users and more sophisticated applications combine to demand more sophisticated handling of traffic in networks. Network vendors have responded with more features and functions, which have dramatically increased the complexity of managing and programming network devices. Device configurations used to be relatively straightforward and required hundreds of lines—now they are very complex and can be thousands of lines or much more.

From a modeling point of view, this can be characterized (and normalized) as increasing the number of capabilities that devices have. These increased capabilities are needed because of the integration of different applications and different users, each of which (in general) have different needs and therefore have spawned new services. These new network services cause a problem, because each new service has its own management approach, which usually requires new commands, new protocols, and new ways to program the service. This causes two immediate problems: (1) these new features and services must interoperate (or at the least, peacefully coexist) with existing features and services and

(2) each of these new features and services are implemented in proprietary ways. Even if there is a standard that is produced to govern the service (e.g., RFC2547bis, which specifies a particular type of VPN[26]), most vendors support this new standard in either a proprietary (that is not necessarily interoperable) manner or by combining support for it with other new features.

This new service will require new management tools, which has placed a terrible burden on IT administrators who must learn each of these new tools and programming models to manage the environment. This causes the IT administrator to duplicate the actions used to configure and/or manage devices on a vendor-by-vendor basis, because these new services require new programming and management methods. Worse, when companies acquire or merge with other companies, the resulting products act differently and are often programmed differently, despite having the same vendor label. I have used networking as an exemplar—this problem exists not just for networking, but for just about anything an IT administrator may need to configure and manage.

The solution is to build a set of common information models that can represent the capabilities of different products of multiple vendors. We cannot stop vendors from implementing new functions in proprietary ways nor would we want to. We can model these new functions using a consistent representation. This enables different functions and different implementations of the same function to be categorized (i.e., placed in a class hierarchy), thereby facilitating their comparison and integration. This in turn lets us use the appropriate mechanism or feature from each device to build an end-to-end service. For example, consider traffic flowing through two routers. One is "DiffServ-aware," which classifies traffic using different DSCPs. The other is not DiffServ-aware, and it classifies traffic using different settings of the ToS byte. These two different implementations of the same function—classification—use different mechanisms to communicate the result. End-to-end management is greatly facilitated by having a model that identifies these two different markings as two implementations of the same function. This enables a mapping to be defined between the values of the ToS byte and equivalent values in the DSCP. Once this mapping is defined, it is possible to build an end-to-end service that has common semantics supported by different mechanisms. Thus, the requirement is for a *set of device-specific functions to be modeled in a uniform manner, so that business rules can be used to define how each different device function can be used to provide the appropriate type and/or level of service.*

2.11.3 Using Models as Part of the PBNM Process

By itself, an information model is not enough to solve this problem. The information model will describe the characteristics and behavior of managed entities. However, this only defines the current state of an object. We need a finite state machine that lets us model the system as a closed loop system. Thus, we need classes to model current state, classes to model the changing of state, and classes to control when state can be changed.

Policy is the mechanism to control when and how state is changed. Policy will provide two essential benefits in the management of heterogeneous network devices:

- Overall simplification of the management process through the use of a closed loop system
- Automating what is currently done manually and improving consistency of configuration changes in the process

The first bullet implies the use of a *layered* information model (i.e., a set of information models; each focused on a particular management domain, that work together to provide a single, unified, cohesive view of the managed environment). Examples of different layers are the business, administrative, system (device-independent), and other layers. This point addresses the holistic nature of management and recognizes the fact that managing a service is more than just taking fault or performance measurements.

The second bullet recognizes that to deploy a configuration change multiple processes need to occur in a specific order. We will see in later chapters that the notion of one or more *workflow processes* is used to control the different processes of approving the configuration, installing it, and verifying that the installation was successful. This is made even more difficult by realizing that workflow processes can be complicated. That's why two things are needed. First, a set of different models (e.g., information models, activity and sequence diagrams, and so forth) must be used to cooperatively specify what the workflow should do and how it should be managed. Second, we need the ability for workflows to contain workflows, so that different processes that have different execution patterns can have their own workflows, yet still be bound within a higher-level process (Figure 2-28).

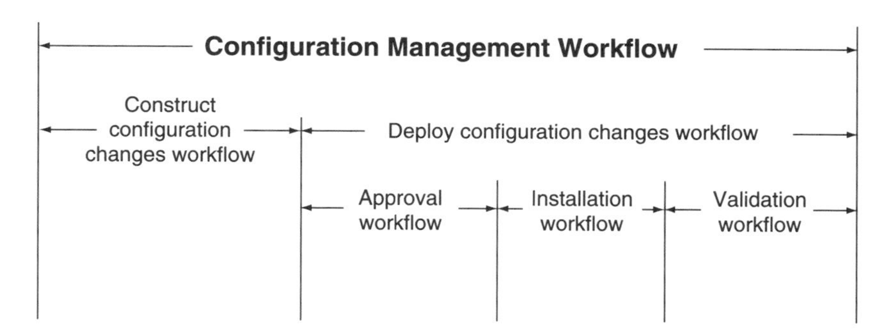

Figure 2-28 Workflows within workflows.

This shows one master workflow, called configuration management, which has two subordinate workflows, construct configuration changes and deploy configuration changes. The former is an atomic workflow, whereas the latter is composed of three subordinate workflows. This flexibility is critical for modeling the different subprocesses that are involved in managing the configuration change process.

Policy controls the different management processes that are represented by workflows. The workflow encapsulates the set of managed entities that must be affected to accomplish a given task. We will see in later chapters that the goal of automating configuration management cannot happen until these different subprocesses are also automated. Policy therefore gives us an extensible vehicle to control what happens when and how.

Thus, an information model should be used to represent not just entities, such as routers, services, and users, but also policies. If policies are an integral part of the information model, we can use them to control when and which managed entities are managed and how they are managed.

2.11.4 Sharing and Reuse of Data

Operational support systems (OSSs) are commonly built using best-of-breed products that provide solutions for different functions needed by the OSS. For example, an OSS may consist of one application for fault and performance measurement, another application for billing, and yet another application for configuration management and service activation. Using a single common model provides two important benefits. First, it lets multiple applications share and reuse the same data. An example of this is using the output of a discovery application to feed an inventory application to dynamically catalog the devices that are currently operational. Second, it enables processes to be built to do a function once and to apply that function in different applications under different contexts. Continuing the preceding example, other applications (e.g., configuration and service activation) could also use the output of the discovery application as input for different processes. Thus, the requirement is for *a set of common information to be used to represent common functions, enabling different applications to share and reuse common information.*

2.11.5 Interfacing with Different Constituents

Most systems define a policy as a single entity; this is incorrect. For example, there are policies to represent business rules, and policies to represent configuring a feature of a device. There is little in common with these two ends of the policy continuum, because they use different grammars to express their function and because they are used by different constituencies. However, they are in reality different views of the same policy. This is shown in Figure 2-29.

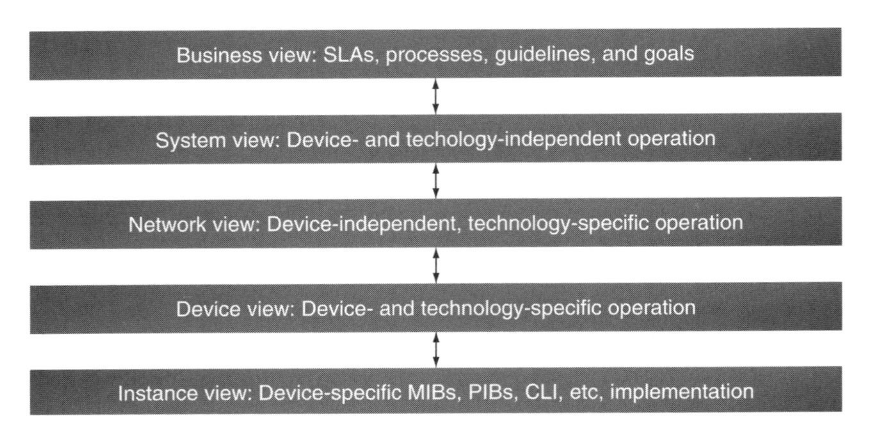

Figure 2-29 The policy continuum.

Each view is optimized for a different type of user that needs and/or uses slightly different information. For example, the business user wants SLA information and is not interested in the type of queuing that will be used. Conversely, the network administrator may want to develop CLI commands to program the device and may need to have a completely different representation of the policy to develop the queuing CLI commands. This will be covered in more detail in Chapter 5. For now, it is sufficient to realize that this is indeed the case. Thus, the requirement is for *policy to be treated as a continuum, where different policies take different forms and address the needs of different users.* Unless there is an information model that can be used to relate these different forms of policy to each other, it becomes difficult (if not impossible) to define a set of mappings that transform the data between each type of policy in the continuum. This is one of the cornerstones of the DEN-ng policy model. Specifically, it provides a layered set of policies with different levels of abstractions and model mappings to translate between them.

2.11.6 Interfacing with Devices and EMSs

Traditionally, device configurations were changed and managed using element management systems (EMSs). The conceptual model of the PBNM system shown in Figure 2-19 can either replace or augment this function.

This decision depends on how integral policy management is to your business and application. A gradual introduction of PBNM systems could be accomplished by having them control one or more aspects of device configuration management (e.g., control the configuring of DiffServ support) or perhaps define different IP address pools and enforce how IP addresses are allocated. Although this simplifies integration (and possibly acceptance) of the new system, it does not offer the complete set of advantages that a true commitment to policy management provides. This latter puts the PBNM system in control of constructing, managing, and deploying device configurations, as well as other associated functions (e.g., user access rights, and permissions to access shared system resources). EMSs do not necessarily disappear in this approach; they are augmenting the function of "pushing the device configuration changes to the device" by wrapping that process in a higher-level workflow process that enables business policies to drive those changes. The main difference is that most EMSs do not use auditable workflows to manage the deployment of configuration changes. This topic will be discussed in later chapters.

Thus, a PBNM system must be able to interface with EMSs. *Rather than being viewed as an alternative to an EMS, a PBNM system should instead be viewed as a higher-level process that guides the use of EMSs.*

2.11.7 Interfacing with NMSs

The purpose of network management is to simplify the management and operation of large networks by maintaining network stability, tuning network performance, and troubleshooting problems that might arise. Network management also aids in strategic planning for network growth.

Network management architectures are implemented using a network management system (NMS) to manage a set of devices. Traditionally, managed devices contain software modules called agents, which gather and store information about the managed device in a repository and provide this information (proactively or reactively) to the NMS. The NMS is a computer system that contains software processes that poll agents in managed devices (automatically or by user request) to check the management information that they contain. It also contains processes that react to alerts from managed devices by executing one or more actions, such as notifying the network administrator, performing event logging, shutting down a system, or attempting system repairs.

PBNM can augment NMSs by providing a framework in which to supply NMS components management information. This includes not just specific details on how a device is currently operating, but other important pieces of management information, such as when a configuration was changed, why it was changed, and who changed it. Thus, the requirement is:

A PBNM system must be able to interface with NMSs. Rather than being viewed as an alternative to an NMS, a PBNM system should instead be viewed as a higher-level process that guides the use of NMSs and facilitates communication between the NMS and other systems in the OSS.

2.11.8 Interfacing with Other Portions of the OSS

The TMN model[27] attempted to provide a framework for telecommunications management. This framework provided for a large variety of functions that are commonly referred to as FCAPS (Fault, Configuration, Accounting, Performance, and Security) management.

TMN is a collection of many standards that define three key areas of communications management:

- An architecture that views "management" as a set of layers and groups of functions. The architecture is recursive and can be used to model multiple operators who may be involved in the service delivery chain.
- A methodology for defining the management behavior of managed devices. This uses an object-oriented modeling methodology known as the "guidelines for the definition of managed objects."
- A set of protocols for management information to be passed between systems. These protocols define a standardized interface at all seven layers of the OSI model with options for wide area and local area networking.

Theoretically, PBNM systems can, and should, interface with all of these different systems. As stated earlier, PBNM ". . . treats the system being managed as a policy-enabled system." Thus, PBNM systems are not limited to "just" configuration management, but rather, are concerned with the overall behavior of the network. Managing this behavior requires PBNM systems to interface with other OSS components. Thus, a PBNM system must be able to interface with different components of the OSS. The PBNM system provides a common lingua franca for communicating between different OSS systems and enables different layers to more efficiently interface with each other.

A PBNM system spans multiple layers of the TMN model, which is in direct contrast to the association of an EMS with the element management layer (and to a lesser degree, the network element layer) of the TMN model. The relationship between a PBNM system and an NMS is less clear, because the definition of an NMS is not as straightforward. This is because most people have successfully disassociated the original meaning of an NMS, which had more to do with managing high-level features of the system, such as faults and alarms, with what they currently refer to as an NMS—a system that can perform higher-level network functions. However, PBNM systems include business functions, which are *not* included in the current (or original) definition of an NMS. Thus, PBNM systems are still fundamentally different than NMSs.

2.11.9 Communication With Policy-Unaware Elements

Many PBNM systems assume that they will use their protocols, object models, and programming methods to communicate with and configure devices that they manage. This assumes that the PBNM system is the "center of the universe." It is exactly this type of thinking that has created stovepipe applications! A common information model should instead be used to ensure that different PBNM systems can share and reuse management information from each other.

The question then is: how are policy-unaware entities controlled, and how do they communicate with and be managed by the PBNM system? This requires the use of a proxy (as shown in Figure 2-19) or a mediation layer that performs a model mapping between the object model used in the PBNM system and the object model used in the policy-unaware entities (or their EMSs). Thus, the PBNM system must be able to interface with both policy-aware and policy-unaware entities.

2.12 Policy Used to Express Business Requirements

Today, the network and the services that it provides exist as their own individual entities, divorced from the operation of the business. One goal of PBNM is to enable business requirements to drive the configuration and management of network services. Although PBNM can help express and integrate the different business, system, and implementation views of the system, the area that has been given the least attention is the link between PBNM and business requirements. This section will examine this link in more detail.

Given the policy continuum shown in Figure 2-29, we need a set of model mappings that will translate the purpose of each policy at its given level to a form that the policy at the next level can use.

This can be better seen by revisiting the definition of policy:

Policy is a set of rules that are used to manage and control the changing and/or maintaining of the state of one or more managed objects.

Therefore, a set of mappings is needed that define equivalent managed objects in each level. This is best explained by examining the example shown in

Figure 2-30. This figure is by no means complete; its purpose is to provide a sample of how the mapping is done.

View	Sample Objective	Sample Objects
Business	*John gets Gold Service.*	Customer; GoldService; GoldApplications
System	*Define three Classes of Services.*	Set of customer-facing services: Gold, Silver, Bronze
Administrative	*Use DiffServ to define traffic conditioning for Gold, Silver, and Bronze; use RSVP to reserve bandwidth when required.*	Define mappings between devices that are DiffServ-aware and *not* DiffServ-aware.
Device	*Pick specific devices and software releases of their operating systems that support the above requirements.*	Define specific type of queuing objects used per device and map their functional differences.
Instance	*Write the appropriate CLI, and monitor using the appropriate MIBs.*	Define objects to represent CLI and MIBs and define mapping between them.

Figure 2-30 Mapping between different entities in the policy continuum.

Each view has its own particular grammar and type of objective, which means that the types of objects that are needed to support policies of one view are at a different level of abstraction than objects of a different view. What we have, therefore, is a set of two parallel mappings—one to translate between objectives and grammar, and one to translate between objects. This duality enables us to use business rules (which by definition do not use networking terms) to manage the construction and deployment of device and system configurations (which by definition use very detailed networking terms). There are two fundamental principles of using business processes to drive configuration management: *individuality* and *process*.

2.12.1 Individuality

Individuality means that different types of configuration changes require different processes. For example, a huge difference exists between changing the SMTP server address of a device and changing how routes are distributed on that same device. Some of the differences include the technical complexity of each change (implying different proficiencies in the personnel that could be assigned to implement the change), different approval processes (because the *business impact* for each change is different), and different guidelines for installing the change. These differences mandate different, customized processes (i.e., workflows) for implementing these changes. The corollary—different configuration changes are not all the same—is significant and is a fundamental principle explored throughout this book.

Because not every configuration task can be handled using the same "template," flexibility is needed to assign different processes to each task, but consistency is also needed to ensure that each task will be handled according to the proper procedure. This is the opportunity for a PBNM system.

2.12.2 Process versus Policy, or Process AND Policy

Many people believe that process management and policy management should exist as separate management efforts and applications. One main point of this book's approach to PBNM systems is to define a richer, more holistic, system where policy and process management can work together to better manage and control network elements and services.

Figure 2-28 showed a sample workflow for controlling different elements of configuring a change to a network device, and it is repeated for convenience here as Figure 2-31 with some embellishments.

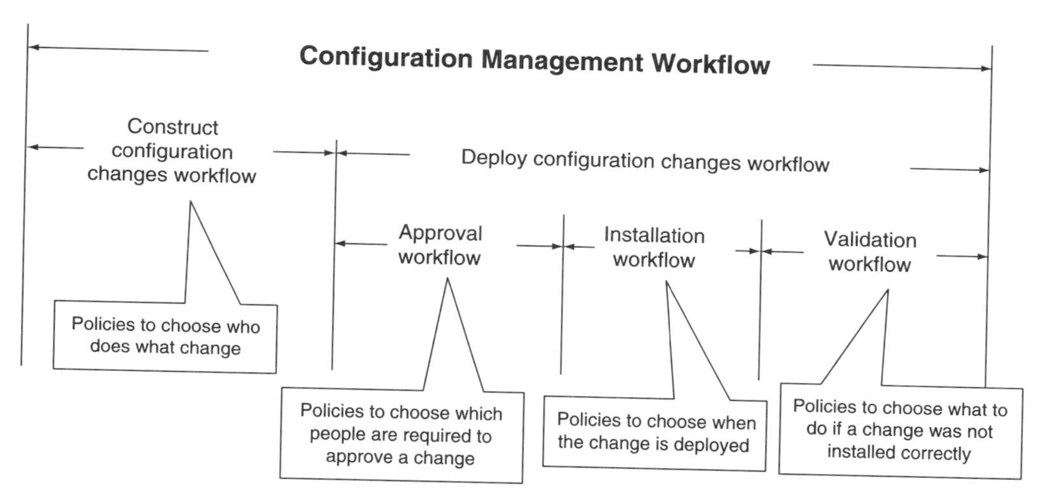

Figure 2-31 The interaction between policy and process management.

As shown in Figure 2-31, policies can be used with processes to better control the different phases of the configuration process. Policies by themselves can define *what* to do, but not *how* to do it. Similarly, processes can define how to accomplish a particular task, but are not decision-making entities in and of themselves. This is because while policies take the form "IF <condition clause> IS TRUE THEN DO <action clause>" (or, in DEN-ng, ON <event clause> IF <condition clause> IS TRUE THEN DO <action clause>), DEN-ng processes take the more restricted form "ON <event clause> DO <action clause>." Thus, policies select which processes to use, and processes perform the requested task. Results of processes are then used to adjust which policies are currently active and enabled, forming a closed-loop management system.

2.13 Summary

This chapter has provided a foundation for understanding PBM. After a brief review of critical object-oriented terminology, a conceptual policy model was introduced to help define the key components of a PBNM system and enables us to focus on the specific terminology needed to better understand that system.

Policy terminology was then introduced and defined. Without a consistent set of terminology, we cannot describe (and certainly cannot build) interoperable policy systems. The terminology was compared against RFC3198; an IETF document that defines several policy terms in the context of specific network management efforts in the IETF. This chapter's policy terminology section began with a detailed comparison of essential terms from RFC3198 and provided enhanced definitions that made the terminology more generic, fixed definitions based on implementation experience, or disambiguated terms that were overloaded with multiple meanings. The second section introduced important new terminology not covered by RFC3198, but which is essential for understanding the design and implementation of PBNM systems.

High-level requirements of a PBNM system were then discussed. The key requirements were:

- Use a set of layered information models to represent different objects at different abstraction layers
- Build policy objects as part of the same information model that represents users, devices, and services
- Form a closed-loop system by using a finite state machine to model the allowable states and state transitions of managed objects and use policy to control when, where, why, and how managed objects change state
- Use business rules to drive network configuration and management through the use of a common information model that includes policy as first-class objects
- Use an information model to solve information overload of management data (both the number of devices and interfaces as well as the plethora of new, yet dissimilar, features of network devices)
- Use the notion of workflow and subworkflows to gather the different managed entities that need to be operated on to implement policy rules
- Interface with all parts of the OSS in a uniform way using policy and a common information model
- Ensure that PBNM systems can communicate with policy-aware and policy-unaware elements equally well

One primary goal of PBNM systems is to enable the network to be operated as profit center instead of a cost center. This concept will be explored throughout the remainder of this book.

2.14 Recommended Further Reading and References

The following is a set of references for this chapter.

1. *http://www.policy-workshop.org/2002/Home.htm*
2. *http://www.noms.org*
3. *http://www.comsoc.org/confs/im/2003/*
4. Strassner, J., *Directory Enabled Networks*, Macmillan Technical Publishing, 1999, New York.

5. Larman, C., *Applying UML and Patterns: An Introduction to Object-Oriented Analysis and Design*, Prentice Hall, 1998, New Jersey.
6. Booch, G., *Object-Oriented Analysis and Design with Applications*, Addison-Wesley, 1994, New York.
7. OMG, *Unified Modeling Language Specification*, version 1.4, September, 2001.
8. http://www.3com.com/other/pdfs/infra/corpinfo/en_US/50066001.pdf
9. http://www.nwfusion.com/links/Encyclopedia/L/725.html
10. TeleManagement Forum, *Shared Information/Data Model, Addendum 1P: Common Business Entity Definitions—Party*, version 3.0, June 2003.
11. Baumer, D., Riehle, D., Siberski, W., Wulf, M., *The Role Object Pattern*, Download PDF from: http://www.riehle.org/papers/1997/plop-1997-role-object.html
12. This is the industry group OSS through Java initiative: http://java.sun.com/products/oss
13. Westerinen, A., Schnizlein, J., Strassner, J., Scherling, M., Quinn, B., Herzog, S., Huynh, A., Carlson, M., Perry, J., Waldbusser, S., *Terminology for Policy-Based Management*, November 2001.
14. *http://www.ietf.org/html.charters/diffserv-charter.html*
15. *http://www.ietf.org/html.charters/policy-charter.html*
16. *http://www.ietf.org/html.charters/ipsp-charter.html*
17. *http://teamlink.tmforum.org/Information %20and%20Data%20Modeling*
18. Moore, B., Ellesson, E., Strassner, J., Westerinen, A., *Policy Core Information Model—Version 1 Specification*, RFC 3060, February 2001.
19. Moore, B., Rafalow, L., Ramberg, Y., Snir, Y., Westerinen, A., Chadha, R., Brunner, M., Cohen, R., Strassner, J., *Policy Core Information Model Extensions*, draft-ietf-policy-pcim-ext-06.txt, November 2001.
20. Snir, Y., Ramberg, Y., Strassner, J., Cohen, R., Moore, B., *Policy QoS Information Model*, draft-ietf-policy-qos-info-model-04.txt, November 2001.
21. Nichols, K., Blake, S., Baker, F., Black, D., *Definition of the Differentiated Services Field (DS Field) in the IPv4 and IPv6 Headers*, RFC2474, December 1998.
22. Yavatkar, R., Pendarakis, D., Guerin, R., *A Framework for Policy-based Admission Control*, RFC 2753, January 2000.
23. This refers to a set of specifications; the top one of which is International Standard 9594-1, ITU-T Recommendation X.500, Information Technology – Open Systems Interconnection—The Directory: Overview of Concepts, Models and Services.
24. Fowler, M., *Role Patterns*, PLoP conference proceedings, 1997.
25. Mr. Fowler maintains a reasonably current web page of his work and contributions. Please see: http://ourworld.compuserve.com/homepages/martin_fowler
26. Rosen, E., Rekhter, Y., Bogovic, T., Brannon, S., Vaidyanathan, R., Carugi, M., Chase, C., Fang, L., Chung, T., De Clercq, J., Dean, E., Hitchen, P., Smith, A., Leelanivas, M., Marshall, D., Martini, L., Morrow, M., Srinivasan, V., Vedrenne, A., *BGP/MPLS VPNs*, draft-ietf-ppvpn-rfc2547bis-01.txt, January 2002.

27. ITU-T, *Principles for a telecommunications management network*, Recommendation M.3010, May 1996.
28. Gai, S., Strassner, J., Durham, D., Herzog, S., Mahon, H., Reichmeyer, F., *QoS Policy Framework Architecture*, draft-sgai-policy-framework-00.txt, February 1999.
29. *http://www.ist-tequila.org*
30. The TeleManagement Forum, SID working group, *Mining Information from the DMTF CIM into the TMF SID*, July 2002.
31. Damianou, N., Dulay, N., Lupu, E., Sloman, M. *Ponder: A Language for Specifying Security and Management Policies for Distributed Systems — The Language Specification, version 2.3*, October 2000.
32. The TeleManagement Forum, *Shared Information/Data (SID) Model — Common Business Entity Definitions-Policy-Addenda 1-Pol*, July 2003.

Policy Management and the Sharing of Data

This chapter describes the need to provide a common representation of information. A novel approach—built around the notion of parallel sets of information models to describe policy and the entities that are managed by policy—is developed to define a coherent information view of the managed environment. The policy and entity information work together to form a complete behavioral description of entities in the managed environment. A combination of a set of models, languages, and a data dictionary are used as the means by which behavior is prescribed for different abstraction levels of the managed environment. This approach enables policy and entity information and semantics to be shared and reused and is fundamental to the design and implementation of PBNM systems.

3.1 Introduction

This chapter defines the need for a common information model to represent policy information and information that policy acts on, monitors, and manages. A common information model is more than just a way of representing information in a standard format—it describes relationships between managed entities and the semantics and behavior of managed entities. This is important because different applications have different needs of the network. Similarly, different network devices have different capabilities, which means two important things: (1) some network devices may be unable to provide a service, because they do not have the necessary capabilities to support that service and (2) two network devices may use a different set of functions to support the same service. This in turn may mean that the same service has different activation and maintenance costs. Finally, different applications place different constraints on which features of what network devices can be used. *This fundamental interplay of capabilities and constraints can be used to define a standard way of describing network elements and services in a vendor-independent manner.*

Without a common way to describe the capabilities and constraints of a network, we will never be able to build and manage agile networks that can quickly meet the needs of varying customers. This is not just limited to service activation; service maintenance and troubleshooting, capacity planning, and virtually anything associated with the network are adversely affected.

The first half of this chapter explains the motivation for expressing information in a common way. Examples of current problems in sharing and reusing network information are examined. The second half of this chapter concentrates on past and current approaches taken, in order to understand what they did right and where they went wrong. It will then introduce a new information modeling approach, Directory Enabled Networks—New Generation (DEN-ng), that was specifically designed to help share and reuse information through the innovative combination of information models and languages.

3.2 The Need to Express Information in a Common Way

Information is knowledge. However, more knowledge than we think is needed is required to properly manage and provision today's networks.

As networks become more and more complex, so must network management and provisioning systems increase in complexity. Why are networks so complex? It is partly because there are many more users running many more applications that generate much more data than ever before. Figure 3-1 shows the huge growth in data between 2000 and 2002. Note that this is in spite of the hard economic times that we have been experiencing.

Another reason is that different applications are trying to use and/or share the same network resources concurrently. The problem is that each application usually has different needs of the network, resulting in conflicting device configurations, conflicting demands on the network, and conflicting service qualities. The problem is no longer bandwidth, but rather how multiple applications can share the same resources in a productive manner.

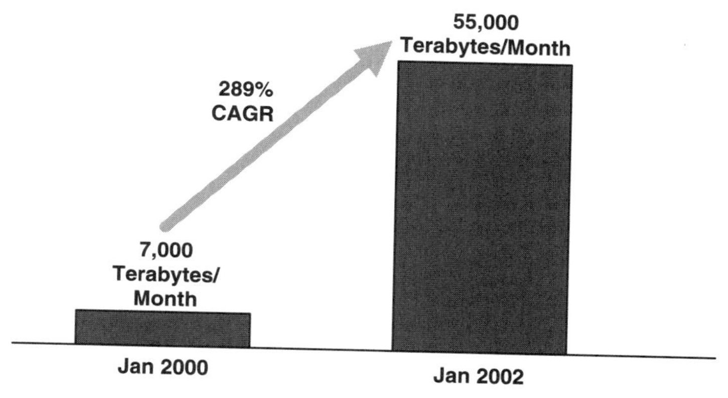

Figure 3-1 The growth of North American Internet Backbone Traffic. (Source: Network World, January 2002.)

Business rules and policies can and should be used to determine which applications get prioritized use of shared resources. Clearly, the only way to make policies do this is to have a common representation of data, which is the subject of this chapter.

3.2.1 Policy Information is Pervasive

People agree that policy information must be shared. Curiously, however, people like to think of policy as a set of specialized information that affects a single area of the system. For example, there has been a lot of focus on "policy servers" built by network and system vendors that control various types of network functions. The emphasis is on building systems that can simplify the programming of how different traffic flows are treated in the network. Usually, this takes the form of a programming tool for constructing a configuration file that defines how a device should treat traffic, although sometimes higher-level concepts, such as business rules, are also included. The problem is that this is an artificial stratification of parts of a policy.

For example, consider the simple policy "John gets Gold Service". In reality, there are many other policies that are used to define this simple policy; some of these are listed below:

- A policy that defines which users get which services when the network is congested
- A policy that defines what applications and capabilities are contained in Gold and other high-level services
- A policy that defines what to do if John can't receive Gold Service when he logs onto the system
- A policy that defines what to do if John receives degraded Gold Service (e.g., Gold Service contains VoIP, but some of John's phone calls are missed or the sound quality is bad)

- A policy that defines how to translate "John" and "Gold Service" to constructs that can be used to program network devices (e.g., source IP address ranges and ports and different types of queuing functions)
- A monitoring policy that periodically inspects the values of key managed entities and adjusts network services delivered by ensuring that these monitored values are within specification

A service level agreement (SLA) can be written that defines an average availability of the service; if the monitored values detect that the service has just gone down and will violate the SLA, then the network operator could choose to switch the customer's service to a new source so that the SLA will not be violated.

Many other policies are needed to implement this service and to exist at different levels of abstraction. Thus, policy can affect many (if not all) of the managed entities in the system. Therefore, policy must use the same artifacts that are used to model and control the entities that they are supposed to manage. We must resist the urge to think of policy as an automated means to change one or more lines in a configuration file. Although that may be one use of a certain type of policy, the PBNM approach defined in this book will always provide the ability to define a continuum of policies that collectively describe how different management actions relate to a set of managed entities.

3.2.2 Management Data, Applications, and Repositories

One of the deficiencies with current management and provisioning applications is their inability to organize information into usable knowledge. A related problem is to filter out excess and/or unwanted information.

Organizing Information into Usable Knowledge

Current management systems find it difficult to organize information into usable knowledge for several reasons. First, think about the vast amount of information present in a network. We can all recall complaining about how hard it is to summarize information or how much harder it is to write a short amount of text that describes a complex idea instead of being able to write a larger amount of text. The same applies to network information. Although the end result may be simple (i.e., an interface is congested or a link is down), the amount of information necessary to determine that fact and the related information that is generated can overwhelm the user unless it is properly correlated and filtered.

Second, recall the simplistic nature of policy-based network management (PBNM) systems designed to date. For example, the IETF/DMTF Policy model of Chapter 2 is shown again in Figure 3-2.

Looking at this architecture, we see two problems: (1) a mismatch in the number of policy levels to computational elements and (2) the lack of model mapping computational elements. We identified five different types of policies in the policy continuum in Chapter 2 (See figure 2-29), but notice that there are only three levels of computation (Console, PDP, and PEP) in the above architecture. Thus, we are either not performing some required model mappings or we

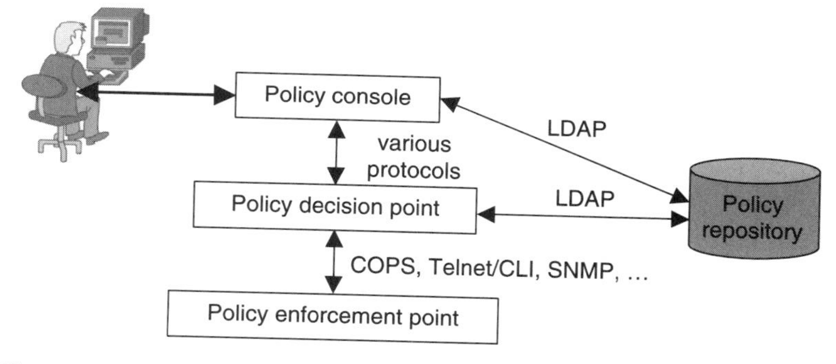

Figure 3-2 IETF/DMTF conceptual model of a PBNM system.

are losing the granularity that we need and making the job of the Console, PDP and/or PEP that much harder. The problem is not that policy information is all stored in one repository, because the repository here is being used passively (i.e., to simply store and retrieve policy information, and not to operate on it). In addition, each of our three computational points must do far more work than it should have to (e.g., conflict detection and resolution as well as model mapping).

For example, look at the Policy Decision Point. Recall that it has the following simple definition: *An entity that makes policy decisions for itself or for other entities that request such decisions.* Nowhere in this definition is anything said about having to perform not one, but two, model mappings (one from the Policy Console to the PDP and another from the PDP to the PEP). Part of the problem is that it is not clear which view each of the computational elements is taking—the Policy Console could take on the business, system, or network Views of the policy continuum, the PDP could be operating at either the network or device views of the policy continuum, and the PEP could be operating at either the device or instance views of the policy continuum, respectively (among others). The result is a "mushing together" of different types of information into a form that is not well-suited to share and reuse the resulting data.

Third, realize that different applications need to share and reuse common information. However, the problem is twofold: (1) not every application has the same view of the same information and (2) not every application needs to share all of the information that is presented to it. To see why, think of two applications; a human resources (HR) management application and a billing application. The common concept of a "user" is present in both of these systems; however, the attributes of the user are very different for each application. The HR application does not need to know anything about the credit history of the user, what applications cost how much, or what applications the user can run. Similarly, the billing application does not need to know anything about the employees that work for the user or the particular employment details of the user.

Now, let's add a third application to the mix: provisioning. Although the provisioning application may know about the user, the network elements that it provisions certainly does not. All they can know are network representations of the user, such as an IP address. Not only do we now need to correlate between possibly different user names (e.g., "johns" versus "john.strassner"), we now also need to match a user name (or equivalent human concept, such as an employee ID) to a nonhuman concept, such as an IP address. The same goes for other concepts, such as services and service qualities. Unless a common information model can serve as a data dictionary to correlate these different names and concepts, applications are unable to share and reuse data.

A final point is that the goals of network device manufacturers and network management developers are usually incompatible. Although software vendors have routinely embedded management code with application code, this is not necessarily true of network device vendors. Most network vendors do not provide the ability to program a "service," such as an IP VPN, with a single command or even a set of commands. Rather, they provide a set of functions (e.g., classification, marking, queuing, dropping, and others) that enable the developer to construct desired functionality by piecing together the appropriate functions. Although very flexible, it means that there is no corresponding management function to monitor and control the service that is programmed by the device. Instead, each of the building blocks used to build the service can (most of the time) be individually managed.

There are two problems with this approach. First, most network devices use different programming models to do different things. The classic example is to use simple network management protocol (SNMP) commands to monitor the functions of a device and command line interface (CLI) to program the functions of the device. If the vendor has not implemented a common information model for that particular product, a mismatch will occur between the semantics of the monitoring function and the semantics of the provisioning function. Worse, if that vendor makes different devices—each of which use slightly different programming models—the above problem is exacerbated because now there is no general way to use different provisioning and/or monitoring functions of different devices. Second, suppose that a vendor has implemented some management functionality by, for example, writing a management information base (MIB) to help manage a set of functions provided by the device. Because different vendors have different functionality in a given type of device (such as a router) and use different programming models, that management function is likely to be implemented in a private MIB and therefore will be proprietary to that particular vendor (as opposed to a standard MIB and implemented by many vendors). As a result, the network operator may have to look at multiple vendor-specific MIBs (each written in slightly different ways that express variations on the same concepts) to manage a service that uses multiple vendor devices.

Compounding this problem is the tendency to have a proprietary set of management protocols and API to manage the device. For example, different vendors support one or more of the following methods to manage quality of service (QoS) commands in network devices:

- Connect through Telnet and issue CLI commands that are specific to that device
- Connect using common open policy service (COPS[1,2]) and manage the device using COPS messages
- Connect using SNMP and issue SNMP commands to manage the device

To provide an end-to-end service of a specific quality across heterogeneous devices, we must develop an equivalence mapping between vendor-specific implementations (i.e., each vendor's CLI is a slightly different programming model with a slightly different implementation). Now, we encounter yet another problem—monitoring the device to prove to ourselves (and to the customer using the service!) that the QoS delivered is that specified in the contract. This is a problem because we have even more methods to monitor a device. No correspondence exists between (for example) a CLI command and a COPS or SNMP command because no common information model equates these different programming models.

We need a data dictionary to portably translate between different synonyms and aliases of the same piece of data. Furthermore, because the data are being used by different applications and exist in multiple views, we need to be able to describe each datum in terms appropriate to the application using it. This is something that almost every information model previously defined has lacked. The DEN-ng and SID information models contain a data dictionary and multiple views (business, system, and implementation) that drive the definition of multiple models and languages that help translate these different concepts into each user of the information. This concept will be covered more in the last section of this chapter and in subsequent chapters.

Migrating Information from Existing Systems into a Shared Information Model

Information in modern OSSs takes different forms. For example, a device interface can be represented in the following different forms:

- Format defined by the OSS, such as EdgePort1
- Format defined by the BSS, such as FinancialGateway
- Format defined by the NMS, such as Rack1-Chassis2-Port17
- Format defined by the EMS, such as a private MIB name
- Format defined by the network vendor, such as fe/0/1/1

Which one of these is the correct name? All of them are, of course, because each name means something significant to its users. Yet, this proliferation of different names that refer to the same managed object is very damaging because it prevents different systems from sharing and reusing each other's data.

Therefore, the purpose of a common shared information model is *not* to replace all of these names, but rather to do two things:

- Define a single name for a managed object that a particular group of users and applications should use
- Define a data dictionary wherein each of these names can be correlated to each other

The first point means that ideally, if there are two EMSs, they should both use the same name—the one defined in the common shared information model. The DEN-ng model defines such common names, but also provides for vendor- and application-specific names. This also enables existing applications to continue to use their own private names (instead of having to be rewritten), but defines the public name that applications should use when sharing the data.

The second point expands on the first by defining synonyms that enable different applications at different levels of abstraction to agree on a common name. Here, the TMF SID specification does an excellent job. Each managed entity in this specification has a data dictionary entry, which defines a set of related concepts and synonyms for this entity. Thus, the SID team is taking the first step in defining fundamental relationships between different managed entities. This step should help speed industry acceptance.

Finally, it migrating management information to a common shared information model helps enable different organizations to interoperate. The shared information model serves as a specification that each management application can map its information to, thus enabling different data from diverse management applications to be shared and reused. This not only ensures that the right data are used, but also prevents wasteful regeneration of the same information. Current systems often have multiple processes that generate the same information, because it cannot be shared.

Controlling the Flow of Unwanted and/or Unneeded Information

Because we have such multiple domains that are so rich with data, we will have to filter unneeded data into just the data that we need. To do this, we need to better understand the sources of the unneeded data and the capabilities of the repositories and protocols used to store and access the data.

The first and most obvious source of data overload is the network itself. Data exist to describe the devices in the network, statistics to describe the performance of network devices and traffic flow between devices, and much more. Applications are not set up to deal with all of this data—they only want a small portion of it. Different applications want different subsets of the data.

The second source of data overload is the data that applications generate. The preceding example illustrated how different applications may have very different views of the same object. Likewise, different types of data have different characteristics (e.g., the speed that their values change) and different uses (e.g., a simple get versus a complicated relational query). This mandates the need for using different repositories with different characteristics to store each type of data. However, unless simple interfaces can be defined that describe what data are exchanged between which applications, applications will not know where to access the data that they need.

In both cases, a common information model is needed that defines not only the data representation and format, but also the data that applications are supposed to transmit and receive. The TMF new generation operational systems and software (NGOSS)[3] is an effort to build a componentized architecture for automating business processes that are part of modern operational support systems. It is a framework whose goal is to enable the rapid and flexible integration of OSSs and business support systems (BSS) in the telecommunications industry. It is a framework because it integrates multiple different viewpoints (e.g., those of the business analyst and system designer) through its object-oriented models and business processes. TMF053 defines a technology-neutral architecture, from which multiple technology-specific implementations can be built. NGOSS is a distributed interface-oriented architecture, which specifies interoperability through a set of common information models and a concept called a "contract."

A NGOSS contract defines the fundamental unit of interoperability in an NGOSS system. Components must use NGOSS contracts[6] to exchange and share data. The concept of an NGOSS contract is shown in Figure 3-3.

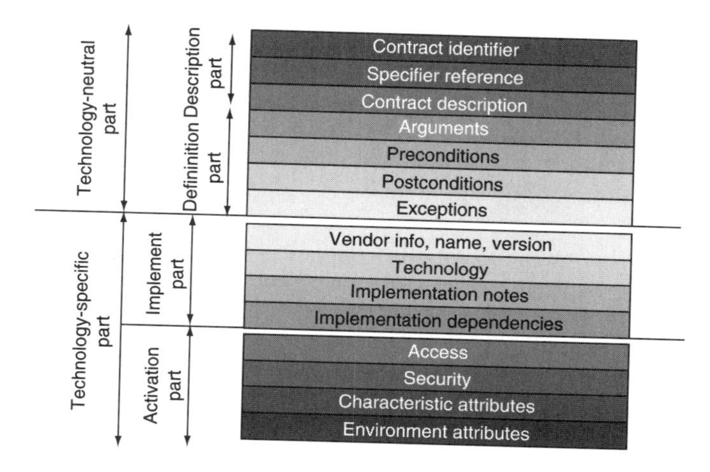

Figure 3-3 Definition of an NGOSS contract.

NGOSS contracts are used to describe and define public shared services in a technology-neutral fashion and then to map this neutral format to one or more technology-specific implementations. More importantly, NGOSS contracts can be defined at the business, system, and implementation levels. This provides a consistent mechanism for specifying public shared services.

A NGOSS contract is a binding agreement to provide information and/or perform an operation. Referring to Figure 3-3, the technology-neutral definition portion defines the semantics of the contract (i.e., what preconditions are necessary to invoke the contract, how it is invoked, and any postconditions that will occur as a result of the contract invocation).

A NGOSS contract is a good example of providing a formal, yet extensible, way to exchange data. The NGOSS contract defined more than just a single object and its data type; it defined the semantics and behavior of the object, which is essential to achieving true interoperability.

Thus, the use of contracts, such as the NGOSS contract, is valuable for defining the specific information that is used and exchanged between different components.

3.2.3 The End-to-End Provisioning Problem

The preceding problems become much more arduous when end-to-end provisioning using heterogeneous devices is attempted. The first problem is obvious:

devices have different features. This means that someone must provide a mapping between different features so that the appropriate feature or features from each device is used. For example, suppose a network has an ATM core and an IP edge (as many networks do). No standard exists to map ATM's QoS features to IP's QoS features. Even if there was, different vendors will use different algorithms to implement various traffic conditioning functions to provide QoS. As a result, we must provide the technology mapping and the implementation mapping for the same QoS to be provided.

Second, devices have different programming models. Anyone who has looked at Cisco's IOS versus Juniper's Junos will find this immediately apparent (IOS and Junos are the network operating systems that run [most of] Cisco's routers and all of Juniper's routers). Not only is the granularity and structure of commands different, the syntax and semantics of commands that perform the same function (e.g., a Cisco ACL and a Junos "Firewall" command) are different. Even worse, because there is generally no common product model among a vendor's devices, vendors tend to develop specialized element management systems (EMSs) for their products (note that Juniper, to its credit, has embraced a single programming model for its M-series routers). As of this writing, Cisco had over 60 different EMSs,[4] multiple operating systems, and a plethora of object models that are incompatible with each other at various levels. One cause of this problem was because of Cisco's prior business strategy, which was to acquire companies and then fit them into its product lineup. Another cause was because of the way that Cisco is structured—highly focused product groups that act somewhat independently.

Realize that so far, only IOS and Junos have been considered. There are many more programming models for many more different types of devices, including access servers, firewalls, and switches. In addition, devices that are not network devices, such as file servers and host machines, must also be taken into account, because they are part of the end-to-end provisioning process. Thus, what you end up with in a heterogeneous network is a collection of device- and vendor-specific EMSs to handle the provisioning of each different type of device.

What you want, of course, is a single model—a unified control plane—that is capable of abstracting the differences of different device feature sets into a single common model. Otherwise, it is impossible to share and reuse data, because that data are named differently, defined differently, and have different semantics.

This last point is very important. When a "stovepipe" application is built, it can manipulate its data without regard to how other applications may want to use it. Now, consider FCAPS (Fault, Configuration, Accounting, Performance, and Security). Assume there are five different applications, one for each category, that are trying to work together to manage devices in a network. The problem lies in each application taking a slightly different view of the same management object. For example, a device interface is usually modeled as different objects for each application, because each application is interested in different attributes of the device interface and because each uses the device interface in a different way. Along the way, common attributes, such as the name of the device interface, get redefined. Thus, management information that should be associated together cannot be associated at all, and new specialized information (e.g., performance data) that a particular application does not have access to (such as the configuration application) cannot be used. This is because, in this example, the configuration application does not have any insight into the structure of the new data, what the names of the objects are, how many there are, and so forth.

We will talk about these and other related problems in Chapters 5 and 6 and will discuss commercial implementations of this in Chapter 10. This reason is one of many why using a common information model that provides such structure is so important for current and future OSSs.

3.2.4 The Operational Support System Dilemma

The preceding problems spill over into the design and implementation of OSSs. OSSs have to perform a wide variety of tasks, and people who build OSSs tend to buy best-of-breed products. This turns into a variation of the preceding problems—unless these best-of-breed products agree on a common information model and the equivalent of an NGOSS contract, true interoperability is impossible. However, this particular variation is much more difficult to implement correctly, because each OSS component has its own design center (i.e., its own basis for how it was built, what it is supposed to accomplish, what data it will import and export, and how data will be exchanged).

This is why an approach such as that used in NGOSS contracts is so important. Interoperability is defined as being able to exchange data and understand what was exchanged. This requires a definition of the semantics and behavior of the objects being exchanged, as well as an understanding for how different objects will interact with each other and the state of the resulting system both before and after the data have been exchanged.

The TMF has adopted NGOSS[3,5,6] as its methodology for rapid and flexible integration of OSSs and BSSs in the telecommunications industry. This approach emphasizes the linkage among the business, system, and implementation layers. This enables business rules and processes to be defined that translate into the design and implementation of a system. The eTOM[7] is a set of business process models that define the enterprise business processes required for a service provider to operate its business, while the system is defined by the TMF053 series documents and the SIM.[9] NGOSS contracts are the interface mechanism between these different layers. Both the business and the system layers used shared information.

3.2.5 Problems in the Sharing and Reusing of Data

Some of these problems are illustrated in Figure 3-4.

Conceptually, two applications—inventory management and configuration management—are sharing and reusing two pieces of the same user object—the user name and its employee ID. The ability for the two applications to do this is made possible by the shared information model.

Paradoxically, this also introduces a problem. Notice that the name of the user, as viewed by each application, is different ("JohnS" versus "JStrassn"). This means that somebody or something will have to map the name "Johns" to

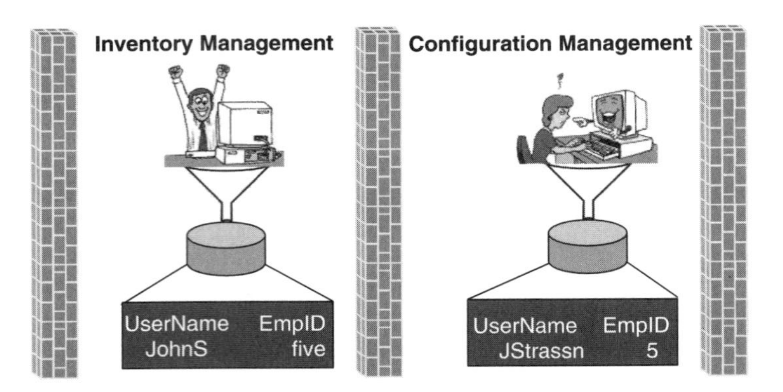

Figure 3-4 Problems in the sharing and reusing of data.

the name "JStrassn." This mapping is not part of the common information model, because it cannot know what private names each of its users give to the entities that it defines. Furthermore, the inventory management application has represented the employee id using a string, whereas the configuration management application has represented the same attribute using a different data type (an integer). The former problem is significant, because names provide more than a mechanism to identify an object—names are used to scope and organize information. This seemingly innocent problem may be indicative of a more serious and much deeper problem—one of being unable to find the desired information because the name is unknown. The latter problem means that, in the best case, an adaptation function must be used to translate one data type to another. In the worst case, multiple queries must be used, if the query language is dependent on data type.

The moral is that we need more than "just" a common information model for OSS component interoperability. We also need a means of mapping different expressions of the same data (e.g., names and data types) to a common format.

3.2.6 Providing an Information View

The business and system views described in the preceding sections need to be tied together. This is not straightforward, because the managed entities (and the level of detail of each entity) that are defined in the business view are different from those defined in the system and implementation views. Importantly, the differences are not just limited to objects, but rather are rooted in different concepts and processes. Just as policy needs to be split into a continuum of policies, we need to establish a continuum of information views (and their attendant mappings between each other) that integrates these different entities.

In addition, each of these views must help guide how the system is implemented. This is best thought of as two separate implementation efforts, because in each case the entities and concepts are different. The problem is how to realize these integration challenges.

The TMF has recently embarked on defining a shared information layer for NGOSS. This work[9] is defined by the shared information and data (SID) modeling team. Their goal is to define a common view of information used in an NGOSS system. The SID is an organized collection of business and system entity definitions, UML models, and a data dictionary.

The SID team has taken a novel approach to defining these business and system entities:

- Rather than building one giant UML model, the team has split up the information into various domains defined by the eTOM and SIM framework documents. These domains define the entities and concepts that are pertinent to that domain. The SID defines the following eight fundamental domains:
  - Common Business Entities, which defines entities that span multiple domains, such as people, organizations, events, and locations
  - Customer
  - Enterprise, which defines the capabilities of the entire enterprise
  - Market & Sales, which defines entities that define various marketing and sales concepts
  - Product
  - Resource
  - Service
  - Supplier & Partner, which defines entities that play the roles of supplier and partner
- For each domain, a business is provided (with a system view being planned to be started in Fall 2003) that documents the various entities in that domain; this takes the form of addenda to the main SID document. Each business and system view of a domain is a separate addendum.
- For each addendum, an explanation of the development of the major entities and examples is provided.
- SID is a federated model that draws from other standards that have been previously published as well as emerging standards. Therefore, each addendum provides not just a UML model, but also a data dictionary. This data dictionary is used to define entities and to define the source of entities (e.g., if they or an idea came from another standard) and can be used to map the SID back to other standards.

The SID can be viewed as the "NGOSS glue." Because the SID defines business and system entities, it provides a common model that is oriented toward the needs of different users of the system. SID's business and system views take care of mapping between these different levels of detail. This provides common terminology and sources of information for NGOSS contracts. The NGOSS contracts use the SID data and bind those data into application-specific uses, which enable implementation to be tied to the business and system needs.

SID is a new type of model; one that I call an "information view." The business and system views define entities of interest to business and system analysts, respectively. The information view is a collection of these views with supporting detail and documentation that enables the business and system analysts to exchange information about the entities and concepts in each management domain. This is exactly what is needed for PBNM. This book defines policies as a continuum of related policies that enable different users operating at different abstraction layers to cross these layers and interoperate. Thus, the policy continuum requires different views of the information that are still connected to each other. This goal is achieved through the SID. DEN-ng provides the system view, as well as the Policy, Resource, and Service domain models.

3.2.7 Expressing Policies

We have already established the need for a set of consistent artifacts to govern the allocation and usage of shared network and system resources by different applications. This is because each application has its own specific needs, but a particular set of applications must be able to peacefully coexist with each other in the same network at any given time. This artifact is, of course, a policy.

Think of the situation if PBNM systems did not exist. Assume that there are two networks: they could be from different service providers or from different groups within the same provider. Different administrators of these systems must first configure the edge of their networks so that each network understands what traffic conditioning to apply to which traffic. The problem is that each change for each device is different (i.e., it refers to a different set of commands, uses different terminology, and so forth). There is no abstraction for the configuration process. So even if you could ensure that the changes were encapsulated in a transaction (which is a big "if"), you must do the same work for multiple devices multiple times. This dramatically increases the chance for something going wrong, whether it was initial operator error or the fact that multiple changes needed to be applied to multiple devices, but were not (for whatever reason), or something else.

Now imagine that we have a set of PBNM systems that all use a common set of policies. This provides a common set of abstractions that can be used to configure different devices. Add to this a mature workflow process that provides the correct transactional and referential integrity, and we have a workable system.

The difference between these two scenarios is that the policies served as a set of abstractions that mapped the different capabilities, commands, and other characteristics and behavior of each device to a common format. Policies are used to aggregate similar configuration commands across dissimilar devices. The end of this section will examine policy information and shared data.

Types of Policies

Several different types of policies need to be represented:

- Policies that correspond to the five different views of the policy continuum defined in Chapter 2
- Policies that correspond to performing different functions (e.g., provisioning, monitoring, and fault correlation)
- Policies that describe what a subject is permitted to do or not do to a given set of targets
- Policies that describe what a particular subject is obliged to do
- Policies that define what actions a grantor can assign to a grantee

- Policies that describe alternate actions to be taken in case the primary set of actions dictated by the policy fail

The last three types of policies will be covered in detail in Chapter 5.

The Policy Continuum as a Means to Specify Policy Coherency

We will talk more about the policy continuum in Chapter 5 and later chapters. It is useful, however, to briefly examine these five different types of policies to understand their requirements on shared data. The policy continuum was defined in Figure 2-29 and is reprinted here as Figure 3-5 for ease of use.

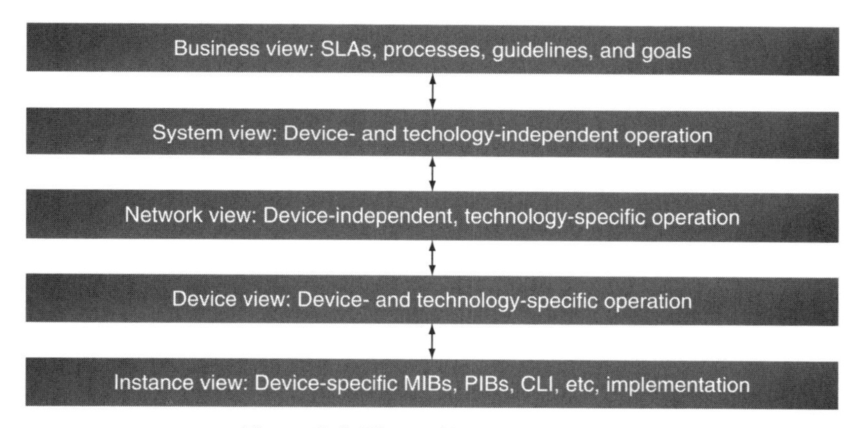

Figure 3-5 The policy continuum.

It is composed of five levels—each of which addresses a different type of user that has a different understanding of the shared entities operating at that particular level of abstraction. For example, the business view provides a description of business entities, such as customer, service and SLA, in business terms. The system view uses these same entities, but adds detail that is not appropriate at the business level. In addition, the system level may introduce additional entities to support one or more business processes defined at the business level.

This is an important point. As a simple example, consider the business definition of a VPN. Businesses view the VPN as a service to be provided to specific customers, and they are not necessarily concerned about the lower-level details of *how* that VPN service is managed and supported. At the system view, however, these things become important. Implementation questions, such as what type of VPN, will be asked to add detail that is necessary to build the VPN. This will lead to more detailed views that focus on the definition of specific entities. For example, it may be decided to implement a BGP-MPLS VPN, as defined by Rosen et al.[10] The model for this shows that the VPN service requires three subservices: one each for BGP, MPLS, and whatever internal gateway protocol (IGP, such as OSPF) is used.

This introduction of data is a result of different users having different roles to play in designing and maintaining the network. Thus, the policy continuum establishes a common lingua franca wherein different users can talk about different aspects of the same managed object or concept in a common way. Furthermore, the lingua franca of the policy continuum enables different actions affecting these different related concepts and entities to be correlated, producing a coherent, cohesive flow of information among the different users of the PBNM system.

This brings to the forefront *why* we have insisted on these different levels of abstraction. A service provider will happily sell a VPN service. A VPN service does *not* require the customer to be aware of which IGP the service provider is using. The IGP is still required to build the service, but it is not a *customer-facing service* (whereas the VPN is). Therefore, no requirement exists to even mention the type of IGP being used at the business level. This also applies to BGP and MPLS in the preceding example.

The other important idea here is "policy coherency." Because different people have different ideas of what a policy is and what it is telling them, we need a means to translate between different levels of abstraction. Referring to the preceding VPN example, we need to be able to tie the high-level specification of the VPN to an approach (i.e., which type of VPN are we going to implement) which has a particular implementation (e.g., the particular CLI commands necessary for a particular router to support this type of VPN). As a result, the shared data must facilitate syntactic adaptation and/or semantic mediation between the different levels.

Syntactic adaptation means that we define a mechanism that translates from one representation using a particular syntax to another representation that uses a different syntax. For example, imagine a simple router whose interface can report on its current speed using a MIB variable. Compare this to a second router, which supports a more robust interface that can report information on its interface using Java objects. The Java object will contain data that is equivalent to that represented by the MIB variable of the first router (in this example, the Speed attribute), but in a different form. More importantly, in the Java object, additional information is present. Some of this information corresponds to other MIB variables that the first router can support (in this example, the Name attribute), and other information does not (in this example, the Configuration object). This is shown in Figure 3-6.

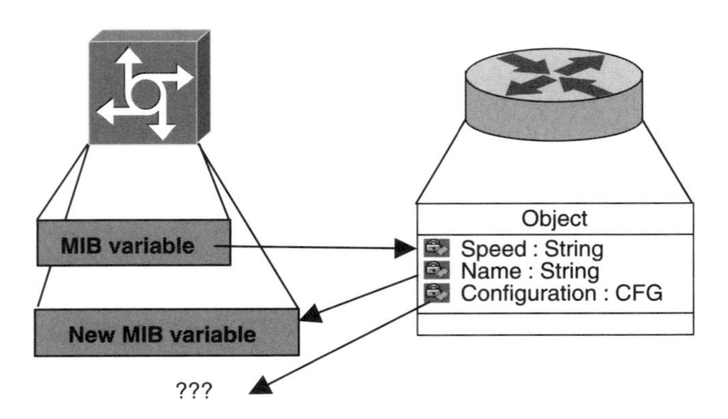

Figure 3-6 Adaptation of two different interfaces.

There certainly is NOT semantic equivalence between these two views, as information is lost between them because the simple router cannot even provide some of the rich information that represents its configuration. In fact, the first interface does not have anything in the form of a single attribute or even a set of attributes that corresponds to all of the information that represents its configuration, whereas the second router does. Yet, we can certainly provide a mapping between the first MIB variable and the Speed attribute of the second router. Furthermore, we can use the common information model to query the first router to realize that the first router has an additional MIB variable that corresponds to the Name attribute of the second router.

Therefore, adaptation is the process of translating of a value provided in one syntactical form to an equivalent value in a different syntactical form. Mediation is (for our purposes) the use of a common information model to enable us to strive to make an equivalent semantic mapping between two concepts. Whereas adaptation is concerned with translating one value to another value, mediation is concerned with translating concepts. In the preceding example, adaptation translates between the MIB format and the Java string format. Mediation causes the developer to realize that additional information can be used to better and more fully describe the relationship between the two router interfaces. The mediation process is used to try and build equivalent mappings to all of the information present in each of the routers by an application. Note, however, that sometimes there will be no way to mediate between different information of different managed entities (as in the preceding example).

Application-Oriented Policies

An application-oriented policy is used to manage and control accomplishing one or more tasks defined by the application. Two common application-oriented policies are device configuration and device monitoring. We will use device configuration and monitoring as examples to determine what different application-oriented policies require of shared data.

Application-oriented policies require two types of shared objects: (1) set of common data that policies apply to and (2) set of policy information used. Depending on the granularity of the policy information, some of data (e.g., policy condition terms or even policy rules) could be shared among multiple applications.

Suppose there is a set of common objects that multiple application-oriented policies apply to. This means that different application-oriented policies will use the same shared data, but in different ways. Conceptually, they require two different views of the same information that applies to the same device. This requires the use of a common information model, because otherwise it is impossible to maintain data coherency. This situation can be thought of as integrating two different pieces of data that each have a small set of features in common, as shown in Figure 3-7.

Object A represents the "complete" object as defined in the information model. For simplicity, just object attributes are shown; relationship and methods are ignored for now (though this thinking applies to them as well). Objects B and C are filtered views of Object A that reflect just the attributes used by

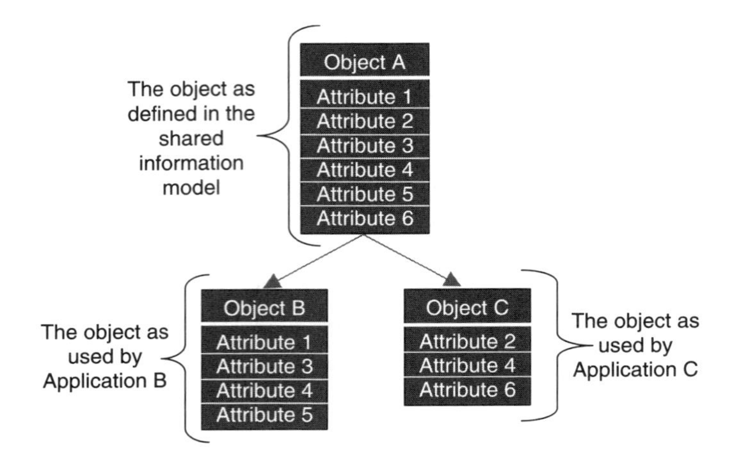

Figure 3-7 Different views of the same object.

Applications B and C, respectively. Let's take a look at the characteristics of these attributes. Notice that there is no dependency on how many or which attributes are used; the only requirement is that an application can define its own attributes to use.

Although some attributes in a view are the same (i.e., Attribute 4), most are different. Attributes that are the same reflect the fact that some information is common and shared. A simple example is the name of the interface, which both applications require. It's immaterial whether the applications supply this attribute using different formats and/or datatypes, because adaptation can be used to map between them without any loss of data. In contrast, other attributes are specific to a particular application. For example, the configuration of the device can be represented by a composite object (e.g., an object that is composed of other objects). Each of these "sub-objects" can have different attributes of different types. This is necessary to capture not just the commands present in the configuration, but also other required semantics. Clearly, this object is specific to the device and to the configuration application. Furthermore, the monitoring application has no use for it. In contrast, the set of statistics for an interface are used by the monitoring application and in general are of no use to the configuration application. Furthermore, the statistical attributes have fundamentally different characteristics than the configuration attributes. For example, the rate of change of these attributes is several orders of magnitude faster than the rate of change of the configuration object.

Common policy information can also be applied to multiple applications, as long as the level of granularity is fine enough. A policy rule is composed of four items: metadata and semantics that define the behavior of the policy, one or more events that trigger the policy, a condition clause, and an action clause. The more generic these terms are, the more likely that other policy rules will be able to use them. These common elements are called *reusable* policy elements and will be further discussed throughout the remainder of this book.

3.3 How to Solve Current Problems in Sharing and Reusing Data

The previous sections have pointed out that there are many complex relationships between managed entities (and the policies that manage them). Therefore, these complex relationships must be taken into account when it is desired to share and reuse policy and/or management data. The most important of these relationships are as follows:

- Different applications have different views of the same object, meaning that different attributes, methods, and relationships of the object are pertinent to each application
- Different policies can apply to different parts of the same object
- The definition of an entity in one level of abstraction often requires the definition of related entities at lower levels of abstraction to be properly understood and/or managed
- Different applications may require adaptation and/or mediation to map different data representations of the parts of the same object that they are interested in to a common format that can be shared and reused
- To obtain reuse, the granularity of representing the policy and entity modeling information must be large enough to be useful, but small enough to encourage different applications to incorporate them

The solution to each of the above problems shares one powerful idea: in that policy and entity information must work together to collectively define a set of views on how to manage and control that entity in the context of a managed environment. Separating policy from the definition of entities that it controls will result in ineffectual management of those entities. More importantly, separating (or even treating differently) policy information from the entities that it controls means that it will be impossible to relate different policies at different levels of abstraction to different entities at correspondingly different levels of abstraction. Therefore, the idea of parallel policy and data continua that together define a coherent information view is fundamental to the design of any information model that is going to be used and even more fundamental to the implementation of PBNM systems. This idea is shown in Figure 3-8.

This figure shows that each of our five views of the environment has a policy-oriented view and a view that enables more and more details about the entities that are being managed by the policy to be related to each other. Referring back to our VPN example, the business view thinks of just the customer-facing service of enabling a user to use the services of one or more service provider backbones to retrieve information. This results in the definition of policy information that (for example) defines which customers can use this service and which cannot and that triggers the definitions of two types of managed entities: high-level business policies that express how this concept will be managed and high-level entity definitions, such as a VPN, that express how this concept will behave and be implemented. This process continues until it culminates in sufficient definition of policies and entities that are required to implement the VPN.

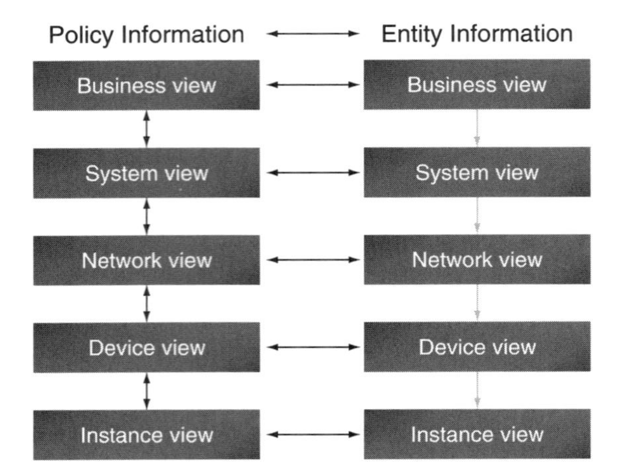

Figure 3-8 Policy and entity continuums.

This process is unique because it provides an inherently tight coupling among what we are trying to manage, how we're going to manage them, and how we're going to implement their management. To accomplish these goals, we must be able to represent policies and the entities that policies control using the same set of tools. This provides a common framework for defining what policies at what levels of abstraction are required to manage what entities at which corresponding level of abstraction.

3.3.1 Solving the Problems of Vendor-Specific CLI

As an example of the effectiveness of the preceding approach, let's compare it to the prevalent method of configuring a device, which is to use vendor-specific CLI directly. Suppose that it is desired to treat traffic originating from a particular subnet in a consistent way across the various ingress points of a network, no matter where that ingress point is. One way to do this easily is to define a reusable policy condition that would detect traffic originating from this subnet. The advantages of this approach are:

- Capturing this test as a common policy condition ensures that it is consistently handled independent of device type and network location.
- Abstracting just the policy condition enables different applications that need to process the traffic differently to use their own specific policy actions (of course, some or all of these could be reusable as well).
- Furthermore, different devices could be triggered in different ways to evaluate this condition—decoupling the condition from how it is triggered (e.g., by an event) facilitates this.

The problem is that different vendors have different CLI commands and different notions to implement this functionality. This is shown in Figure 3-9.

A good analogy is the Tower of Babel. Each vendor has a slightly different language (in our example, IOS versus Junos). For many vendors, different products

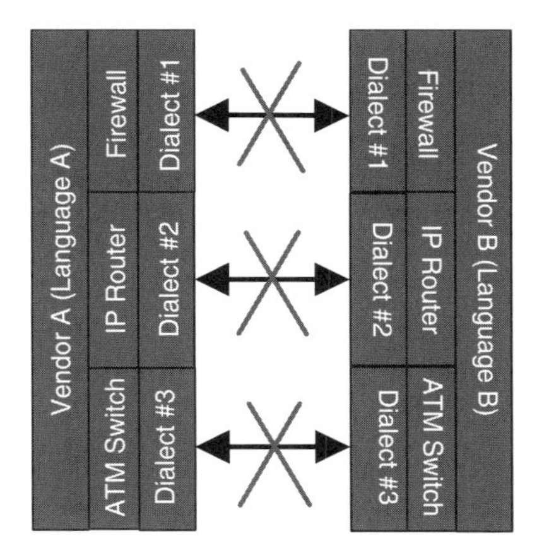

Figure 3-9 The Tower of Babel of vendor configuration languages.

have different dialects of the same language. This corresponds to the changes to the CLI commands that are present in different products made by that vendor or even for different releases of the same product. The problem is that the poor network operator needs to know the syntax and semantics of each dialect of each language. Think how hard it is for you to learn a foreign language. Now imagine that you need to learn not just many foreign languages, but multiple dialects of them, to do your job. This is too difficult to do and cannot scale.

For example, programming Cisco devices would most likely make use of access control lists. If Juniper devices are also used, then we run into an immediate problem, because Juniper does not even have anything called an access control list. Rather, they have the concept of a firewall that implements similar functionality. Second, the structure of the commands is different for each vendor. Continuing with the MPLS VPN example, we find that for Cisco and Juniper devices, some concepts are the same and just require slight syntax variations. Figure 3-10A shows the command format for specifying a route distinguisher (which enables distinct routes to be created to the same IPv4 address prefix).

The only difference in the Cisco versus Juniper implementation is the actual name of the command. This is complicated, however, by the fact that Cisco and Juniper use different ways to configure interfaces. Cisco has distinct configuration modes that you enter, which apply different commands to different semantic levels in the configuration (e.g., globally or to a specific interface). This is shown in Figure 3-10B—the router bgp command is applied at a different level than the neighbor command. Juniper has a similar concept, but it does not force the user to enter different command modes. Rather, the configuration takes the form of a single text file that defines different areas to use to control different functions. To see this, look at Figure 3-10B, which shows how BGP could be used to configure the communication between a PE and a CE router in each case.

The syntax of the commands are now quite different, having different command names and parameters.

These examples were chosen for simplicity to illustrate the point. Note, however, that there are a variety of other differences, ranging from different options

Feature	Cisco	Juniper
Route Distinguisher	rd (as-number:number I ip-address:number);	route-distinguisher (as-number:number I ip-address:number);

A

Cisco	Juniper
Router(config)# router bgp *autonomous-system* **Router(config-router)# neighbor { *ip-address* I *peer-group-name*} remote-as *number*** **Router(config-router)# neighbor *ip-address* activate**	**[edit]** **routing-instances {** ***routing-instance-name* {** **protocols {** **bgp {** **group *group-name*; {** **peer-as *as-number*;** **neighbor *ip-address*;** **}** **}** **}** **}** **}**

B

Figure 3-10 Example of syntactically similar CLI commands (A) and an example of syntactically different CLI Commands (B).

(e.g., an attribute that can be set) on a protocol to concepts that have no direct analogy, that exist between not just these two vendors, but in general between all vendors. These differences are independent of the type of device or the service being built. For example, similar differences can be found in the command terminology used for firewalls and access servers.

The power of the common information model is that it can be used to abstract vendor-specific differences in languages and programming models, thereby ensuring that the same function is programmed in similar ways. Instead of worrying about the syntactical differences between an access control list command and a firewall command, the desired functionality of the command can be defined in a vendor-independent way. As long as a means exists to map this vendor-independent form to a vendor-specific form, life is simpler, and configurations are less error-prone.

3.3.2 Solving the Problems of Differing Programming Models

Although some vendors use a single programming model to configure and monitor their devices, many do not. For example, CLI might be used to configure devices, and SNMP may be used to monitor them. The advantage of PBNM systems is similar to the above benefit of abstracting different CLIs, except that it is now applied at a higher level of abstraction. Instead of different forms of the same command, different application functions (which involve sets of commands) are abstracted. This means that it will be easier to identify common policy information and common data that policies operate on, even if different programming models are used for different functions. This is where the real power of a common information model surfaces, because by using the same model for two (or more) different tasks, we ensure that the model can represent common information of both those tasks. In the specific example of configuration and monitoring, we are able to define configuration functions that have direct management function corollaries, and vice versa, even if there is no direct link between commands in the different programming models (such as between the implementation of SNMP and CLI commands for most networking vendors).

3.3.3 Communication Between OSS Components

Many integration issues hamper communication between different OSS components, such as a provisioning application and a billing application (Figure 3-11).

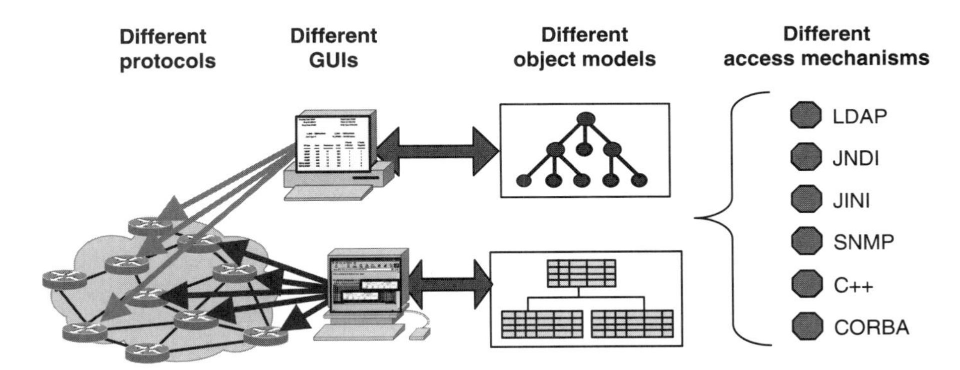

Figure 3-11 Data sharing problems.

As previously mentioned, different vendor products generally require dedicated EMSs. Figure 3-11 represents this by two different GUIs. The GUIs will use different protocols to configure and monitor the devices that they are managing. More importantly, because the GUIs represent different products, they each have their own different object models that control their functionality. This is significant, because each object model defines its own application-centric view of the piece of the network that it is managing, not of the entire system. The problem is not the many different access mechanisms that are used to extract data from each network management application, but rather the different object models embedded within each application. This greatly impedes the exchange and reuse of data.

3.4 Existing Approaches to Sharing Information

This section will survey existing approaches for sharing and reusing information, setting the groundwork for recommending our PBNM development approach.

3.4.1 The DEN Approach

Before DEN, a sizable gap existed between the functionality that devices offer and the functionality that application services want to use. Specifically, there was no common way to what the applications needed from the network and what the network could provide. For example, consider a network that supported two different applications—distance learning and Voice over IP. Both of these applications need QoS. However, each application needs a very different type of QoS, because each application thinks of QoS in slightly different ways. Although both expect "better" handling of their traffic compared to other applications, the distance learning application defines "better" using different terms and different semantics than the VoIP application.

Furthermore, each application expresses its needs for QoS in very different ways, using application-specific terminology that maps to application-specific functionality. For example, distance learning, VoIP, and Oracle all need QoS; either because best-effort delivery will compromise the service quality and/or because best-effort delivery will not ensure that the application receives the prioritized access and use of shared resources that it requires. However, the QoS that is required by (for example) VoIP is fundamentally different than the QoS required by most other applications (e.g., Oracle), because voice traffic has different requirements of the network than (for example) data traffic.

Figure 3-12 depicts this problem as a "conceptual gap." The gap arises from the realization that developers of these applications do not think in terms of queuing and other important traffic conditioning functions that the network provides. Yet, these applications require and depend on these and other networking functions to operate and provide satisfactory end-user experiences. Similarly, network engineers think in terms of low-level traffic conditioning functions, not in terms of high-level application functions. This result is shown in Figure 3-12.

There is currently no direct support for distance learning in existing network devices (e.g., if you look in a particular vendor's programming model, such as CLI or SNMP, you will find that there is no command or set of commands that tell the device that a distance learning application is sending traffic and how to

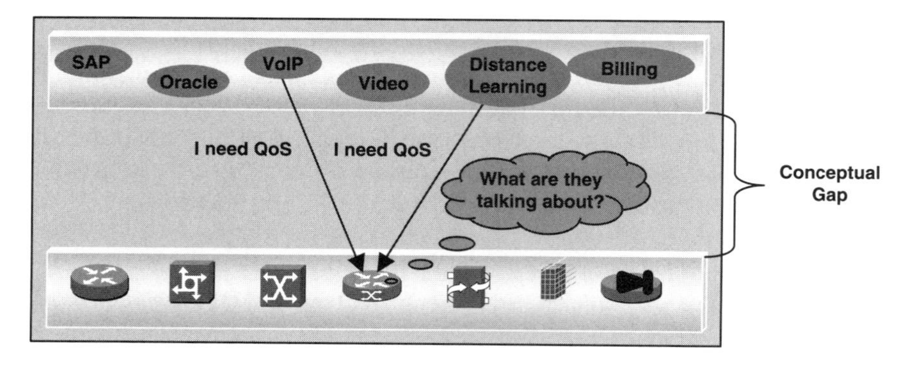

Figure 3-12 The lack of abstraction in binding application needs to network services.

treat the traffic). Consequently, there is no direct way for the device to support such applications. We must instead rely on a set of highly skilled network engineers to devise a way to detect when this traffic is being sent and, upon its detection, mark it in a special way. This marking tells devices in the network to perform a set of common traffic conditioning functions when packets corresponding to that application are received. The network engineers then program appropriate individual features for each device that carries traffic to support the specific traffic conditioning needs of distance learning. This usually translates into a set of complex CLI commands that are interspersed at various places in the configuration of the device. This makes it very hard for less skilled engineers to maintain and change this code.

Earlier attempts at simplifying this problem encouraged the network engineer to think of the device functions as a set of building blocks that can be assembled to provide the necessary traffic conditioning for distance learning. Although this does simplify the task of configuring the network devices, it in turn causes two problems. First, if different devices have different capabilities and hence different commands, the building blocks will be different. It is therefore quite difficult to ensure that the appropriate traffic conditioning is being provided by each device, because now mappings between a different set of functions for each device must be done to ensure that each device is treating the distance learning traffic in the same relative manner compared to other traffic from other applications. *DEN adds value by providing a framework to normalize these different building blocks.*

A worse problem is that other important applications, such as VoIP, also have no direct support in the CLI of the device. The preceding problem is now greatly exacerbated. Conceptually, VoIP, distance learning, and other applications must be gathered into different classes of service (e.g., Gold, Silver, and Bronze), so that they can receive the same relative level of service that is better or worse than the service that other applications receive. However, what happens when both distance learning and VoIP ask for special traffic conditioning functions from the same device? Although some devices have the capability to accommodate this by varying the drop probability of the traffic, this will not solve the problem of trying to get multiple devices from multiple vendors to provide the same features; rather, it is simply another function or set of functions that must be accounted for and mapped to appropriate mechanisms in other devices. *DEN adds value by abstracting the traffic needs of distance learning along with the needs of the other types of traffic that are using the network into a set of common terms.* This abstraction is critical for determining how the network devices should be provisioned to support the applications that are using it.

These are the main reasons why DEN was invented by the author. DEN can be thought of conceptually as providing a set of information layers that reduce the gap shown in Figure 3-12. DEN provides a set of abstractions that enable applications that require network services to express their needs to the network in a form that the network can understand. Similarly, DEN enables network services to be defined in a way that applications can understand.

The gap between the needs of applications running on the network and the services needs to be bridged, and DEN bridges this gap by developing intermediate network service abstractions that can be matched to intermediate application service abstractions. The network service abstractions mask the various lower-level device mechanisms, while the application service abstractions translate the needs of the applications into a form that can be communicated to the network. DEN does this by defining a common set of terminology (e.g., "Gold" service) that applications can use to define which network services they need. Relating this to our previous example of VoIP and distance learning, both applications need QoS, but they need different types of QoS. DEN enables the differences in QoS that each application needs to be abstracted into a common set of services (i.e., gold, silver, bronze, etc.) that can then be translated into technology- and device-specific functions. This approach is shown in Figure 3-13.

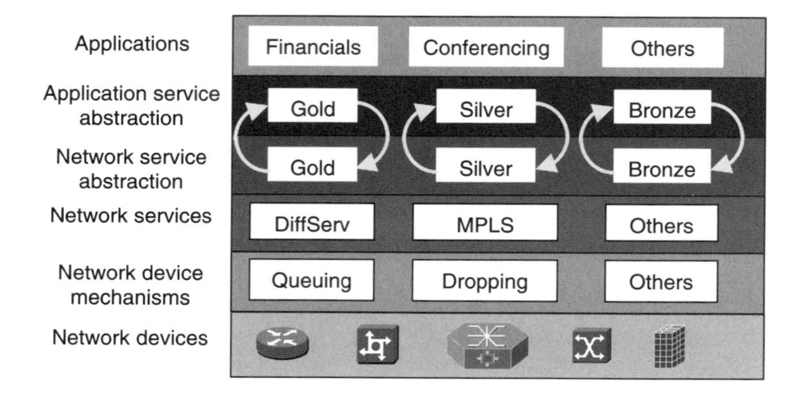

Figure 3-13 DEN service abstraction.

This approach is sound in principle. However, when DEN was merged into the DMTF CIM, the effort to develop this approach was stopped, and instead work focused on building out more of the CIM models. In particular, the effort to build higher layer abstractions, such as the set of Gold, Silver and Bronze services that map both to applications and to low-level network commands, was stopped, as were other abstractions, such as roles. This is because the DMTF wanted to concentrate more on developing the models themselves. The problem is that this renders those models academic works that are very hard to use in real life. As we will see, the DEN-ng effort is once again picking up this abstraction effort.

3.4.2 The IETF Approach

The IETF policy effort started as a Birds of a Feather (BOF) meeting in the 42nd IETF meeting.[11] The strategy was to reuse the basic DEN policy model[12] to define the overall representation of policy and then concentrate on filling out the details to define how this could be used to manage services defined by the integrated and differentiated services working groups.[13,14] The basic approach is shown in Figure 3-14.

The policy core information model[15] was built to define the structural representation of policy in terms of a set of classes and relationships that bound dif-

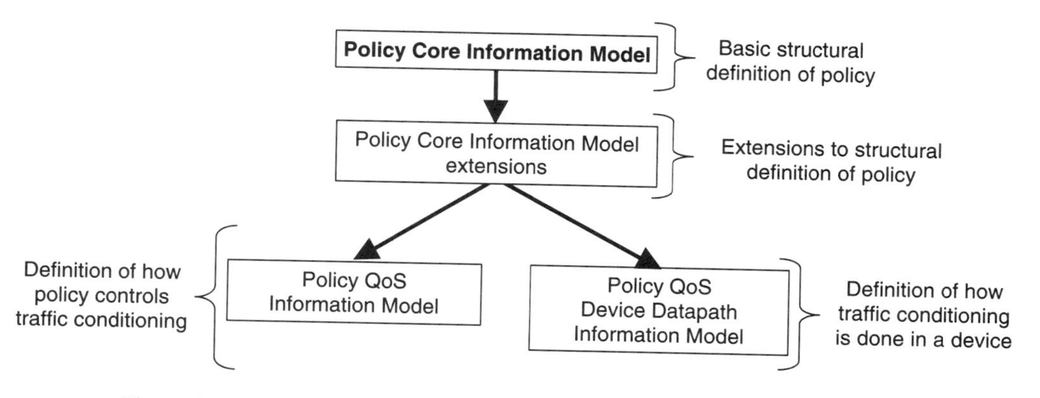

Figure 3-14 The approach of the IETF policy framework working group.

ferent classes together to define the semantics needed to represent policy. For example, a policy rule can aggregate one or more policy conditions. More importantly, the structural representation of policy rules, conditions, and actions was defined independent of the content of policy. This is critically important, because it was recognized that policy would necessarily take different forms to accommodate the needs of different applications at the device level.

For example, security policy has its own established set of terms and functions that differ from equivalent concepts in networking. However, if there was no common way to organize these different application-specific concepts, then there would be no method to apply different application-specific policies to the same device. The need to apply multiple discipline-specific policies to control different functions of the same device was recognized as being crucial for the design of policy systems in the policy framework working group of the IETF. Therefore, the first RFC (Policy Core Information Model, or PCIM[15]) produced by this working group was a basic model that defined how policy was represented in a manner that was independent of device and application.

One of the first proposals that was worked on was the QoS Policy Information Model, or QPIM.[16] This document defined an object-oriented information model for representing policies that administer, manage, and control access to two specific types of QoS—integrated and differentiated services—and how each type of QoS could be controlled in a device. QPIM was defined by refining the class definitions defined in PCIM. In other words, QPIM started with the basic class definitions defined in the PCIM and subclassed them to define and refine functionality specific to representing QoS as defined using integrated and differentiated services.

In this way, QPIM did not change the basic structure of representing policy, but rather defined content that seamlessly fit into that structure. This methodology of adding specific content based on a general representation served as a model for the development other types of policy information models (e.g., those having different content, such as the IPsec[17] model).

As originally written, QPIM defined a set of extensions to PCIM to represent QoS policies. Examples of these included the extension of the "policy repository" concept, the specification of a policy condition as a triplet (variable, operator and value) of objects, as well as other extensions. The working group thought that these extensions were of general use to applications other than QoS and tasked a group of authors to construct policy core information model extensions (PCIMe) to externalize these extensions and gather them into a single document that other information models could use. The PCIMe defined several extensions and changes to the existing PCIM classes and relationships. The QPIM was then changed to appropriately reference the new PCIMe.

The final piece of the QoS puzzle is the QoS Device Datapath Information Model (QDDIM).[19] The purpose of this document is to define an information model to describe the QoS mechanisms available in a network device. These mechanisms control how traffic is conditioned through the forwarding path of a network device. This document is intended to be used with the QPIM; the QDDIM defines the QoS mechanisms (e.g., the classification, marking, metering, dropping, queuing, and scheduling functionality) that can then be managed and configured using policies defined by the QPIM. Together, the QPIM and the QDDIM describe how to write QoS policy rules (based on the common representation of policy as defined in the PCIM and PCIMe) to configure and manage the QoS mechanisms present in the datapath of device. All of these documents will be described in more detail in the book.

In summary, the IETF approach focuses on the development of information models that use a generic framework (as specified by PCIM and PCIMe) that forms the basis for developing specific content (as defined in QPIM and QDDIM). However, this work is currently stalled at the information modeling layer. Although there is a model mapping of PCIM to a directory model that uses X.500 or LDAP,[20] no other data model mappings have been developed. For the IETF approach to be used, the IETF must define a set of data model mappings into one or more technologies.

There are, however, several problems with this approach. Compared to the Ponder[21] work, there is no definition of high-level policies (e.g., authorization, obligation and delegation policies) and no means to trigger policies (e.g., with the use of an event). In fact, there is no language for specifying policies; meaning that important issues, such as conflict detection, cannot be handled. Compared to DEN-ng, the IETF approach uses instance-based, rather than specification-based, reuse. In other words, the IETF approach does not provide a means of abstraction, such as roles and templates, which DEN-ng does, to build object instances. Instead, it provides a means to model the generic structure of policy rules, conditions, and actions (through PCIM and PCIMe), the specific structure of QoS policy rules, conditions, and actions (through QPIM), and the current state of device traffic conditioning mechanisms (through QDDIM). This is an incomplete picture, because it doesn't address anything more than the current state of the device. Finally, the association of policy with consumer (subject) and target objects is not specified in usable detail. Although this work forms the basis of part of the information model for representing policy objects, it is incomplete in its definition and representation of policy information and objects that policy control for PBNM systems.

3.4.3 The DMTF Approach

The DMTF originally was concerned with developing an object-oriented information system to model the instrumentation of desktop systems. Over time, it has developed a more generic model to manage information. However, its roots still show in its approach and the level of detail that it provides in its models.

The DMTF started by providing its own metamodel, which it calls a metaschema, and an accompanying language (Managed Object Format, [MOF]) for expressing modeling constructs, such as how to define classes, attributes, and relationships. MOF is a human-readable, textual format for describing richly structured management information, such as object definitions and relationships. MOF is constructed so that automated tools can process it (e.g., compilers). One important tool of MOF is the concept of *qualifiers*. Qualifiers can be thought of as directives for defining an artifact, such as an attribute, method, method parameter, class, or instance. For example, if a class has the qualifier ASSOCIATION, then that class is used to implement an association.[22]

With respect to policy, the DMTF has taken the IETF model and fit it into its overall set of models. This provides the opportunity to link the elements of the IETF model to the elements defined by the DMTF. In particular, the definition of subject and target of the policy could now be defined in greater detail. Unfortunately, subject and target are still only defined in vague detail. For example, the definition of how policy is related to a particular entity is incompletely specified in several relationships. Let's examine one of these—the PolicyRuleInSystem association. The purpose of this association is to relate a policy rule to a System. The main problem with this association is that in CIM, the System class is used to represent a network device, such as a router (DEN defined specific base classes to represent the physical and logical aspects of network devices, but the DMTF rejected this approach and used the more general system class). For networking, policies usually do not apply to an entire device—they instead apply to a particular component of the device, such as an interface. It is impossible in both the current version of CIM (2.6 as of this writing) and the next planned version of CIM (2.7) to specify a port of a network device in sufficient detail as to write policy to configure that portion of the network device. Hence, the developer would have to subclass this association (or worse, invent a new one) and build a new entity (or more likely, a set of entities) that represent the components inside the System object which the policy affects.

In general, the same problems with the IETF model hold true with the DMTF approach. One notable difference is that the CIM has defined an event model. Unfortunately, the event model conflicts with the notion of events and states defined in the UML metamodel. In addition, although the DMTF did define events, it did not relate them to policy.

3.4.4 Language Approaches

A generic language provides a very flexible approach to defining policy rules and expressing how they control the behavior of different managed entities. Two popular approaches are: (1) to define a formal policy language and (2) to use a general purpose scripting (or even programming) language to model and enforce policy. The latter is beyond the scope of this book and by definition is vendor- and/or application-specific. However, the former will form a portion of the proposed approach for building a PBNM system. Therefore, it is helpful to briefly survey some examples of such languages.

The Ponder[21] language is a declarative, object-oriented language for specifying different types of security and management policies using role-based access control (RBAC). It does this by grouping policies into roles and relationships and then defining configurations of roles and relationships as management structures. The power of Ponder is its melding of a declarative language (for ease of use and simplicity of policy construction) with an object-orient model (for extensibility of what can be modeled and how entities can be modeled). However, Ponder is really oriented toward high-level management expressions that involve RBAC and therefore has a natural bias toward security policies. It needs to be expanded to handle more generic policy structures and applications. In particular, there is no specification of low-level policies that could be used to control (for example) traffic conditioning functions. In addition, there is no detailed class hierarchy as of this writing. Viewed in this light, we come to the somewhat surprising conclusion that Ponder addresses high-level policy specifications, whereas the IETF information models are oriented toward defining lower-level functionality. However, this is of no bother, because our goal is to have a continuum of policies and models that work together at different levels of abstraction. Therefore, Ponder serves as a valuable tool for formulating high-level policies.

The trust policy language (TPL), developed by IBM, uses an XML-based syntax to define access control to resources or services. A TPL policy rule consists of a list of required certificates that specify which groups a user can belong to, and a function using various certificate fields to determine whether access control is granted. Unfortunately, TPL is concerned only with defining access control of security policies. The PolicyMaker and KeyNote languages are similar in scope. Again, their concepts can be studied, and ideas for building a more generic policy language can be gleaned from better understanding the advantages and limitations of these languages.[23,24]

Lucent's policy definition language (PDL) is aimed at network management. It is based on defining policies using a declarative language. Policies rules take the form:

Event *causes* action *if* condition

PDL addresses the triggering of policies by defining a "policy defined Event (PDE)" as follows:

Event *triggers* PDE *if* condition

This enables high-level network conditions to be defined, and policies invoked when a particular type of event occurs (either by the system or by policy itself. The PDE has a standard set of parameters, including a timestamp and a URL. It also defines a time window in which a set of events can occur and how these events relate to a set of policies.

However, PDL lacks several things. The most important of these are the inability to aggregate policies or to build composite policies, little support for reuse of policies, and an over-reliance on the event-condition-action triplet instead of object-oriented class hierarchies and additional programming language support. This latter problem causes the generation of overly complex PDE hierarchies, in which a PDE triggers one or more additional PDEs, until eventually actions are performed. It is unclear how these complex hierarchies can be resolved and managed to guarantee that the desired effect will indeed occur. Although there is some limited support for conflict resolution, this area in general needs a lot more attention, especially because it is prone to generating and using complex hierarchies.[25]

In summary, languages enable implementations of varying levels of abstraction to be expressed in a standard form defined by the capabilities and programming constructs of the language. The single most important impediment to using languages has been the misconception that a single language must be usable by all consumers of the policy continuum. Instead, this book recommends a continuum of languages (and a continuum of models) that work together to define the managed objects and semantics necessary at each abstraction level of the policy continuum.

3.5 How to Express Information in a Common Way

This section will examine different ways to express two types of information: common policies and common objects that the policies apply to. This section dispels the current theories that information must be represented using "just" information models or "just" languages. Instead, by using both methods in a complementary fashion, data *and* semantics can be shared and reused.

3.5.1 Using Information and Data Models

Information models are independent of repositories, languages, and access protocols. Therefore, we need to define model mappings for policy-based applications to use the content of information models. In general, two types of mappings are of interest (Figure 3-15). The first mapping is used to translate data from an information model to a particular data model. In general, many

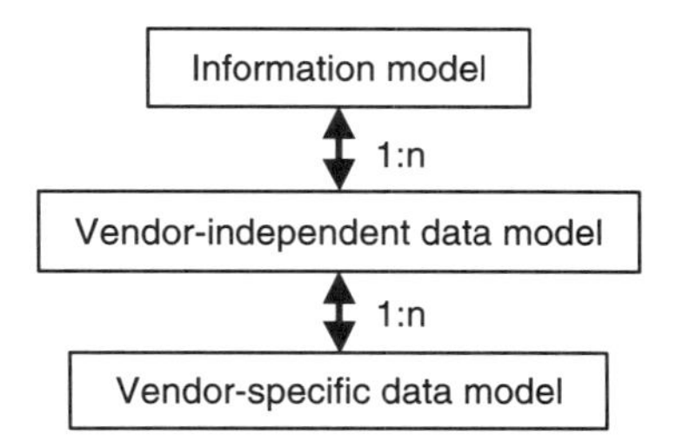

Figure 3-15 Model Mappings.

repositories are used in a single system; each having its own characteristics making them suitable for a particular type of data. Each different repository will require its own mapping because of the fundamental differences that exist in the way that each repository stores and access its data. Storing data is a function of how the data is organized and what data structures are used. Accessing the data is a function of the protocol used and the capabilities of the protocol. This is why using a single information model is so important—as long as one information model is used, an object (or parts of an object, as described in Figure 3-7) can be represented in different repositories in many ways but still maintain coherency between these representations.

The second mapping is used to translate from a generic form that is used for a particular type of repository to vendor-specific implementations of that type of repository. For example, if the type of vendor-independent data model chosen is a directory, then there may be different mappings to different vendor implementations (e.g., Novel NDS versus Sun iPlanet). This layer is sometimes necessary because there are significant implementation differences in these (and other) directories, despite the fact that standards for directories (and other repositories) exist. This is caused by three facts: (1) standards can never be precise enough to ensure that each feature is implemented exactly the same way; therefore, implementation differences are caused by the inability of the standard to precisely define functions at a level guaranteed to ensure interoperable implementations; (2) vendors will always seek to emphasize their added value, and therefore will not implement all or parts of the standard in favor of what they perceive to be a "better" feature, and (3) each vendor will offer its own additions that are not part of the standard that can offer better performance compared to the performance obtained by using the standard because the vendor can more easily optimize their implementation instead of a standard implementation.

Information models are unparalleled in their ability to express complex relationships between different data in an easy to understand graphical format. This format lends itself to building a model to capture common data and semantics and then refining the model to accommodate application-specific behavior.

The advantage of a graphical format is its ability to express complex concepts pictorially, enabling the human mind to more easily grasp these concepts. Clearly, languages can represent the same information, but not in as elegant and compact a form. The difference revolves around the level of abstraction being used. In a language, the user needs to search through the entire file to find all related classes, attributes, methods, and semantics. In an information or data model, these relationships are portrayed graphically, making it easier to see and understand the relationships.

Information models place a heavy emphasis on the use of relationships. For example, consider the definition of relationships that define how different managed objects interact with each other. Relationships can be defined in languages and in models, but take different forms. Relationships as defined in languages often have a structure that requires them to be implemented as a set of related functions in multiple places of multiple source files. This requires the reader to first find each related function (which is no easy task for a complicated project!) and then understand its implementation by looking at its code. In an information model, a single picture can graphically represent the same information (though admittedly at a higher level of abstraction). This enables the reader to better understand the purpose of the relationship (e.g., the set of managed objects and how they are related) by looking at one artifact (the picture of the model) instead of many (the different code files).

UML[26] uses a special symbology (called adornments) to define different semantics. The type of relationship (e.g., whether it is a simple dependency relationship between two objects or a more complicated relationship, such as a composition) describes semantics that can be easily captured and shown in a graphical format. For example, an association is shown as a line connecting managed objects, whereas an aggregation is shown with a diamond at one end (to denote the aggregate object), and a composition is shown with the diamond filled in.

The power of a standard symbology is that it defines common concepts in a standard way. If the graphical format is UML, then the format itself is extensible in pre-defined ways (e.g., tagged data, stereotypes, and the object constraint language).[26] Although languages can convey similar semantics by using the same naming and signature (i.e., the set of input and output parameters, their data types, etc.) of related functions, *these must be manually enforced and are not a natural construct of the language*.

UML also distinguishes between associations and association classes. Although both represent relationships, the latter uses a class to implement the relationship. This enables the relationship to have attributes and methods and even respond to events. This is an example of UML serving as a guide to define in a standard graphical format the semantics of the information that must be implemented.

Management- and policy-based applications rely on their ability to correlate current state information with predefined models of how those managed objects should behave. UML models are very desirable because of their inherent ability to represent information in a standard and an inherently extensible fashion. This enables policy application developers to extend their original models to accommodate the changing state of objects that they are managing and to model new and modified behavior that is exhibited over time. More importantly, the states of each managed object can be visually related to each other.

The advantage of using information and data models is that they are visually oriented. The adage "a picture is worth a thousand words" is very applicable to modeling. People are sometimes scared off by the seeming complexity in a UML model; in reality however, it is mostly because of the lack of familiarity with UML. True, UML models have the ability to contain additional information that some other models (such as a MIB) cannot easily represent. But if this information is worth representing, then it isn't going to disappear, and why not represent it in a single model, rather than requiring the developer to read and understand multiple documents that each have different notations? The power of information and data models lies in their ability to succinctly describe the characteristics and behavior of a system. Furthermore, the visual aspect of these models often makes it easier for jump-starting people who are new to policy-based management. Finally, the format of an information or data model often helps people develop extensions (as compared, for example, to scrolling through a large text file trying to find all of the dependencies in various modules, tables, and code that will be affected), because they can clearly see how different objects relate to each other.

However, information and data problems do have a major problem—the lack of standardization in mapping between an information model and a specific type of data model. The only significant work that has been done in this area to date is in the mapping of an information model to a directory. Remember that it isn't enough to just have a common representation of managed entities—we also need to have a common representation of the semantics of an entity and how it interacts with other managed entities. Another major problem with information models is that they "just" represent information, not how to use that information. The simplest example of this is conflict detection. Although information models can easily represent the entities necessary to model a policy rule, they cannot be easily used to detect and resolve conflicts between two or more policy rules. This is because there is nothing in the information model that enables these conflicts to be modeled and resolved, as conflicts are run-time artifacts. As we will see, this is one of the primary benefits of using a language for describing policy.

3.5.2 Using Languages

Languages may be independent or dependent on applications, vendors, platforms, repositories, and other factors. We will ignore the (obvious) use of languages for specific applications and instead focus on how languages can be used in a generic way.

There are several possible uses of languages. First, a language provides a concise definition of the policy (and associated information). This format is usually directly machine-readable or easily translatable to a machine-readable form. A machine-readable format can simplify implementation by identifying concepts and language constructs that are required in the system implementation. Some tools exist to generate code in one or more languages from models; but that only serves to reinforce the point that models are useful for the definition of objects and object instances.

Second, models are used to define information and how managed objects are related to each other in analysis and definition time, not in run-time. Consequently, they cannot be used to detect and resolve run-time problems. Languages can be used for both design- and run-time development. A PBNM system must therefore use a combination of models and languages to implement policies. In fact, a language is a very natural way to represent most policies and the information that they manage. The basic approach is straightforward—if we assume that policies consist of conditions and actions, it follows that the policy is a type of logical expression. Conditions and actions can be viewed as terms of the expression. Languages are a natural means for representing expressions. Furthermore, languages can be used to elegantly solve the goal of translating a policy from one domain (e.g., for use by business analysts) to another domain (e.g., for use by system administrators).

But languages have more appeal than this. Consider the following two policies:

- Engineers get Gold Service
- FTP Traffic gets Bronze Service

Although these two policies are correct, they could conflict with each other (i.e., what happens when engineers send FTP? Do they get Gold or Bronze Service?). Humans may or may not catch conflicting policies such as these. More importantly, models cannot represent all of the different permutations of conflicting policy elements. They instead define the overall representation of the policy and rely on a run-time tool to do this. Although OCL[26] can express some obvious conflicts (such as the preceding example), it cannot be used to detect different *effects* of policy at run-time (e.g., one policy may call for a certain number of queues to be built to condition traffic, whereas a different policy may call for a different number of queues to be built). However, if these policies were both represented using a language, it would be simple to catch this conflict because they represent conflicting expressions.

As important as this is, the real power of languages lies in their ability to connect the different expressions, or views, of a set of policies that make up the policy continuum. For example, Figure 3-5 showed five different views of the same policy. The policy continuum is realized through the mapping of information that is expressed using different grammars that are optimized for the different users of each continuum level. This use of languages is complementary to models and occurs at the analysis and design phases of a PBNM system.

The reason that languages need to supplement the use of models in defining the policy continuum is that each client of the policy continuum operates using different concepts and terminology that require different grammars. Although models can express much of the information, they cannot translate that information between different levels of abstraction because they cannot capture all of the semantics and implementation effects of those semantics. The solution is to build a set of languages that can be used in each layer of the continuum. Each language in the set is targeted specifically to satisfy the needs of one or more users of the policy continuum, while providing the coherency to ensure that each concept at one level of the continuum can be translated to other levels of the continuum that need that information. Therefore, the PBNM system not only provides different views of the same policy but also provides a set of different policies that collectively implement the policy continuum.

This set of languages forms a common specification of the overall policy. They enable specific information corresponding to each level of abstraction to be exchanged between heterogeneous components to enable them all to perform their function in providing, securing, distributing, and administering policy. The language enables multiple vendors to interpret a given policy in the same way, while enabling vendors to provide value-added services.

However, although languages can be used to concisely express the relationships between the different terms of a policy, most languages lack spatial and pictorial representation to illustrate how the policy is formed. It is much easier for a human to look at pictures. Because policies tend to be complex, pictures of how the policies are constructed may help the user understand and even define the policy. Put another way, the use of graphics better enables people unfamiliar with policy to understand policy more quickly and is fundamentally easier (especially for non-programmers) to use because diverse information and relationships between managed objects can be represented in a single picture (instead of multiple functions that are defined in different places of a file or perhaps in different files altogether).

The other major problem with using a language is the structure and grammar of the language itself. It is impossible to build a language that everyone likes. Developers will compare it to the language(s) that they program in, and start complaining about real and/or perceived limitations of the policy language compared with the language that they program in. Administrators and other people who are not developers will demand user-friendliness and verbosity; two characteristics that are typically not found in languages oriented toward developers, and consequently they will complain about how hard it is for them to express their needs in a language that wasn't built to express their needs.

The key point to realize is that a policy language will be used not only by developers, but also by business planners and administrators. Therefore, we must define a set of policy languages to implement the translation of policy information between different levels of the continuum (Figure 3-16).

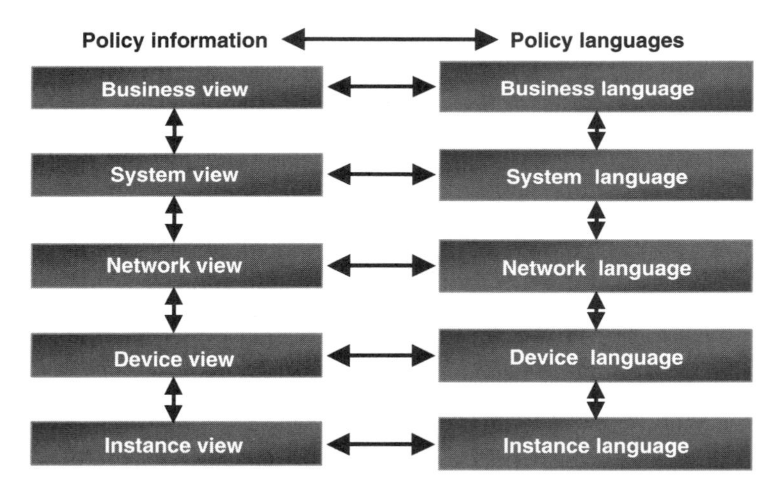

Figure 3-16 The structure of our ideal PBNM language.

This book recommends that a continuum of languages be used to match the continuum of policies that are defined. This approach solves the problem of trying to use a monolithic language to address the needs of different users of the language by constructing our PBNM language as a composition of different "sub-languages." Each sub-language addresses the specific needs of the users at that particular level of the continuum and of course enables specific information pertinent at that level of the information to be efficiently conveyed. The idea is for each language at each layer of the policy continuum to easily and efficiently present its inputs and outputs to other languages residing at other layers of the policy continuum. The effect is to create a single composite language out of a set of languages that are each optimized for use at a particular level of the policy continuum. This concept will be discussed in more detail in later chapters.

3.5.3 What About a Policy Protocol?

The least flexible, but most easily achievable approach, is to embed policy within a protocol or other communication mechanism. On the other hand, implementations are easily interoperable, and their behavior is deterministic. There are several examples of policy embedded in a protocol, including the IETF policy models that can be embedded in the IETF's COPS protocol.[1] Policy itself, in the IETF approach, is defined by constructs called policy information bases (PIB)s which are written in a subset of SNMP's structure of management Information (SMI).[27]

Unfortunately, this approach is inherently limited, because typically policy rules, conditions, actions, and variable bindings are predefined and specifically enumerated in a protocol. This does not leave much room for flexibility and expansion. More importantly, because these approaches are targeted at low-level implementation, they cannot address the needs of the policy continuum. Business analysts do not think in terms of protocol and the lower-level content of information that can be carried by COPS (and other similar protocols). The use of such an approach by itself causes irreparable harm because it prevents many important users of the PBNM system from understanding and expressing their needs.

Most importantly, this approach by itself simply cannot work. This is because protocols simply move data between a source and a destination. This misses the point of the common information model, which is specifically designed to make *all* data available in a standard format and organization. Protocols are inherently limited in what they can express. Finally, different applications and especially different devices are built to support a set of different combinations of protocols. It is impossible to define a single protocol to serve all needs of all people. A good example of this is the proliferation of protocols for intra-domain networking, such as OSPF and IS-IS. Both are standards because both are appropriate for different applications. Trying to use a single protocol (or worse, multiple protocols) for all policy needs implies the need to build complicated proxies that translate between these protocols. The translation process is very difficult, because not only do data expressed in each protocol need to be adapted to the capabilities of other protocols, the *semantics* of each protocol (i.e., what it can and cannot express) must be translated because, in general, each protocol is limited by the type of information that it can carry and the operations that is supports for accessing and managing that information. For example, the structured query language (SQL) of relational database management systems has powerful inherent capabilities (e.g., joins) that enable richer and more powerful information relationships to be expressed than the capabilities of the lightweight directory access protocol (LDAP).

Without the common information model, all the protocol would do is to ship incompatible data to different users of the data. Therefore, a specific protocol for PBNM systems is not recommended. Rather, the PBNM system is defined as a policy continuum that consists of UML models, data dictionaries, and languages and uses whatever protocols necessary to ship the information between different entities using the information.

3.5.4 The Recommended Approach

As can be seen, each of the preceding approaches has advantages and disadvantages. It is therefore important to realize that no one solution meets the needs of all applications and environments. This is why a combination of a set of languages and a set of models are used; each bound to a particular layer of the policy continuum. Languages provide conciseness and the ability to directly express semantics that enable each layer of the policy continuum to translate its information and concepts to other layers of the policy continuum that need that information. More importantly, languages are machine readable and lend themselves to sophisticated uses, such as conflict detection. They lack a visual way to express complex relationships of managed objects. Models enable new users to quickly understand the content and help guide the design and implementation of the PBNM information through its rigorous structure.

To ensure that different models and languages operating at different levels of abstraction have a common source of understanding, a data dictionary is added so that all of the metadata and semantics of each entity are appropriately captured. It can also be used to translate between different users of the same or different level of the policy continuum. For example, the data dictionary defines a set of synonyms and aliases that a given entity is known by, so that different users know that the same entity is being referenced even though it has a different name. This approach is taken by the TMF SID team and by DEN-ng.

3.6 The New DEN-ng Approach

DEN-ng is the next version of the original DEN standard. It is a federated model (meaning that it takes content from other sources) and was constructed to meet the needs of the TMF NGOSS architecture. Its policy model will be covered in detail in Chapters 6 and 7. The DEN-ng model uses the following ground rules to construct its content:

- It uses the UML metamodel to ensure that a standard set of tools and building blocks are used in its construction
- It is extended using methods defined in the UML (e.g., stereotypes, tagged data, and OCL)
- It uses a *layered* set of models; one that defines common concepts and others that build on these concepts but focus on a particular management domain
- It recognizes that not everyone is fluent in UML, and so each model specification consists of the following:
    - An English description of the use case(s) that motivated the definition of the entities defined

    - Business and system views of entities for each management domain, consisting of UML models and explanatory text and examples
    - A comprehensive data dictionary that defines the description, source, synonyms, aliases, semantics, and definitions of classes, attributes, methods, and relationships
- Best software practices (e.g., the extensive use of patterns, which is unique for information models)
- The use of a finite state machine for defining the life cycle of managed entities (this means that DEN-ng requires three types of classes—classes to model state, classes to model the changing of state, and classes to model when and how state can be changed)

DEN-ng was built to integrate different processes and entities. It uses the TMF eTOM to define business entities and processes, and other TMF documents (e.g., TMF053 and GB921) to define entities and processes. DEN-ng then defines a set of intermediate entities that are required to ensure that the business entities and processes can be integrated with the system entities and processes. Conceptually, it is an extensible UML-based framework in which other models can be placed and, as we will see, languages can be attached to.

This unique area of DEN-ng cannot be overemphasized. Most other information models concentrate on a particular level of abstraction, such as networking technology, business process modeling, or system modeling. DEN-ng concentrates on a concept and then models the business and system views of that concept. Thus, a DEN-ng network service is fundamentally different than the same network service in other information models because it contains more than just the network definition of how to implement the network service. For example, DEN-ng includes:

- The relation of that service to business entities, such as a product and a SLA
- The definition of which users and applications can manage and configure that service
- The definition of which users and applications can use that service
- The list of other services that this service depends on
- The list of other resources that this service depends on
- The definition of the network service in different networking abstractions, according to the policy continuum

Figure 3-17 summarizes the different DEN-ng models. There is a common core model to enable each of the individual domain models to interact with each

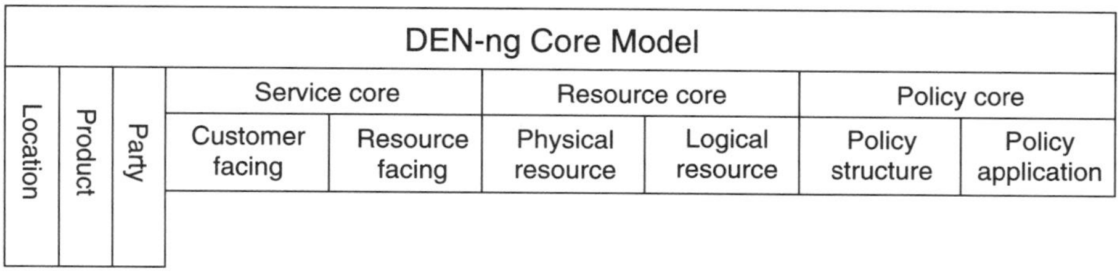

DEN-ng Core Model
Location
Product
Party
Service core
Resource core
Policy core
Customer facing
Resource facing
Physical resource
Logical resource
Policy structure
Policy application

Figure 3-17 Conceptual overview of the DEN-ng model.

other. Each of the more complicated domain models has its own core model to enable each of the more detailed domain models to fit together better and to enable them to be worked on in parallel.

The policy model of DEN-ng does all of the above to define its set of entities. It is seamlessly linked to the other DEN-ng domain models through a set of roles and explicit relationships. For example, a policy can be used to define which users can access a service, which users can modify a service, how to reconfigure the service if an SLA is violated, and so forth.

The DEN-ng policy model reuses basic concepts from the IETF model, such as requiring the common structure of a policy rule, but adds events, constraints, and additional semantics that are not present in the IETF model. The DEN-ng policy model differs from the IETF policy model in that the DEN-ng policy model consists of a series of layered models that contain related entities defined at different layers of abstraction. For example, a VPN is defined as a service object at the business layer, but is then refined into a set of objects and relationships at more granular layers, along with policies that are built to manage the VPN. This will be discussed more in Chapters 9 and 10.

More importantly, the DEN-ng policy model has been heavily influenced by the design of Ponder and has included a set of languages that are used in conjunction with its models. This is another reason that the data dictionary of DEN-ng is so important—without it, it would be difficult, if not impossible, to bind the model information to the language constructs.

DEN-ng is being implemented by Intelliden in its product line, which will be discussed more in Chapter 11. DEN-ng is also the focus of several major industry efforts. British Telecom, for example, is working on an innovative architecture called Transform that leverages DEN-ng. Other TMF companies are examining it and the SID. (Remember that DEN-ng is one input to the SID, but DEN-ng is progressing in parallel with the SID. This enables its more aggressive components to be worked out by the DEN-ng authors and then submitted as a finished project to the SID team, which then integrates it with the rest of the work. In point of fact, DEN-ng and the SID proceed in parallel, with DEN-ng focusing more on policy, network elements and network services, and the SID focusing more on business concepts.)

3.7 Summary

This chapter has described the need to provide a common representation of information. It has emphasized that an information model is more than just a common representation of data and is instead a powerful means to express objects, their semantics, and the interaction that an object has with other objects, in a common way. The information model is organized as a set of layered models; each aimed at a particular abstraction layer as defined by a policy continuum. This ensures that all users of policy can deal with common terminology and concepts and can relate their definitions and data to other information in the managed environment.

A brief survey of previous approaches was provided. This concluded that information models by themselves are not enough to ensure that common data can be shared and reused. Languages enable implementations of varying levels of abstraction to be expressed in a standard form according to the capabilities and programming constructs of the language. As with models, a continuum of languages is required to enable different users of the policy continuum to define objects and policies at appropriate levels of abstraction.

This combination of models and languages requires a common methodology to translate between the different concepts and entities defined by the model and the language. A data dictionary was defined for this purpose. It ensures that all of the metadata and semantics of each entity are appropriately captured. It can also be used to map between different synonyms or aliases that different users of the policy continuum give to the same entity in the policy continuum.

This combination of models, languages, and a data dictionary enable a PBNM system to cope with three difficult problems: (1) the proliferation of vendor-specific CLIs and programming models, (2) the inability of different OSS components to share data and (3) the reuse of information. Previous approaches have failed because of the lack of a set of abstractions that serve to map between different vendor capabilities and programming models. The unique combination of models and languages that are bound to different layers of the policy continuum ensure that multiple users with different foci (e.g., a business user, a system analyst, and a network engineer) can use the same system because appropriate views, using appropriate terminology and concepts for each of these different users, are defined. A data dictionary guarantees that the different terminology and concepts inherent in using different levels of abstraction (the policy continuum) and different methodologies (models and languages) are mapped to each other. This produces a framework in which the differences of vendor-specific CLIs and programming models are replaced with a common set of management abstractions.

The DEN-ng model embodies the preceding principles, but also ties these to the need of the TMF NGOSS approach. In particular, this means that the entities and processes modeled by DEN-ng are in turn shaped by business, system, and implementation views of the managed environment. The policy and entity information work together to form a complete behavioral description of entities in the managed environment and how they are controlled.

3.8 Recommended Further Reading and References

1. Durham, D. (ed.), Boyle, J., Cohen, R., Herzog, S., Rajan, R., Sastry, A., *The COPS (Common Open Policy Service) Protocol*, RFC 2748, January 2000.
2. Chan, K., Seligson, J., Durham, D., Gai, S., McCloghrie, K., Herzog, S., Reichmeyer, F., Yavatkar, R., Smith, A., *COPS Usage for Policy Provisioning (COPS-PR)*, RFC 3084, March 2001.
3. TMF, *NGOSS Architecture Technology Neutral Specification*, TMF053, May 2002.
4. Please see the home page for Cisco's Element Management Systems (note that this page lists four EMSs and one OSS, the rest being NMSs—the problem is

that all have EMS functionality, but most do not have NMS functionality): *http://www.cisco.com/warp/customer/44/jump/network_management.shtml*

5. TMF, *NGOSS Architecture Technology Neutral Specification (Glossary), V.3.0*, TMF053a, November 2002.
6. TMF, *NGOSS Architecture Technology Neutral Specification (Contract Specification), V.3.0*, TMF053b, February 2003.
7. TMF, *eTOM: The Business Process Framework, V.3.5*, GB921, June 2003.
8. TMF, *Shared Information/Data (SID) Model, Business Entity Definitions-Policy, V.1.0*, GB922, Addendum 1-POL, July 2003.
9. TMF, *Shared Information and Data model V.3.0*, GB922, July 2003. This is a members-only working group of the TMF, with the following members-only site. *http://teamlink.tmforum.org/Information%20 and%20Data%20Modeling*
10. Rosen, E., Rekhter, Y., Bogovic, T., Brannon, S., Vaidyanathan, R., Carugi, M., Chase, C., Fang, L., Chung, T., De Clercq, J., Dean, E., Hitchen, P., Smith, A., Leelanivas, M., Marshall, D., Martini, L., Morrow, M., Srinivasan, V., Vedrenne, A., *BGP/MPLS VPNs*, draft-ietf-ppvpn-rfc2547bis-01.txt, January 2002.
11. The proceedings of the first policy framework working group meeting are documented in *http://www.ietf.org/proceedings/98aug/index.html*
12. Strassner, J., *Directory Enabled Networks*, Macmillan Technical Publishing, September 1999.
13. Please see the home page of the IETF IntServ working group: *http://www.ietf.org/html.charters/intserv-charter.html*
14. Please see the home page of the IETF DiffServ working group: *http://www.ietf.org/html.charters/diffserv-charter.html*
15. Moore, B., Ellesson, E., Strassner, J., Westerinen, A., *Policy Core Information Model—Version 1 Specification*, RFC 3060, February 2001.
16. Snir, Y., Ramberg, Y., Strassner, J., Cohen, R., Moore, B., *Policy QoS Information Model*, draft-ietf-policy-qos-info-model-04.txt, November 2001.
17. Jason, J., Rafalow, L., Vyncke, E., *IPsec Configuration Policy Model*, draft-ietf-ipsp-config-policy-model-05.txt, February 2002.
18. Moore, B., Rafalow, L., Ramberg, Y., Snir, Y., Westerinen, A., Chadha, R., Brunner, M., Cohen, R., Strassner, J., *Policy Core Information Model Extensions*, draft-ietf-policy-pcim-ext-06.txt, November 2001.
19. Moore, B., Durham, D., Strassner, J., Westerinen, A., Weiss, W., Halpern, J., *Information Model for Describing Network Device QoS Datapath Mechanisms*, <draft-ietf-policy-qos-device-info-model-07.txt>, February 2002.
20. Strassner, J., Ellesson, E., Moore, B., Moats, R., *Policy Core LDAP Schema*, draft-ietf-policy-core-schema-14.txt, January 2002.
21. Damianou, N., Dulay, N., Lupu, E., Sloman, M., *PONDER: A Language for Specifying Security and Management Policies for Distributed Systems, The Language Specification, v2.3*, October 2000.

22. For more information on the DMTF CIM, please see: *http://www.dmtf.org/spec/cims.html*
23. For more information on IBM's Trust Policy Language, please see: *http://www.alphaworks.ibm.com/tech/TrustEstablishment*
24. For more information on IBM's Trust Policy Language, please see: *http://www.crypto.com/papers*
25. For more information on Lucent's Policy Definition Language, please see: *http://www.bell-labs.com/user/jlobo1*
26. OMG *Unified Modeling Language Specification, Version 1.4*, September 2001.
27. McCloghrie, K., Fine, M., Seligson, J., Chan, K., Hahn, S., Sahita, R., Smith, A., Reichmeyer, F., *Structure of Policy Provisioning Information (SPPI)*, RFC3159, August 2001.

Policy Operation in a PBNM System

This chapter will describe how policies are used in a PBNM system. It will first address basic communication issues, including how policies are organized, assigned, coordinated and distributed. Then, basic management issues will be discussed. This focuses on understanding how policies are applied, and the effect of applying policies to a system.

4.1 Introduction

This chapter describes how policies operate in a policy-based network management (PBNM) system. There are three main issues to understand: (1) policy communication, both among the components of a PBNM system and to the entities that are controlled by the PBNM system, is crucial to delivering timely policy services; (2) it is important to understand how policies are applied to managed entities and the effect of applying policies to the target of the policy as well as the overall system; and (3) policies are themselves complex entities. Therefore, it is important to understand how policies operate (i.e., knowing the physical and logical requirements of a policy, when the policy is applicable, which managed entities the policy will affect, and what happens when a particular policy is applied)

4.2 Policy Communication

A policy system must be able to communicate effectively to outside entities and between its constituent components. This section will describe the most important issues in policy communication.

4.2.1 General Communication Issues

There are several approaches to distributed communications. These approaches aim to solve the "N-squared" complexity of N independent systems talking to each other in a point-to-point fashion. For PBNM systems, two of the more popular approaches are publish-subscribe messaging systems and distributed communications systems.

Publish-Subscribe Message Buses

A publish-subscribe messaging system helps coordinate the exchange of information among a set of distributed components. The theory is remarkably simple. Assume that some system components generate information that other system components need. Furthermore, assume that not all system components need this newly generated information. The publish-subscribe approach defines each system component as a publisher of information, a subscriber to information, or both. Publishers send information onto a bus, and subscribers take information from the bus. The bus is the mechanism used to manage the "N-squared" complexity of intercomponent communications.

More technically, a publish-subscribe messaging bus binds distributed program components at runtime. This can be used either to build processes or to construct applications. The message bus enables different components to execute concurrently and communicate with other components via the message bus. Furthermore, components can dynamically connect to or disconnect from the message bus.

Fundamentally, there are two types of messages. Requests are messages that are sent to either a particular component or a set of components to request that an action be performed. They can be either synchronous or asynchronous.

Similarly, although there are many different ways of matching published messages to clients that subscribe to specific messages, the two most popular are subject-based (also called topic-based) and content-based message buses. The difference is that the former delivers messages based on their address, whereas the latter delivers messages based on the content of the message.

Commercial publish-subscribe systems include Talarian[1] (which has been bought by Tibco), Vitria,[2] Tibco,[3] and MQseries.[4] Research systems include Siena,[5] Wisen,[6] Gryphon,[7] and Scribe.[8] PBNM can use these types of message buses to incrementally update policy information and to signal information to coordinate the creation, modification, deletion, validation, and installation of policies. The messages themselves are used to enable, disable, and change policy rules and information.

Newer Approaches in Distributed Communications Systems

One of the more promising approaches is Jini.[9] It is an architecture that enables systems (which used to be called "communities," but are now referred to as Jini Networks by the Jini Community) to be constructed from objects and networked together. It provides an infrastructure for delivering services in a network and for creating spontaneous interaction between programs that use these services, regardless of their hardware and/or software implementations. Any type of network composed of services (e.g. applications, printers) and clients can be assembled, disassembled, and maintained on the network. Services can be added to or removed from the network, and new clients can find existing services—without administration.

Jini services forms the common integration point that helps tame the N-squared complexity of intercomponent communications. The Jini approach recommends building a set of simple services to perform a complex function, instead of a monolithic, complex service to perform that same complex function, which makes the Jini approach inherently more distributable than other approaches. In this approach, legacy, non-Java systems are wrapped via surrogate or proxy techniques and presented to the *community* of Jini services and consumers as *just another Jini service.*

Brief Summary of Important Jini Concepts

Jini provides a lookup service that allows other services connected by the communication infrastructure to be found. This Jini Lookup Service (JLS) helps locate and resolve a needed service. The service interfaces are mapped onto objects that implement the service. A client wishing to use a service uses a lookup protocol to locate a lookup service and then uses the lookup service to find the desired services. Objects in a lookup service may include other Jini lookup services or encapsulate other non-Jini services. This mechanism bridges the various lookup services and allows clients using them to access the Jini community.

The Jini infrastructure provides a mechanism—called discovery/join—for Jini technology-enabled devices (e.g. disk drives, printers, routers, computers) to discover the appropriate lookup service and join the appropriate community. When an entity joins a community, its services are added to that community's lookup service. Likewise, when a device leaves the community (by removal or lease expiration), its services are deleted from the lookup service.

Jini supports distributed events and event notification. This allows interested services to get notification when particular events occur within certain objects. This is *not* a traditional publish-subscribe system, but rather, events are used to activate services.

Services are provided in a Jini community on a lease-based access, which means that a consumer is provided access to a service for the lease period. This lease is negotiated as part of the service protocol. A service is requested for a period of time, granted for some period, and after that amount of time, if the lease is not renewed, the resource is released. This is a very powerful concept. Granting a lease to the user of a distributed resource allows the resource owner to determine if the resource is no longer required and reclaim it, even in the event of partial or complete network failure. This is one of the fundamental concepts behind distributed garbage collection. Individual servers are free to implement leasing according to whatever policy is appropriate for the resources they provide.

Each Jini-enabled device has enough information stored on it to enable hot-plugging. There is no need to install drivers or configuration. Its services are available instantly. This enables the concept of "service-on-demand." A client looking for a particular service may connect to any Jini network (i.e. community) and look in the directory for the service it wants, use the service, and then disconnect.

Jini approaches systems integration by defining almost everything as a Jini service. This approach differs significantly from the approach taken in message bus systems. Message bus systems focus on using adaptation and mediation among systems that were not designed from the outset for integration with other systems. Adaptation is a mechanism to convert one entity that has a particular syntactic representation to another entity that has a different syntactic representation to enable interoperation of the two. Effective interoperation may still require the application of a mediation layer.

Mediation is the process of translating from one level of representation and/or abstraction to another level of representation and/or abstraction. Often, mediation of two entities is achieved by translating the two entities to a common third form. Mediation enables entities to interoperate in a loosely coupled way.

Jini promotes the building of specialized, focused components that *play well with other systems from the outset*. Rather than building complex adaptation and/or mediation tools, Jini systems embrace the common Jini service model, abstracting everything as a service. This can be supplemented by wrapping non-Jini components with (for example) Jini proxies so that they appear to be Jini services and therefore conform to the common Jini model.

An Example—the Fine Grain NGOSS Catalyst Project

One powerful example of using Jini was the Fine Grain NGOSS Catalyst project in the TMF.[10,11] Fine Grain uses a Jini-based architecture to integrate new applications and services that support Jini natively as well as older applications that are not Jini-enabled. (Note that newer network devices are starting to support Java and Jini.)[12,13] Thus, in Fine Grain, everything is a service, whether it is hardware or software. Each service is represented by a Java Interface. Communication can be through remote invocation, publish-subscribe, or Java messaging.

The Fine Grain NGOSS essentially replaces message buses with the common Jini model of network services accessed through common APIs. Although used where appropriate, information buses are only one of the interaction approaches used among a range of options, which also include remote, synchronous invocations and shared tuple-space repositories. Furthermore, where asynchronous messaging buses are used within Fine Grain environments, they are presented to their clients in the common framework of Jini services that are discovered like any other Jini service.

The computational part of implementing the OSS was built as a set of Jini services that exchanged information through a JavaSpace,[14] which of course, is another Jini service. The particular commercial implementation used was IntaMission's IntaSpace.[15]

One of the most important features of Fine Grain was improving system survivability. This is typically a weak point in existing OSS systems, because each OSS component presents a different interface and is supported by different communication mechanisms and APIs. In Fine Grain, everything is a Jini service, and thus has consistent interface and set of APIs. Therefore, all services can be monitored and automatically restarted on failure. We will talk more about Fine Grain in Chapter 11. Fine Grain services can therefore be described as:

Jini Services plus an Information Model (subset of DEN-ng) plus Policy

Conclusions

In PBNM systems, every component (even devices) must be able to provide and consume information. This effectively makes the system a closed-loop system, and more importantly makes communication possible between policy components and the devices that they control.

There is no one "right" method of implementing communication. This is more of a function of the needs of the organization implementing the PBNM system and existing capabilities of the components being integrated. However, a PBNM system requires a much richer and more robust communications infrastructure than previous approaches because it is critically important that when policies are changed in the system (e.g., a new policy is deployed or an old policy is revoked) that all affected components know about that change and can react appropriately. Examples of both communication approaches will be provided in Chapters 8 and 11.

4.2.2 Assignment of Policies

Policies can be assigned to virtually any type of entity. For example, policies can be assigned to change the particular type of traffic conditioning, such as the queuing and dropping algorithms used, based on measured parameters (e.g., the number of packets received, what percentage of traffic exceeded its allowed limit, and other factors). Policy can be applied at a much higher level of abstraction, such as to control how a user logs on to the system.

The literature has so far treated policies as disparate entities, which is unfortunate because most high-level policies are in fact related to low-level policies. In the preceding examples, the user could be a subscriber who is assigned a particular set of services based on his or her identity according to a service level agreement (SLA). This SLA must be translated to a form that can be used to condition that user's traffic. Therefore, it affects a set of network devices.

Figure 4-1 shows how different policies at different levels of abstraction must work together to ensure that the user in this example receives the services that he or she has contracted for.

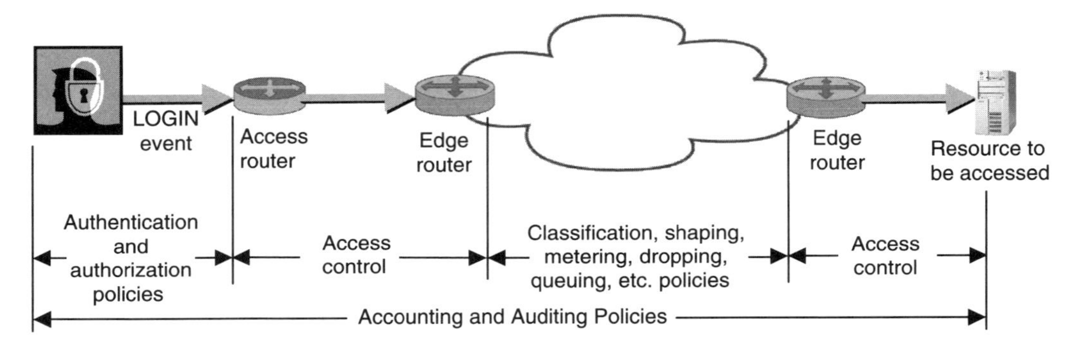

Figure 4-1 How different policies apply in a system.

The preceding policies span a several different entities, including:

- User
- Services that the user is authorized to perform
- Paths connecting the user to network and system resources
- Resource to be accessed, such as a router
- Mechanisms within a router that condition traffic
- QoS and SLAs
- Access control
- Accounting and auditing of actions done

These are just the major entities—all entities have not been listed to keep the example simple. Obviously, if all of these entities have to be managed just so the user can connect to a resource, how can this possibly the scale? The answer is through the use of abstraction.

To Users

Clearly, one cannot manage any appreciable number of users individually. Users need to be organized into groups that have the same privileges and access rights. Exceptions can then be dealt with by either defining subgroups or on an individual basis.

We will use *roles* to organize users. By associating roles with particular groups, the same access rights and privileges are inherited by every user in a given group. Roles correspond to positions that people play in an organization. They also mirror different types of users that subscribe to services (e.g., users can have the roles of gold, silver, and bronze for best, above average, and best effort services, respectively).

We accommodate situations in which a user plays multiple roles through combining roles. These role-combinations (see Chapter 2) are implemented via *role-selectors* (again, see Chapter 2). Policies can thus be applied to groups that have one or more associated roles. For example, we may have the roles sales representative, sales engineer, and sales director. Organizationally, these are shown in Figure 4-2 as:

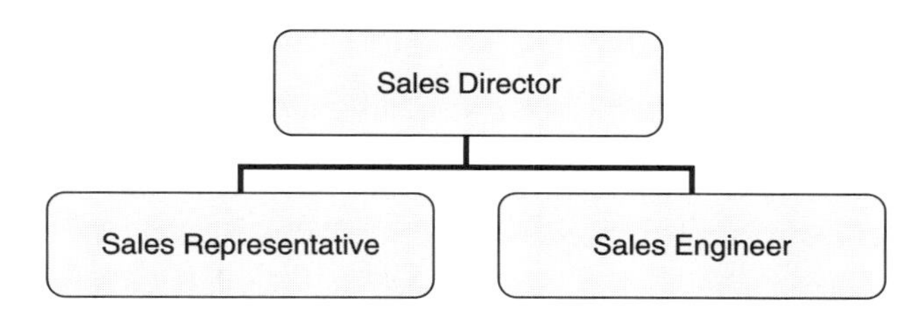

Figure 4-2 A fictional sales organization hierarchy.

The preceding figure represents a reporting hierarchy. From a service perspective, sales director must be able to do anything that either a sales representative or a sales engineer can do. Furthermore, we know that the sales engineer is there to support the sales representative. In other words, the sales representative will initiate and maintain direct contact with the customer, whereas the sales engineer will play a supporting role. We therefore end up with the following roles (Figure 4-3):

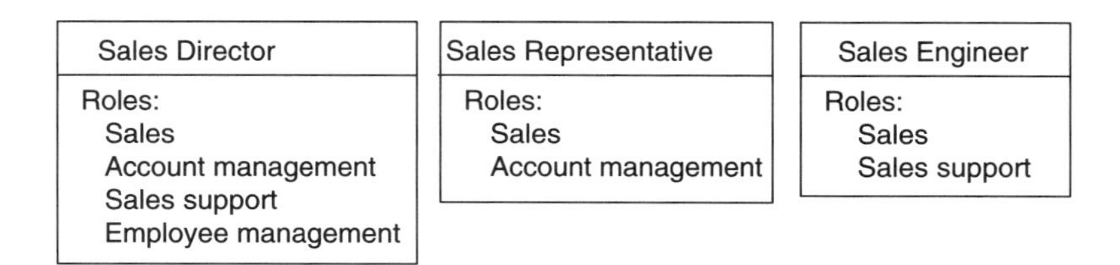

Figure 4-3 The roles for our fictional sales organization.

As expected, the sales director has all roles that the sales representative and the sales engineer have. In addition, the sales director has the employee management role, which provides managerial authority to the sales director. Because the sales engineer, sales representative, and sales director all have a sales role, they can be easily differentiated from other types of employees, such as those in the marketing and engineering departments. This enables the administrator to use the sales role to implement programs, such as *Sales Force Automation*, which enable all people in the sales organization to have the same basic set of privileges, access to resources, and even laptop configuration.

Thus, by associating different types of policies with different roles, we can assign different behavior to people who have different roles. By associating people in groups, we can avoid dealing with individual people and instead can manage a significantly smaller number of groups. Furthermore, this abstraction frees us from worrying how to accomplish a particular task if a specific individual is unavailable if that task can be performed by anyone with that role.

To Network Devices

Different routers have different purposes in the network. For example, core routers need to excel at forwarding large numbers of packets as quickly as possible, whereas edge routers must be able to classify different packets and flows very quickly. Access routers connect users to the edge of the network and must deal with different types of protocols that mobile and remote users employ to connect to the network.

One way of dealing with this complexity is to use the concept of roles. This enables us to apply different types of policies to different devices as a function of their purpose in the network. It also greatly simplifies configuring the network. For example, instead of having to painstakingly find all BGP edge interfaces that peer with a particular service provider, we can assign the role-selector "BGP + Edge Interface" to select just these interfaces. This allows us to (for example) easily change the BGP peering policy (e.g., change one of the BGP path attributes that is being passed). Without our role-selector, it would be much more difficult to find *all* of these interfaces; let alone ensure that they are all changed at the same time.

Thus, by associating devices and device interfaces with roles, we can apply different types of policies to different devices and device interfaces as a function of the purpose of that device or device interface. This tool is invaluable for managing large, complicated networks.

To Mechanisms Within Network Devices

Networks are becoming too large to have their individual devices (let alone their interfaces and sub-interfaces) individually managed. Furthermore, the complexity of traffic is such that individual traffic flows and paths cannot, in general, be managed. Rather, we need to rely on more sophisticated aggregation mechanisms, such as Differentiated Services,[16] which defines how different traffic can be aggregated to receive common traffic conditioning. A large organization might run hundreds of applications. Each application could in general spawn a set of different traffic types that together constitutes the application. From an efficiency point of view, each particular traffic type usually can receive conditioning commensurate with its own particular needs.

The good news is that this enables the system and network resources to be more efficiently utilized. The bad news is that this generates too many types of flows to deal with on an individual basis; both because of limitations of the protocol (i.e., there are only six bits in the DiffServ field) and because of limitations of the router (e.g., not enough queues). Furthermore, how could a service provider manage hundreds of different classes of service?

Therefore, we must first classify traffic so that we can aggregate similar traffic together and concentrate on conditioning traffic aggregates instead of individual traffic flows (Figure 4-4). Conceptually, the following is performed in the router:

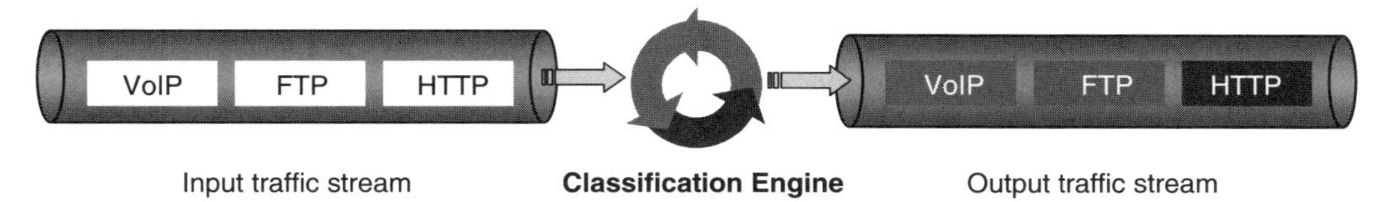

Figure 4-4 Classifying traffic inside the router.

In the preceding figure, three different types of traffic (corresponding to multiple applications) enter the ingress interface of a router. Typically, the first place the traffic is directed inside the router is the classification engine. This software separates the traffic into multiple streams; each of which can potentially receive different conditioning. The result is shown on the right hand side of the preceding figure—the traffic has now been separated and marked according to the relative priority of each traffic stream. This enables different conditioning (e.g., queuing, dropping, etc.) to be performed on an individual traffic stream basis.

Classification is just one type of mechanism in the router; many others perform various functions on the packets that they receive. Some of the more important mechanisms include policing, shaping, dropping, queuing, and scheduling. Policing is a mechanism wherein the input rate of traffic can be limited. Shaping smoothes the output rate of traffic by buffering flows that exceed their rate. Policing and shaping are discussed in more detail in Figures 4-5 and 4-6, respectively. Dropping refers to the ability to drop packets (either randomly or specifically) to tell certain types of applications to slow its transmission speed. Queuing refers to an algorithmic way to sort the traffic and then prioritize the sorted traffic for sending on an output link. Scheduling is an algorithm that selects traffic from multiple queues and interleaves them.

Depending on the sophistication of the router, this can translate into one overall traffic conditioning policy that attempts to take care of all of these different mechanisms, or more likely, a set of policies, where each device mechanism policy is used to adjust a particular function of the router. If we use multiple device policies to control different functions inside a router, each policy must be cognizant of the effect that other policies have on the overall resources of the router. For example, if fancy queuing (e.g., class-based weighted fair queuing) is enabled, this will require a certain amount of resources from the router. Depending on the amount of memory, CPU processing speed, and many other factors, the router may not have enough processing capability remaining to perform other functions. Furthermore, some device mechanisms directly affect other device mechanisms. As a simple example, if the classification function is not correct, none of the traffic conditioning performed will be correct either (unless there is no congestion in the router, in which case it does not matter if we classify packets at all).

The solution is to use a combination of roles (to identify the function of the mechanism and abstract it to a common terminology) and a detailed information model that represents the interdependencies of these different mechanisms. If the interdependencies are properly represented, then the policies that control them will be properly represented.

Between Heterogeneous Devices

The preceding discussion focused on a router. Clearly, many more types of devices can be used in the network. All of these devices must work together to provide a user with the services that he or she requires. Although roles help to an extent, they do not solve the mismatch between device functions.

Directory-enabled networks–now generation (DEN-ng) uses the notion of *capabilities* to abstract functions. For example, priority queuing and weighted fair queuing can both be abstracted into a "queuing" capability, even though these queuing algorithms do very different things and affect traffic flows very differently. By being able to characterize what different types of queuing do, we can mix and match these functions as appropriate. More importantly, when devices have different capabilities, it enables us to perform a "best match" to choose functions that complement each other.

To Domains and Sub-Domains

People control network devices. Usually, people guard these precious assets very closely. It is very rare that one group will allow its devices to be managed by another group's administrator. Hence, our PBNM system must be able to deal with groupings of devices that belong, in an administrative sense, to a particular user or group.

DEN-ng defines a "NetworkDomain" as such a grouping. DEN-ng NetworkDomains provide a means of grouping together a set of devices that share common characteristics. NetworkDomains can be nested, which enables hierarchies to be formed. For example, this enables devices that are under common administrative control to be efficiently modeled. NetworkDomains are associated with users and groups of users that play various roles, such as *administrator*.

Between the Network and the Environment

We live in an environment in which networks change (e.g., routes are recomputed when a link goes down), the number of users change, the application mix changes, and different business rules go into effect which re-prioritize which applications receive preferential treatment and preferred access to shared system and network resources.

Most current implementations do not take this into account, but rather continue to try and apply policies to managed entities independent of whether they are appropriate at a particular time.

Clearly, if the environment changes, feedback must be available to enable the PBNM system to compare the current state with the desired state of the entities that it is managing. Therefore, our PBNM system must be able to incorporate feedback, so that the current state of a managed entity may be tracked against its desired state.

However, sometimes additional constraints come into play that restrict the set of capabilities of a managed entity that can be used. DEN-ng models these as *constraints* and uses OCL to define how they relate to managed entities. This ensures that only those capabilities that are appropriate and allowed to be used can be used at any given time.

4.2.3 Coordination of Different Policies

Most vendors build specialized policy systems. That is, the policy system does not know about or use all of the functionality of a given device; rather, it provides management and control over a small percentage of the functionality of a device. For example, a vendor might build a security server, a QoS server, and an IP address management server (for example, see Cisco products[17-19]). Building specialized policy servers such as these and others is fine, as it enables the vendor to more easily build a best-of-breed product for each of these three functions.

The problem is that these three policy servers may need to talk to the same interface of the same device. If each policy server uses different terminology and different commands, how can one ensure that the commands given by one policy server do not conflict with the commands given by another policy server?

Equally important, how does one ensure that a set of discipline-specific policies (such as the three mentioned) are applied at the right time and *in the right order* to the appropriate devices?

Although roles can help select the right policies, we also need to rely on the concepts of *subject* and *target* as defined in Chapter 2. If we can develop policies that use roles to identify the subject and target of the policy, then we can start synchronizing their application. Of course, this also requires an efficient communication mechanism to ensure that the policies are distributed to the appropriate system components and requires a common understanding of each policy even if it uses terminology that is specific to a particular discipline. This is handled by the DEN-ng information model.

4.3 General Management Issues

Management is a very complex task. The application of policies depends on the nature of the subject and target of the policy, the effect of applying the policy, and the purpose of the policy. This section will first provide a high-level categorization of policies and relate them to different management tasks. This categorization enables a PBNM system to more quickly determine the type of policy that it needs for a given situation. This section will then discuss how policies are applied.

4.3.1 Policy Categorization

Policy-based management reflects the complexity of information technology. In the real world, many types of configuration and operational problems must be solved (or prevented). Furthermore, the different types of entities being managed are quite large because the number of vendors, systems, software, and devices deployed each has different capabilities. To this we must add the fact that the environment changes dynamically and in an unpredictable fashion. For example, it is impossible to anticipate the number of users or types of applications that the users will use at any given time. Instead of having to over-engineer the network and waste valuable resources, PBNM systems provide the ability to quickly fine-tune the capabilities of the network in accordance with the specific requirements of the current environment.

One can imagine building policies for almost anything. Policies can run the gamut from generic to specific. High-level business rules, such as those governing how employees are authenticated, affect many devices at many different levels. For example, one such business rule could affect a Radius server (for dial-in authentication), an access server (to aggregate users that are dialing into the network), a directory server (for storing credentials), and many other devices. Each affected system will have a different implementation of this single policy. Therefore, it is helpful if we employ a categorization of policies so that we can more easily organize policies and bind them to target entities. All categories defined in these subsections are included in the DEN-ng categorizations except where noted.

By Meta-Policy

Policies can be categorized according to their function. One example of this is a meta-policy (i.e., a policy about a policy). Meta-policies can be used to control the behavior and semantics of a policy, how and when it is applied, and many other factors that affect its use and operation. For example, a meta-policy may specify what to do if a conflict with another policy is encountered or if an exception occurs in the execution of a policy rule. Meta-policies could also be used to define mappings between high-level and lower-level policy rules.

By Technology

Organizing policies by technologies is perhaps the simplest and most obvious way to categorize policies. In this approach, policies are organized by the specific problem domains that they address. A very general policy rule or SLA can cover many technologies, whereas other rules are specific to a problem domain and area of expertise. Examples of the two extremes are:

- Ensure that a given SLA (which defines performance metrics, such as the percentage that a customer's connection is available and the latency incurred on that connection) is met
- Ensure that a specific application does not get more than x kbps of bandwidth at any given time.

Assume that the first example spans a network of multiple technologies (e.g., an ATM core, a Frame Relay distribution network, and an IP edge network). This policy not only affects each device used by the customer's connection, but also requires the different expressions of QoS in each technology to be equated with each other. This is because there is no standard mapping between the different QoS configurations of ATM, the QoS specification in Frame Relay (committed information rate), and either the ToS (Type of Service) byte or DiffServ field in IP. Thus, a policy is needed that equates these different expressions of QoS and adjusts them as appropriate. Here again, we see the utility of capabilities. Each of these different mechanisms is elegantly represented by the same type of capability. This enables us to better match diverse functions and ensure that they work together and that each delivers the appropriate service.

The second example can be implemented using a variety of device-specific features, such as policing versus shaping. Policing is defined as a feature that limits traffic flow to a configured bit rate. Although most policing commands allow the specified traffic rate to exceed (often called bursting), this rate is for a short period. In addition, most policing facilities have no buffering capability. In other words, packets that cannot be transmitted are simply dropped, as shown in Figure 4-5.

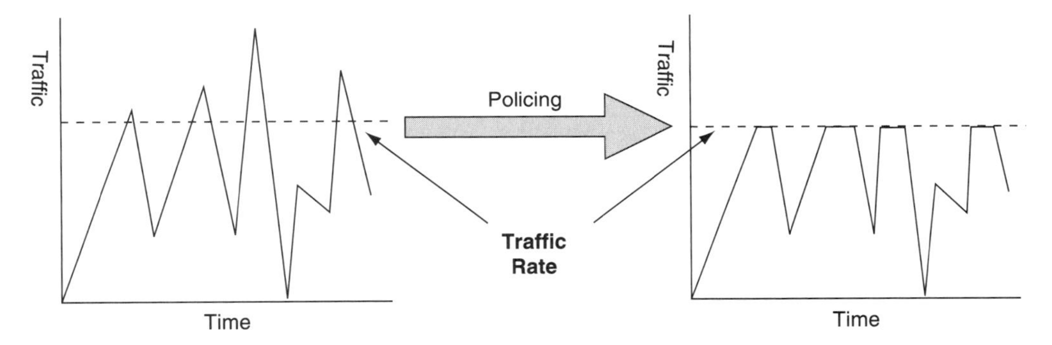

Figure 4-5 The effect of policing on network traffic.

In contrast, traffic shaping is used to buffer the output so that the output flow is regulated to a defined rate. It comes into play when the amount of traffic to send exceeds the output capacity of the link.

The important difference between traffic policing and traffic shaping is that policing drops packets that exceed the configured rate, whereas shaping tries to buffer those packets.

Figures 4-5 and 4-6 are two examples of two common functions implemented by most network vendors. The added value of PBNM solutions is that even though different network vendors will implement these two functions using a different set of commands, an information model can normalize the differences between them. Furthermore, the *same* policy can be used to manage either of these functions, regardless of how each function is implemented in a particular network device.

Policies are not limited to networking. They can be defined for accounting and billing, security, storage management, and many other categories. The point of this category, however, is to define what specific technology (or technologies) this policy covers.

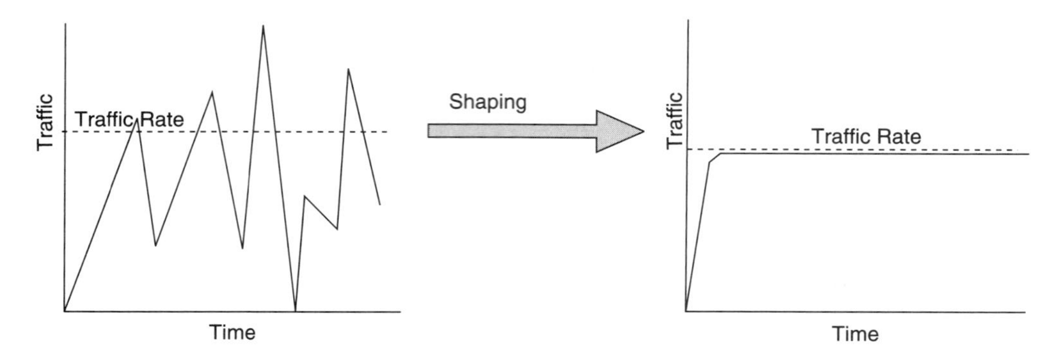

Figure 4-6 The effect of traffic shaping on network traffic.

By Subjects and Targets

It is advantageous to organize policies to correspond to administrative hierarchies when they are applied in business organizations. Similarly, policies should be grouped according to the domains that they are going to be applied to or according to the person that maintains them and applies them.

All of these goals can be accomplished by organizing policies based on the subjects and targets that they apply to. This enables policy rules to be applied to the particular set of managed entities by a given set of people. Thus, one may have a general policy rule that governs how configurations are applied to a particular type of network device, such as a customer-premise router. This policy may define a standard set of commands that all such customer-premise routers contain. A second policy may also be used to modify customer-premise routers; but it could be limited to a particular set of people (e.g., support technicians) that are allowed to modify specific parts of the configuration. Although both examples apply only to changing the configuration of customer-premise routers (and not other types of routers), the first can be executed by any member of the support organization, whereas the second can only be executed by more senior personnel.

Finally, the concept of the policy continuum is based on varying scope of administrative coverage. General, high-level rules are likely to have more global coverage, whereas device-specific, low-level rules are device- and perhaps even component-specific.

Using the IETF's PolicyKeywords Attribute

RFC3060[20] defines a class attribute, PolicyKeywords, to classify policy groups and rules by their purpose and/or intent. This classification is useful in for searching, selecting, and grouping policy rules and groups of policy rules. This attribute is an array, which enables a particular policy rule to belong to multiple categories.

RFC3060[20] define the following keywords: "UNKNOWN", "CONFIGURATION", "USAGE", "SECURITY", "SERVICE", "MOTIVATIONAL", "INSTALLATION", and "EVENT". Their definitions, along with comments relating these definitions to the definitions used in DEN-ng, are defined briefly below. (If there is no difference between the DEN-ng definition of the keyword and the RFC3060 definition of the keyword, then no comments are provided.)

- Configuration policies define the default (or generic) setup of a managed element. *Comment*: configuration also implies changing the configuration of a managed element. Therefore, DEN-ng splits this keyword into two keywords: "Default" (for the RFC3060 definition) and "ConfigChange" to represent changes to the configuration of a managed entity.
- Usage policies control the selection and configuration of entities based on specific "usage" data. *Comment*: the DEN-ng definition is: "monitor a given set of metrics and then reapply Configuration Policies based on the current value of that metric"; it is felt that this is a clearer definition than that given in RFC3060, although the intent is the same.
- Security policies is a general category that consists of authentication and authorization mechanisms, permitting or denying access to managed elements, encryption, and other policies that affect the security of the system. They address the configuration and selection of these mechanisms. *Comment*: this category is kept for backwards compatibility and to do some high-level pruning of policies; DEN-ng defines its own hierarchy of more specific security policies. This is because the subject and/or target of these different types of security policies often differs, as well as recognizing the fact that the policies themselves are different. In addition, DEN-ng defines more granular security policy categories, consisting of Authentication, Authorization, Accounting, and Auditing.
- Service policies characterize network and other services. *Comment*: this category is kept for backwards compatibility and to do some high-level pruning of policies; DEN-ng defines its own hierarchy of more specific service policies (e.g., customer-facing versus resource-facing, or internal services). Having a policy this broad (e.g., "service") renders it almost useless, except for simple operations (such as separating service policies from other types of policies).
- Motivational policy is a type of policy that "motivates" satisfying a policy goal. *Comment*: high-level business policies often fall into this category. While the engineer wishes for more specific definitions, it is important to remember that policies are defined by the user.
- Installation policies define what is permitted versus disallowed on a system or component and the mechanisms for installation. *Comment*: this definition unfortunately is overloaded. Two types of policies are defined here—policies that control what is and is not permitted to be installed and policies that govern how a component is installed. Therefore, DEN-ng separates this category into two categories.
- Error and event policies enable automated response to events. *Comment*: these two concepts are "mushed" together in RFC3060, but the actual policy specification only has event. In DEN-ng, events and errors are separated. This is because errors are not necessarily propagated as events, and events are used for many other things than just to communicate that an error occurred. Furthermore, errors are specifically handled through exceptions. While it could be argued that an error is a form of an event, it is a special

type of event. Thus, DEN-ng defines error policies, exception policies, and event policies with suitable hierarchies of each.

Comparison with Ponder Policies

Ponder provides four basic types of policies, called authorization, refrain, obligation, and delegation (see Chapter 5). Brief definitions of these four types of policies are as follows:

- An **authorization policy** is a type of security policy whose purpose is to define what activities a subject is permitted to do (or not permitted to do) to a set of target objects. Authorization policies act on behalf of clients that have already been authenticated.
- A **refrain policy** is used to define a set of actions that subjects are not permitted to execute on target objects, regardless of whether they have been authorized to do so. They restrain the set of actions that a subject may perform and are similar in function to negative authorization policies, but with one important difference—refrain policies are subject-based, whereas negative authorization policies are target-based.
- An **obligation policy** defines the set of actions that a subject must do (i.e., the subject interprets the policy and performs actions on the target).
- A **delegation policy** are used to permit subjects to grant the same or a subset of privileges, which they possess (because of an existing authorization policy), to other subjects (called grantees). Delegation policies cannot *add* new privileges to the grantee.

Table 4-1 compares the Ponder policies with those described above (RFC3060 policy keywords and DEN-ng categories).

Table 4-1 shows that Ponder can represent the types of policies defined in the IETF PolicyKeywords class attribute. It also shows a difference in approaches. The IETF concentrated on building a high-level classification of policy that

Table 4-1 Mapping of IETF policy categories to Ponder policy types

	Types of Policies			
IETF Keyword	Authorization	Refrain	Obligation	Delegation
Configuration (default) (DEN-ng)			X	X
Configuration (change) (DEN-ng)	X	X	X	
Usage			X	
Security	X	X	X	X
Service	X		X	
Motivational			X	
Installation (permissions) (DEN-ng)	X	X		
Installation (for a component) (DEN-ng)	X	X	X	
Event (DEN-ng)	X	X	X	X
Error (DEN-ng)			X	
Exception (DEN-ng)			X	

deliberately used noncontentious terminology. Although it is good to have agreement, classification requires more specific terminology to be useful. Ponder took the opposite approach to policy categorization; it defined generic policies that fit a large number of needs.

DEN-ng uses a combination of attribute-based classification (based on extensions of the IETF model), enhancements to Ponder, and a class hierarchy that was organized according to some basic classification rules (we will explore this more in Chapter 6). The result is a rich and malleable classification toolkit that enables the developer to set up policy hierarchies (and more importantly, relationships between policy rules and managed entities) that are customized to the specific applications and environment that the PBNM system is being used in.

4.3.2 How Policies Are Applied—Outsourced, Internal, or Interactive

Another useful classification is how the policy is implemented. An outsourced policy model implements policy by having components request policy-related decisions to be made by other components of that same framework. The policy decision-making function is located in a component that is physically separate from the component where the policy is executed. A well-defined, real-time interaction exists between components in an outsourced policy model. This has the advantage of centralizing the decision function, so that different types of policies can take advantage of a single decision process. The disadvantage is that it takes time to send a policy request to a decision-making entity and to receive an answer from the decision-making entity. Therefore, it may not be applicable for all types of policy decisions.

An internal policy model implements policy by keeping the decision-making process co-located with the policy request, execution, and enforcement processes. In this model, devices are configured to execute policy before the events that will prompt decisions. Devices use the appropriate preprovisioned mechanisms to implement the appropriate functionality when different traffic is received. Thus, no real-time interaction exists between components in this model. This is the exact opposite of the outsourced policy model, and so its advantages and disadvantages are reversed.

An interactive policy model implements policy by installing policy expressions within appropriate components of the system. These are complete and self-contained expressions of policy information and the rules that define what and how to evaluate to make a policy decision. Interaction between policy components is defined on a system-specific basis. This policy requires some type of interaction (human or otherwise) before a policy decision can be enforced. The advantages and disadvantages of this model are both that it is interactive. That is, interactive policies can be more responsive and more easily fine-tuned to the needs of the user. However, they are by definition slower, because they require communication with a human and are often subject to interpretation.

None of these three models is the "best" or "correct" way to implement policy. The "best" or "correct" way depends on the nature of the environment in which the PBNM system is being deployed. However, for PBNM to be optimally deployed, each model should be used if it is appropriate to do so. Stated another way, a robust PBNM solution is one that can support outsourced, internal, or interactive policies.

4.3.3 An Application of Policy—Provisioned versus Signaled QoS

QoS often takes two forms: provisioned (using internal policies) or signaled (using outsourced policies). Policies used in applications that require provisioned QoS (e.g., VoIP) are predefined policies that proactively execute to try to keep behavior at a status quo. For example, traffic conditioning can be prespecified according to a set of SLAs. These SLAs can also be used to identify the traffic that needs to be conditioned (e.g., traffic with this source port originating from this IP address range must receive preferential treatment). These policies are static in nature, which means that a set of them must be used to accommodate the dynamic nature of the network environment. They affect predetermined points in the network and provide prespecified behavior on those points.

Policies used in applications that require signaled QoS react to a specific event and are sometimes called "on-demand" policies because they enable resource requests to be asked for only when needed and to be satisfied on-demand. A good application of signaled QoS mechanisms is to provide guaranteed service. Because of the network changing dynamically, the only way that a guaranteed service can be provided is usually to over-provision parts of the network. This is wasteful and unduly complicates the network design when guaranteed services are not required. Furthermore, it may not always be viable to overprovision the specific part of the network that needs to supply the guaranteed service.

Signaling enables the specific needs of the application to be communicated to each node that will provide the service. Signaling uses requests to reserve resources in each affected network device, whereas provisioning configures the network device to act a certain way regardless of what traffic is flowing or will flow. Signaling also enables additional information, such as user and application identification, to be communicated between system components. For example, user and application identification data are not present in an IP header—only IP source and destination addresses and ports are provided. This additional information can help improve the manageability of the network. In addition, signaling mechanisms traverse the identical network path that they seek to control. This enables very specific information to be communicated to the network devices along this path, enabling not just guaranteed services, but more optimal use of network resources.

However, these information comes at a price. First, the signaling results in extra network traffic. Second, additional setup time, formulating the request, processing of the response, and releasing the resource when it is no longer needed incurs additional time and overhead. Third, and most important, it requires state to be kept in each node that is participating in the request. Therefore, signaling should only be used when, without it, the application cannot provide good end-user experience (e.g., in VoIP—imagine losing every fifth call or not hearing every fourth word in a conversation!) and/or for flows that are relatively long in duration (e.g., at least one minute).

Therefore, many times the network will be engineered so that signaled traffic, such as RSVP, will be *tunneled* over a provisioned network (such as a DiffServ cloud). This provides the best of both worlds—signaling can be used to carry per-state information, and DiffServ can be used to enable the network to provide scalable services. This is being pursued in the IETF RAP[21] and ISSLL[22] working groups.

Policy plays a critical role in both signaled and provisioned QoS. In each case, policy is used to configure the network functionality. Policies can also be used to direct the network to implement alternative behavior. Alternate behavior is invoked in response to events and can be either limiting or not, depending on the nature of the event and who is changing the state of the system. For example, if the CEO needs additional bandwidth, he or she should get it. In contrast, if a link or application fails and customers can no longer receive high quality video, then a policy could specify that all premium customers receive low-quality video, while all other (non-premium) customers receive SMS updates, and all customers (premium and otherwise) get a rebate because their service level agreement was violated.

4.4 Policy Operation

The following factors must be checked to ensure that a particular policy can be executed.

4.4.1 Physical System Requirements

Policies require system resources to execute. First and most obvious is the amount of memory required by the policy. Second, the overall complexity of the policy (e.g., how many rules does it consist of, how much additional information does it require, etc.) must be evaluated. Some policy decision components have finite limits on the number of policy rules (or the number of conditions and/or actions within a rule) that can be loaded at any single time.

4.4.2 Logical System Requirements

Policies also require logical resources to execute. For example, the entity executing the policy must be able to understand the policy. If it cannot, then it must pass the policy to an interpreter or proxy that can either translate the policy to a form that the managed entity can execute, or execute the policy and return the result of the policy execution.

Another factor that must be evaluated is the performance impact of executing the policy. For example, if an outsourced policy is meant to affect a particular traffic stream, a set of packets will pass through the device before a decision can be received and a suitable policy implemented. Three questions need to be asked in this case: (1) does it matter than some packets got through (clearly, if it does, an outsourced policy should not have been used), (2) how many packets can get through (i.e., without the policy being applied to them) before the application is adversely affected, and (3) what is the effect on the overall performance of the application (e.g., even for an internal policy, if it contains too many rules, then the device will not be able to forward packets as fast as expected). These three questions enable one to determine whether or not a particular type of policy is appropriate for a given system.

4.4.3 Applicability of the Policy

Different devices have different capabilities. This not only applies to devices that are implementing policy decisions, but policy servers that are making decisions.

Devices that implement policy decisions must have the functions required to enforce a policy decision. The use of capabilities enables different functions of different devices to be equated with each other, so that a common policy server can control them. Similarly, matching the role of a policy with subject and target roles helps ensure that a particular policy is meant to be applied to a given target.

However, just because a policy is intended to be executed on a particular device doesn't mean that it is appropriate to execute that policy on that device at a given time. Several additional factors must be taken into consideration. First, we must determine if the device is able to evaluate the policy. This means that instead of blindly applying a policy to a device because the role of the policy matches the role of the device, we need to determine the resources required for the device to implement the policy. For example, a new policy may have a lot of conditions, and the device may not have the capacity to store and/or evaluate these additional conditions.

Therefore, it is important that the environment that policies are going to be used in is analyzed to ensure that the policy is still applicable. The problem is that the *state* of the managed entities that the policy will affect has most likely changed. This basic fact drives the need for both a common information model *that models multiple states of its managed entities* and a robust distributed communication mechanism.

The DEN-ng information model holistically combines three types of classes: classes to model the state of a managed entity, classes to model how the state of an entity can be changed, and classes that model the control of when the state of a managed entity is changed. By using a finite state machine and these three types of classes, DEN-ng achieves closed loop control of its managed entities.

4.4.4 Effect of Applying Policy

All policies will affect some object. Abstractly, this can be categorized as a set of actions that either changes the state of the object to a new state, or keeps the state of the object in its current state. Regardless, executing actions can either apply just to that object, to a set of objects related to that object, or to that object and a set of related objects.

Policies are built and then applied to a system. Therefore, before they are applied, each policy should be checked to ensure that there are no adverse affects of applying the policy. There are two aspects of this. Multiple policies can conflict with each other, meaning that until that conflict is resolved, the conflicting policies cannot be deployed. On the other hand, a policy can be proven to not conflict with other policies, yet it still cannot be safely run. This is a measure of the feasibility and satisfiability of applying a policy. Finally, policies can be designed to be coordinated. For example, different policies, such as a QoS policy and a security policy, may need to be combined to achieve the desired control over user-generated traffic. We therefore must ensure that such policies can be safely combined.

This section will cover all of these topics in more detail.

Effect of Applying Policies on the Environment

Policies can be designed to affect a single object. This is akin to changing one or more attributes in an object. The effect of the policy is localized to that single object. Two types of policies fall into this category: direct action policies alter the state of the data model of that object, whereas indirect action policies leave the data model unaltered and instead change the life cycle model of that object. For example, a policy can be written that will change the threshold at which packets stop to drop. This has fundamentally affected how that dropper behaves. Compare this to a policy that periodically backs up data from a laptop onto a file server. Nothing in the data model of the laptop has been altered. However, from a life cycle point of view, the data has changed state. Put another way, the difference between these two policies is that the former is a direct action on the object that changes its behavior, whereas the latter is an indirect action that alters how the external world views that object.

Policies can also be designed to have a larger effect. This effect can be applied to the target object and/or other objects that have relationships to the target object. For example, a policy could control how routing updates are controlled. Each routing update changes the routing table to the router that receives the update, but also changes how the rest of the network behaves because now, additional routers know about the new routes being advertised in the update. This policy changes both the target object (the router) and additional routers that the target router communicates with in the network. Compare this to a policy that acts as a surrogate for another set of objects. For example, a policy can be sent to one object that tells that object to send a set of policies to other objects. Here, the first object is acting as a surrogate for the second set of objects. Its data model may not change, but the policies that it sends may affect the values of instances of the data models of the second set of objects.

Conflict Detection

Conflict detection and resolution is a complex topic. This treatment will expose all of the major problems and challenges to be solved, but will not be an exhaustive treatment of the topic. There are a host of references on conflict detection.[23,26]

The simplified environment described in the IETF architecture doesn't explicitly address conflict detection, much less conflict resolution. The architecture shown in Figure 4-7, which is being proposed in the TMF and other fora, does.

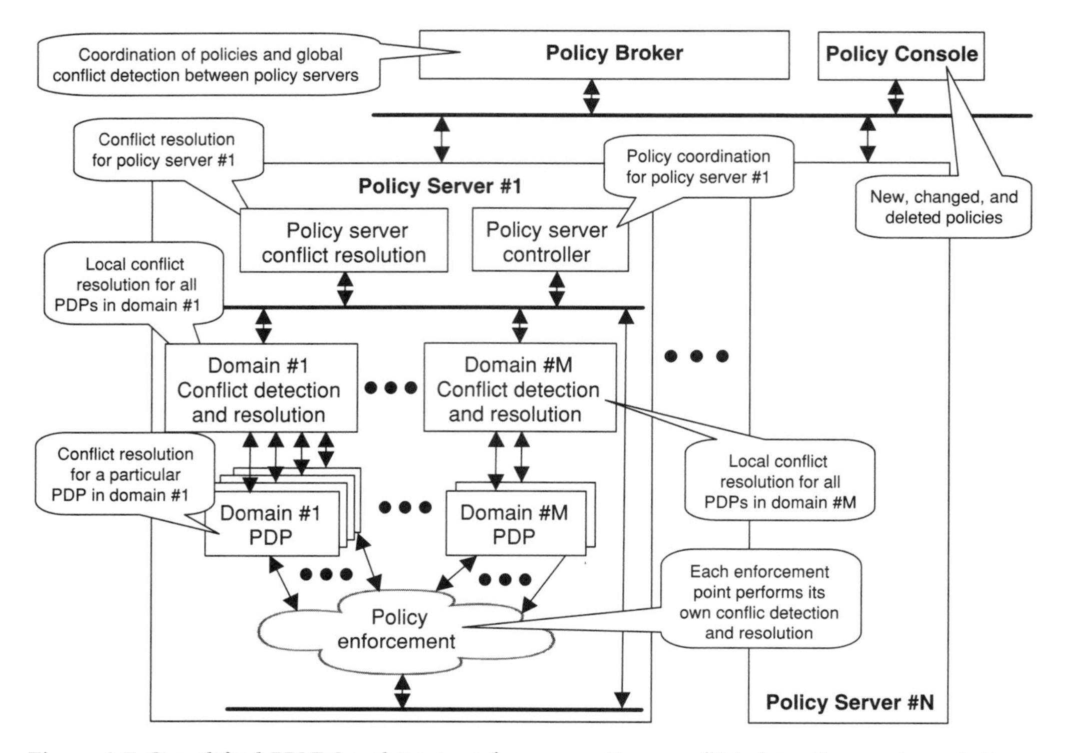

Figure 4-7 Simplified PBNM architecture for supporting conflict detection and resolution.

This architecture explicitly identifies five different levels of conflict detection and resolution: globally between policy servers, globally across the different components of a particular policy server, locally for all PDPs in a domain of a policy server, local to a particular PDP within a specific domain of a given policy server, and local per device (i.e., a PEP) in a domain. The following sections explain this architecture in more detail.

Types of Conflicts. Recall that a policy conflict was defined in chapter 2 as:

A policy conflict occurs when the conditions of two or more policy rules that apply to the same set of managed objects are simultaneously satisfied, but the actions of two or more of these policy rules conflict with each other.

Our simplified model of a policy rule is a triplet, consisting of three clauses: event, condition, and action. Events are used to trigger the *evaluation* of a policy rule; conditions specify if the policy rule is *applicable*, and actions define what to do. Therefore, three main types of conflicts can occur in a PBNM system:

- Conflicts between different policy rules
- Conflicts between the actions within a policy rule
- Conflicts between policy rules within a policy rule (e.g., sub-rules within a rule)

The most common type of conflict is between different policy rules. For example, imagine the following simple set of policy rules:

1. John *always* gets Gold Service
2. John *always* gets Bronze Service

Clearly, this is untenable, as the above two policies are specifying that John always gets two different types of services. This type of conflict can be easily detected at the time of entry. Now, imagine the following set of policy rules:

3. John *always* gets Gold Service
4. FTP *always* gets Bronze Service

This appears to be OK, as the subjects are different. However, what happens if John sends FTP? Does he get Gold or Bronze? (With my luck, probably *no* service, but that is an entirely different problem.).

How did this conflict occur? Note that in the above two rules, a computer won't see that there could be a conflict between these policy rules because the subjects (John and FTP) of the two policy rules are different. Furthermore, the event portion ("*always*") doesn't help distinguish between these two policy rules. Therefore, the conflict arises because John could send FTP traffic. In other words, the conflict arises because the actions cannot be concurrently executed because the subjects conflict with each other. However, the root problem is that *the subjects are at different levels of abstraction.*

Thus, the first thing that we need to do is to ensure that each of the three components of a policy rule is specified at a common level of abstraction with respect to the same component in other policy rules. For example, assume that policies 3 and 4 above are both used to define what class of service a given subscriber gets. Class of service can be specified either explicitly (e.g., by specifying John as the subject of a policy) or implicitly (e.g., by specifying the application as the subject of the policy; this works because users run applications). So, if we rewrote the above policies as:

5. In Engineering, the user John *always* gets Gold Service
6. In Engineering, the application FTP *always* gets Bronze Service

we now have a valuable hint that these policies may conflict with each other. This is because they both occur in the Engineering policy domain. We can then check to see if we have a policy conflict. Indeed, we do, because the user John can use the application FTP. Thus, we see that the condition clauses are simultaneously satisfied, but the action clauses cannot both be satisfied simultaneously.

The simplest, and most rare, form of policy conflicts is when two actions are specified within a policy rule that cannot both be satisfied. At first, one might think that this is the fault of the system used to enter policies in the first place. While that is probably at least partially true, remember that there are multiple levels of policy in our policy continuum. Certainly, at the lowest level (the device instance), this type of rule shouldn't occur. However, at the highest level, where it is desirable to let users specify policies in abstract terms, this could very well occur.

Fortunately, this type of policy conflict is usually easily detectable by simple static analysis.

The final category of policy conflict is when policy rules are made up of other policy rules. Examples of this type of policy rule abound in networking. One example is to allocate bandwidth. For example, the following example is taken from Snir et al.:[27]

- *For input traffic . . .*
  - *IF (protocol is of type UDP) THEN (guarantee 30% of available bandwidth)*
  - *IF (protocol is TFTP) THEN (guarantee 10% of available bandwidth)*
  - *IF (protocol is NFS) THEN (guarantee 40% of available bandwidth)*
  - *IF (protocol is of type TCP) THEN (guarantee 40% of all available bandwidth)*
  - *IF (protocol is HTTP) THEN (guarantee 20% of available bandwidth)*
  - *IF (protocol is FTP) THEN (guarantee 30% of available bandwidth)*

The above example shows that for a single policy rule about input traffic, there are two nested policy rules, one to allocate bandwidth for UDP applications, and another to allocate bandwidth for TCP applications. Furthermore, each of these two nested policy rules has its own set of two nested policy rules. As explained in Snir et al.,[27] this is necessary to correctly specify the bandwidth allocation for this set of 4 protocols. In particular, note that this policy is fundamentally different than structuring all of the six rules as a single if-then-else statement. This is because different percentages of the bandwidth are being allocated based on first, whether the protocol is UDP or TCP and second, based on the particular type of protocol that this flow is.

The problem with using nested policy rules is simply one of complexity. The more complicated the policy rule specification is, the more complicated the checking must be to ensure that no conflicts arise with the rule. Furthermore, there is a tendency in building nested policy rules to not use the same level of abstraction in specifying the subjects and targets of each policy rule at different nesting levels. Unless this is done, we're back to the problems detailed in the first example. Even though this is conceptually a single policy rule (that contains nested policy rules), the nested policy rules will *act* as if they are separate policy rules if care is not taken to ensure that the conditions and actions of all nested policy rules are at the same level of abstraction.

Conflict Detection at the Device Instance Level. The lowest level of conflict detection is at the device instance level. This is necessary because there is always a translation between the programming model used in a particular device and the next higher level in the policy continuum.

Remember that each device in general has its own specific programming model. A simplified programming model may be described as follows:

- *Vendor*
  - *Type of device (e.g., router or switch)*
  - *Model of device*
  - *Model of managed entities in device (e.g., how much and what type of memory)*
  - *Version of network operating system in each managed entity in the device*

- *Set of features used in the network operating system in each managed entity in the device*
- *Version of network operating system in the device*
- *Set of features used in the network operating system in each managed entity in the device*
- *Type of communication mechanism used*

Note that the above model is a programming model; other models may also apply to the device. For example, the above model describes logical functions of the device; physical composition is also important, and may be implemented using an entirely different model.

Given this level of complexity, it is impossible to specify policies at the device-view (i.e., the next higher level in the policy continuum) that can apply to all devices without another translation level. The problem then becomes that each device or entity acting on behalf of that device must be smart enough to determine if a newly received policy can be executed. If it can, then that device (or entity acting on behalf of that device) must determine if executing that policy will produce conflicts with any other deployed policies in the device.

It is important to remember that at this level, we are talking about functions within a device. Even if two devices appear to have the same functionality, in all probability they don't. This is because of device-specific hardware differences, which in turn cause software implementation differences. For example, in order for high-speed routers to condition traffic at line speed, much of their decision logic must be implemented in hardware. This implies the use of application-specific integrated circuits (ASICs). This then requires the underlying algorithm to be slightly modified, so as to take advantage of the particular features and capabilities of the ASICs of the device.

Therefore, since different devices will in general have different implementations of the same function, the devices themselves need to validate any policies that they receive. In particular, side effects of using a particular feature should be taken into account. For example, fancy queuing provides the ability to give different classes of traffic different levels of service. However, this comes at the price of increased resource utilization. Turning on fancy queuing could prevent the device from running other functions.

DEN-ng models this as a set of capabilities and constraints. Capabilities are a means of representing different functionality in a standard way, and are implemented in DEN-ng via a set of classes and relationships that normalize particular types of functions. Constraints enable specific restrictions to be placed on particular functionality. For example, a device may have a wide range of different encryption options available, but a particular business rule may restrict the encryption options that can be used to a smaller number. Taking advantage of a particular set of capabilities might place additional constraints on other capabilities of the device. However, this is beyond the scope of this book.

Conflict Detection Within a PDP. Networks tend to be built using sets of devices with similar capabilities. Such sets of devices can be managed by a single PDP. Alternatively, a single PDP that understands how to provision a particular feature often is able to manage that particular feature of multiple devices. For example, many of the popular QoS "policy servers" (which are really PDPs in the architecture shown in Figure 4-7) can provision QoS features for many different vendor devices. However, most such policy servers do not enable *all* of the particular domain features to be used. The user should check to ensure that the mechanisms used by a policy server include those that the user wants to use. Otherwise, the alternative is for the user to modify the results of the policy server. This is very dangerous, because now the deployed configuration is different than what the policy server thinks it should be, and so it becomes out of sync with the device.

In either case, the PDP has four primary functions with respect to conflict detection. The first is to ensure that a given policy doesn't conflict with other policies that are already deployed in the devices that the PDP is managing. When possible, it is desirable for the PDP to first download all possible policies that may be invoked, so that it can run static analysis on those policies to check for conflicts.

The second function of the PDP is to ensure that a policy can be executed by a particular device. For example, a policy may specify the creation of six queues, corresponding to six different classes of service. Some devices that the PDP is managing may have limits on the number of queues that they can create and support. In this case, there is a conflict with the *capabilities* of these two devices that the PDP controls.

The third and fourth functions of the PDP are to perform feasibility and satisfiability analyses. These are described later in this chapter.

The Need for Multiple PDPs. As defined in Chapter 2, a policy server is an entity that contains multiple policy decision points (PDPs). This is an important point and needs to be reinforced. Most network vendors currently build separate PDPs (which they often call policy servers) to focus on provisioning and managing distinct functions of the device. For example, a router can have a large number of distinct commands. Each of these commands fall into one or more categories (e.g., security and QoS might be separate commands, whereas functions like an access control list could be used by multiple functional categories). Network devices, such as routers and switches, in general have large numbers of different commands. Building a PDP to handle all of these different commands is very difficult; consequently, vendors separate the management of different functions by category and group them into distinct PDPs.

In addition to each PDP being responsible for managing a set or subset of device functions, each PDP is also responsible for managing all or a portion of a set of network devices. For example, a particular PDP may only be responsible for making IP QoS decisions. Therefore, it may only manage the subset of devices in the network that participate in delivering QoS. Furthermore, since not every interface of a device participates in QoS, the PDP may be restricted to managing just those interfaces that do use IP QoS. Multiple PDPs that perform the same function may also be required for redundancy, locality of control with their managed entities, or numerous other reasons.

Finally, multiple PDPs may be needed for redundancy, to implement hot or warm standby, to provide localized functionality, or for a host of other reasons.

This drives the need for another layer in the control hierarchy: the Local Conflict Detection entity.

Local Conflict Detection Among a Set of PDPs. The policy domain conflict detection and resolution (PDCDR) entity is responsible for examining a policy in the context of all other policies currently operating in the set of PDPs that it controls to determine if there are conflicts on a *specific object* between this policy and any other policy already defined. Here, an "object" could be a network device, an interface of a network device, or even a device mechanism, such as a classifier or queue within a network device. The set of policies to be examined is governed by the particular domain that the PDCDR is managing. Referring to Figure 4-7, there are M domains, and therefore M PDCDRs.

For example, suppose that PDP No.1 configured a device to use priority queuing, because the only traffic that was flowing through the device at that time was voice traffic. Now suppose that additional non-voice traffic, such as FTP, enters the same interface that was carrying the voice traffic. Clearly, priority queuing is a bad choice, since the FTP traffic will swamp the voice traffic. Furthermore, when FTP is recognized in the router's classifier, undoubtedly a policy rule will fire that wants to change the queuing for FTP. The problem, therefore, is that if we change the queuing (from priority to, for example, some variant of weighted fair queuing), the voice traffic will be adversely affected. We instead need to classify the traffic into two different types of traffic, and recognize that each needs its own type of treatment. If the router cannot handle different types of queuing on the same interface, then we need to ensure that each traffic type is given a separate egress interface.

Note that when the FTP policy rule fires, it doesn't conflict with other policy rules designed to handle FTP traffic. Rather, it conflicts with a *policy rule that is already running* that handles a different type of traffic.

The PDCDR checks for policy conflicts that apply to all network devices that are controlled by the set of PDPs that it controls. Furthermore, the PDCDR detects local conflicts of a device-specific nature. For example, a device-independent policy may specify a particular amount of bandwidth to be reserved. As we will see, all the Global Conflict Detection and Resolution (GCDR) entity can do is to specify the amount of bandwidth to be reserved, and ensure that the value to be reserved is within the capacity of a particular device. However, when this policy is enforced at run-time, some of the bandwidth for that link may have been previously allocated by other policies that were implemented earlier. The GCDR has no way of determining this. Similarly, individual PDPs may not be able to determine this, because multiple PDPs can affect the same device. Thus, the PDCR is the logical place for performing this check.

Domains. The definition for a policy domain (repeated from Chapter 2 for convenience) is:

A policy domain is a collection of entities and services that are administered in a coordinated fashion using a set of policies. The policies are used to control the set of services and entities according to a common methodology, such as a finite state machine.

In the preceding figure, there are M domains. Each domain gathers together a set of PDPs that manage a particular function or set of functions, such as QoS. This enables policies to be built on a domain-specific basis. It also enables conflict detection and resolution to be done at more granular levels, where first conflicts are detected and fixed at a specific domain level before the rest of the system is considered. This helps make conflict detection and resolution more scalable.

Global Conflict Detection. Referring to Figure 4-7, we see that different PDPs in different domains could change the same interface of the same device. For example, a QoS PDP and a Security PDP could both try and change the same interface of the same network device. This causes two problems.

First, if the PDPs are organized by function, then by definition one PDP will not be cognizant of the actions of another PDP. In the above example, the QoS PDP would be oblivious to the actions of the Security PDP, and vice-versa. Unfortunately, there's nothing to prevent one PDP from sending commands that conflict with the commands already sent by another PDP.

Second, suppose that the PDPs are organized by vendor, and that each PDP controls roughly the same set of functions. Usually, vendors build PDPs using a particular GUI and data model. This means that in general, commands from each PDP will use a different grammar. Therefore, the PDPs themselves won't be able to recognize conflicts between different PDPs because each has its own grammar.

Both of these problems point to the same solution: a common entity is needed to perform conflict detection and resolution among a set of PDPs and/or PDCDRs. We may therefore define Global Conflict Detection as follows:

Global Conflict Detection is the process of examining a policy in the context of all other policies of the Policy System to determine if there are conflicts between this policy and any other policy already defined. This process does not take into account any actual device or network state but only the assertions of other policies. Furthermore, it does not target any single object.

Whereas the PDCDR checks for policy conflicts that apply to all network devices that are controlled by its PDP or policy proxy, the GCD checks for policy conflicts that do not apply to any specific network device—rather, these conflicts apply to different components of the network as a whole and are independent of any specific network device.

Global conflicts are conflicts based on the properties of the policy itself; they are not based on properties of the specific devices (or their interfaces) to which the policy might apply. Two policies conflict globally with each other when all their conditions are satisfied, but one or more of the actions of one policy conflict with one or more of the actions of another policy.

The set of PDCDRs handle conflict detection and resolution within their respective domains. However, multiple PDPs from different domains could be needed to collectively change the same interface of the same device. The GCD monitors all policies from the PDCDRs as well as all new, changed, or deleted policies in the system as a whole. As such, it is the first item that gets checked when a new policy is submitted to the system. Since it is a global check, it ensures that this new policy doesn't conflict with other policies that are already installed in the system. It thus represents a simple yet important optimization, ensuring that only those policies that do not conflict with other policies at a global are processed for further conflict detection.

Policy Broker. The policy broker controls how different policy servers interact with each other. It therefore has two different functions.

The first function is to ensure that conflicts don't exist among different policy servers that are trying to work together. For example, suppose that there are two IP networks that pass traffic between each other. Assume that both networks are using the Differentiated Services[16] paradigm to provide Gold, Silver, and Bronze classes of service. They distinguish each class of service using a unique DiffServ Code Point, or DSCP. These two networks will pass traffic to each other through their edge routers, as shown in Figure 4-7.

The problem is that Network A is set to pass egress traffic marked with a DSCP of 10 for Gold Service. Unfortunately, Network B has defined Gold Service to be DSCP 17. The result is that while the networks can talk to each other, the traffic that they exchange will be inappropriately marked.

One of the purposes of the Policy Broker is to recognize this situation and ensure that the edge routers of each network change their marking so that they each have a common understanding of Gold, Silver, and Bronze traffic. Thus, the Policy Broker might say that the Edge Router for Network A can mark its egress traffic with a DSCP of 10 as long as that traffic is not destined for network B. If it is destined for Network B, then it will instead mark that traffic with a DSCP of 17. The alignment of DSCPs in Figure 4-8 is an example of coordinating multiple policies so that they can work together.

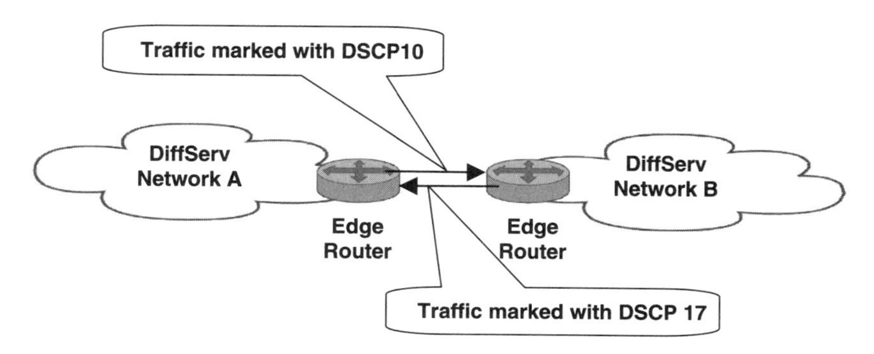

Figure 4-8 The need for a policy broker.

The other purpose of the Policy Broker is to coordinate the application of different policies in different policy servers. For example, an SLA may have been newly installed. This SLA consists of several related business goals that each has different effects on the configurations of devices that support the SLA. The SLA can only be properly supported if all of the changes to all of the affected devices are implemented in concert. This requires an entity such as the Policy Broker that can control policy decisions in each policy server that is used to support the newly installed SLA.

Conflict Resolution

Conflict resolution is a vast and complicated subject, full of ongoing research efforts. This is because policies are themselves so different in nature that it is impossible to define a single methodology that will guarantee resolution of any possible policy conflict. Furthermore, there is a difference between static and run-time conflicts. Static conflicts can be detected before they are implemented in a device, whereas run-time conflicts can only be found when they are deployed and something goes wrong. While there is a set of tools that can be used to resolve static conflicts, run-time conflict resolution is an open research area. The following is a selection of some of the more popular methods of conflict resolution.

Obeying Inherent Dependencies Among Policies. Sometimes, policy conflicts arise because of the order in which policies are applied. If this is the case, reordering their application can remove the policy conflict.

This type of conflict detection is amenable to model-driven analysis. Order is one form of dependency. If the policies are understood well enough to be modeled, then associations can be defined that characterize dependencies between the order in which they are applied.

The above is a simple example of generalized dependencies that exist between and among policies. Dependencies can be defined that don't just specify ordering of policies—they can specify additional constraints, such as realizing that a particular policy can only be executed before or after another policy has started or finished execution. Information models are well-suited for properly representing such dependencies, as they can represent them using associations, aggregations and compositions. UML also enables each of these to be represented using a class or not. Classes enable these relationships to have attributes, methods, and respond to events. The result is a much richer, more realistic model of the behavior of the system.

Priority. Priority is a convenient means for specifying the order of execution among different policies as well as for temporarily overriding a specified or implied execution order. The former case is in reality the same as above. The latter case is interesting, because sometimes, environmental constraints are such that planned actions cannot be taken. In this case, it is critical that the policy server have the flexibility to reorder policy rules.

A good description of the use of priority is given in Moore et al.[28]

Subject and/or Target State. Sometimes, policies should only be executed when the subject and/or target is in a particular state. Conflicts arise when the subject and/or target state is ignored, and conflicts can be resolved by ensuring that the subject and/or target is in the correct state before the policy is executed.

Time. Time is an especially useful tool for resolving policy conflicts. The simplest example is to use time as a mask that enables or disables a policy. For example, suppose we had the following conflicting policies:

- John *always* gets Gold Service
- John *always* gets Bronze Service

By using time, we can resolve this conflict as follows:

- John gets Gold Service *only between 8am and 5pm*
- John gets Bronze Service *only between 5pm and 8am*

The policies no longer conflict with each other because time has been used to limit the application of a particular policy. A good example of the use of time may be found in Moore et al.[20]

Events. Similarly, events can be used to distinguish between when two policies are applicable. By specifying different events, policies that would otherwise conflict with each other can be separated and applied under completely different conditions. Note, however, that often specifying different events is, by itself, not enough. For example, if the time component of two policies is set to "always", then the two policies may not start out in conflict, but may end up in conflict.

Safety Issues

There are two main safety issues in applying a policy, called feasibility and satisfiability. These two concepts are different than conflict detection and resolution, in that the policy that is about to be enforced does not conflict with any other policy. Rather, these are two different checks on whether it is possible to enforce this policy in the current system.

Feasibility. Feasibility is defined as follows:

Feasibility checking is a set of checks that compares the available services of the network with respect to the full set of policies that want to use those services. Feasibility checking will most likely require post-policy deployment checking that is sensitized to the particular network elements involved as well as the nature and effects of the deployed policies.

Feasibility checking assumes that none of the deployed policies conflict with each other. Rather, feasibility checking is concerned with whether or not it is possible and practicable for the network to support the deployment of a particular policy or set of policies when another set of policies are already deployed at a given time. For example, suppose that a set of policies have already been deployed on a network, and now a new user logs onto the network. This new user is a premium user and requires the deployment of new policies to properly condition that user's traffic. Can the network support this new service? In other words, is it feasible for this policy to be deployed given the current network conditions?

Sometimes, this can be determined by static analysis. Most of the time, however, feasibility checking will require post-policy deployment checking that is sensitized to the particular network elements involved as well as the nature and effects of the deployed policies. In other words, the policies will need to be deployed and then measured to see if the network can support the enforcement of these policies concurrently.

Satisfiability. Satisfiability is defined as follows:

Satisfiability checking is a set of checks that ensures that the resources needed by a particular policy, in isolation from all other policies, are available in the entities to which this policy applies.

Again, it should be emphasized that this is not a matter of searching for conflicts—there are by definition no conflicts in this case. Rather, this is a matter of determining whether or not the resources required by this policy can be used by this policy without adversely affecting other deployed policies.

For example, suppose that a policy requires that a certain set of paths through the network provide a certain specific queuing behavior. Suppose further that on one of the paths at one of the interfaces, no advanced queuing mechanisms are available. This would mean that the needs of the policy are not satisfied. Thus, the policy itself is not satisfied, implying that this policy cannot be implemented in these devices. Note that the policy didn't conflict with other policies—rather, it was unable to execute because its pre-conditions were not met.

Policy as a Closed Loop System

The power of a PBNM system lies not in the power of individual components of that system, but in the system as a whole. This power can only be harnessed if the different components of a PBNM system can work together in the form of a closed loop system.

In such a system, each component has its own purpose to play. However, there is a marked difference between the PBNM system that provides policy-driven network services, and the administration of the PBNM system.

Figure 4-9 below illustrates a simple system in which a set of users are accessing a common resource.

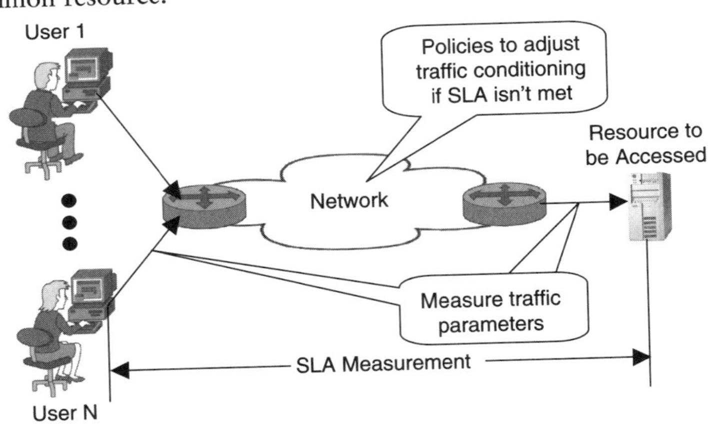

Figure 4-9 A simple policy-based system.

Figure 4-10 shows conceptually how the PBNM system is used to control the services in Figure 4-9:

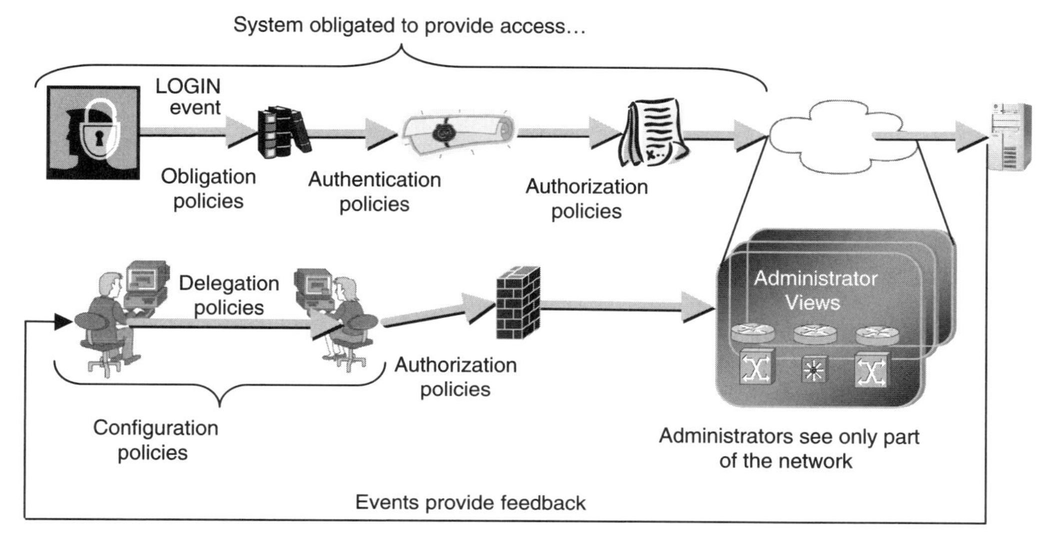

Figure 4-10 Different management paths for the simple policy-based system.

Figure 4-10 shows three distinct management paths for our simple policy system. The first controls how users log onto the system and gain access to system resources. Obligation policies define the set of actions that a subject must do. Authentication policies determine how to prove that a user is indeed who that user purports to be, based on a set of factors (e.g., different authentication methods may be used for LAN versus dialup logon). Authorization policies determine what activities a subject is permitted to do (or not permitted to do) to a set of target objects. These policies are defined in more detail in chapter 5.

The second management path consists of the policies used to configure the network to deliver contracted services. Delegation policies are used to permit subjects to grant the same or a subset of privileges, which they possess (due to an existing authorization policy), to other subjects (called grantees). This enables different administrators to perform the same subset of tasks on the same device. Configuration policies are used to build network services that provide access to shared system resources that are available using the network. Networks, however, are complicated entities, and devices that make up a network are usually owned by different groups within a company, or even different companies. In general, an administrator will not have the ability to see, let alone change, every device in the network. This is shown conceptually in figure 4-10 through the use of administrator views. Each view requires an authorization policy to be run to ensure that the administrator has the proper credentials to view and operate on the entities contained in that view.

The presence of administrator views is another important reason why the GCD and especially the Policy Broker are needed in a PBNM system. Without these critical entities, there would be no way to coordinate the application of policies by different administrators to different devices.

The third and final management path is the feedback loop. One of the most important features of a PBNM system is to be able to *measure and monitor* the delivery of the services that it is controlling, and to subsequently *adjust* the system configuration to ensure that contracted services are met. This can't be done without the use of feedback. This is one of the reasons why DEN-ng uses the {event, condition, action} triplet instead of the simpler {condition, action} doublet used by the IETF and the DMTF.

Thus, we see that even for a simple system, many different policies must be used and coordinated in order to ensure that services that are contracted for are properly delivered.

4.5 Musings on Implementation

It is difficult to describe thoughts on the ease of implementation for any arbitrary PBNM solution, due to the variety of variable factors that are present in such implementations. However, some general conclusions on how easy or difficult it should be to implement different areas of PBNM solutions, along with their overall effect on network management, can be provided. This is the focus of this section.

4.5.1 Communications Issues

The vast majority of PBNM solutions either provide an API or some type of communications bus to enable the different components of the PBNM solution to communicate with each other.

If the PBNM solution is not heavily distributed in nature, then often an API is provided. A good check for potential buyers of these systems is to see if the functionality offered by the API is the same as the functionality offered by the GUI component of the policy server. If it isn't, then care must be taken because the product will in effect change into two products—one operated by the GUI and one operated through the API. If the functionality provided by the API and the GUI appear to be the same, then hopefully the API being exposed is the same as the API used within the product itself. If it isn't, then you're back to the two product situation, because changes made with the API will probably be different than changes made with the GUI. Or perhaps there will be other limits, such as performance or functionality, which will be different.

However, a more robust solution is provided if *in addition* to the API, the PBNM solution provides some type of messaging system to enable it to drive and be driven by external entities. If the PBNM solution is using a particular type of messaging bus, check and see whether that bus supports connectors to other types of messaging buses.

One of the most important design criteria to be applied when using messaging buses is the type of information that will be transported on them. Most systems use events. Events can come in many varieties—lightweight notifications that simply tell the client that an interesting event occurred, slightly more heavyweight request-response pairs of events, and even more heavyweight events which carry as their payload data to be processed and acted upon by consumers of the event.

Another design factor that should be included is how the event is delivered. Most messaging bus companies support several different means of delivering events, ranging from ultra-high reliability (usually involving a database to treat the event as a transaction) to high reliability (for example, using TCP and requiring specific acknowledgment of the receipt of an event) to best-effort delivery. Along these same lines, an essential feature is determining whether the messaging bus will ensure that a message is delivered only *once* and not several times.

The final design factor to take into account is the type of events and messages that will be sent. Many systems use varying notions of what an event is. One way to mitigate this is to use *self-describing* events. In this approach, the event itself contains all of the information needed to interpret it and act on it. This is really the first step towards building a formal object model that describes the different types of events and messages in the system, what objects they relate to, and what entities produce and consume them. It is strongly recommended that a formal object model for events and messages be built. Otherwise, there is no way to guarantee that the semantics of a particular type of message will be interpreted the same way by different components.

4.5.2 Assignment and Coordination of Policies

This book advocates a Policy Continuum, which enables different expressions of policy that manage the same managed object to be related to each other. It also advocates coordinating different policies so that they achieve a common goal. This was illustrated in Figure 4-1, where different policies worked together to manage end-to-end traffic behavior.

It cannot be over-emphasized that the only way that either of these two approaches can work is through the use of a detailed object-oriented information model. This is because assigning and coordinating policies is all about the notions of a *Policy Subject* and a *Policy Target* (definitions for these terms were provided in Chapter 2). Therefore, the information model should be used to both define policies as well as to relate a policy to another managed object. Without the use of a structured tool such as an information model, it will be very difficult to catch all instances of when a policy should be related to one or more managed objects.

In the case of coordinating the order of when different policies are applied, the information model provides the basic definitions to be used, but additional tools (such as a state machine and a sequence diagram) will also be required.

4.5.3 Scalability of PBNM Solutions

One of the least understood benefits of PBNM solutions is scalability.

The first requirement for scalability is not necessarily the *number* of network devices that are to be controlled. Rather, the first requirement should be the *number of different functions* that are being controlled. Referring to Figure 4-7, this is because different functions are assigned to different Policy Domains, each of which can have multiple PDPs. Thus, the first requirement is to determine how many Policy Domains and PDPs are required.

An important input to this decision is whether different devices are being used. Clearly, if multiple devices from multiple vendors are being used, then it is almost certain that multiple programming models will be used. This will almost certainly require different PDPs to handle the different vendor devices. However, it is important to be aware that the use of different devices, and especially different versions of the network operating system in those devices, is also likely to cause different PDPs to be used. The next decision is one of fail-over and redundancy. The architecture described in Figure 4-7 was designed to solve these and other problems. The final decision, one that is unfortunately overlooked by most policy vendors, depends on the number of policy decisions and actions that can be performed during a particular time period.

4.6 Summary

This chapter focused on how policies are applied and operate. Two types of communication approaches were described. Publish-subscribe systems are simple and efficient, and enable different PBNM components to seamlessly communicate with each other. Furthermore, they optimize communication between components. Jini takes the view that everything is a service, and enables services to be networked together. It advocates building complex services from a set of simpler services. Whereas publish-subscribe systems rely on adaptation and/or mediation to interface with other systems, Jini promotes the building of components that can easily interface with other systems using Java interfaces to simple services. In either case, the communication mechanism is used to perform various policy operations, including the enabling, disabling, and changing of policy rules and information.

One area in which existing systems could be improved is how policies are assigned to system components. The policy continuum defines relationships between policies at different levels of abstraction, but unfortunately, most current systems don't relate different types of policies to each other. Instead, they are treated as disparate policies. This makes end-to-end management exceedingly difficult. This chapter examined several different abstraction mechanisms, and showed how they can be used to relate different policies to each other. Principal among these mechanisms was the use of roles. While roles provide a significant benefit in and of themselves, their potential is only fully realized when they are used with an information model. This is because roles by themselves cannot represent relationships and dependencies between managed entities. Both of these can be effectively represented using an information model.

The importance of capabilities and constraints was emphasized. Capabilities enable us to represent the functions that are available in a managed entity, and constraints provide us a means to characterize restrictions on using those functions. These must be incorporated into policies for efficient control and management of network services.

Policy coordination was discussed. A classification hierarchy was defined that helps organize policies. However, it is important to realize that network vendors build specialized PDPs that are designed to manage a particular type of feature, such as QoS. End-to-end services, however, require multiple such features (e.g., security, IP address management, and numerous others) to be coordinated to deliver the appropriate service.

Policy operation involves four main factors. Physical and logical resources must be present in order to evaluate and execute the policy. Similarly, physical and logical resources must be available in the policy target to enable it to implement and enforce the policy. Third, policies must be checked for conflicts with other policies that have been already deployed in the system before they are applied, to ensure that there are no adverse affects of applying the policy. Finally, in order to ensure that a policy can be safely run, feasibility and satisfiability checks should be done.

Some behavior is difficult to control. Policies can be designed to be coordinated, so that the application of a set of policies achieves the desired behavior. For example, a QoS policy and a security policy may need to be combined to achieve the desired control over user-generated traffic. Control and coordination cannot be accomplished using the simplified IETF/DMTF architecture, because it doesn't contain the necessary building blocks to detect different types of conflicts. It also doesn't provide for any type of distributed communication mechanism. This chapter defined and explained an architecture that is suitable for these types of problems.

This chapter also covered an often neglected subject in policy: safety. Safety was defined as ensuring that it is both feasible to implement a policy as well as that all pre-conditions necessary for that policy to execute can be satisfied.

Finally, it was emphasized that a PBNM system is only effective if it forms a closed loop system. It is not sufficient to simply apply a policy and walk away. Rather, one must monitor the policy to ensure that it is operating as desired.

4.7 Recommended Further Reading and References

The following is a set of references for this chapter.

1. *www.talarian.com*
2. *www.vitria.com*
3. *www.tibco.com*
4. *www-3.ibm.com/software/ts/mqseries/txppacs/ma0d.html*
5. *www.cs.colorado.edu/users/carzanig/siena*
6. *www.ics.uci.edu/IRUS/twist/wisen98*
7. *www.research.ibm.com/gryphon*
8. *www.research.microsoft.com/~antr/PAST/scribe.pdf*
9. *www.sun.com/jini*
10. This was a public presentation at the 2002 JavaOne conference. The session was T3269.
11. This is a TMF member's only document that describes the Interface Implementation Specification of the Fine Grain NGOSS Catalyst Project. It is specified in the following document:

TMF839v1.5 and in the following (members-only) web site: *www.tmforum.org/sdata/documents/TMFC1379%20TMFC1000%20TMF839v1[1].5.pdf*

12. This is an example of a next-generation multi-service optical device. It merges the delivery of SONET, Fast Ethernet, and Gigabit Ethernet. More importantly, it uses a Java-based management system, and for the Fine Grain project, Turin built a Jini interface to their management software. Please see: *www.turinnetworks.com*
13. This is an example of another next-generation device. It is aimed at delivering *last-mile* Ethernet services over existing copper and fiber infrastructure. It used Jini-based management software in Fine Grain. Please see: *www.hatterasnetworks.com*
14. Please see the following for the JavaSpaces Service Specification: *wwws.sun.com/software/jini/specs/jini1.2html/js-spec.html*
15. This is a very interesting example of a commercial implementation of a JavaSpace. It features some notable extensions, the most important of which is the concept of *evolution*. Please see: *www.intamission.com*
16. Please see the following RFCs:
    Nichols, K., Blake, S., Baker, F., Black, D., *Definition of the Differentiated Services Field (DS Field) in the IPv4 and IPv6 Headers*, RFC2474, December 1998.
    Blake, S., Black, D., Carlson, M., Davies, E., Wang, Z., Weiss, W., *An Architecture for Differentiated Services*, RFC2475, December 1998.
17. Cisco's Secure Policy Manager—please see: *www.cisco.com/warp/public/cc/pd/sqsw/sqppmn/*
18. Cisco's QoS Policy Manager—please see: *www.cisco.com/warp/public/cc/pd/wr2k/qoppmn/index.html*
19. Cisco's IP Address Management Policy Manager—please see: *www.cisco.com/warp/public/cc/pd/nemnsw/nerr/index.shtml*
20. Moore, B., Ellesson, E., Strassner, J., Westerinen, A., *Policy Core Information Model—Version 1 Specification*, RFC3060, February 2001.
21. This is the Resource Allocation Protocol working group. Please see: *www.ietf.org/html.charters/rap-charter.html*
22. This is the Integrated Services over Specific Link Layers working group. Please see: *www.ietf.org/html.charters/issll-charter.html*
23. Damianou, N., *A Policy Framework for Management of Distributed Systems*, Ph.D. Thesis, February 2002.
24. Lupu, E., *A Role-Based Framework for Distributed Systems Management*, Ph.D. Thesis, July 1998.
25. Dunlop, N., Indulska, J., Raymond, K., *Dynamic Conflict Detection in Policy-Based Management Systems*, Policy 2002 Conference, June 2002.
26. Fu, Z., Wu, F., Huang, H., Loh, K., Gong, F., Bladine, I., Xu, C., *Ipsec/VPN Security Policy: Correctness, Conflict Detection and Resolution*, Policy 2001 Conference, January 2001.
27. Snir, Y., Ramberg, Y., Strassner, J., Cohen, R., Moore, B., *Policy QoS Information Model*, draft-ietf-policy-qos-info-model-04.txt, November 2001.
28. Moore, B., Rafalow, L., Ramberg, Y., Snir, Y., Westerinen, A., Chadha, R., Brunner, M., Cohen, R., Strassner, J., *Policy Core Information Model Extensions*, draft-ietf-policy-pcim-ext-06.txt, November 2001.

Representing High-Level Policies

This chapter will describe the structure of high-level policies and the functions that they must perform in a PBNM system. These policies are aimed at the business and system levels of the Policy Continuum, and are based on extending the Ponder system. The extensions are based on incorporating concepts from the DEN-ng policy architecture (described in Chapter 7) and the DEN-ng approach to modeling information. This chapter connects the introductory concepts presented in Chapters 2 to 4 with the advanced concepts in Chapters 6 to 8.

5.1 Introduction

The purpose of this chapter is to define the detailed requirements for using high-level policies in a PBNM system. These requirements will be derived by first analyzing the types of high-level policies that a PBNM system uses. Fundamental to this is the need to describe the primary building blocks of policy—domains and roles. Because the Ponder language is arguably the most robust of the several languages in current use, this chapter will start by describing the most important features of Ponder language.

Next, the four fundamental types of Ponder Policies will be described. The use of events and constraints in the Ponder language will also be described. This sets the stage for describing how DEN-ng has enhanced the Ponder language. Important enhancements, such as access control, exception control, and events, will be described.

Grouping is a fundamental operation in a PBNM system. Ponder and DEN-ng groups will be compared and analyzed. Finally, an example of how high-level business policies are specified and used will be provided. The relationship between policies and the underlying information model will be emphasized.

This chapter will begin to use UML in detail. Readers unfamiliar with UML should read Appendix 1.

5.2 High-Level Policy Specification

Chapters 2 and 3 emphasized the need for different types of policies to link business and system processes and procedures together. There is also the need to define different policies to control different types of functions within a device. This section will concentrate on specifying policies at different levels of abstraction, so that these links can be made.

The approach used begins with the work from Sloman et al.[1] and use concepts from the Ponder language to capture high-level policy requirements. As pointed out in Chapter 3, however, Ponder has a natural bias toward security. Therefore, this section will examine the fundamental concepts of Ponder and then change and enhance them to build a more robust high-level language.

Fortunately, Ponder relies on the use of roles. As will be seen in Chapter 6, roles are one of the main components of the DEN-ng policy model. This is very important because it enables a natural linking between Ponder and DEN-ng to be built.

5.2.1 Domains

Grouping is one of the fundamental organizational concepts of the human mind. Clearly, for large systems that have large numbers of objects, policies are useful for managing sets of objects. Domains are collections of objects that have been grouped together for one or more management purposes. This enables multiple policies to apply to the same sets of objects.

Both Ponder and DEN-ng use domains. However, there are important implementation differences between them. These are discussed in the following sections.

Ponder Domains

Domains enable objects to be grouped together, so that policies can be applied to all or some of those objects. Domains organize the objects that they contain according to one or more characteristics or rules, such as object type, geographical location, or other defining characteristics.[2] In Ponder, domain membership is implicit (i.e., the domain does not *contain* objects, rather, it merely holds references to object interfaces).

Domains can be nested to form a hierarchy, as shown in Figure 5-1a. This hierarchy has an interesting consequence, shown by way of example: Objects in container D can inherit the policies and behavior present from container B as well as container C (and container A).

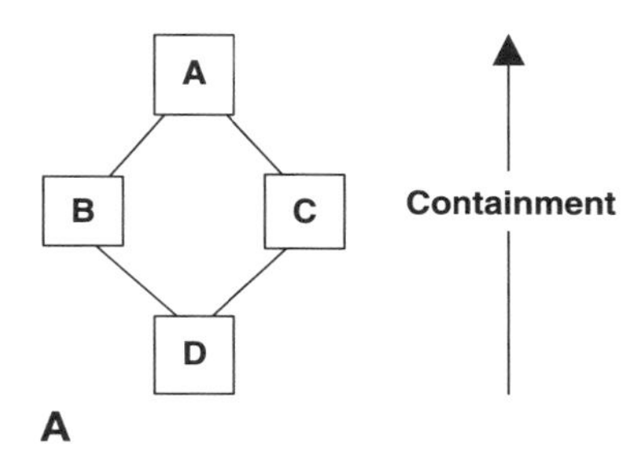

Figure 5-1a Nested domains forming a hierarchy.

Domains are generalized containers. Therefore, not only can they contain managed entities, they can also contain policies. The object is to group sets of policies that apply to sets of managed objects. Ponder enables objects to be members of multiple domains. In other words, domain memberships can overlap. This is shown in Figure 5-1b.

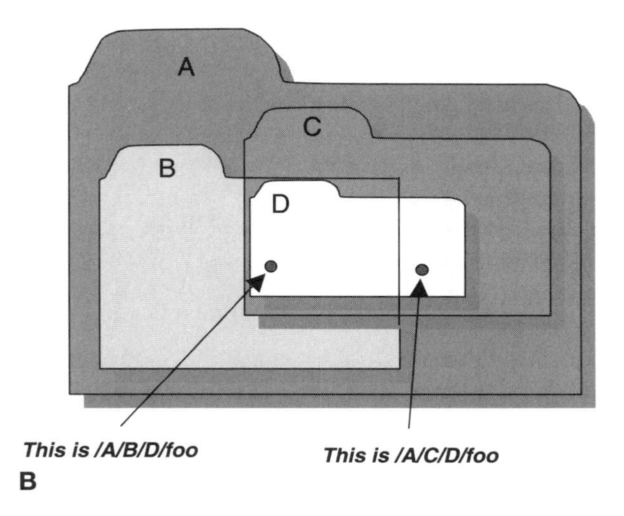

Figure 5-1b Containment and naming in Ponder domains and sub-domains.

This figure shows two objects that are each called *foo*. They can exist concurrently in container D because they have different identities. This is because one is identified as a child of D, which is a child of B, which is a child of A (denoted as /A/B/D), while the other is a child of D, which is a child of C, which is a child of A (denoted as /A/C/D).

We can now see why Ponder places such a heavy emphasis on security. Defining access control enables the system to control who (or what) is authorized to perform particular management actions. Otherwise, it becomes impossible to control who is doing what to which objects! (The reader will notice that this organization is similar to how a directory is organized, including the concept of each container having its own access control.[3])

Ponder defines a "domain scope expression" that enables domains to be combined from which objects can then be selected. This enables a policy to be applied to multiple objects from multiple domains. This is important because this enables domains to be used to organize information and policies to be used to apply common behavior to sets of managed objects. For example, it may be necessary to group different users into different administrative domains to enable their respective managers to access their personal and financial details. If they all have something in common, then it is desirable to be able to write a single policy that acts on a collection of the appropriate users from each domain.

However, the real power of domains lies in their utility to abstract individual objects. It is simply impractical to specify policies for each individual object in a large system. Instead, this enables us to specify a policy or set of policies on a per-domain basis. The key point is that *domain membership can change without requiring policies to change.*

DEN-ng Domains

DEN-ng domains are similar to Ponder domains but are more restrictive. In DEN-ng, domains are *directed acyclic graphs*, meaning that a strict hierarchy is formed with no ability for one domain to belong to multiple domains from multiple trees. As a simple example, the nested hierarchy shown in Figure 5-1a is not legal in DEN-ng because domain D belongs to two different domains that have different paths.

The reason that this was done was to simplify containment and scoping. Experience has shown that Ponder has too much flexibility. This restriction also naturally mirrors the limitations found in directories, one of the primary targets of DEN-ng. However, as we will see, DEN-ng adds two interesting twists to domain usage. The first is the use of roles to populate domains. The second is the use of policies to control domain membership.

5.2.2 The Use of Roles in Policies

Roles are critical in both Ponder and DEN-ng. DEN-ng roles are influenced by Ponder but contain significant enhancements and differences in semantics. The first subsection will summarize how roles are used in Ponder. This will be followed by comparing the definition and use of roles in DEN-ng with that of Ponder.

The Definition and Use of Roles in Ponder

In Ponder, roles are used to associate a set of policies to a set of policy subjects. Recall that policy subjects were defined in Chapter 2 as follows:

A policy subject is a set of entities that is the focus of the policy. The subject can make policy decision and information requests, and it can direct policies to be enforced at a set of policy targets.

The idea is to realize a set of common semantics that should be governed by the policies. For example, positions within an organizational hierarchy, such as division manager, department manager, and technical lead, are assigned a natural set of decreasing privileges and authority to make decisions. By specifying organizational policies in terms of positions rather than persons, the assignment of a new person (or removal of a person) to (or from) a manager position can be done without re-specifying the policies defining that position.

Position is one simple concept that can be easily implemented using a role. We will discuss the use of roles more when we examine the DEN-ng role model.

This aspect of treating a role as a container is further emphasized in Ponder. Ponder roles can also specify the set of policies that apply to a managed entity that is the subject of a policy. Ponder defines domains as containers that can be associated with one or more roles. A domain contains a set of authorization, refrain, obligation, and delegation policies whose common subject is a role.

A role-based framework for management of distributed systems has been defined in Lupu.[4] This work emphasizes the possible correspondence between organizational hierarchies and roles. Conceptually, it views a role as a position in an organizational hierarchy. The set of authorization policies (which define the rights associated with that position) and the set of obligation policies (which define the duties of that position) are then linked to the role. It also includes relationships that model the ways roles of an organization can be related to each other.

Lupu suggests that roles and relationships can be specified as container classes that contain policy templates, instead of actual policy instances. Policy templates are parameterized policies that may have one or more elements unspecified or left abstract. Reuse of role and relationship specifications is also achieved through inheritance. Inheritance allows a role class to inherit all of the attributes, methods, constraints, and relationships of its superclass, as well as be specialized with particular rights and duties that differentiate this role from other role classes. Thus, inheritance enables incremental refinement of the role structure for specific management needs through repeated subclassing.

The Definition and Use of Roles in DEN-ng

Ponder views a role as an intelligent container, which can contain a number of policies that can then be associated with a particular implementation of a management structure, such as an organizational hierarchy. This use of roles is really as a *role-selector*, which was defined in Chapter 2 as:

A role-selector is a means of grouping together a set of objects, so that a set of policies can be applied to them.

In the IETF, the notion of a role attribute was developed by Moore et al.[5-6] DEN-ng uses this concept (as well as the concept of role-selectors) and expands on these concepts, as we will see.

In order to collect objects, it is convenient to give objects an attribute that can contain the name of a role. This is called a *role attribute* and was defined in Chapter 2 as follows:

A role attribute is a fundamental characteristic of an object that is used to define the purpose or function of that object.

The management application can then apply a role-selector function to populate a container based on the value of the role attribute of a set of objects. This is a very powerful notion—it enables objects located in different domains to be collected into a single container based on their role, so that common policies can be applied to them.

DEN-ng also makes frequent use of the role object pattern as defined in Baümer[7] and of *role objects*, which were defined in Chapter 2 as:

A role object is an object that is not meant to stand on its own; rather, it is meant to supply a combination of common and unique functionality that can augment the basic definition of another object. The unique functionality may be supplied in the form of additional attributes, methods, and/or relationships.

Note that this latter is not specifically identified in Baümer[7] but is a characteristic of DEN-ng and the work in the SID.

We therefore have three types of "roles" in DEN-ng. These are *role-selectors*, *role attributes*, and *role objects*. DEN-ng containers can collect objects that have a particular *role attribute*. *Role-selectors* are functions that collect and group these objects into a container based on the value of each object's *role attribute*. *Role objects* are used to split a managed entity into a core, or fundamental, part and a variable part whose attributes, methods and relationships define the *role* that the managed entity will play (this will be discussed more in subsequent chapters of this book). The use of role objects enables the model to be relatively independent of technological advances. For example, a router that has the ability to act as a firewall is modeled by defining a firewall role and associating that role to that particular device.

Figure 5-2 shows how role attributes, role-selectors, and role objects interrelate in a managed environment that uses DEN-ng. (Note that this is a simplified view of the DEN-ng *ResourceRole* model only; every DEN-ng domain model uses roles.)

Different managed entities can be organized in a hierarchy. Figure 5-2, shows three generic containers: ManagedCollection is a generic DEN-ng container that forms a portion of an administrative hierarchy. The role abstractions shown in Figure 5-2 enable different policies to apply to different parts of the same device. Figure 5-2 has been simplified to make it easier to understand.)

Imagine a user defines two application-specific containers, named Edge Routers and Core Routers. Further assume that these role attributes that can be matched with a hierarchy of role objects. This is done using an appropriate role-

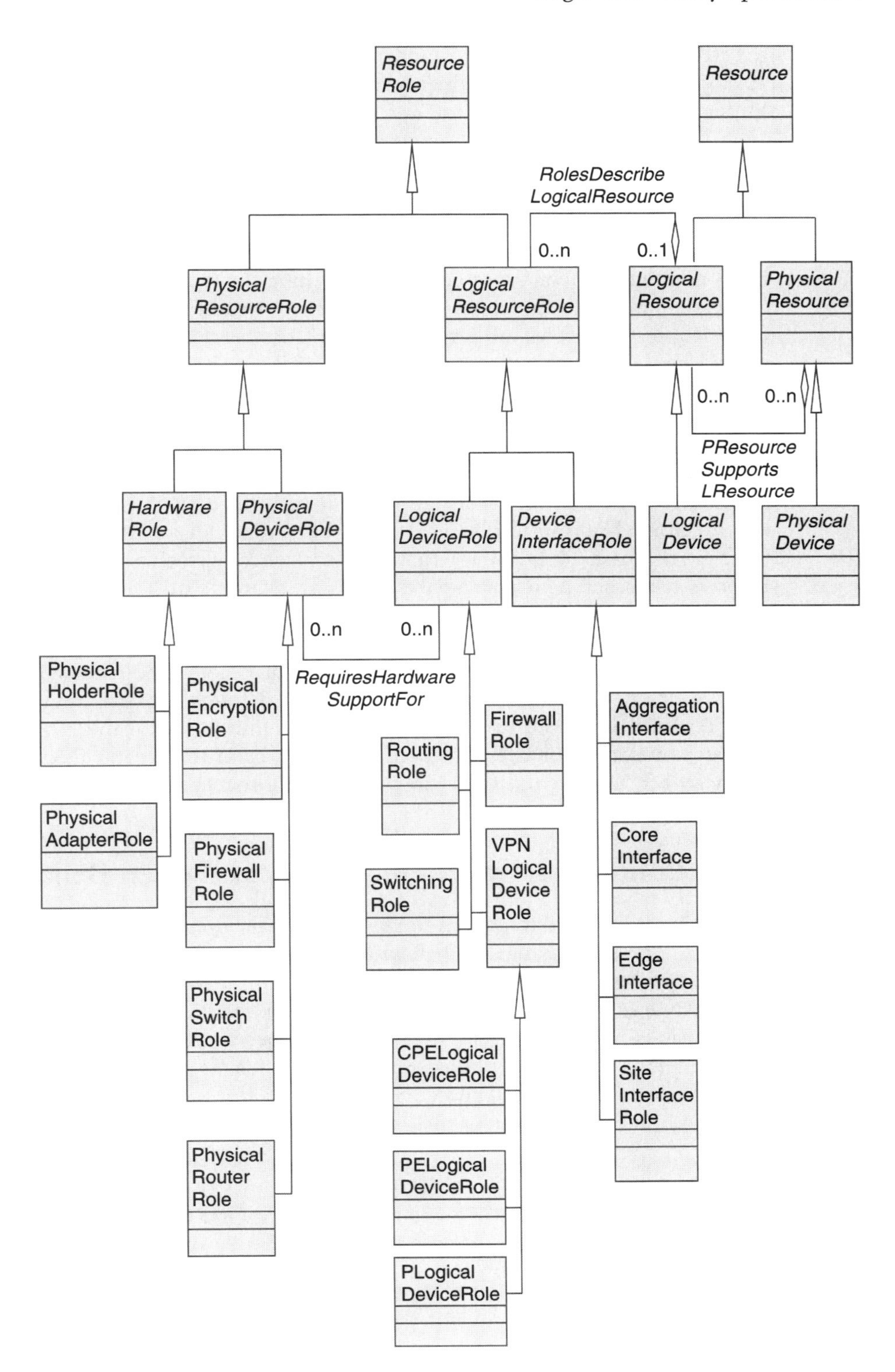

Figure 5-2 DEN-ng *ResourceRole* interaction.

selector. In DEN-ng, it is important to remember that there are physical, as well as logical, roles that a given device can play. These are represented by the PhysicalDeviceRole and LogicalDeviceRole classes. Each of these has their own class hierarchy. For simplicity, Figure 5-2 shows just some simple examples of physical and logical roles taken from the DEN-ng model.

This correspondence can be detected in a number of ways, depending on the repository being used. For example, if the repository is a directory, then at the minimum two different correspondence methods could be used. The first would be to populate the container based on matching the type of *object* Class, which is an attribute of all directory entries. Here, we would match (for example) the Core Routers container object class to the CoreInterfaceRole object class. The second method would be to define a role attribute. For example, this attribute could be an enumerated integer, where again we would seek to match the value corresponding to core routers to the value corresponding to devices that had a CoreInterfaceRole.

This matching function is a role-selector. While it is conceptually represented as an aggregation in the information model, in this case, it is a query made of the directory. The result of this query will be to place all routers (or router interfaces in the case of DEN-ng) that have a role of CoreInterface in the Core Routers container. This correspondence is an example of a *model mapping* (as defined in Chapter 2).

We can therefore see that the use of DEN-ng roles is richer than those defined in Ponder. As we examine the different types of DEN-ng policies, you will see that the generic "domain-scope-expression" for defining subjects and targets has been replaced in the corresponding DEN-ng definitions with the more specific "constraint-expression." This is used to remind us of the greater flexibility that is inherent in the use of OCL in the DEN-ng approach.

5.3 Basic Policies for Specifying Business-Driven Behavior

Basic policies are used to govern the behavior of the system. The idea is to apply different policies to change the behavior of the system. Ponder defines four types of basic policies, which are covered in the following four subsections. These will be used to develop the first level of translation (from high-level business-oriented specifications of behavior to lower levels of specification) in our PBNM system. Note that all basic policies can be specified as parameterized types to enable them to correspond to classes defined in (for example) an information model.

5.3.1 Authorization Policies

An authorization policy is a type of security policy. Its purpose is to define what activities a subject is permitted to do (or not permitted to do) to a set of target objects. Authorization policies that permit actions are called positive authorization policies, while those that do not permit actions are called negative authorization policies.

Note that authorization policies act on behalf of clients that have already been authenticated. Ponder does not specify authentication policies, but DEN-ng does. These are described later in this chapter.

Authorization policies are designed to protect the target objects from unauthorized actions being performed on them. They do this by setting the access control of the target objects. This is a key point—it is the *target* that executes the policy and therefore determines what actions it will allow and enforce.

Negative authorization policies are used to either prohibit certain actions from being performed on a set of targets, or revoke privileges that were previously assigned. This flexibility enables administrators to express high-level access control in terms of both positive and negative policies. This is important, as this mirrors how people naturally express policies.

Note that having positive and negative authorization policies may result in conflicts between the two. However, such conflicts are easily detected through static analysis of the policy specification (i.e., if the subjects and targets, and one or more actions, are identical for any positive and negative authorization policy, then we have a conflict).

Finally, default positive (or negative) authorization policies can be defined and enforced. It is easier to start with a default negative authorization policy (since that fully protects the target) and relax these concerns by adding positive authorization policies as required. The positive authorization policies thus act as a set of criteria that, if matched, provide access to the target; if none of the positive authorization policies are matched; then the default policy will deny access. Thus, it is very easy to set up layered access control with this approach.

In Ponder, all policies can be specified as parameterized types from which instances can be created. Positive authorization policies define the set of actions that subjects are permitted to perform on target objects. Their syntax is simple and is expressed in Figure 5-3.

Here, **bold face** words are Ponder reserved words. The terms "auth+" and "auth–" define positive and negative authorization policies, respectively. Square brackets define optional elements, and braces define repetitive elements. Ponder is very free-form in nature, as it does not impose any specific ordering on its terms (e.g., one policy can define subject on the second line and target on the third line, and another policy could define subject on the third line and target on the second line).

```
inst ( auth+ | auth- )[pathname] policyName {
   subject [<type>]          domain-scope-expression;
   target [<type>]           domain-scope-expression;
   actionList                action-list;
   [ when                    constraint-expression; ]
}
```

Figure 5-3 The syntax of a Ponder authorization policy.

Constraints are expressed using a subset of OCL (Object Constraint Language[14]) as defined in OCL. This is a formal language used to express constraints. Constraints are used to specify invariant conditions that must hold for the system being modeled. OCL is defined such that when OCL expressions are evaluated, they do not have side effects. This means that their evaluation cannot alter the state of system using them. Consider the simple example of a Ponder Authorization Policy in Figure 5-4.

```
auth+      /users/administrators/jsmith/allocateBandwidth {
   subject /users/groups/executiveGroup;
   target  /realms/sites/site1 OR /realms/sites/buildings/building4;
   action  enable(videoconference, bandwidth=1M, priority = 5);
   when    time.between(1400, 1700); }
```

Figure 5-4 An example of a Ponder authorization policy.

This example is the authorization of a video conference. The name of the policy is *allocateBandwidth* and is located in the path */users/administrators/jsmith*. The policy subject is the *executiveGroup*, and the policy target is any person that is a member of the *executiveGroup*. The action is to enable the video conference with a bandwidth of 1M and a priority of 5 between 2pm and 5pm for any user in the *executiveGroup* that is located in *site1* or *building 4*.

We are now in a position to evaluate the effectiveness of Ponder authorization policies. Several shortcomings of Ponder will be identified; these will be further addressed later in this chapter.

A Ponder authorization policy can be applied to one or more domains, so our need for applying policies to multiple objects in multiple domains is easily satisfied. Access to the target is controlled by the action list, so the Ponder authorization policies have also captured the ability to apply access information. Note, however, that a Ponder policy does not define how access control is expressed.

Ponder authorization policies take the view of protecting the target. So, while they can accurately model access control from the target's point-of-view, they cannot model more complex interactions where the subject and the target jointly need to determine what a subject can or cannot do to a target. Note that this cannot be done simply by building two authorization policies (one for each direction of the interaction) because Ponder does not contain any means to relate the two authorization policies to each other. In addition, authorization policies are specified from the point-of-view of the target and therefore do not take into account the requirements of the subject.

The largest shortcoming is the lack of defining events that trigger the evaluation of authorization policies. Ponder does not specify this, leaving it up to the implementation. This is not a prudent course of action. Because the purpose of policy is to control behavior, it becomes very important to ensure that policies themselves are properly applied. This means that in addition to events, we need the concept of a state machine that defines when and how policy is to be applied. Otherwise, the *misapplication* of policy could potentially do more harm than good because it could cause an unwanted state transition of one or more managed entities.

We will use these observations to design an enhanced framework for these (and other) policies in our PBNM system. These enhancements, and the affect that they have in the different syntaxes and semantics of these policies, will be explained later in this chapter, when the DEN-ng versions of these policies are described.

5.3.2 Refrain Policies

Refrain policies are used to define a set of actions that subjects are not permitted to execute on target objects, regardless of whether they have been authorized to do so. Put another way, refrain policies limit the set of actions that a subject may perform, regardless of whether the subject has been previously authenticated and authorized to perform those actions.

Refrain policies are similar in function to negative authorization policies, but with one important difference—refrain policies are subject-based, whereas negative authorization policies are target-based. For example, suppose that the decision to enforce access control involves one or more attributes of the subject. An authorization policy cannot be used because it works by interpreting target properties. As another example, an obligation policy may be operating on a subject. The only way to restrict actions that the subject could normally perform would be to use a refrain policy.

The syntax of refrain policies is the same as that of negative authorization policies and is repeated in Figure 5-5 for convenience.

```
inst refrain policyName {
  subject [<type>]          domain-Scope-Expression;
  target [<type>]           domain-Scope-Expression;
  action                    action-list;
  [when                     constraint-Expression; ]
}
```

Figure 5-5 The syntax of a Ponder refrain policy.

Consider the simple example in Figure 5-6 of a Ponder Refrain Policy. This policy prohibits video conferences from being enabled to any site on a Monday.

```
refrain     /users/administrators/jsmith/assignBandwidth {
   subject  /users/groups/executiveGroup;
   target   /-;  //any target
   action   enable(videoconference);
   when     time.day(Monday); }
```

Figure 5-6 An example of a Ponder refrain policy.

5.3.3 Obligation Policies

Obligation policies define the set of actions that a subject must do. In contrast with authorization policies (which are target-based), obligation policies are subject-based (i.e., the subject interprets the policy and performs actions on the target.

Obligation policies are triggered by events. Obligation policies represent a potentially powerful tool that can be used in a variety of novel ways, as they represent the ability to embed predefined actions to control behavior that is executed when an event happens.

This is the foundation for the {event-condition-action} triplet that has been referred to in Chapters 2 and 3. In order for an entity to be obliged to do something, that entity must know when to perform the action. This can be simply and elegantly specified by an event. The event acts as a trigger that determines when the actions must be performed.

Events can be simple or complex in nature and may require external entities and/or services to be involved. A simple event is a single occurrence of something significant. A complex event is a set of events that, when taken together, signify a more abstract event that has taken place. Note, however, that both can affect one or a set of Policy Targets.

For example, suppose that a system controller detects when boards are inserted into or removed from the system. This is a simple event because a single significant occurrence happened at a discrete time that the system detected. Another example is when a timer expires, changing the state of the system (as in the BGP finite state machine). This, too, is a single event. A third example of a simple event is when a user logs onto the system. There is a single event that identifies this action, but it almost always invokes an external service to communicate that event to another entity that will take action. The first two examples will most likely only directly affect a single managed entity. However, the third example will almost surely affect a set of managed entities, as well as launch a set of dependent policies (e.g., to determine what services the user is authorized to use, and to start the accounting process so that the user is correctly billed for services that he or she uses).

An example of a complex event could be the generation of an alarm alerting an operator that the number of packets dropped for a particular interface has exceeded a predetermined threshold. This is a complex event, because it consists of a set of simpler events (e.g., the device updating the MIB counter, an external service monitoring the MIB counter, and realizing that the counter has exceeded its predetermined value, the sending of that information to a controlling entity, and finally the generation of an alarm by the controlling entity). This example may or may not affect more than one managed entity, depending on how the PBNM system is programmed to respond to alarm events. The syntax of an obligation policy is illustrated in Figure 5-7.

Consider the simple example in Figure 5-8 of a Ponder Obligation Policy.

This simple obligation policy defines the actions to be taken whenever an event that requests a video conference is received. To keep the example simple, the only action specified in this policy is to reserve bandwidth. Note that *how* this action is to be performed is not specified by the Ponder policy.

```
inst ( oblig) policyName {
   on                          event-specification;
   subject [<type>]            domain-Scope-Expression;
[ target [<type>]              domain-Scope-Expression; ]
   do                          obligation-action-list;
   [ catch                     exception-specification; ]
   [ when                      constraint-Expression; ]
}
```

Figure 5-7 The syntax of a Ponder obligation policy.

```
oblig videoConferenceSetUp {
  on       request( videoConference, source, bandwidth );
  subject  /devices/routers/InternetGatewayRouter;
  target   /devices/edgeRouters;
  do       target.reserve( bandwidth ); }
```

Figure 5-8 An example of a Ponder obligation policy.

Note that the specification of an event, subject, and action list are required. The keyword **on** is used to specify an expression that refers to a single or a set of events. Events can be related to each other using a set of pre-defined operators. These operators do not provide a full Boolean expression; rather, they are used to relate event occurrences to each other (e.g., the operations "event 1 occurs before event 2," "event 1 occurs n times," and "event 1 or event 2 occurs in either order" are examples of the types of relationships that can be specified).

A refrain policy can also be used when the subject is permitted to perform the action but is asked to refrain from doing so when particular constraints apply. Refrain policies are similar to negative authorization policies, except that refrain policies specify actions that must not be performed by the subject, whereas negative authorization policies specify actions that are not permitted to be done by the subject to the target.

The target element is optional because an obligation action can specify further actions that the subject must take. The optional catch clause is used to specify an exception that is executed if the actions specified by the obligation policy fail to execute correctly. A simple set of concurrency operators is available to control whether actions are executed sequentially or in parallel.

5.3.4 Delegation Policies

Delegation policies are used to permit subjects to grant a set of privileges to other subjects. These other subjects are called *grantees*, while the subjects doing the granting of privileges are called *grantors*. The set of privileges that a grantor has are determined, for example, by executing one or more authorization policies on behalf of the grantor. One of these privileges may in fact specify what privileges a grantor can grant to a grantees.

Delegation policies cannot *add* new privileges to the grantee. An important point is that delegation policies do *not* transfer access rights from subjects to

```
inst deleg+ ( associated-policy-name ) policyName {
 grantee [<type>]              domain-Scope-Expression;
 [ subject [<type>]            domain-Scope-Expression; ]
   [ target [<type>]           domain-Scope-Expression; ]
   [ action                    action-list; ]
   [ when                      constraint-Expression; ]
   [ valid                     constraint-Expression; ]
   [ hops                      int-value; ]
}
```

Figure 5-9 The syntax of a Ponder delegation policy.

grantees; subjects continue to retain their access rights after a delegation is performed.

The syntax for delegation policies is shown in Figure 5-9, and a simple example of a Ponder Delegation Policy in Figure 5-10.

```
deleg+ (allocateBandwidth) delegateBandwidthAllocation {
   grantee   /users/groups/adminGroups/execAssistants;
   target    /realms/sites/site1;
   action    enable( videoconference, bandwidth=1M, priority=5 );
   when      time.between( 0800, 1700 );
   valid     time.duration( 24 ); }
```

Figure 5-10 An example of a Ponder delegation policy.

This example is associated with the allocateBandwidth policy defined in Figure 5-4. It states that the subject of the allocateBandwidth authorization policy (which was any executive) can delegate the ability to enable a video conference on policies from the domain /realms/sites/site1 to grantees in the domain /users/groups/adminGroups/execAssistants. (In other words, an executive is enabling his or her executive assistant to enable a video conference; note, however, that the original allocateBandwidth policy also had as a target /realms/sites/buildings/building4, which has been omitted in this policy.) The when clause specifies that a delegation may be performed between 8am and 5pm, and the valid clause specifies that this delegation is valid for the next 24 hours.

A delegation policy is always associated with an authorization policy; this is enforced by requiring the delegation policy to specify its associated authorization policy as a formal parameter. Grantee is the only required part of the delegation policy. Note that the subject, target, and action lists must be a subset of their respective definitions in the original authorization policy.

Delegation policies can be constrained in three ways:

- Specifying time periods over which the delegation should be valid before it is revoked
- An OCL expression that can define a constraint based on either system attributes or attributes of a subject, target, grantee, or action
- Maximum number of cascading delegations allowed (maximum number of delegation hops or levels)

These constraints are specified by the *when*, *valid*, and *hops* clauses, respectively.

Negative delegation policies forbid delegation of certain actions. Only positive delegation policies contain delegation constraints; they make no sense in negative delegation policies. The syntax for a negative delegation policy is thus the same as that for a positive policy, without the ***valid*** and ***hops*** clauses, which are used to specify delegation constraints.

5.3.5 Definition of Events in Ponder

Events define certain significant occurrences, based on the value of an attribute, that are used to trigger obligation policies. Events can be parameterized, which enables event expressions to be defined that relate specific values to generic expressions.

Ponder defines several types of events.[1] The syntax of events is of the following form:

Event-name [(formal-parameter-list)] = event-expression

An event expression can take one of the following forms (e1, e2, and e3 represent events):

- basic event (i.e., the name of the event with any optional parameters, as in *foo [(. . .)])*
- e1 && e2—occurs when both e1 and e2 occur irrespective of their order
- e1 | e2—occurs when e1 or e2 occurs
- e1 -> e2—occurs when e1 occurs before e2
- e1 + time—occurs a specified period of time after e1 occurs
- {e1 ; e2}!e3—occurs when e1 occurs followed by e2 with no interleaving of e3
- n * e—occurs when e occurs n times

5.3.6 Constraints in Ponder

Constraints are used to limit the applicability of basic policies. Like events, constraints can be parameterized. Unlike events, constraints can be used with all policies.

Ponder differentiates between time- and state-based constraints. While both of these constraints lend themselves to static analysis (in order to detect and resolve conflicts between policies), their use is very different. Time-based constraints have the effect of defining a window of time in which the policy is active and can be applied. Thus, time-based constraints can only conflict with each other if there is an overlap in time between the two policies. In contrast, state-based constraints can only be in conflict if they relate to states of the same managed entity.

5.3.7 Filtering Policies in Ponder

Filtering policies are used in Ponder to transform the values of information returned from actions. This enables generic policies to specify generic information that can subsequently be refined by more explicit filters. For example, a generic authorization policy may define access for a set of users to a set of servers. Different filtering policies may then be used in conjunction with this *single* authorization policy to specify different access permissions that different users have on different servers.

Filtering policies act *after* the action in the authorization list to which they apply is performed. If the action succeeds, then a decision is made on whether the results need to be transformed. Filters can only be applied to positive authorization actions. Filter policies have the syntax shown in Figure 5-11.

```
actionName { filterName }
filter = [ if condition ] {
// can be 1 or more of these expressions
  { ( in parameterName = expression ; | result = expression ; ) }
}
```

Figure 5-11 The syntax of a Ponder delegation policy.

The filter is associated with a particular action, not a particular policy. The filter takes the form of specifying a condition; if that condition is true, then it changes either the input expression to the action or the result of the action (Figure 5-12).

```
inst auth+ VideoConf {
        subject                 AGroup;
        target                  BGroup;
        action                  VideoConf(Bandwidth, Priority)
        if (Time.after("1900"))
           {in Bandwidth=3; in Priority = 1;} // executed if condition is satisfied
           {in Bandwidth=2; in Priority=3;} // default filter if condition isn't satisfied
}
```

Figure 5-12 Example of a Ponder filter policy.

This policy enables members of the AGroup to set up a video conference with members of the BGroup. The video conference action takes two parameters, bandwidth and priority. At this point, the filter triggers. If the current time is after 1900, then the bandwidth and priority parameters are set to 3 and 1, respectively. Otherwise, the bandwidth and priority parameters are set to 2 and 3, respectively.

5.4 DEN-ng Additions to the Ponder Framework of Basic Policies

This section will briefly describe the additional features of DEN-ng policies that were designed to complement the features present in Ponder basic policies.

Note that this section is concerned with linking languages, like Ponder, to information models, like DEN-ng. This approach is one of the fundamental principles of the DEN-ng policy model. Note, however, that the IETF and the DMTF have both rejected this approach.

5.4.1 Generic Enhancements to All Types of Policies

Ponder forms an excellent basis to start the design of a high-level policy language. First, however, it is appropriate to describe generic enhancements made to Ponder to meet the generic requirements that we have described so far in this book. The results of these requirements will be summarized into a single template that will be used by all DEN-ng policy expressions.

Access Control

Ponder does not define a policy to create or modify access control privileges. Rather, Ponder assumes that access control is already predefined by an external system.

This misses the opportunity to use policy to control *how* access control is defined, deployed, and updated. For example, an organization may define a set of business rules that describe different authentication and authorization policies to be taken by the *same* user depending on how that user connects to the network. As another example, a policy can define when active certificates should be revoked based on a set of metrics. Clearly, without the use of policy, this will be significantly more difficult to implement. Individual exceptions, such as described in the first example, could only be accommodated by keeping track of each individual exception. This cannot scale for even small organizations.

There are a number of requirements that must be satisfied in order to provide a robust access control solution. Principal among these include the following:

- Defining a set of access control schemes that meet the needs of different users and applications; this enables different needs of different users and applications to peacefully coexist in the same managed environment.
- Identifying which access control scheme is in force for which entities at which time, as a function of which environmental conditions or other constraints are in effect; this recognizes that access to a given resource may change as a function of environmental conditions, such as time of day, amount of congestion, or login method.
- Identifying which domain and groups of entities (such as a user) a given access control scheme applies to; this enables a consistent set of permissions to be applied to a set of targets through the use of domains and groups.
- Identifying which roles a given access control scheme applies to—this provides the flexibility of associating different access control mechanisms with the same entity and distinguishing them on a role-by-role basis.
- Identifying the collection of access control information that collectively defines the overall access permissions that are applied to a managed entity; this enables sophisticated access control to be built up through logical operations on domain and group membership.
- Defining the policy or set of policies that are used to determine what behavior will be applied when different access control permissions are

applied to the same managed entity; this enables different decisions to be applied to different managed entities as a function of what role the managed entity is playing, as well as other factors, such as the current value of different attributes that define the managed environment.
- Identifying which entity roles can modify access control information and schemes for which managed entities; this ensures that only certain users can define access control information for certain managed entities based on the role they play in the organization

There are two important benefits that this collection of requirements provides. First, the most common complaint of network technicians is that when they use a given Element Management System and an administrator is assigned, that administrator now has global privileges for all devices that are governed by that Element Management System. This is not desirable and does not reflect the real world. Network resources are very precious resources, and usually, one group will not allow technicians or administrators from another group to manage their resources.

Furthermore, a given organization's network technicians often have a different variety of skill levels. It is mandatory that policies be established that determine which user or group of users have which privileges to execute which functions on which devices. The requirements listed ensure that these goals are met.

Second, access control should not be viewed as a static choosing of permissions that are collected into a group. Rather, it is a rich mechanism that can dynamically compute permissions and obligations based not only on the identity of the user, but also on the type of service being requested by the user and the role that the user is playing. This enables administrative authorities to tailor the service granted based on the current environment. The requirements listed cannot be satisfied without using policies to control the generation and application of access control. This will be further discussed in later chapters of this book.

Controlling Policy Evaluation and Execution

Ponder provides a rich framework for defining basic system building blocks in the form of its four basic policies. However, it does not provide a framework in which it is guaranteed that policy execution will be safe. In order to do this, we need additional metadata that describes when and how policies can be evaluated and executed. This metadata in essence relates an individual policy to other policies that exist in the system.

Since we are using a state machine to model the life-cycle aspects of managed entities, it is desirable to define policies in such a way as to enable them to be used in the same state machine. The approach will be to define policies to enable them to control when the state of a managed entity can change. The easiest way to do this is to extend the Ponder policy definitions such that each policy has a set of pre- and post-conditions that are used to define the state of the system before and after the policy has executed, along with a set of exceptions that could be encountered.

- The pre-conditions define the set of conditions that must be satisfied in order to invoke the policy with reasonable assurance of its successful completion.
- The post-conditions define the state that the system is left in if the preconditions are met and no exceptions are encountered in the execution of the policy.

- Finally, the set of exceptions that can be encountered specify abnormal behavior during the execution of the policy. An exception occurs if the pre-conditions are satisfied, but the post-conditions cannot be met.

DEN-ng adds these three types of information, plus generic metadata, to the definition of all of its policies. This will be discussed in more detail in Chapter 6.

Event Interaction

In Ponder, events are only formally defined to trigger obligation policies. Specifically, the Ponder syntax specification says: "Basic Common Elements are those that can be specified in Authorization, Delegation and Refrain policies—Events make no sense in those policies."

In DEN-ng, events can be used to trigger all policies (i.e., each type of DEN-ng policy is a triplet of the form {event-condition-action}). This offers some significant usage enhancements. For example, in Ponder, obligation policies must be used to control the execution of authorization, delegation, and refrain policies. It is unclear why this level of indirection is necessary. Even more important, exceptions can be abstracted as events that must be handled. For example, an action may be executing as part of a policy. If that action generates an exception, then the exception can be handled directly by allowing other policies to be invoked directly by the exception. Conceptually, this corresponds to the exception generating an event that either directly or indirectly (e.g., through a publish-subscribe message bus) invokes one or more policies to handle the exception. This proves to be a generic method that can be used for handling all types of policies.

Exceptions

DEN-ng is based on using a finite state machine to control, in a closed-loop fashion, the entities that it is managing. Part of this control is based on ensuring that managed entities are in the correct state at any given time. Exceptions can be used to signal to the rest of the system that something abnormal has occurred.

Two types of abnormalities are currently defined. A generic abnormality can be signaled by any ManagedObject. In addition, action clauses within a policy can be used to signal abnormalities.

In either case, the approach used in DEN-ng is to define an explicit set of policies that are invoked when a particular type of exception is raised. This provides a generic framework that enables the PBNM system to react to any type of exception in a standardized way. Application-specific exception handling can then be defined as specialization of the standardized exception handling. This helps foster interoperability between multiple PBNM systems.

Therefore, DEN-ng defines exceptions at three different levels: as a signal that something abnormal has occurred within any ManagedObject in the system, as part of any action that is a part of any policy, and as part of each policy definition. These three types of exceptions work together to provide different levels of granularity to enable the user to specify, manage, and monitor the particular state of any given managed entity. They also provide the framework for a standardized yet extensible response mechanism to exceptions to be defined.

Relationship Between Policy Model and an Information Model

In Ponder, there is a high-level relationship between the types of policies that are defined in the language and an object-oriented class hierarchy, defined as an information model, to further define the semantic relationship between policies and to guide the implementation of policies. This relationship is shown in Figure 5-13.

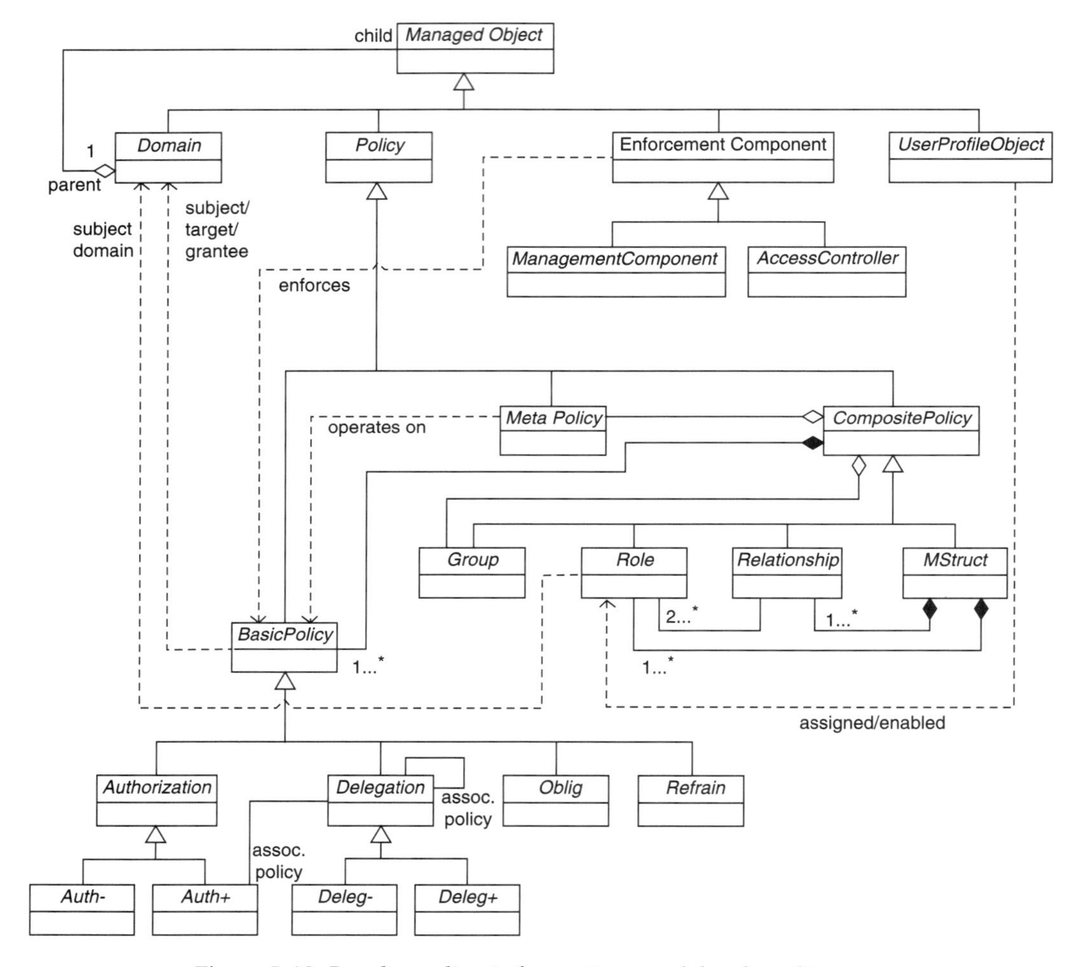

Figure 5-13 Ponder policy information model—class diagram.

This class diagram is a good start. However, it is not self-contained. For example, the Group class does not have a superclass defined for it. Furthermore, one of the principle reasons to use an information model is to model the definitions and relationships between different managed objects. Ponder defines everything as a "ManagedObject." This is important because it facilitates the binding of a policy to a set of subjects and targets.

However, the Ponder information model stops short of defining *how* to bind a policy to a subject or target. There are also a number of inconsistencies (e.g.,

BasicPolicy should participate in a shared aggregation relationship, not a composition relationship, with CompositePolicy) and questions (why roles cannot be assigned to and used by BasicPolicy?). Finally, the model in Figure 5-13 does not describe different types of ManagedObjects and how policies affect them. Clearly, a policy to control when a new Card is inserted into a Device are different syntactically and semantically than policies that control how to update routing information. These are answered in the DEN-ng policy information model, which is discussed in Chapters 6 and 7.

Version Control

Ponder has no notion of versioning of its policies. This presents a number of difficult problems. For example, if a particular policy is deployed in many places and that policy is subsequently updated, it is difficult to ensure that every instance of that policy has been correctly updated to the new version.

DEN-ng includes a version header in its definition, which simplifies determining what version an existing policy instances is. This also enables fast, efficient searches to be conducted to update new versions of a particular type of policy when they are available.

Summary of DEN-ng Generic Additions to Policy

The following fields are specific to DEN-ng policies, and are applicable to all DEN-ng policy definitions unless otherwise noted:

- **Version:** *The version of this policy.*
- **Pre-Conditions:** *The set of conditions that must be satisfied in order to invoke the policy with reasonable assurance of its successful completion; conceptually, this defines the state of the subject and target of the policy before the policy executes.*
- **Trigger-Conditions:** *The set of conditions that define the triggering, or beginning of evaluation, of this policy; this is usually a set of events but is not limited to just events (e.g., a previous policy could directly invoke a new policy).*
- **MetaData:** *A set of data that describes the policy; it is used to provide information about this policy and how it is intended to be used.*
- **Post-Conditions:** *This defines the state that the system is left in if the pre-conditions are met, and no exceptions are encountered in the execution of the policy; conceptually, this defines the state of the subject and target of the policy after the policy executes.*
- **Exceptions:** *This defines the set of exceptions that can be encountered. Exceptions specify abnormal behavior encountered during the execution of the policy. An exception occurs if the pre-conditions are satisfied, but the post-conditions cannot be met. This provides the policy designer with finer-level control over the target entities* ***during*** *the execution of the policy.*

This results in the DEN-ng policy template in Figure 5-14. Note the inclusion of the subject-role-list and target-role-list fields. These expand on the notion of

```
inst ( xxx+ | xxx- ) [pathname] policyName {
   version                      version-expression;
   pre-conditions               constraint-expression;
   trigger-conditions           constraint-expression;
   subject-role-list            constraint-expression;
   target-role-list             constraint-expression;
   metadata                     metadata-expression;
   post-conditions              constraint-expression;
   exceptions                   exception-list;
}
```

Figure 5-14 The generic DEN-ng policy template.

subject and target as defined in Ponder to include the broader concept of roles as defined in DEN-ng. By making each of these fields a constraint expression, roles can be used to match and/or constrain the matching of what the policy is about and to what it applies.

Finally, two important differences between Ponder and DEN-ng are (1) the absence of optional fields in the DEN-ng template, and (2) the lack of free-form grammar in DEN-ng (i.e., in Ponder, the order of the expression components is not fixed, whereas in DEN-ng, the order of each statement in each component of the expression is fixed). This makes for a more consistent grammar and simplifies the design of the compiler. This helps to produce more optimized code to implement these policies.

5.4.2 DEN-ng Access Control Policies

DEN-ng access control policies revolve around associating permissions with roles, as opposed to specific users or devices. This provides an inherent flexibility in the system specification because users or devices can dynamically be assigned to or removed from roles. This powerful level of indirection is shown in Figure 5-15.

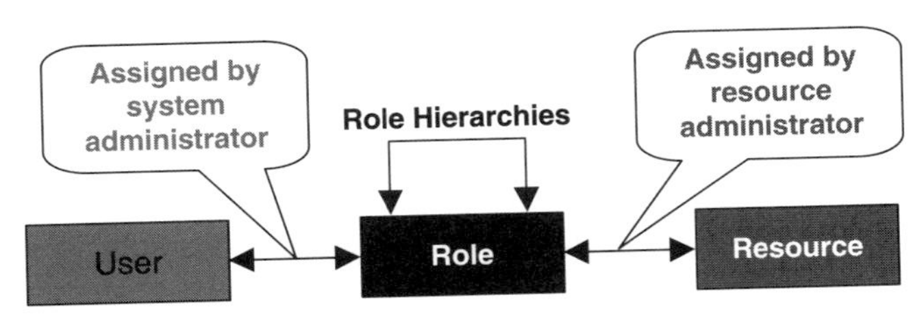

Figure 5-15 Role-based *access control as used in DEN-ng.*

In DEN-ng, users do *not* have discretionary access to objects in the system. Instead, access permissions are associated with roles, and users (as well as resources) are assigned a set of roles. This enables policies to be written based on authentication and authorization capabilities and obligations. It also ensures that the same concepts are handled in the same way. For example, conceptually there is no difference between a user authentication process or a resource authentication process. The DEN-ng role abstraction enables both of these processes to be handled in exactly the same manner.

Roles can be hierarchical. Users acquire access rights and privileges based on their role *membership*. This means that the set of rights and privileges that a given resource has is a function of the groups that the resource belongs to and the roles that are assigned to each of those groups.

Role-based access control provides the following extensible features:

- Authorization beyond typical "all or nothing" security.
- Flexible permission assignment through the use of *role, person,* and *resource hierarchies.*
- Privileges are inherited (inheritance filters are used to provide greater flexibility).
- Privileges can be cumulative or exclusive.

An important concept in DEN-ng access control policies is the concept of a *session.* This is shown in Figure 5-16 (note that the same concept applies to resources as well):

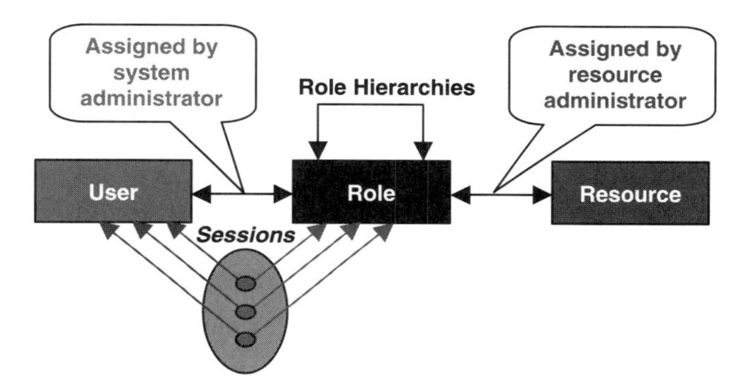

Figure 5-16 Session role-based access control as used in DEN-ng.

Sessions enable roles (and hence access control) to be assigned on a transitory basis. Role assignment can be computed on a dynamic basis, or retrieved and reused from static definitions. This enables certain privileged users, such as administrators, to always have the same set of access control privileges while enabling other users (e.g., "clients" of the system) to have their access control governed by a set of metrics that change as a function of (for example) the environment. For example, a subscriber may have increased access to functions not normally available by paying a per-session fee.

In either case, a session can be used to define a logical container that defines how access control is computed. The session itself is treated as a named object, which facilitates its tracking and manipulation. For example, different sessions can be identified and have the same or a different set of operations (such as accounting and auditing) performed on specific objects within that session or the session as a whole.

Privileges must be defined using the system's concept of objects that define access control. Examples are certificates, public and private keys, and other forms of credentials. Credentials are used to verify that the sender is, in fact, who the sender purports to be. Each type of operation requires its own set of functions that are used to create, read, update, and delete (CRUD) credentials of that type. For example, certificates require a CRL, or Certificate Revocation List, to be checked to ensure that the credential is still valid and has not been revoked. These requirements will be added into DEN-ng authentication policies.

5.4.3 DEN-ng Authentication Policies

Ponder has no authentication policies—the best that Ponder can offer is to use Ponder obligation policies to point to (or invoke) an external authentication mechanism. Furthermore, Ponder authorization policies assume that both the subject as well as the target has already been authenticated.

DEN-ng has taken an alternate approach. The DEN-ng approach is that authentication is a fundamental service, and policies to govern authentication are mandatory for any system. In order to determine what actions a policy subject is authorized to perform, one or more DEN-ng authentication policies are executed.

DEN-ng authentication policies are intimately related to the concept of roles. It is important to be able to abstract behavior to higher-level concepts, such as roles. This enables one authentication method to apply to multiple entities without specifically naming any of the entities. One can write:

"All sales engineers shall use a DES card to authenticate to the system when on the road."

and have such a policy affect *all* of the sales engineers in the corporation, regardless of where they are based, or what their specific title (e.g., grade of sales engineer) is. This holds true if new sales engineers are added because they are assigned the role sales engineer when they join the corporation.

The simplified syntax for a DEN-ng authentication policy is in Figure 5-17.

```
inst ( authen+ | authen- ) [pathname] policyName {
   version                      version-expression;    // DEN-ng policy template
   pre-conditions               constraint-expression; // DEN-ng policy template
   trigger-conditions           constraint-expression; // DEN-ng policy template
   subject-role-list            constraint-expression; // DEN-ng policy template
   target-role-list             constraint-expression; // DEN-ng policy template
   privilege-list               constraint-expression; // permissions
   privilege-definition         enumerated type;       // type of credentials
   privilege-support            enumerated type;       // CRUD for credentials
   (activate | deactivate) when constraint-expression;
   metadata                     metadata-expression;   // DEN-ng policy template
   post-conditions              constraint-expression; // DEN-ng policy template
   exceptions                   exception-list;        // DEN-ng policy template
}
```

Figure 5-17 The syntax of a DEN-ng authentication policy.

The DEN-ng authentication policy is modeled after Ponder, though this is supplemented with the eight additional fields defined in Figure 5-14. These eight fields are part of the generic DEN-ng policy template. Subject and target definitions in DEN-ng are similar to those of Ponder, except that there is a more powerful use of roles in DEN-ng than there is in Ponder. While roles do not have to be used in DEN-ng policies, they provide an important abstraction that in practice has proven to be extremely useful in simplifying the complexity of system management.

Also, the definition of when to apply this policy has been changed. In Ponder, this clause is optional and consists of the simple keyword *when*. In DEN-ng, this clause is made mandatory for the sake of consistency, and its semantics have been increased by using the *activate* and *deactivate* keywords and the ability to have an expression compute when the policy should be activated or deactivated. This enables the exact same policy to be easily enabled or disabled via these two keywords. This provides greater and more flexible semantics than simply having a positive or negative authentication policy. These three changes (subject and target definitions and when to apply the policy) are common for all DEN-ng policies and will not be discussed further for any policy.

Finally, the authentication enforcement mechanism must be able to communicate with the calling application, so that it can tell the calling application whether an action requested is allowed or not. DEN-ng defines three possible responses: YES, NO, and PARTIAL. YES and NO state that the action is or is not allowed, respectively. PARTIAL is used to indicate that additional dependencies are involved. For example, a system may implement a multi-step authentication mechanism; PARTIAL can be used to indicate that the first part of the multi-step authentication mechanism has been satisfied, but additional steps must still be satisfied before the requested operation(s) can be performed on the target resource(s). A simple example of a DEN-ng Authentication Policy is shown in Figure 5-18.

```
authen+ /users/salesGroup  logonPolicy( user, timestamp, accessDevice) {
   version                 2.1;
   pre-conditions          connectionType == PPP;
   trigger-conditions      event( logonRequest );
   subject-role-list       user == salesRole;
   target-role-list        accessDevice;
   privilege-list          /scripts/privileges/salesGroup;
   privilege-definition    secure token;
   privilege-support       /scripts/privileges/support/updateToken;
   activate when           event( credentialsReceived );
   metadata                null;
   post-conditions         run authorizeSalesGroup( user, timestamp );
   exceptions              /scripts/logonScripts/badLogon( user, timestamp ); }
```

Figure 5-18 The syntax of a DEN-ng authentication policy.

This example is used to authenticate a member of the salesGroup when that user tries to log onto the network. The three parameters of this policy—the user, the time of the logon request, and the particular accessDevice that received the logon request—are used throughout the policy.

Since sales people can be present in the office or travel on the road, there are in general at least two different types of authentication policies—one that is used when the user is connected through Ethernet at his or her local office and one when the sales person travels. This distinction is captured on the pre-conditions clause where the type of connection is specified as PPP. (Of course, other connection types could also be allowed; these would be supported through additional values for the connectionType attribute.)

This policy is triggered when a logon event is received by the network and applies to all users that have the role *salesRole*. The target is the accessDevice that received the logon request—it needs to be told to let traffic from this user enter the network. The set of operations that the user will be granted if the logon request is successful is defined in a script. The type of authentication mechanism specified is a secure token (e.g., a special device that produces a unique entry in response to a challenge from the access device). The updateToken script is used to support the use of this mechanism.

The authentication policy will be activated when an event denoting the receipt of the credentials (in response to the challenge by the access device) is received. When this event is received, the script *authorizeSalesGroup* is executed. If the challenge is not correct, an exception is raised, which triggers the execution of the *badLogon* script. There is no metadata associated with this policy.

5.4.4 DEN-ng Authorization Policies

DEN-ng provides some additional semantics for authorization policies compared to Ponder authorization policies. The first is that some applications need a richer authorization decision. Specifically, they need access rights to be determined as a function of the subject and the target, not just the target alone. This means that the decision on what can and cannot be done is changed to the combination of what the target will allow and what the subject is capable of doing. For example, if a user who is not an administrator wants to invoke a function restricted to administrators, it is easy to see that this decision can't be allowed regardless of the target.

Furthermore, sometimes things are restricted because of the current operating environment or some other data. Consider the business rule that prohibits access to certain servers over non-secured lines. In this case, the fact that the user has authenticated properly and is therefore normally entitled access to these servers is overridden by the special provision that users must access these servers through a secure mechanism.

These enhancements can be accommodated by including a list of constraints that express how the subjects and targets interact. The third enhancement is the realization that, in general, some set of pre-conditions must have been satisfied in order to enable a policy to be triggered.

The final enhancement is the recognition that some policies fail to execute correctly for any number of reasons. In order for such policies to be well-behaved (in the computational sense), a specification of what to do if the policy fails to execute properly must be specified. Therefore, DEN-ng modifies the basic syntax of authorization policies (Figure 5-19). An example of a DEN-ng authorization policy is in Figure 5-20.

```
inst ( author+ | author- ) [pathname] policyName {
   version                         version-expression;
   pre-conditions                  constraint-expression;  // 2nd and 3rd enhancements
   trigger-conditions              constraint-expression;
   authorization-function          constraint-expression;  // 1st enhancement
   subject-role-list               constraint-expression;
   target-role-list                constraint-expression;
   actionList                      action-list;
   metadata                        metadata-expression;
   post-conditions                 constraint-expression;
   exceptions                      exception-list;
   (activate | deactivate) when    constraint-expression;  // 4th enhancement
}
```

Figure 5-19 The syntax of a DEN-ng authorization policy.

```
author+ /users/salesGroup   authorizationPolicy( user, timestamp, accessDevice) {
   version                   1.5;
   pre-conditions            event( authenticationSuccessful );
   trigger-conditions        event( authenticationSuccessful );
   subject-role-list         user == salesRole;
   target-role-list          accessDevice;
   privilege-list            /scripts/authorization/salesGroup;
   privilege-definition      /scripts/authorization/enforceUser;
   privilege-support         /scripts/authorization/support/updateSalesGroup;
   activate when             event( authenticationSuccessful );
   metadata                  null;
   post-conditions           run startAccountingWithLogging( user, timestamp );/
   exceptions                scripts/authorization/authorizationError( user, timestamp ); }
```

Figure 5-20 An example of a DEN-ng authentication policy.

This example is used to define the set of services that a member of the salesGroup is authorized to use (after that user has been successfully authenticated). This example is triggered by the receipt of a successful authentication event. When that event is received, a script is run that reconfigures the network to provide this user with the services that this user is authorized to use. In addition, a script is run to an accounting process (e.g., to determine the charge to assess the user's department), as well as to start logging the activities that the user is doing (as part of a general business policy).

5.4.5 DEN-ng Obligation Policies

DEN-ng makes two important enhancements to Ponder obligation policies:

- Actions can include a priority that enables a finer-level of granularity to be specified than just "a follows b."

- Exceptions are enhanced, in that each action of an action-list can throw an exception if it doesn't execute correctly (in addition to a more general exception thrown by the obligation policy itself, which is the only exception that Ponders supplies).

This leads to the syntax for DEN-ng obligation policies in Figure 5-21.

```
inst ( oblig+ | oblig- ) [pathname] policyName {
   version                      version-expression;
   pre-conditions               constraint-expression;
   trigger-conditions           constraint-expression;
   subject-role-list            constraint-expression;
   target-role-list             constraint-expression; // 1st difference–see below
   do
      obligationActionList      action-expression-list;// 1stand 2nd enhancements–see above
   metadata                     metadata-expression;
   post-conditions              constraint-expression;
   exceptions                   exception-list; // 2nddifference–see below
   (activate | deactivate) when constraint-expression;
}
```

Figure 5-21 The syntax of a DEN-ng obligation policy.

There are two fundamental differences in syntax between Ponder obligation policies and DEN-ng obligation policies. First, in Ponder, targets are optional. For consistency, DEN-ng makes the target clause mandatory. One can still specify that the subject is also the target but rather than doing this implicitly by not having a target clause (as is done in Ponder), DEN-ng uses the keyword *self*. Second, in Ponder, the handling of exceptions is made optional. DEN-ng mandates the handling of exceptions in order to enforce consistency among all types of policies. (Note that the *catch* and *when* clauses of Ponder are subsumed into this pseudo-syntax.)

5.4.6 Events in DEN-ng

DEN-ng events are identical to Ponder events in their definition; the difference is in their application. Ponder restricts events to being used as part of an obligation policy. In DEN-ng, events are integrated as a fundamental part of all policies. This is because every DEN-ng policy is made up of three clauses: an event, a condition, and an action clause. This is shown in the UML snippet in Figure 5.22 (a more complete policy model will be shown in Chapter 6):

A PolicyRule contains an event clause, a condition clause, and an action clause. These are specified by the *IsTriggeredBy*, *PolicyConditionInPolicyRule*, and *PolicyActionInPolicyRule* aggregations, respectively. In particular, the event clause is a *set* of events, which in the above figure are represented by the Policy EventSet object. The reason that this set of events is an object is to enable it to be reused by other policies as an identical trigger (or part of a trigger).

A PolicyStatement is a formal Boolean expression. Both PolicyConditions as well as PolicyActions can be formed using a PolicyStatement through the *PolicyStatementInPolicyCondition* and *PolicyStatementInPolicyAction* associa-

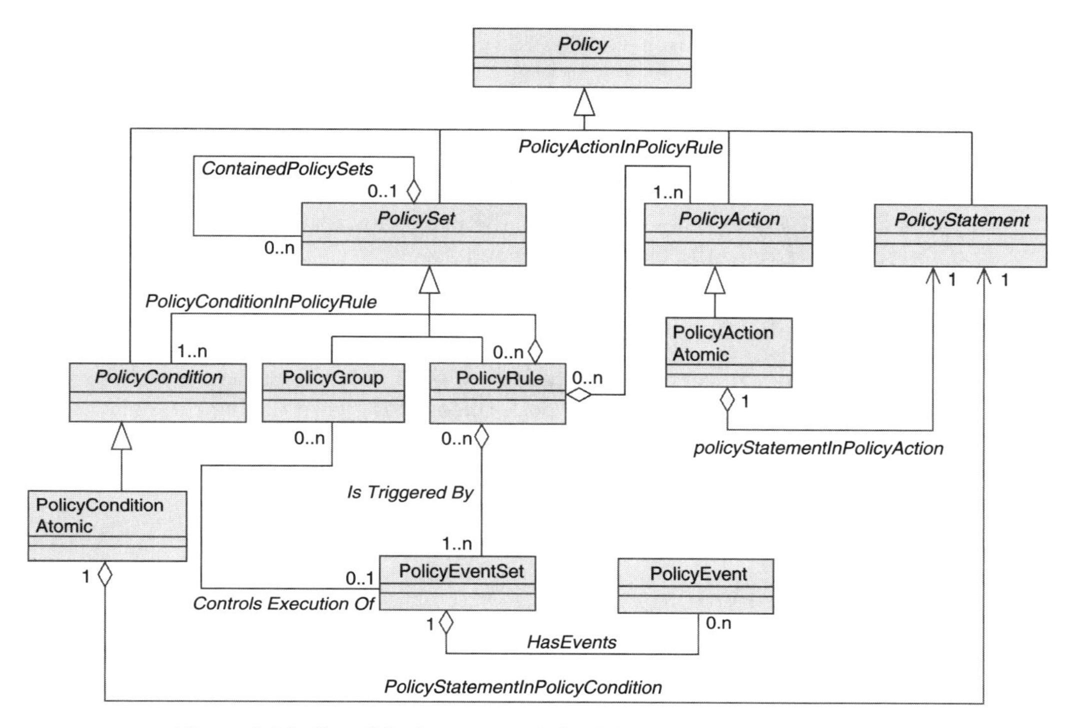

Figure 5-22 Simplified excerpt of the DEN-ng core policy model.

tions, respectively. Note that each of these associations has a cardinality of 0..n to 0..n, meaning that a PolicyConditionAtomic (or PolicyActionAtomic) can be made up of a PolicyStatement but does not *have* to be made up of a PolicyStatement. A PolicyStatement is made up of a variable, an operator, and a value. Because these are classes, the developer has inherently more power available in defining a condition or an action clause than in Ponder.

Finally, the *ControlsExecutionOf* association is used to distribute a set of events to a set of PolicyGroups. This is a very powerful mechanism and enables different managed entities to respond to the same set of changing conditions.

These policy definitions all include the definition of when an event (or set of events) triggers their evaluation. This is specified in the *trigger-conditions* part of the policy specification.

5.4.7 Constraints in DEN-ng

In DEN-ng, constraints are used to restrict the evaluation of the event, condition, and/or action clauses. This is a much richer use of constraints than how Ponder uses constraints, which limits the applicability of a Basic Policy by restricting the "when" portion of the clause.

DEN-ng constraints are always specified in OCL. This enables a standardized language and representation to be used in their definition. Not only does this enhance interoperability, it also enables different vendors to build their own constraint extensions and to be able to understand other vendors' constraint extensions.

One final difference between DEN-ng and Ponder constraints is Ponder's notions of time- and state-based constraints. In DEN-ng, time-based constraints are formally modeled as time conditions that can be included as part of a formal definition of a PolicyRule. State-based constraints are implied in the definition of a DEN-ng PolicyRule—the state is the projection of the event, condition, and action clauses onto the Finite State Machine.

5.4.8 Filtering Policies in DEN-ng

DEN-ng has no concept of policies used to filter data. The filtering and/or transformation of any result values is performed by the entity using the policy. This decision was made because such filtering actions are not only application-specific, they are session-specific. Because there is little chance of their reuse, it was felt that their inclusion only complicated the grammar.

5.5 Composite Policies

Large systems require abstraction mechanisms that enable policies to be defined such that they apply to a group of resources. This enables a particular set of policies to be applied to one or more sets of resources.

Ponder calls such policies "composite" policies, and DEN-ng keeps that name. Composite policies enable policies to be grouped together using any appropriate mechanism (e.g., roles, or ownership by an administrator or organizational unit) and applied to a set of ManagedObjects.

Ponder differentiates between groups, roles, relationships, and management structures. Groups define the scope in which contained policies apply. Roles are associated with domains that specify the common subject for all policies inside the domain. Relationships group two or more roles together. Finally, management structures group sets of role relationships to model hierarchical structures, such as organization hierarchies. Figure 5-23 shows the Ponder class hierarchy for defining composite policies.

DEN-ng takes a different approach in two accounts. First, several of the constructs here have fundamentally different meanings. For example, in Ponder, a MetaPolicy is used to "specify constraints, over a set of policies, on the permitted types of policies or their policy elements" and are specified using OCL. DEN-ng has no such construct—instead, it would model the policies as a group and then specify constraints on these policies using OCL. Second, DEN-ng reasons that each of these four types of grouping mechanisms expresses one aspect of collecting and organizing subjects and/or targets. The DEN-ng class hierarchy is therefore different than the Ponder class hierarchy and will be discussed in detail in Chapters 6 and 7.

Perhaps the largest difference is in fact the class hierarchy. Referring to Figure 5-23, it is unclear why the Group, Role, Relationship, and MStruct classes are subclasses of the CompositePolicy class. Here are the differences compared with the DEN-ng model:

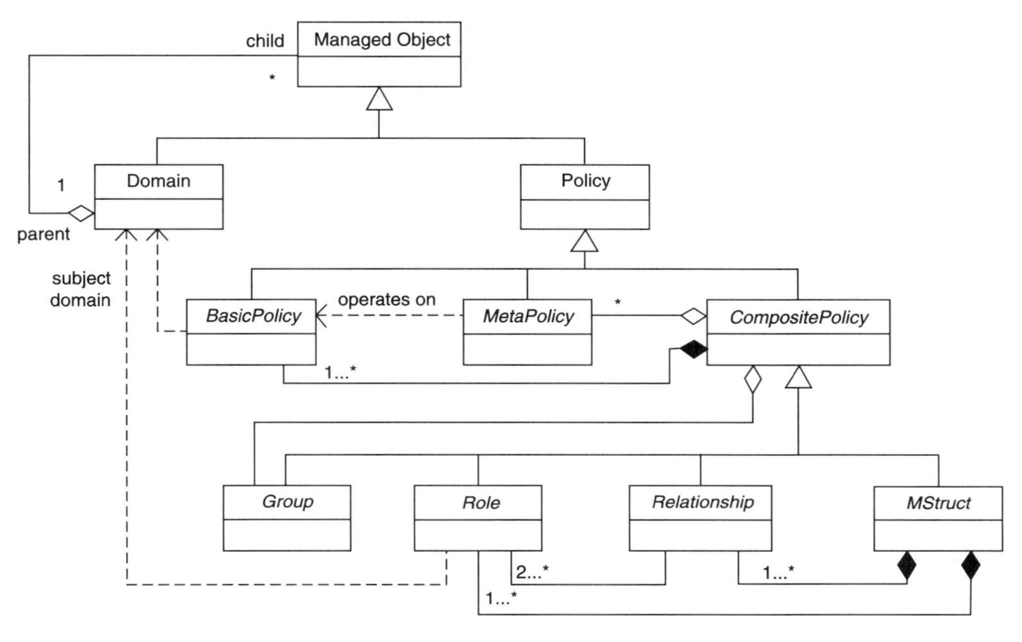

Figure 5-23 The Ponder composite policy class hierarchy.

- Groups are a general mechanism of collecting and organizing entities. While CompositePolicies are certainly groups, groups should be specified as an entity independent of a policy. Rather, a policy should be able to be associated to a group (if the policy is to act on members of the group) or aggregated by a group.
- Similarly, roles are not a subclass of a CompositePolicy—roles exist as a separate entity. If it is desired to apply a policy to a role, then this is an association relationship, not an inheritance relationship.
- Relationships are defined in the UML metamodel and are fundamental constructs that have nothing to do with CompositePolicies. In the Ponder specification, Relationships are defined as "relationships specify policies pertaining to the relationship rather than the individual participating roles." DEN-ng provides this functionality in a more elegant and extensible means through use of UML.
- The Ponder specification defines an MStruct as "a management structure [that] defines the configuration of roles and relationships in organisational units in terms of the required instances of the roles." This is provided in DEN-ng by first defining separate class hierarchies to represent organizational units, roles, and policies and then defining relationships that associate them. Policies, of course, can be applied to the entity and/or the relationship.

However, it is appropriate to discuss the four grouping mechanisms specified above and compare them to their DEN-ng counterparts. This will be done in the following four sections.

5.5.1 Groups

A Ponder group is used to define a syntactic scope for specifying a set of related policies to be instantiated at the same time. It is best thought of as a simple packaging construct to group related policies together. This grouping can either be for ease of administration (e.g., organizing the policies in a certain way to simplify their maintenance) or ease of application (e.g., for reusability, or to associate a specific set of policies that are going to be used with a specific set of managed entities in one unit). A group can contain zero or more basic policies, nested groups and/or meta-policies in any order. In Ponder, a meta-policy is used to specify constraints on the policies within the scope of the group; as we will see, DEN-ng has a different notion about what a meta-policy is and how it can be used. Figure 5-24 illustrates the Ponder group syntax.

```
inst group  groupName {
   { common-element-definition }
   { basic-policy-definition }
   { group-definition }
   { meta-policy-definition }
}
```

Figure 5-24 The structure of a Ponder group.

In DEN-ng, groups have a similar structure, but have additional semantics compared to those of a Ponder group. Part of these additional semantics are derived from the IETF policy model,[5-6] whereas others are extensions needed to implement the NGOSS concept of a contract,[8-9] and others arose from implementation experience. Most importantly DEN-ng uses a group to contain an object, not just a policy.

The additional semantics for DEN-ng PolicyGroups will be covered in more detail in Chapter 6, when the subject of containment is covered in more detail. However, there is an important implementation detail that concerns groups in general, and how they are implemented in data models. Briefly, the idea is that a third dimension to a group is defined—one that can be used to help the implementation of the group. The idea is to associate the semantics of the operations to be performed on objects that are contained in a group with the capabilities of the repository in which the group is stored. If this is not done, then different applications that use different repositories cannot interoperate. For example, consider two applications—one uses a directory, and the other uses a relational database. The latter has capabilities, such as triggers and stored procedures, which can be used to implement and enforce policy semantics. However, the directory does not have these (or similar) capabilities defined as part of the standards that specify how a directory operates. Therefore, unless these extra semantics are specified in the policy, they cannot be the same. In this particular example, the directory implementation will have to use some other type of software to provide these semantics. Thus, it is important that a group be able to specify where these semantics are defined. Figure 5-25 shows a simplified DEN-ng PolicyGroup.

In particular, note that a PolicyGroup can control the execution of policies that it contains through the receipt of one or more of a set of policies.

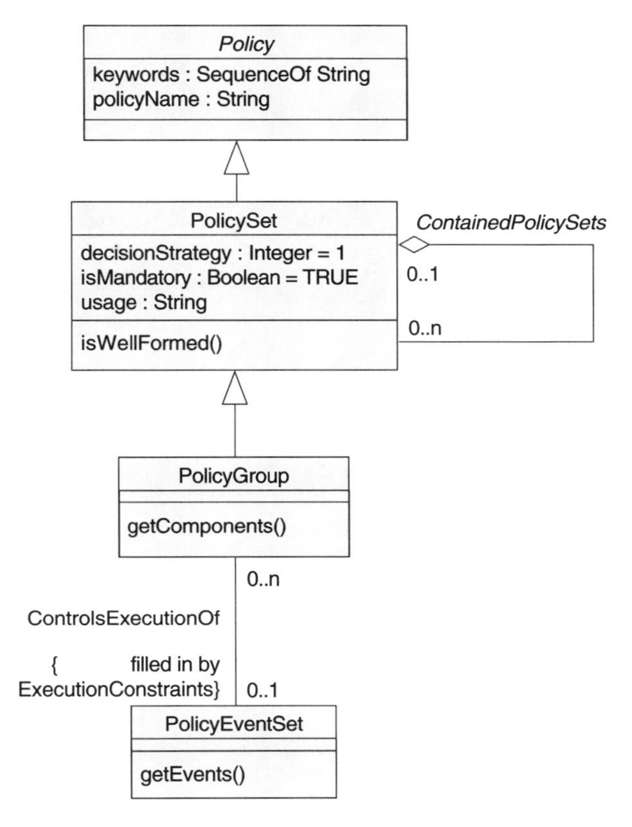

Figure 5-25 The structure of a DEN-ng PolicyGroup.

Furthermore, an OCL expression defines how the set of events affect the PolicyGroup, and is dynamically defined by the particular set of events and PolicyGroup. This leads to the definition of groups in Figure 5-26.

```
inst group groupName {
   access-control          constraint-expression;
   role-combinations       role-combination-list;
   common-elements         entity-reference-expression;
   basic-policy-list       basic-policy-list;
   group-list              group-list;
   metadata                metadata-expression;
   priority                priority-expression;
   decision-strategy       decision-strategy-expression;
   execution-strategy      execution-strategy-expression;
   constraint-list         constraint-expression;
   container-actions       action-expression-list;
}
```

Figure 5-26 The structure of a DEN-ng group.

The access control field is used to define special privileges that control what users can view, create, add, modify, and delete the rest of the fields in the above specification. This access control affects the other fields in the CompositePolicy and is independent of how any other access control policies (these simply defined which users could operate on the CompositePolicy) are defined.

Role-combinations (see Chapter 2 for the definition of role and related terms) enable the group to function as a "smart container." For example, an application can define a procedure that puts objects that have a particular role into a specific container that contains only objects of that role.

The next three fields (common-elements, basic-policy-list, and group-list) are identical to those defined in Ponder. However, Ponder's "meta-policy" field is replaced by the last six fields because, taken literally, a "meta-policy" is a "policy about a policy." There are a number of semantics that a policy can define about another policy—these six fields define the most important of them. Metadata represents a field that defines information needed to properly implement the desired semantics of the container. Priority enables different groups to be prioritized relative to each other. For example, if a selector chooses a set of groups and possibly individual PolicyRules, then an execution strategy (e.g., which policies will be evaluated in what order) must be chosen. Furthermore, a comparison method is needed to define such an execution order. In DEN-ng, PolicyRules and PolicyGroups can each have a Priority, which can be used to tell the system if it should be retrieved, searched, or if the policies contained in them should be executed and in what order. Decision-strategy defines how policy rules contained in this group will be evaluated (e.g., by executing the first PolicyRule that satisfies the conditions and then stopping, or by finding all matching PolicyRules and then executing their actions). Execution-strategy defines whether the presence of one or more failures halts execution of the rest of the actions or not. (Priority, decision-strategy, and execution-strategy are enhancements[6] and will all be discussed in detail in Chapter 7). The constraint-list is identical in function to the "meta-policy-list" defined in Ponder—it provides the ability to constrain how objects in a group interact with each other. Finally, the container-actions clause enables the system to attach one or more actions that should be executed when a specific operation is executed. This enables advanced system functions to be attached to groups.

Thus, we see that DEN-ng groups are geared specifically for policies and executing policy rules that are contained in a group, whereas Ponder groups are much more generic in nature. We also see that an effort has been made in DEN-ng to explicitly define as much of the semantics of executing a PolicyRule as possible. This was done to help and encourage interoperability.

5.5.2 Roles

In Ponder, roles are used to group a set of policies that have a common subject. Ponder is focused mostly on automating organizational hierarchies. This is done by associating a role with a position, enabling the assignment of a new person to the position without copying the policies associated with that position multiple times. Ponder also can be applied (at least in theory) to other managed entities, such as routers and router interfaces. The problem is that Ponder defines "just" the policies and does not define associations between the policies and the managed entities. Ponder treats a role as defining the set of authorization, obligation, refrain, and delegation policies that are applicable to a particular subject.

In DEN-ng, roles are much richer. Refer to Chapter 2 for the definition of a DEN-ng role and related terminology. Roles will be discussed in more detail throughout the rest of this book. For now, there are two important points to remember about DEN-ng roles:

- Both users, as well as resources, can have roles.
- There are three types of DEN-ng role objects—role attributes, role objects (implemented as hierarchies of role classes), and the use of the role object pattern.

Consequently, DEN-ng has a much richer use of roles than Ponder.

5.5.3 Role Relationships

In Ponder, a role relationship is used to specify the rights and duties of the related roles toward each other. This in turn specifies resources shared by each role, as well as the semantics that describe how each role relates to each other role. In other words, role relationships provide a means for defining policies that are part of the interaction between the roles.

Role relationships are not defined as such in DEN-ng. Instead, a two pronged approach is used.

The first is to define the concept of an interaction.[10] Here, an Interaction is defined to be an arrangement, contract, communication or joint activity between two entities. The second step is to define each entity as having a Role such as a PartyRole (Party is the DEN-ng and SID abstraction for an individual or an organization) or a ResourceRole.

The advantage of the DEN-ng approach is that each relationship is represented by objects that represent the role that each entity is playing as well as separate objects that represent their interaction. Each of these objects can have their own specific relationships with each other. Most importantly, each object and relationship has their own semantics that don't appear in the Ponder definitions.

5.5.4 Management Structures

Ponder defines a grouping concept called a Management Structure (MStruct). A management structure is a composite policy that contains the definitions of basic policies, groups, roles, role relationships, meta-policies, and other nested management structures into various groups that parallel organizational structures. Their purpose is to provide a mechanism for the application of policies to large groups of managed entities.

DEN-ng has no specific class that parallels a management structure. Rather, it models similar types of structures as part of the appropriate domain models. For example, organizations and groups of users are modeled in the Party Domain, whereas PolicyDomains and AutonomousSystems are modelled in the Policy and Network Domains, respectively. There is also a generalized Collection that can be used to build the Ponder equivalent of an MStruct, if desired.

5.5.5 Other High-Level Policies

The approach used in DEN-ng is to provide a library of generally useful policy "templates" to represent common application functions that desire to be controlled by policy. Interoperability cannot be fully achieved unless common representations of data and entities, as well as common semantics defining CRUD operations, are provided. For example, if two different applications authenticate a given user in different ways, the differences in the authentication mechanisms used by each application impose differences in the semantics of determining whether the user was really authenticated or not.

DEN-ng defines a set of common application functions, such as accounting and auditing, which can be called to perform common tasks required in a PBNM system. Each of these policies uses the generic template shown in Figure 5-14. This ensures that common semantics exist for all DEN-ng policies. However, the treatment of these policies is beyond the scope of this book.

5.6 Use of High-Level Policies in a PBNM System

The high-level policies described in this chapter are suitable for specifying control of managed entities from the business and system points-of-view. The link between the business and system points-of-view is arguably the most critical (though not the hardest to implement) of any of the links in the Policy Continuum. This is because the business world has become disconnected from the worlds of system and networking design and implementation. Unless a firm link between the business, system, and networking worlds is established, the network will always be a cost center and never be a profit center.

Conceptually, high-level policies are used as shown in Figure 5-27. This figure shows that the high-level policy rules presented in this chapter serve as a set of translation mechanisms that enable application-specific business rules and procedures to be expressed in a common format. This approach is taken because vendors are not going to scrap their current products just because a new standard comes along! They will only do so if customers demand it, which in turn requires a smooth transition mechanism between existing systems and DEN-ng systems.

The approach shown in Figure 5-27 enables different applications to use their own particular representations of business rules and procedures, as long as each application-specific format can be mediated to the common DEN-ng format. The mediated DEN-ng business format is directly compatible with the DEN-ng system format, which could then be mediated to application-specific system views of this information. This general pattern repeats for all five levels of the Policy Continuum.

5.7 High-Level Policy Example

To better understand the architecture in Figure 5-27, a simple example of how the architecture is used follows.

Suppose that our system offers VPN services to some of its customers. In order to establish if a user is a valid customer in good standing or not, two sep-

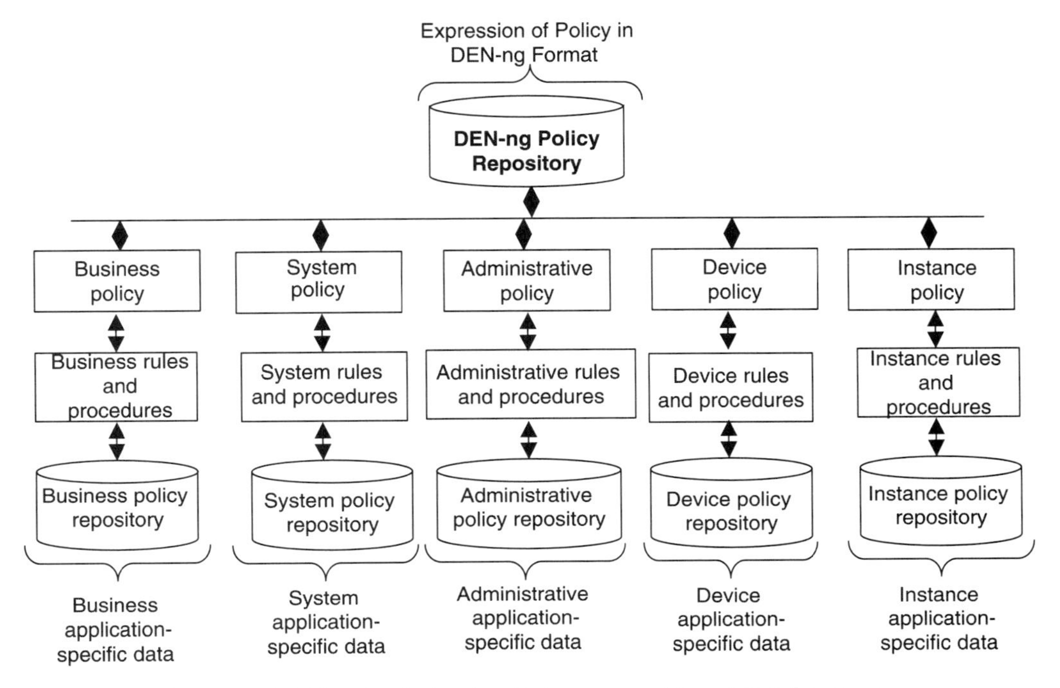

Figure 5-27 How high-level policies are used in a PBNM system.

arate applications are used—one for validating the login of a user to the system, and one for billing users of the system. These two systems are separate in order to enable different vendors to be able to supply these applications. This also better mirrors how OSSs are designed. Naturally, each vendor has used their own view of the data to design their own application-specific repositories. This is shown in Figure 5-28.

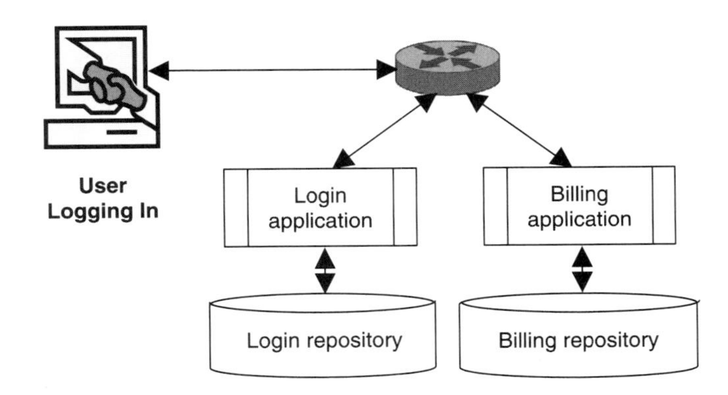

Figure 5-28 The integration of disparate applications into a PBNM system.

Assume that each application specifies its own business policy. Clearly, we need these to be coordinated. But how can they be if they are specified differently? For example, suppose that the business policy of the login application is defined as follows:

> *Users logging in through dial-up connections must use one-time tokens to authenticate (1)*
> *Users can only have a single login session at any one given time (2)*

Let's also suppose that the business policy of the billing application is defined as follows:

> *Customer traffic should start being billed as soon as a successful session has been established (3)*

Clearly, the two login rules, as well as the login and billing applications, depend on each other. Unfortunately, they do not directly refer to each other, and they use different terminology. The solution to this problem is to use the underlying information model to define these two policies as being related through their policy subjects.

5.7.1 Approach

The policy rules defined in this chapter can be used to transform each of the above rules to a common format where they can be associated with each other. The first step is to define a set of mappings from application-specific terms to common terms.

This involves equating different representations of the same entity (e.g., "user" in rule [1] and "customer" in rule [3]) to a common representation in a policy rule as well as in the appropriate information and data models.

In general, the mapping of business concepts between business applications will either use common definitions (e.g., a "session") and/or common concepts defined in an information model (e.g., an authentication method) to integrate different applications with each other. These common concepts (whether they come from a common information model or not) can be stored in the common DEN-ng Policy Repository in Figure 5-20. As we will see in Chapter 6, the Policy Repository stores many different types of information in addition to policy rules, conditions, and actions. Thus, complete policies can be easily assembled by retrieving information from a single location. It also enables efficient reuse of policy and other information stored in the Policy Repository.

5.7.2 Equating Policies in the Business View

We can start to transform policy rule (1) and (2) into a single DEN-ng Authentication Policy through substituting the values in Figure 5-29 into the generic DEN-ng authentication policy template.

The name of this policy is standardDialupPolicy. It is stored in the dialup container, which is in the authentication container, which is in the policies container, in the Policy Repository.

```
inst ( authen+ ) /policies/authentication/dialup standardDialupPolicy {
  version                1.2;// the version of this policy
  pre-conditions         session( user ) == NULL; // any necessary pre-conditions that must be
                         //satisfied before this policy can be evaluated
  trigger-conditions     event(logonRequest( dialupUser )); // event causing this policy to be evaluated
  subject-role-list      AuthenticationSystem; // the focus of the policy
  target-role-list       DialupAccessRouter; // the target of the policy
  privilege-list         < >; // existing permissions
  privilege-definition   2; // type of credentials is one-time token
  privilege-support      /scripts/support/otp-policy; // policy for CRUD operations on credentials
  activate when          event( logonRequest ) is received; // when this policy should go into effect
  deactivate when        event( logoutRequest ) | event( disconnectUser ) is received; // when this policy
                         // shouldbe removed from the system
  metadata               < >; // any additional semantics to be provided
  post-conditions        run standardDialupBillingPolicy if successful; // any necessary post-conditions
                         // that must be fulfilled if the policy succeeds or fails
  exceptions             dataNotFound, loginUnsuccessful, timeOut; // list of possible exceptions
}
```

Figure 5-29 An example of a business-level DEN-ng policy.

The version of this policy (1.2) indicates that the second minor revision to the first version of this policy is being used. Pre-conditions are any conditions that must pre-exist before the policy can be invoked. In this case, this policy is insisting that this user should not have an existing session. This corresponds to the business requirement that a dialup user can only have one active session at a time, which satisfies rule (2). The trigger conditions are how this policy is going to be evaluated. In this case, the trigger condition is the receipt of the login event itself.

The subject-role-list is the Policy Engine that is evaluating this policy rule. However, the target-role-list is the set of entities that will receive and implement this policy. In this basic stage of executing the policy, a single router (DialupAccessRouter) has been identified as the focus of implementing this policy. Note that this will change soon because if this policy successfully executes, then the target-role-list of the next policy can be different.

The three privilege commands define the user object as having no pre-existing privileges (meaning that they must be determined by this policy) and that a one-time-token policy should be used for the actual authentication process. Note that a script is used to support the use of the one-time-token.

The activate and deactivate commands define when this policy should go into effect and when it should be removed from the system. Care must be taken to ensure that a policy does not conflict with other policies when it is deployed into the system. Similarly, when a policy is removed, the Policy Server must check to ensure that entities that are using that policy will not be adversely affected. These factors will be covered in more detail in Chapters 7 and 8.

In this case, there are no other standard metadata available or that is applicable to this policy. However, there is a post-condition, which happens to be the link to invoking the billing policy. Finally, a list of exceptions is provided, so that the rest of the system can be aware if something doesn't evaluate correctly in this policy.

```
inst ( authen+ ) [/policies/billing/dialup] standardDialupBillingPolicy {
   version                  1.1; // the version of this policy
   pre-conditions           session( dialupUser ) == customer; // any necessary pre-conditions
   trigger-conditions       event( sessionInitiated ); // event causing this policy to be evaluated
   subject-role-list        BillingSystem; // the focus of the policy
   target-role-list         DialupAccessRouter; // the target of the policy
   privilege-list           < >; // not applicable
   privilege-definition     < >; // not applicable
   privilege-support        < >; // not applicable
   activate when            event( sessionInitiated ) is received;// when this policy should go into effect
   deactivate when          event( logoutRequest ) | event( disconnectUser ) is received;// when this policy
                            // should be removed from the system
   metadata                 < >; // any additional semantics to be provided
   post-conditions          < >; // any necessary post-conditions
   exceptions               dataNotFound, badCredit, timeOut; // list of possible exceptions
}
```

Figure 5-30 A second example of a business-level DEN-ng policy.

Policy rule (3) is similarly transformed (Figure 5-30). The two high-level business policies shown in Figures 5-22 and 5-23 have equated the terms "user" and "customer." In addition, the first step in implementing these policies has been taken, as additional structure and semantics are added to the policy rules defined in (1)—(3) above.

The next step is to represent our "dialupUser" entity in our information model. This is done in DEN-ng through representing the dialupUser as a subclass of the DEN-ng PartyRole object. This is shown in Figure 5-31.

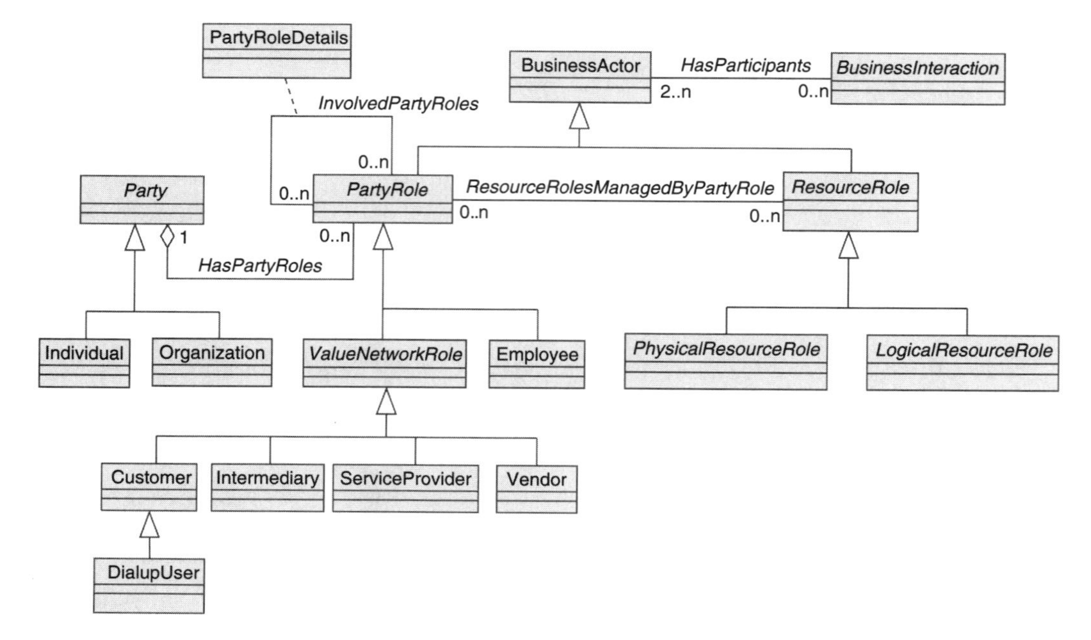

Figure 5-31 Excerpt of the DEN-ng party model.

Referring to the model in Figure 5-31, we see that a DialupUser can be modeled as a subclass of Customer. We can use this fact to link Customer to a number of important other entities through the BusinessInteraction class, as well as to other managed entities in our environment that come into play as we progress down the Policy Continuum.

5.7.3 Equating Policies to Other Levels of the Policy Continuum—An Approach

An exhaustive mapping of this seemingly simple business policy is beyond the scope of this book, as are mappings to the other levels of the Policy Continuum. Later chapters discuss parts of the DEN-ng models. In particular, Chapters 9 and 10 provide several examples of policies to manage different classes of service as well as policies to control QoS. These policies could be associated with the simple business policies that we have described in this chapter through the use of the PartyRole object.

The following discussion will help set the stage for these discussions by showing how we can associate different managed entities at varying levels of the Policy Continuum with each other through the DEN-ng information model.

One of the most important relationships to establish is which Services a given Customer can utilize. There are several ways to establish this relationship, but one of the easier ways is through the information model.

Reviewing our example, we have an MPLS VPN Service. We'll assume that this Service is actually part of a set of Services that the Customer has subscribed to. Figure 5-32 shows a simplified excerpt from the DEN-ng Service Model. This shows how an MPLS VPN Service is modeled.

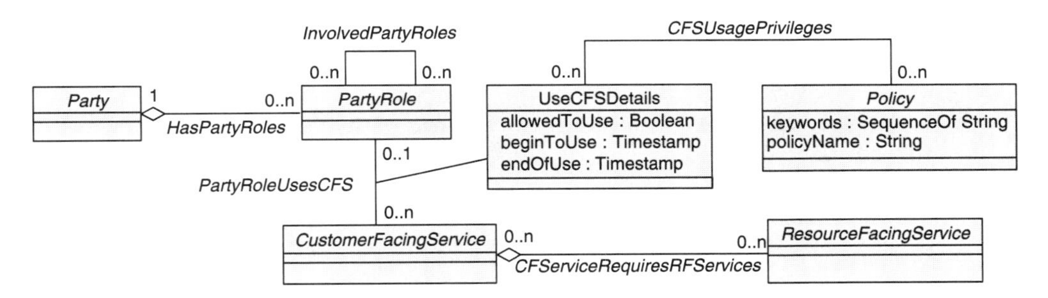

Figure 5-32 Relationship between customer and service.

This simple extract from the DEN-ng Service Model shows how Customer, which is a specialization of PartyRole, can use a particular CustomerFacingService, such as an MPLS VPN Service. A CustomerFacingService is an abstraction that defines the characteristics and behavior of a particular Service as seen by the Customer. The difference between it and a ResourceFacingService is that a CustomerFacingService is a Service that is part of a Product, which a Customer can purchase. A ResourceFacingService is not part of a Product. In fact, Customers shouldn't be aware of ResourceFacingServices at all.

DEN-ng uses the composite pattern[15] extensively. This pattern is a simple three object pattern and is shown in Figure 5-33.

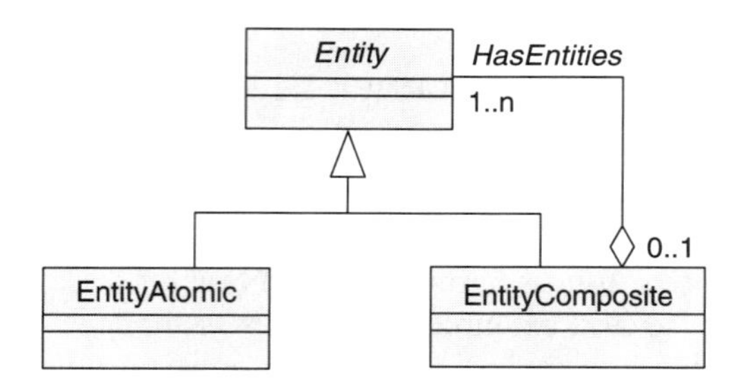

Figure 5-33 The entity composite pattern used in DEN-ng.

The DEN-ng variation of this pattern says that an Entity can be refined to be either an atomic (i.e., non-divisible) entity or as a set of entities. These are represented as EntityAtomic and EntityComposite, respectively. The aggregation *HasEntities* enables an EntityComposite to be made up of one or more Entities, meaning that an EntityComposite can be made up of one or more EntityAtomic and/or EntityComposite objects. This is a very powerful notion and is used in all of the eight DEN-ng domain models.

Both CustomerFacingService as well as ResourceFacingService uses the Composite Pattern. Because these are Service classes, the names of the atomic and composite subclasses have been changed to reflect their use and are called CustomerFacingServiceAtomic, CustomerFacingServiceComposite, Resource FacingServiceAtomic, and ResourceFacingServiceComposite, respectively. Figure 5-34 shows how an MPLS VPN is represented by subclassing both the CustomerFacingService, as well as the ResourceFacingService.

The *CFService RequiresRFServices* aggregation is used to define the set of ResourceFacingServices that are required for a particular CustomerFacing Service to operate correctly. This means, of course, that every subclass of CustomerFacingService inherits this relationship, which enables the appropriate ResourceFacingService subclasses to be associated with the CustomerFacing Service subclass that requires them.

A ServicePackage defines the concept of bundling a set of different Customer FacingServices, so that a Customer can purchase a single Product. However, this purchase will involve the specification of a set of services. This enables a single Product to be offered to the customer, even though in reality the Product consists of a set of different Services that must work together to provide the functionality that the Customer needs. In our case, the Customer will be able to choose one of four different ServicePackages—Platinum, Gold, Silver, or Bronze. Each of these ServicePackages contains one or more ServiceBundles, as indicated with the *ServicepackageSpecUsesServiceBundleSpecs* aggregation. A Service Bundle is the base class for defining the different classes of service that a cus-

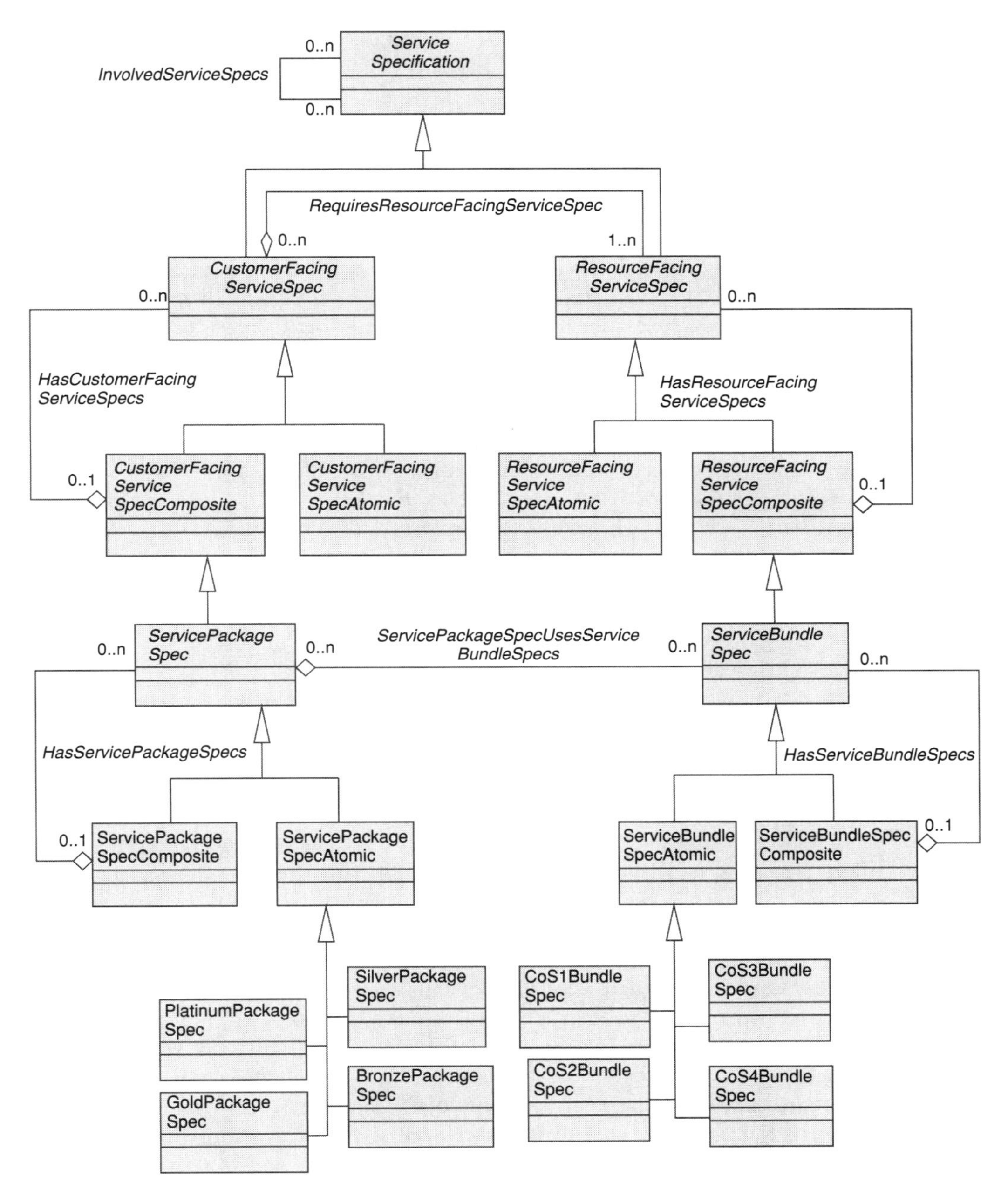

Figure 5-34 Relationship between CustomerFacingServices and ResourceFacing Services.

tomer can subscribe to. It is thought of as a "bundle" to enable the needs of different sets of services to be grouped together. The "bundle" conveys the concept of grouped services that are related, enabling a customer to upgrade to a higher-level Service (or downgrade to a lower-level Service) as desired.

The use of classes that contain the word "Spec" in them denotes classes that serve as templates. This ensures that offered Services will be defined and implemented in a consistent fashion. These classes will be described in more detail in Chapters 9 and 10. ServicePackages and ServiceBundles represent *instances* of their associated ServicePackageSpec and ServiceBundleSpec entities.

For example, suppose that the customer has chosen the SilverPackage. This particular package provides for better-than-best-effort delivery of mission-critical services, as well as best-effort delivery of data and web traffic. These two different class of service specifications are implemented using two different ServiceBundles (CoS3 and CoS4, respectively).

Now assume that the Customer upgrades to the GoldPackage. This package provides better delivery of mission-critical traffic, and so changes CoS3 to CoS2. However, the GoldPackage also enables the Customer to use VoIP, which requires CoS1. Each of these different ServiceBundles corresponds to a set of different network configuration commands. They are all, however, types of ResourceFacingServices, and the Customer is blissfully unaware of their existence. Clearly, none of the ResourceFacingService subclasses are services that a customer will buy directly. However, each of these is important to realize the MPLS VPN Service.

In our case, we want to link a Customer to a particular ServiceBundle which includes an MPLS VPN Service. This can be done because Service uses the composite pattern, which means that any Service (or subclass of Service) can be a part of another Service. MPLSVPNService and ServicePackage are both CustomerFacingServices, which means they both inherit this aggregation.

We also want to be able to use policy to control the services that the network offers. The issue of identifying a particular user from amongst a set of users is a complex one, and is highly dependent on the set of protocols and information available. This is covered briefly in Chapter 9. This is a general discussion that links a PartyRole to a Device, which has various features available (Figure 5-35).

Figure 5-35 shows that a Service depends on one or more PhysicalResources, as well as utilizes one or more LogicalResources. In DEN-ng, PhysicalResource is an abstract base class for describing different types of hardware that constitute a Product. It specifies the fundamental hardware characteristics that are required to implement a particular type of PhysicalResource. Its two subclasses represent the fundamentally different views of physical resources—as a complete, atomic device (e.g., the "InternetGatewayRouter") or as a set of hardware that make up a device (e.g., the cards, power supplies, etc. that are needed to make the "InternetGatewayRouter" operational).

We can continue linking managed objects of interest, like the software that runs on a device, to parts of this model (An example of this will be given when we look at Intelliden's product in Chapter 11). Similarly, we can continue to develop refinements of our policy language to better manage and control these and other managed objects. This work is quite detailed, and is currently

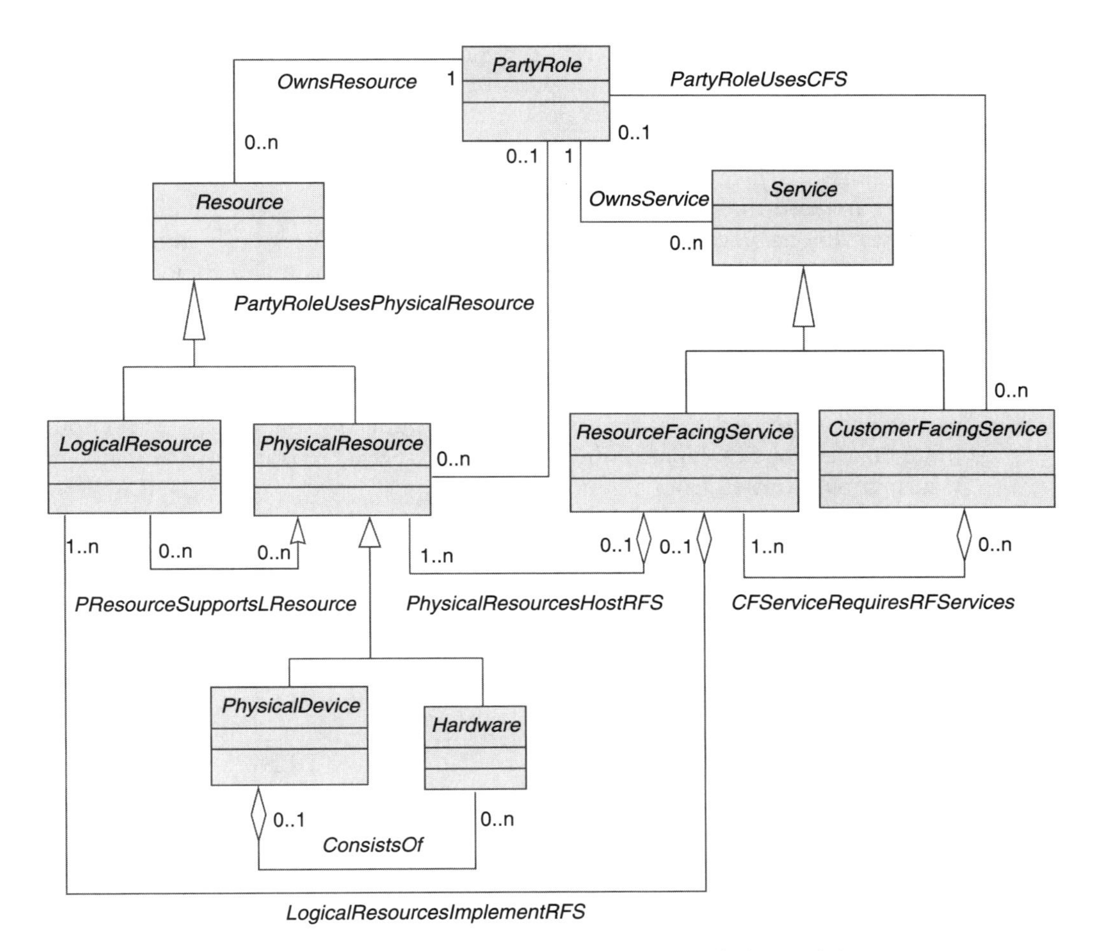

Figure 5-35 Relationship between customer and physical devices.

continuing in the TMF[11] in both the Red Team[12] and the SID teams.[13] The Red Team is responsible for overarching architectural issues regarding the TMF's NGOSS program, while the SID team is responsible for information and data modeling.

5.8 Summary

This chapter has presented a new high-level language, which is part of the DEN-ng effort, which is suitable for capturing business rules. This was based on the Ponder language. New extensions were made in the area of domains and roles. DEN-ng domains were made more restrictive to simplify containment and scoping. DEN-ng roles combine the power of role attributes, role-selectors, and role objects. This enables three different types of abstractions to be used to simplify system design, enabling devices or parts of devices to be managed via policy.

The four basic types of Ponder policies were defined so that the DEN-ng extensions to them could be better understood. Authorization policies define what activities a subject is permitted to do or not to do to a set of target objects. Refrain policies are used to define a set of actions that subjects are permitted to perform on target objects (regardless of whether they have been previously authorized to do so or not). Obligation policies define the set of actions that a subject must do when a particular event or set of events occurs. Delegation policies are used to permit subjects to grant a subset of privileges possessed to a set of grantees.

Three additional important Ponder concepts—Events, Constraints, and Filtering Policies—were then discussed. Events in Ponder are expressions that can be parameterized. Constraints are used to limit the applicability of the four basic Ponder policies. Finally, filtering policies are used to transform the value of information returned from actions.

Given this background, the DEN-ng enhancements were then discussed. The first enhancement was the definition and use of a common template that all types of DEN-ng policies could use. This template included the following concepts that are not addressed in Ponder: the ability to create or modify access control privileges, the notion of pre- and post-conditions (to help guarantee the safety of executing policies), the use of events to trigger any policy's execution (not just obligation policies, as in Ponder), and a definition of exceptions (which are missing in Ponder) that enables policies to be defined for any problem encountered. Finally, version control was also added to DEN-ng policies.

DEN-ng authentication policies were then described. They focus on using a rich definition of roles. It was emphasized that in DEN-ng, users do not have discretionary access to system objects; rather, access permissions are determined based on the role of the user and the role of the devices and/or services that the user is trying to access. To make this more robust, the concept of a session was introduced. The tradeoff is additional time to compute access permissions for increased security.

DEN-ng authentication policies (which are missing in Ponder) and authorization policies were then discussed. The difference between DEN-ng and Ponder authorization policies is mainly in additional semantics provided by DEN-ng. Two important examples of this are the specifying of access rights and making authorization a function of the current operating environment.

DEN-ng obligation policies are enhanced in three important ways compared to Ponder—event specifications can use a full Boolean syntax, action specification can include a finer-level of granularity, and the use of exceptions use is more powerful.

There is a marked difference in events, constraints, and filtering in DEN-ng compared to Ponder. Events are part of all DEN-ng policy rules. Constraints are used to restrict the evaluation of the event, condition, and/or action clauses of a policy rule, which is much richer than Ponder's usage (i.e., to limit the applicability of a policy rule by restricting the "when" portion of the condition clause). Finally, DEN-ng doesn't have any notion of a filtering policy—the entity executing the policy is responsible for filtering received information.

Both Ponder and DEN-ng define composite policies. However, in DEN-ng, emphasis is placed on using grouping constructs that may optionally have constraints specified to build composite policies. As will be seen in Chapter 6, the DEN-ng model has some rich grouping constructs and makes frequent use of the Composite pattern.

The final section of the chapter discussed how to use these high-level policies in a PBNM system. An example was given that traced some of the policies that are active in a policy-enabled MPLS VPN product. Two examples of mapping how business rules could be translated to DEN-ng business policies were given. It was shown that different business policies can be related to each other through the DEN-ng information model. Excerpts from various DEN-ng models then were able to associate a PartyRole to different types of business processes, Services, Products, and Resources. Thus, we saw that the combination of the Policy Continuum in conjunction with an innovative policy language and an information model were able to completely specify policies that can bind Customers to Services to Products to Devices to Software.

5.9 Recommended Further Reading and References

The following is a set of references for this chapter.

1. Damianou, N., Dulay, N., Lupu, E., Sloman, M., *PONDER: A Language for Specifying Security and Management Policies for Distributed Systems, The Language Specification, v2.3*, October 2000.
2. Sloman, M., Twidle, K., *Domains: A Framework for Structuring Management Policy*, Chapter 16 in Network and Distributed Systems Management 433-453, 1994.
3. *Information Technology—Open Systems Interconnection—The Directory: Models*, ITU-T Recommendation X.501, 1999.
4. Lupu, E. C., *A Role-Based Framework for Distributed Systems Management*, Ph.D. Thesis, Department of Computing, Imperial College, London, UK, July 1998.
5. Moore, B., Ellesson, E., Strassner, J., Westerinen, A., *Policy Core Information Model—Version 1 Specification*, RFC 3060, February 2001.
6. Moore, B., Rafalow, L., Ramberg, Y., Snir, Y., Westerinen, A., Chadha, R., Brunner, M., Cohen, R., Strassner, J., *Policy Core Information Model Extensions*, draft-ietf-policy-pcim-ext-06.txt, November 2001.
7. Baümer, D., Riehle, D., Siberski, W., Wulf, M., *The Role Object Pattern*, Download PDF from http://www.riehle.org/papers/1997/plop-1997-role-object.html
8. TMF, *NGOSS Architecture Technology Neutral Specification*, TMF053, version 3.0, April 2003.
9. TMF, *NGOSS Architecture Technology Neutral Specification (Contract Specification)*, TMF053b, version 3.0, February 2003.
10. TMF, *Shared Information/Data (SID) Model—Addendum 1I—Common Business Entity Definitions—Interaction*, GB922, version 3.0, June 2003.
11. TMF web site is *www.tmforum.org*
12. Strassner, J., editor, *NGOSS Architecture Technology Neutral Specification (Behavior and Control Specification)*, TMF053c, version 1.0, February 2003.

13. Strassner, J., editor, *Shared Information/Data (SID) Model—Addendum 1POL—Common Business Entity Definitions—Policy)*, GB922, version 1.0, July 2003.
14. OMG, *Unified Modeling Language Specification,* version 1.4, September 2001.
15. Gamma, E., Helm, R., Vlissides, J., *Design Patterns-Elements of Reusable Object-Oriented Software,* Addison-Wesley, November 2000.

Motivation for the DEN-ng Policy Model

This chapter will first describe the motivation for creating a new policy model. Then it will introduce a central theme of this book—how the business, system, and networking worlds can be related.

6.1 Introduction

This chapter defines the motivation for why a new policy model is being built. Particular attention will be paid on understanding the functional requirements that need to be solved in order to better manage current and next generation networks, as explained in the earlier chapters of this book. Of these, the ability to connect the business and network viewpoints is arguably one of the most important goals of this new policy model. This forms the use case from which this new policy model is built.

The new DEN-ng policy model will be defined in Chapter 7.

6.2 Motivation

There are currently three main policy models in the industry: (1) the DEN model,[1] (2) the IETF model,[2-5] and (3) the DMTF model.[6] Of these, DEN was used as the basis for both the IETF and the DMTF models. Recently, the IETF and DMTF models have diverged, mainly because the DMTF model has additional infrastructure that is not present in the IETE model.

Numerous additions have been proposed to these models, all of which are in various states of completion (see, for example, the conferences in the reference section at the end of this chapter, each of which has numerous papers that extend one or more aspects to these models, and reference 12). There are other efforts, such as Ponder,[13] that imply additions and/or changes to one or more of these models.

6.2.1 Why Build Another Information Model?

The question, therefore, is why should yet another model be built? The driving purpose is that the use cases have changed. The original DEN model was a bit ahead of its time—a model for extensibly representing policy was proposed without accepted architectures and use cases in place. In defense of the original DEN model, this was done mainly because corporate marketing engines had already started preaching policy, and people felt that it was necessary to formally define what one could and could not do with policy. The IETF and DMTF efforts took the original DEN model and refined it, so that its original design (which was admittedly focused on policy for network devices) could be extended to policy for different types of systems. As we will see, the DEN-ng model incorporates ideas from the original DEN and IETF models, but diverges significantly from the DMTF CIM model in its representation of resources and services. Thus, we can represent the conceptual derivation of these models as shown in Figure 6-1 below.

These previous models focus on policy as a domain that is isolated from the rest of the managed environment. Policy is therefore somewhat "detached" from the other domains; as a result, only weak links exist between managed objects in the policy domain and managed objects in other domains (e.g., services and resources). In contrast, the main use case for the DEN-ng policy model was to define a policy model that was closely integrated with the rest of the managed environment. This would enable it to be used to help define the architecture of

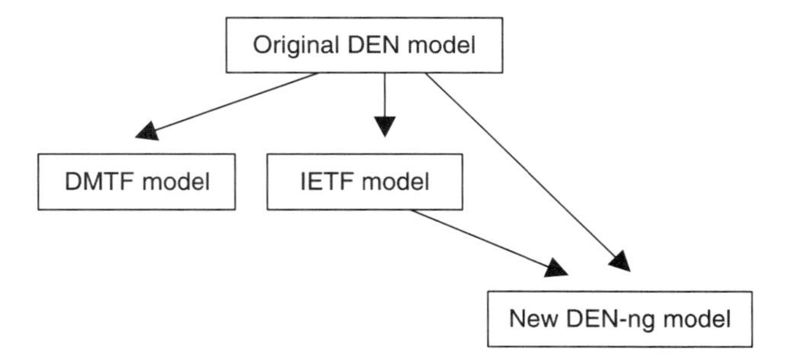

Figure 6-1 Relationship Between the DEN, DMTF, IETF, and new DEN-ng policy models.

the system. In other words, it wasn't enough to define an information model that could represent policy—what was needed was an information model that defined how policy interacted with the rest of the managed environment. This had three important consequences.

Build Closer Relationships Between Policy and Other Domains

First, it was apparent that building a policy information model in isolation of other information models wasn't going to work. The original DEN specification, as well as CIM, had many different domains in addition to policy. Each domain had its own model(s). However, there was little effort made in associating the policy model with these other domain models. In addition, the original DEN and CIM models didn't specify in enough detail how policy could be applied to a managed object. The DEN-ng model took a different approach. It built out the policy model as one of the last domain models, and then concentrated on associating appropriate parts of the policy model with appropriate parts of other domain models. This means that there is an inherent relationship—almost a *binding*—between the DEN-ng policy model and other DEN-ng domain models. **This is critical to show how a policy can be implemented.**

Define How Policy Affects Other Managed Entities

Second, the existing models concentrated on representing policy. They either did not address, or addressed in a very superficial manner, how policy affected other managed entities. The difference here is subtle but important. Current policy models concentrate on defining the structure of a policy rule, what its condition terms are, and so forth. Although there was a lot of talk about policy changing a value in a device configuration file, the details of *how* that was accomplished were left unspecified. For example, the IETF and DMTF models do not specify the concept of a device interface or physical port. If the device uses CLI to change its configuration, how then can policy be used if these fundamental concepts are not modeled? The DEN-ng policy model fixes this unfortunate situation by developing other domain models alongside the policy model and ensuring that appropriate elements in the policy model can be associated with appropriate elements in other models. DEN-ng has as its goal the ability to translate policy expressions *directly* to running code, something that cannot be done with existing models.

Define Static and Dynamic Characteristics of Policy

Third, the original models (and almost all current additions to those models) are still thinking of policy in a very static way. That is, they use policies to express the static configuration of target devices. Most concentrate solely on the network layer, and do not provide an information model for representing business entities and how they affect target devices. In fact, there is very little literature on detailed information models that are designed with business views in mind, and even less literature describing how business information models can be linked to information models of other domains. For example, how does a changed SLA affect device configuration files? Clearly, the SLA defines how traffic should be treated, but when it is changed, the policy of treating that traffic is changed. How is that accomplished? Or how does a customer, who has just bought a new product with a higher class of service, get that service installed and running? These are two simple examples of linking the business world, with its set of entities and concepts, to the system and networking worlds, which have different expressions for those concepts. Policy is required, but unless the representations are equated, the business, system, and networking domains will always remain disconnected. This adversely affects service activation and deployment.

6.2.2 The Role of Standards

The IETF concentrates on protocol development and, to a lesser extent, APIs and architectures. One might wonder why policy is in the IETF. The IETF realized that policy was essential to making it easier to configure and manage devices. However, it doesn't mesh well with the existing focus of the IETF, and work in the policy framework working group is ending.

The mission of the DMTF is to "lead the development of management standards for distributed desktop, network, enterprise and Internet environments." The mission of the TeleManagement Forum (TMF) is to provide a "program of member-driven collaborative activities that help identify, create, develop, and implement real world solutions that automate and streamline telecom operations."[12] Thus, whereas both the DMTF and the TMF are concerned with standards for information and data models, the DMTF seeks to standardize how entities are managed, and the TMF focuses on the automation of telecom operations, which includes the standardization of how entities are managed.

The Difference Between Information Model Standards and Other Standards

There is a fundamental difference between a standard describing an information model and a standard describing another concept, such as a protocol or an interface. An information model, as its name implies, is all about representing information in a managed environment. Thus, an information model seeks to describe the behavior and characteristics of all managed entities that are relevant in that environment. Other standards concentrate on describing the behavior and characteristics of "just" entities that are relevant in their domain.

For example, a standard written about the operation of a protocol does not usually include a model of that protocol. Instead, descriptions of protocol messages and other parts of its operation are given. Even if a model of the protocol is provided, models of other managed entities that run that protocol, such as network devices and users, are not provided.

In contrast, an information model will in principle contain the same information describing the operation of the protocol along with how different entities use and interact with that protocol.

The Granularity of Information Defined in an Information Model

This does *not* mean that an information model should represent every single aspect of every single system in minute detail. However, it does mean that the models have to be inherently extensible, so that different vendor-specific implementations can be represented on a common framework.

For example, consider the configuration of the Border Gateway Protocol (BGP). This is a widely deployed protocol supported by many different router vendors. The IETF has produced at the time of this writing over 25 Request for Comments (RFCs)—the first stage of a standard) on or involving this protocol. Yet, interoperability between different vendors is still very difficult:

- Each vendor strives to differentiate its products from its competitors by offering different features. If these have different side effects and different resource requirements, how can they be used together?
- Although each vendor supports a subset of the standard BGP features, the programming model each vendor uses is different. This means that a network engineer must learn a different configuration language for each vendor used in the network. How many foreign languages can you speak? More network vendors than that likely supply equipment in your network.
- Syntax and semantic differences between different network operating system versions must be learned.

The above example is not specific to BGP; it is applicable to any standards-based protocol, and applies to enterprises as well as service providers. Furthermore, network vendors will always operate this way—they will implement a part of the standard and then provide value-added features to differentiate their products. Standards, such as protocol RFCs, will therefore never be able to address this problem.

However, this doesn't mean that standards are useless. Protocol RFCs can never achieve interoperability because the standard cannot be written to a sufficient level of detail to define implementation, and because vendors do need the freedom to innovate. Instead, what is needed is to *standardize the concepts that enable different implementations to interoperate*. Although this cannot be provided in a protocol RFC, it can in an information model that represents how that protocol operates and interacts with other managed entities.

6.2.3 Scope of Currently Available Information Models

Both the IETF and the DMTF approaches do not attempt to represent business entities and objectives. Although a few other approaches do,[16] only the Shared Information and Data (SID) model being built in the TMF has addressed building a set of models designed to support business, system, implementation, and runtime views that are closely tied to an overall architecture. This architecture is called New Generation Operations Systems and Software (NGOSS). The NGOSS architecture is defined in reference 17 and contains some very important concepts.

One of the main goals of the DEN-ng policy model was to meet the needs of an accompanying architecture. The architecture and the information model together provide the required semantics to share and reuse information between different architectural components. This ensures that systems can be built from the information model, and that they will behave as described in the information model. The DEN-ng policy model accomplishes this goal in two fashions. First, it uses the different layers of the policy continuum to define different abstractions that must be modeled. This enables the different users of policy to work with and express concepts in their own terminology, rather than having the terminology and concepts of a static model given to them, never to be changed. Second, the DEN-ng policy model was developed by starting with defining business concepts, and using those concepts to help define the system aspects of the DEN-ng information model. Once these business and system views were defined, an implementation view was built that focused on ease of implementation.

This development cycle closely matches the development cycle espoused by NGOSS, and is shown in Figure 6-2 below.[18]

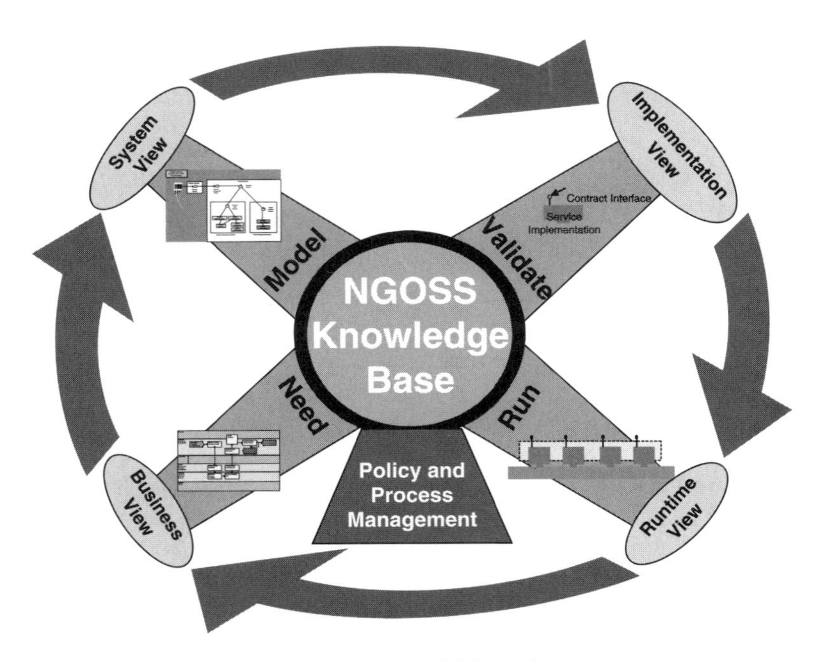

Figure 6-2 The TMF NGOSS architecture.

Figure 6-2 shows a conceptual view of the NGOSS architecture. The NGOSS knowledge base is a collection of information and data models, specifications, contracts, code, and supporting documentation that collectively and cooperatively describe how to build an NGOSS system. The four quadrants represent the business, system, implementation, and runtime views. The behavior of the system is driven by policy and process management functions.

The TMF has developed a set of principles and procedures to coordinate each of these four quadrants. This takes the form of the various architectural specifications (e.g., TMF053 series), the eTOM and SID documents, the contract work, and other elements that form the NGOSS knowledge base. The DEN-ng policy model was built to fit into this approach. Any comprehensive architectural approach requires a set of documents that each focus on one or more specific details of the overall architecture. This describes the TMF053 series: the overall approach is explained in the main document, and its annexes focus on one or more aspects of that architecture. This is the primary reason why DEN-ng was targeted at NGOSS—it was the most comprehensive architecture to date and presented an opportunity for an information model to be the basis of many different facets of that architecture, not just quality of service (QoS) or security.

A key objective of the NGOSS methodology is the development of models that focus on particular characteristics and procedures within an NGOSS system. These are characterized by the four viewpoints shown in Figure 6-2. The viewpoints are in turn tied together through the use of common shared information and a common means to exchange that information: contracts. Put another way, the purpose of the SID is to *unite* the business and system views and to produce a single model that can be implemented more easily. A by-product of this approach is that the combination of the SID and contracts promote interoperability.

The DEN-ng policy model was developed using an iterative top-down, bottom-up approach. Business concerns were first considered, which provided a high-level structure for and helped define key concepts of the overall policy information model. This model was then augmented by considering concerns necessary to actually apply it in real life. Once these business and system views were defined, an implementation view was built that focused on ease of implementation (Figure 6-3).

It should be noted that this approach to building a policy information model is significantly different than previous approaches.

There were other reasons to develop a new information model which are detailed in the following subsections.

UML Compliance

The IETF and DMTF models are not UML compliant[19] and are therefore not based on a universally accepted standard (if you are a TMF member, please see reference 20 for more details). UML is standardized by the object management group. The IETF resisted adopting UML for several reasons: UML is a large and complex specification; the contents of the UML specification are largely outside the interests and expertise of the IETF; and the IETF does its work using ASCII, making it impossible for any reasonably complex UML diagram to be depicted.

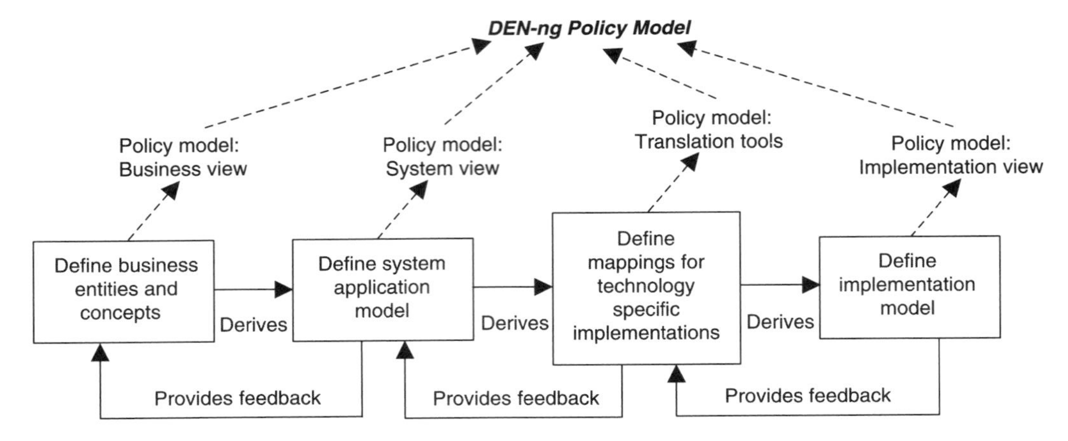

Figure 6-3 Design approach for building the DEN-ng policy model.

The DMTF uses its own "simplification" of UML. Unfortunately, in so doing, it completely redefines the UML meta-model. This makes the DMTF CIM incompatible with any product derived from the UML meta-model. For example, the DMTF mandates that all associations are to be built as classes. This can be seen from looking at the DMTF "meta-schema" (Figure 6-4) which is the DMTF's replacement for the UML meta-model.

Referring to Figure 6-4, note that an association is defined as a subclass of a class. This means that an association *must* be implemented as a class.

In the UML meta-model, there is a fundamental difference between an Association and an AssociationClass. A small excerpt of the UML meta-model is shown in Figure 6-5.

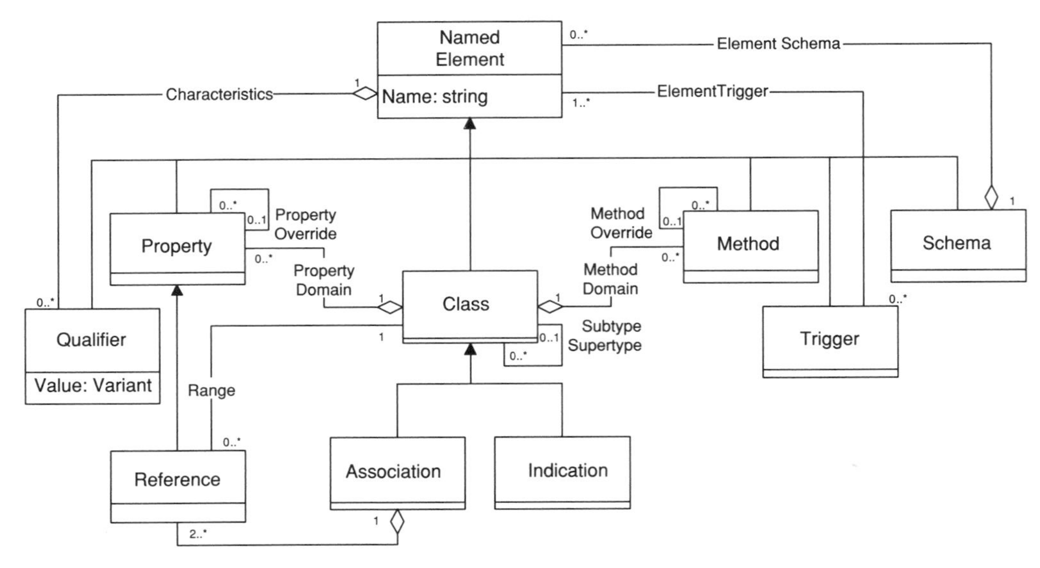

Figure 6-4 The DMTF "meta-schema."

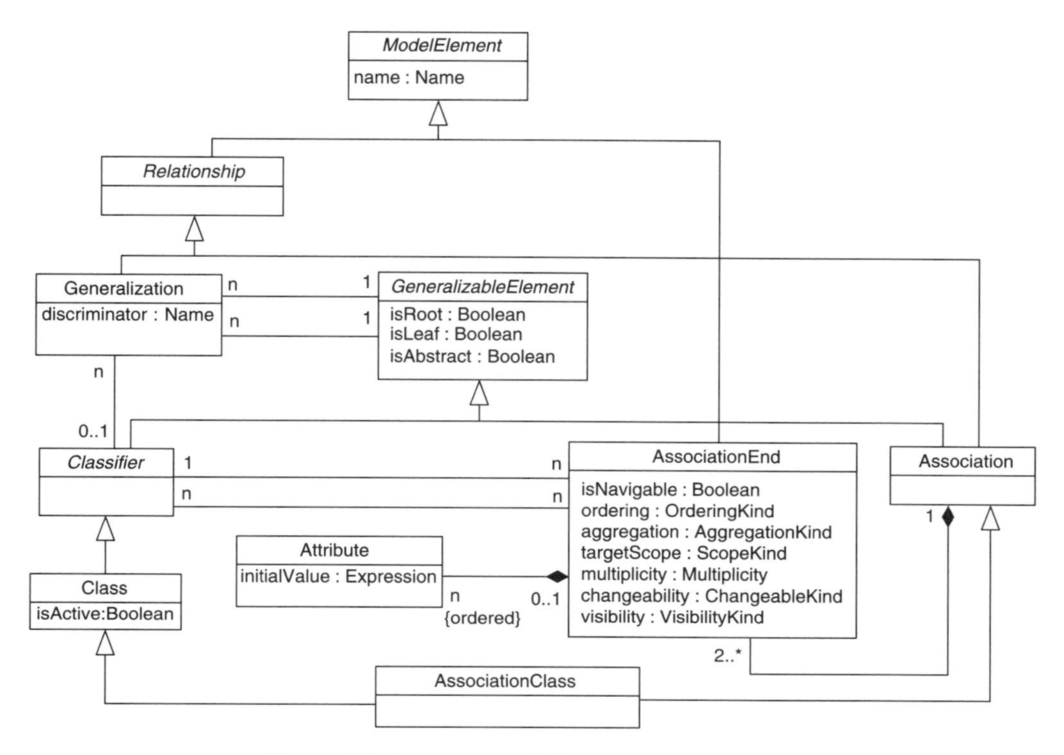

Figure 6-5 An excerpt of the UML meta-model.

The most important differences between associations in the CIM meta-schema versus the UML meta-model are summarized as follows:

- The UML meta-model defines *two* different meta-objects: an Association and an AssociationClass. Therefore, not all associations are classes. In contrast, the CIM "meta-schema" defines all associations as classes.
- The UML meta-model defines a class called an AssociationEnd, which provides special semantics for Associations as well as AssociationClasses. The CIM does not have this construct.
- In the UML meta-model, an Association *does not* inherit from a classifier, whereas an AssociationClass *does* inherit from a classifer. This means that the semantics between an Association and an AssociationClass are very different.

The result is that the *semantics* of a "CIM association" are very different from either a true "UML Association" or even a "UML AssociationClass." The ramifications of this are significant: interoperability is comprised at a very fundamental level, because an association in the CIM is not the same as an association in either the UML meta-model or the DEN-ng information model. Thus, although the models may have the same or similar relationships represented, the semantics of these relationships are different.

There are many other areas where CIM is not compliant with the UML meta-model and/or UML specification; however, this is beyond the scope of this book.

DEN-ng is derived directly from the UML meta-model and hence fits in with related efforts such as the OMG's Model Driven Architecture[22] work. This is not to say that the DEN-ng policy model will ignore other models. In fact, the opposite is true—the DEN-ng policy model is in fact a federation of multiple ideas from multiple models, all integrated using a novel framework. However, it was important to be able to produce a "better" policy model that was also UML compliant.

Lack of Specificity in Representing Resource Entities

Most previous information models simply lacked detailed representation of important managed entities. Here are two quick examples.

The ITU's M.3100[23] specification is widely accepted. Quoting in part from its abstract:

This Recommendation provides a generic network information model. The model describes managed object classes and their properties that are generic and useful to describe information exchanged across all interfaces defined in M.3010 TMN architecture.

Yet, simple concepts like a physical port and a logical device interface are not represented in this model.

As another example, the DMTF CIM also doesn't have the concept of a logical device interface. Although it has a concept of a port, it is lacking some critical attributes (such as supporting the naming of the port using the vendor's terminology) that make it impossible to use the CIM to build device configurations.

The M.3100 model and the CIM do contain some interesting information. However, their lack of architectural focus shows up in the contents of the model. Since one of the major applications of policy is to control the configuration of individual network device commands, this means that neither of these models are suitable for use for representing how policy can control network device configurations.

As another example, policies that are supposed to be able to configure mechanisms that affect traffic flowing into or out of a device interface simply cannot be adequately expressed in the CIM. This is somewhat odd, since the CIM went to the trouble of working on a QoS model that represented internal device QoS mechanisms.

In contrast, the DEN-ng policy model was built to emphasize the dependence and association between policy entities and other managed entities. With respect to QoS, the DEN-ng PhysicalResource, LogicalResource, and Service models all were built to show how policies could be used to support different use cases, including controlling the changing of a network device's configuration.

The close relationship between the DEN-ng policy model and other parts of the DEN-ng model will be shown at a high level throughout this book.

Using the Policy Continuum

The policy continuum is used to ensure that different users and applications that are using policy can communicate with each other. If policy is going to be used at an enterprise level (throughout a system), it is mandatory that different users and applications either have the same view of policy or can easily translate what they mean by policy to another user operating at another level of abstraction. This also enables different types of users having different roles and responsibilities to use different expressions of policy—expressions that are more closely suited to their needs and application.

The policy continuum is only used by the DEN-ng and SID models.

Integration of Different Information Models

Another fact that has been impeding the use of information models in the industry in general is that most of the other efforts don't seem to combine policy with other information models. For example, the CIM took no initiative to talk about different types of policies, such as those that affect human versus device operation, and how they interact with people, devices, and other types of managed entities.

This was a particular focus of the DEN-ng model and was facilitated by its use of roles and patterns. As the following chapters will shows, roles and patterns are common constructs that help discover abstract information and hidden relationships. DEN-ng uses roles and patterns to help "connect" its different domain models.

Incorporating State and RunTime Information

Another important use case was to be able to use the policy model to incorporate environmental and other runtime characteristics. Most current policy information models represent the current state of a managed object; they have little if any facilities to represent how that state changes over time. Examples of this are the IETF and DMTF models, along with most of the security models proposed.[24,25]

In contrast, DEN-ng recommends the use of a finite state machine to represent the life cycle of a managed object. This enables policy to be used to adjust network devices in order to meet the changing requirements and business objectives of the system in which it is being used. This requires the use of a finite state machine (or a similar artifact) to tie together the different state representations of managed objects. This enables policy to take the important role of *controlling which state transitions to implement when*.

Integrating the Business, System, and Implementation Worlds

Chapter 1 stated that the " . . . the true breakthrough was when PBNM was defined as a means for business rules to drive the configuration of the network. This brought forth the promise of changing the network from a cost center to a profit center." A good example of this line of thinking is provided in Strassner (2002).[26] This paper states that "business driven device management is a new paradigm that enables business rules to be used to manage the construction and deployment of network configurations." The DEN-ng model is used to implement this approach.

Approaches such as this greatly emphasize the importance of policy. Without policy, the network will never be profitable. Unfortunately, most information models either concentrate completely on the expression of business concepts, or on the expression of system concepts. Furthermore, most do not carry either of these forward to the implementation stage, to ensure that the models can be easily and efficiently implemented.

DEN-ng (and the SID) are both focused on this effort, as shown in Figure 6-1.

Summary: Reasons to Build a New Information Model

These, then, are the main reasons for building a new policy model:

- UML compliance
- Lack of specificity in representing the characteristics and behavior of resource-managed entities
- Providing multiple views of policy with the policy continuum, so that different types of users can use it equally effectively
- Associating the policy domain with other domains so that different aspects of managed objects could be managed and controlled using different types of policies
- Using a finite state machine to represent different states of managed objects, and policy to control when, where, and how a state transition was accomplished
- Providing policy to model business, system, implementation, and runtime aspects of managed objects

6.3 The New DEN-ng Approach—Functional Requirements

This section will define the set of important functional requirements for the DEN-ng approach in general and specifically how they relate to the DEN-ng policy model.

6.3.1 Using Standards

One of the best examples of how standards can be used to the good of everyone is the Internet. The Internet has always operated under a philosophy of openness, which is the primary reason it has risen to such a level of importance. The Internet is based on open, non-proprietary technology that everyone can share. This enables equipment from different vendors to interoperate.

Even if all of the equipment in a network comes from a single vendor, standards are still needed if devices are to interoperate. This is because of the MAD world of Mergers, Acquisitions and Divestitures. Many networking companies have been built using a philosophy of acquisition. When one networking company buys another networking company, the devices from the company being bought likely do not use the same operating system, or implement the same functionality, as the devices from the buying company. Thus, the customer can get many different devices that operate in fundamentally different ways from each other. Without standards, the customer is forced into "swivel-chair management," relying on different management programs using different management methods to manage devices from the same vendor.

Industry standards enable products from different vendors to interoperate. This allows customers to choose the products that best suit their needs, while being assured that these products will operate properly in their networks.

Unfortunately, protocol standards are treated very differently than information modeling standards. Most networking companies do not use an object-oriented information model to design their products, and those that do usually do not publish it. Although this is unfortunate, it is not the end of the world! Management software can in effect produce a façade to hide this and instead present a standards-based interface to program or report on the device. Thus, it is not necessary for network vendors to adopt information modeling standards like DEN-ng; rather, it is necessary for management software vendors to start using the same standard.

Although it may seem that conformance to such information modeling standards would hurt the business of management software vendors, the opposite is true. If all management products adhered to industry standards, each management product would be able to communicate to other management products, forcing differentiation to be achieved in other areas. As a result, the customer (e.g., the service provider or enterprise that is building and running the network) would have a much wider variety of high-quality products to choose from when managing the network. If vendors did adhere to such information modeling standards, their products would be more interoperable. End users would benefit through increased plug-n-play, and management vendors would have an easier time incorporating that vendor's products into its offering. Ultimately, however, the customer will be the main driver for determining whether network vendors adopt such standards.

This is why the catalyst programs of the TMF are so important. This is a perfect way for vendors to get used to the power of the SID and DEN-ng models, and hopefully will create demand from service providers and enterprises for more vendors to implement a standards-based information model.

6.3.2 Managing Devices from Different Vendors

Network vendors provide added value through hardware and software. People building networks want the freedom to choose best-of-breed products that are best suited to accomplishing the task needed. Fundamentally, networks are all about connecting people and resources together.

PBNM solutions have not been realized to their fullest because most PBNM solutions are built to manage a particular set of functionality from a single vendor. Sometimes, multiple vendors are supported, but then support is almost always limited to the same functionality from these vendors (e.g., the ability to build a VPN using different devices from a small set of vendors).

The problem with this approach is that as the business grows, the services that need to be supported by the network also grow. This means that the organization will have to purchase additional management products to manage these new services. However, what happens if a customer purchases old and new services? Then the different management products that are being used to control these services must each know that this service is being used by the same customer. For example, each management product must be careful to avoid changing the configuration of the device that both management products are using in a noncompatible way. Without a common information model (which includes a common policy model), how can this be accomplished?

6.3.3 Abstracting Different Device Functionality

Network devices have become very specialized. Although large, general-purpose routers (as an example) exist, almost every major network vendor now makes many different types of routers for very different purposes. For example, Cisco[27] currently defines 19 different types of router product families. Each product family has one or more models, and every model can run a large number of different versions of IOS and different networking and other types of Cards.

Furthermore, different versions of IOS have different features available. If the network consists of different devices that are running other operating systems, these will have different features available as well. Therefore, a common set of abstractions is needed that enables different features to be equated to each other. This is necessary for two reasons. First, common behavior across different devices can only be achieved by building a mapping between these different features, so that the same relative traffic conditioning can be supplied by heterogeneous devices. Second, if the same policy is to be used to manage and control these features, then an abstraction of different features is required, or else the number of policies will increase, thereby making policy control much more difficult.

Thus, more than a robust networking model is needed that can build these abstractions—an equally robust policy model that can be used to control the network model is also required.

6.3.4 Representing Different Constraints

Almost all examples of current information models do not use constraints. Most information models that are not based on UML do not even provide constraints, and those that are usually do not make use of this feature. Instead, they try to represent constraints using static subclasses. For example, as this book was being written, the DMTF considered modifying their policy model to introduce the notion of an *UnconditionalPolicyRule*, from which *PolicyRule* derives. The purpose of *UnconditionalPolicyRule* is to represent an implied true condition. However, this has two problems. First, it directly contradicts the existing policy models,[2-4] and it even contradicts commonly accepted terminology (e.g., see the definition of *PolicyRule* in Westerinen [2001][28]). Second, it violates object-oriented programming, since a subclass should be an extension of its parent. In other words, a subclass should inherit the attributes, methods and relationships of its superclass and provide additional behavior. In this example, the subclass (*PolicyRule*) is actually a constraint of the base class (*UnconstrainedPolicyRule*).

The root problem is the lack of constraints. If constraints were available, there is no reason to ever have an *UnconditionalPolicyRule*, as the definition of a *PolicyRule* now becomes lost. Instead, *PolicyRules* always are defined with conditions, and a constraint is applied to set the condition to "true" without being evaluated. This is the approach taken in the DEN-ng policy model.

Compare the above dilemma to an excerpt of the DEN-ng *PolicyRule* model, shown in Figure 6-6.

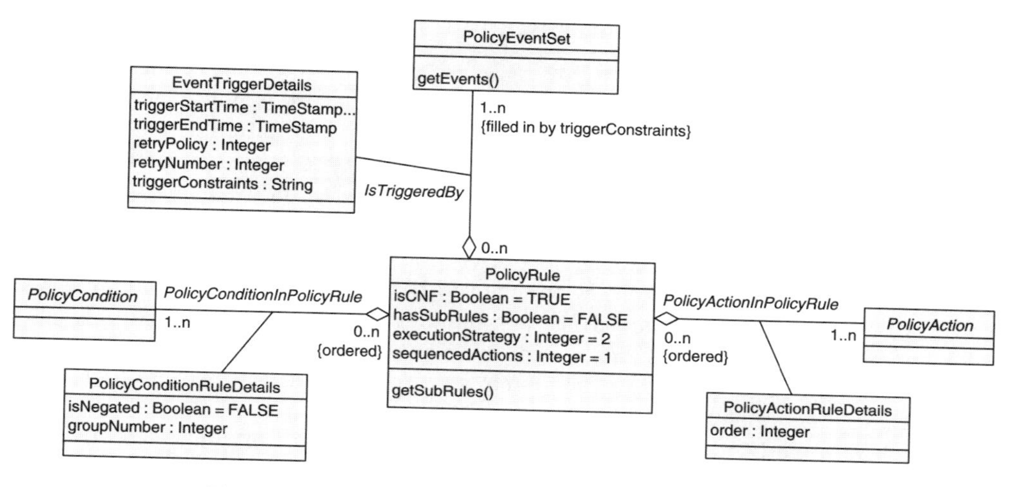

Figure 6-6 Simplified DEN-ng model of a policyrule.

Figure 6-6 shows several different types of constraints that apply to a DEN-ng *PolicyRule*. At the bottom of the figure, the "{ordered}" annotation, which appears twice, signifies that the *PolicyConditions* and *PolicyActions* that are aggregated by the *PolicyRule* must both be presented in an ordered fashion. Second, both of these aggregations are implemented as classes. This provides a convenient point for adding additional constraints, as is done in the *IsTriggeredBy* aggregation. Finally, looking at the *IsTriggeredBy* aggregation, the annotation "{filled in by triggerConstraints}" is a simple expression that means that additional semantics will be placed on this event, as dictated by the triggerConstraints attribute of the EventTriggerDetails class.

The degenerate form of a *PolicyRule* that was desired in the DMTF example can easily be achieved by adding an Object Constraint Language (OCL) expression to the *policyConditionInPolicyRule* aggregation that defines whether a particular *PolicyCondition* should be ignored or not. This is a much better and more extensible way of handling this unusual condition, since it does not alter the structure of the *PolicyRule* class. Rather, it establishes *on an individual PolicyCondition* basis whether a particular *PolicyCondition* should be ignored or not.

This approach can be taken because DEN-ng is derived from the UML metamodel, and therefore uses OCL to express constraints. OCL is a language used to express constraints that do not have any side effects. That is, when a constraint is evaluated, it cannot alter the state of the corresponding executing system. Constraints are typically used in one of two ways:

- to specify invariant conditions that must hold for the system being modeled, and/or
- to specify application-specific constraints

Constraints are very important, since they enable the same basic design to be altered to suit environmental, policy, and other types of changes.

6.3.5 Applying Different Policy Domains to the Same Device

The current literature largely ignores treating different policy domains. However, this is a naturally occurring case in network engineering. Consider a network that runs OSPF. There are in principle four different types of OSPF routers: internal routers, backbone routers, area border routers, and AS boundary routers. An OSPF autonomous system consists of multiple areas and a backbone. Each area is a contiguous group of hosts and networks, as well as routers that communicate with those networks. The backbone consists of networks not included in any area, routers attached to those networks, and routers attached to more than one network. DEN-ng handles this situation by assigning each of these types of routers one or more roles. Assigning a router multiple roles enables it to be known to the system as a router that fulfills more than one of these functional duties.

An area border router attaches to multiple areas, running one copy of the routing algorithm for each area to which it is attached. An area border router is used to distribute topological information about each of its attached areas to the backbone; then, the backbone distributes that same information to other areas. Thus, an area border router is part of multiple policy domains, and policies can be applied that control how routes in one domain are distributed to other domains. Each policy domain has its own set of roles. This enables a natural and extensible equivalence to be defined between a policy domain and its roles. More complicated examples, such as bandwidth brokers, also exist.

DEN-ng extends the work of Ponder to define a domain. As stated in Chapter 5, DEN-ng domains are *directed acyclic graphs*, meaning that a strict hierarchy is formed with no ability for one domain to belong to multiple domains from multiple trees.

DEN-ng adds two interesting twists to domain usage. The first is the use of roles to populate domains. The second is the use of policies to control domain membership. Thus, DEN-ng models the different interfaces belonging to different network domains as being contained in different policy domains. This enables different policies to be used to control interfaces in different domains. This is a good model of the real world, since it enables the developer to isolate one set of interfaces using a role and apply appropriate policies to just those interfaces, and not all of the interfaces of the target device.

6.4 The New DEN-ng Approach—Connecting the Business and System Viewpoints

One of the most important parts of the new DEN-ng design is to provide a means whereby the model can support different viewpoints. This section will explore how the business and system viewpoints are defined, connected, and supported in the DEN-ng policy model.

6.4.1 Business Interactions

The SID defines an interaction as an arrangement, contract, communication, or joint activity between one or more parties playing a certain party role.[29] An interaction may be in regard to one or more product offerings, product specifications, and/or one or more products. An interaction is further defined by one or more places. In addition, one interaction may reference another interaction (Figure 6-7). (Note that differences between the SID and the DEN-ng interaction models will be shown by signifying SID-specific classes as prefixed with SID, so that they can be more easily compared with the DEN-ng classes in subsequent figures; the SID model of course does not prefix its classes with "SID.")

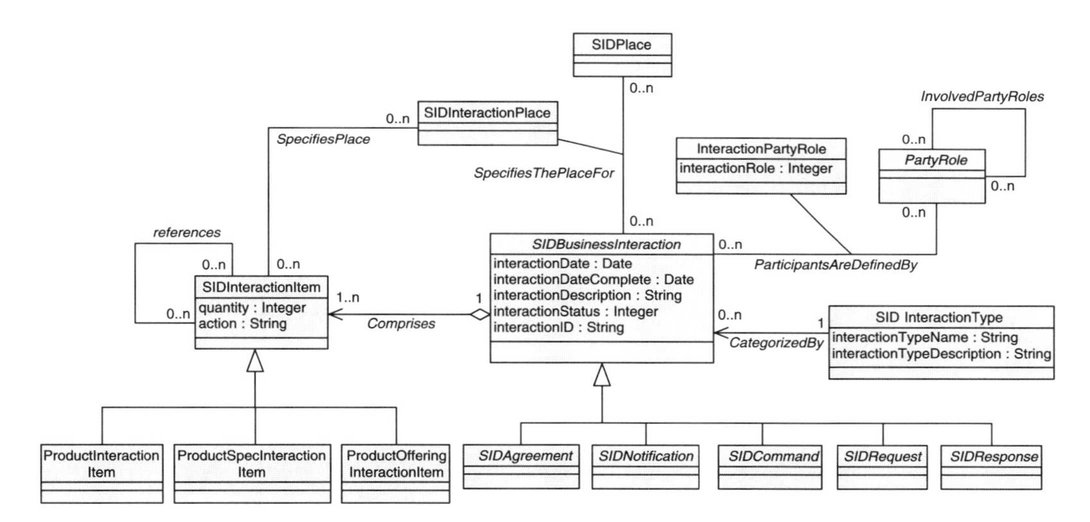

Figure 6-7 Simplified SID interaction model.

Figure 6-7 defines a *BusinessInteraction* as having a start and end date, an optional description, a current status (for tracking the state of the *Interaction*), and a unique ID for identifying instances of the *Interaction*. *Interactions* can be optionally categorized by their types. Each *Interaction* is made up of one or more *InteractionItems*, which specifies a quantity and an action and can be defined to operate on a *Place*. Finally, each *Interaction* defines a set of participants, each of which are characterized by its particular *PartyRole*. (SID and DEN-ng use the role object model to represent a *Party* in terms of roles that the *Party* has. A *Party* is an abstraction that represents an *Individual* or an *Organization*. *PartyRoles* can be defined to represent the different business, organizational, and other functions that a *Party* has. Thus, a *Person* can have the roles of *Employee*, *Manager*, and *SupportSupervisor*).

The DEN-ng interaction model is close but slightly different than the SID interaction model (Figure 6-8). The most significant difference is that the DEN-ng interaction model enables services, physical resources, logical resources, and

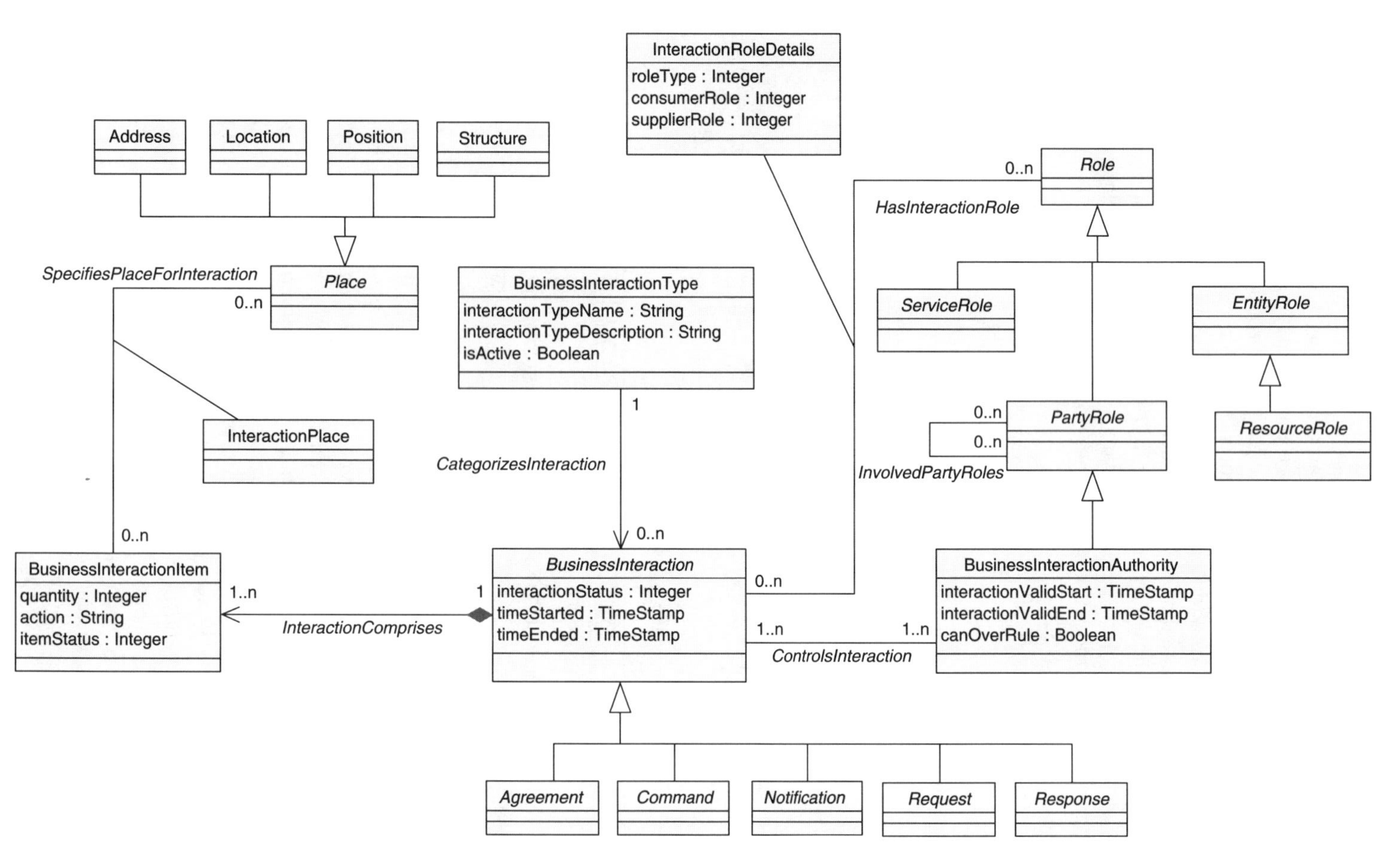

Figure 6-8 Simplified DEN-ng interaction model.

party roles to interact with each other. The *Interaction RoleDetails* class enables this *Interaction* to be characterized in terms of consumers and suppliers. An *InteractionAuthority* is defined, which defines the semantics of how the *Interaction* is to take place, the time period for which the *Interaction* is valid, and whether one *Authority* can overrule another. *InteractionType* has an attribute (*isActive*) for easy filtering of active versus inactive types of *Interactions*. *InteractionItem* has a status attribute for each item. There are additional subclasses of *InteractionItem*, but they are too numerous to show and do not add significant value to understanding Figure 6-8. Finally, there are two changes to *BusinessInteraction*. First, the semantics of its status attribute have changed. Since each *InteractionItem* has its own individual status, the *interactionStatus* attribute represents the status of the entire *BusinessInteraction*. Second, the start and end times have a more accurate data type. The differences in naming between Figures 6-7 and Figure 6-8 arise because DEN-ng is built to be implemented. Thus, every artifact (class, attribute, method, and relationship) has a unique name. This is not yet true of the SID model.

Both the SID and DEN-ng interaction models are much more complete than what has been shown. Also, the DEN-ng and SID interaction models are due to converge in the near future (meaning that differences between the models will be resolved). This is a general statement regarding any difference found between a DEN-ng model and a SID model in this book.

6.4.2 Service Level Agreements, Objectives, and Specifications

All types of business interactions potentially require policy. We will concentrate on the agreement sub-type of business interaction, since it can be used to model a service level agreement (SLA). The SLA involves a number of other business entities, such as products, services, and resources and/or their specifications. The TeleManagement Forum (2002) defines an SLA as follows:

A service level agreement (SLA) (also known as a Service Level Guarantee) is a type of agreement that represents a formal negotiated agreement between two parties designed to create a common understanding about products, services, priorities, responsibilities, and so forth. The SLA is a set of appropriate procedures and targets formally or informally agreed between parties in order to achieve and maintain specified Quality of Service.

There are three distinct types of SLAs.[30] The most common of these are called customer SLAs, which are designed to govern the *ProductOffering* supplied to the customer. This type of SLA is usually subject to a formal contract between the customer and the operator. It is written in terms that the customer can understand, and is based on the end-to-end delivery of the components of the product.

The second type is called a supplier SLA, which the SID team calls a "party SLA." This is to emphasize that different *Parties* or *PartyRoles* can participate in this type of SLA. The purpose of this SLA is usually to ensure that various performance objectives are met for the resource and service components that are provided by a third party. It is usually derived from the contractual agreement between the supplier and the operator.

The final type is called an internal SLA. This is a resource-focused SLA, which is aimed at managing the set of services and service components that are defined by one of the other two types of SLAs. It is subject to agreements between organizational functions within the operator's business.

The DEN-ng SLA model is more involved than the SID SLA model; therefore, only the former will be shown and described.

Figure 6-9 shows that the three different types of SLAs all have *SLAItems* (since their common superclass, *Agreement*, has a composition relationship with *AgreementItem*, which is in turn the subclass of *SLAItem*). This enables individual components of an SLA to be defined for each type of SLA.

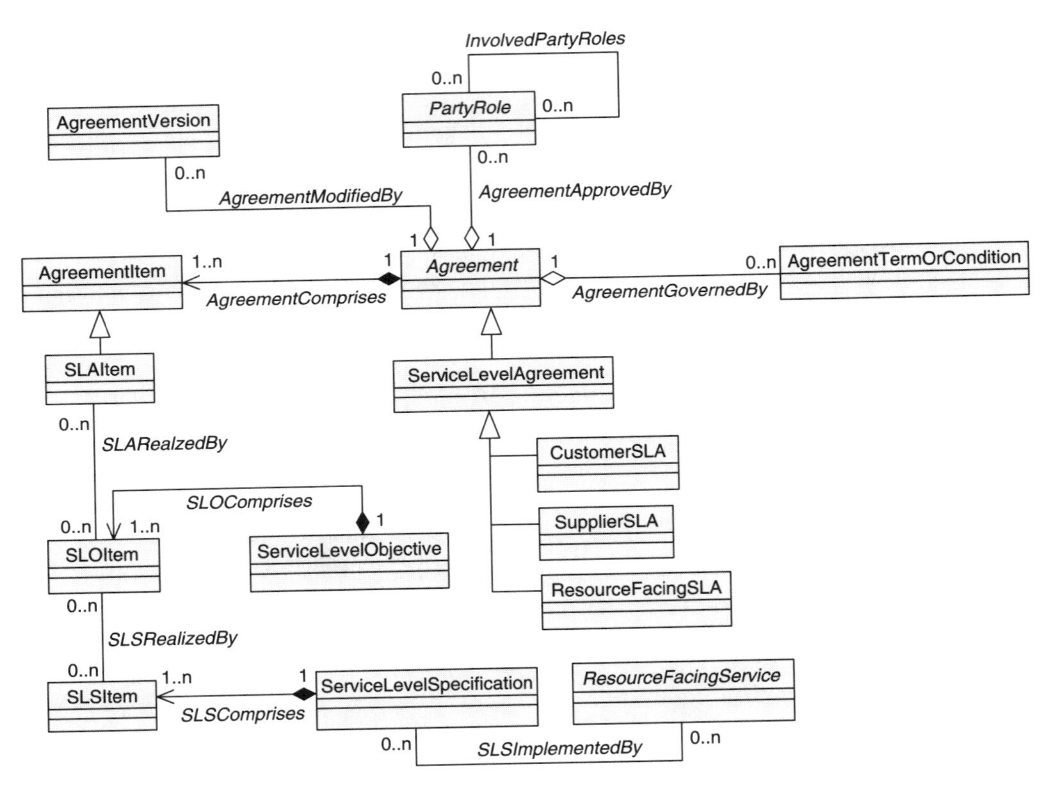

Figure 6-9 Simplified DEN-ng SLA model.

SLAs are expressed in terms of *ServiceLevelObjectives (SLOs)*. Conceptually, an SLA embodies a set of either pre-defined and/or negotiated SLOs. An SLO is a means to translate the business requirements of an SLA into system requirements. It is defined in terms of parameters and metrics, thresholds, and tolerances. In addition, certain consequences are associated with not meeting the SLOs. Each *SLAItem* is related to zero or more *SLOItems*.

SLOs are expressed in terms of *ServiceLevelSpecifications (SLSs)*. Conceptually, an SLO contains a set of pre-defined or negotiated set of SLSs, and consequences that occur, if the objectives are not met. An SLS is a means to translate the business and system requirements of an SLO into an implementation. Each SLS is related to zero or more *ResourceFacingServices*. A *ResourceFacingService* is the base class for defining *ResourceFacingServices* in the DEN-ng service model. A *ResourceFacingService* is an abstraction that defines the characteristics and behavior of a particular *service* that is not visible to the *customer*, but nevertheless must be present in order for the *CustomerFacingService* (e.g., a service that is directly visible to the customer) to work. Thus, for network services, an SLS service has the requirements for deriving one or more *ResourceFacingServices*.

Recall that in DEN-ng, an SLA can be between several different types of entities (e.g., people, physical resources, logical resources, and services). Therefore, SLOs and SLSs also have associated business entities that can be used to define the level of quality, performance, and other metrics that are associated with the *Product*, *Service*, or *Resource* that an SLAItem defines. In addition, SLOs and SLSs have system entities, and SLSs have implementation entities. The advantage of the DEN-ng model is that it provides a mapping between the sets of different entities (e.g., business, system and implementation) that make up SLAs, SLOs, and SLSs.

6.4.3 Relating Business Interactions to Products

Business interactions are related to products at a fundamental level; they define how products and their components are used, ordered, and so forth. The DEN-ng model in Figure 6-10 shows how it models these interactions.

Figure 6-10 shows that the composite pattern is also used with *Product*. *ProductComponent* (also called a *ProductItem*) is the atomic part, and *ProductBundle* is the component part. A product can have either physical Resources and/or CustomerFacingServices. Note that a *Product* doesn't have a direct association with either Logical Resource or a ResourceFacingService, since the *Party Role* that buys or uses the *Product* is not directly aware of either of these. Each *Product* can have zero or more *Services* as well as zero or more *Resources* associated with it. This presents a business view of how a *Product* is used. For example, suppose that a particular service provider decides to offer an MPLS VPN as a *Product*. It provides a *Connectivity Service* and different *QoS Services* for different applications being used. It also provides a customer premise *Router* (which is modeled as a *Physical Resource*, but programmed as a *LogicalResource*) so that the customer can connect to the *VPN service*. The different *Services*, along with the *Router*, are each different *Products* (i.e., *ProductComponents* and *ProductBundles*) that are part of the *Product*.

Each *Product* can also have its own way of interacting with the service provider. The service provider may decide to offer a special upgrade of one or more of its *Services*. There will be a set of terms and conditions that govern which customers this upgrade is offered to. Similarly, new equipment could be offered by a manufacturer that can replace the *Router*, or new versions of the operating system of the *Router* may be offered to provide more functionality to enable more sophisticated traffic conditioning. These will also have terms and conditions associated with them, but will be significantly different in nature than

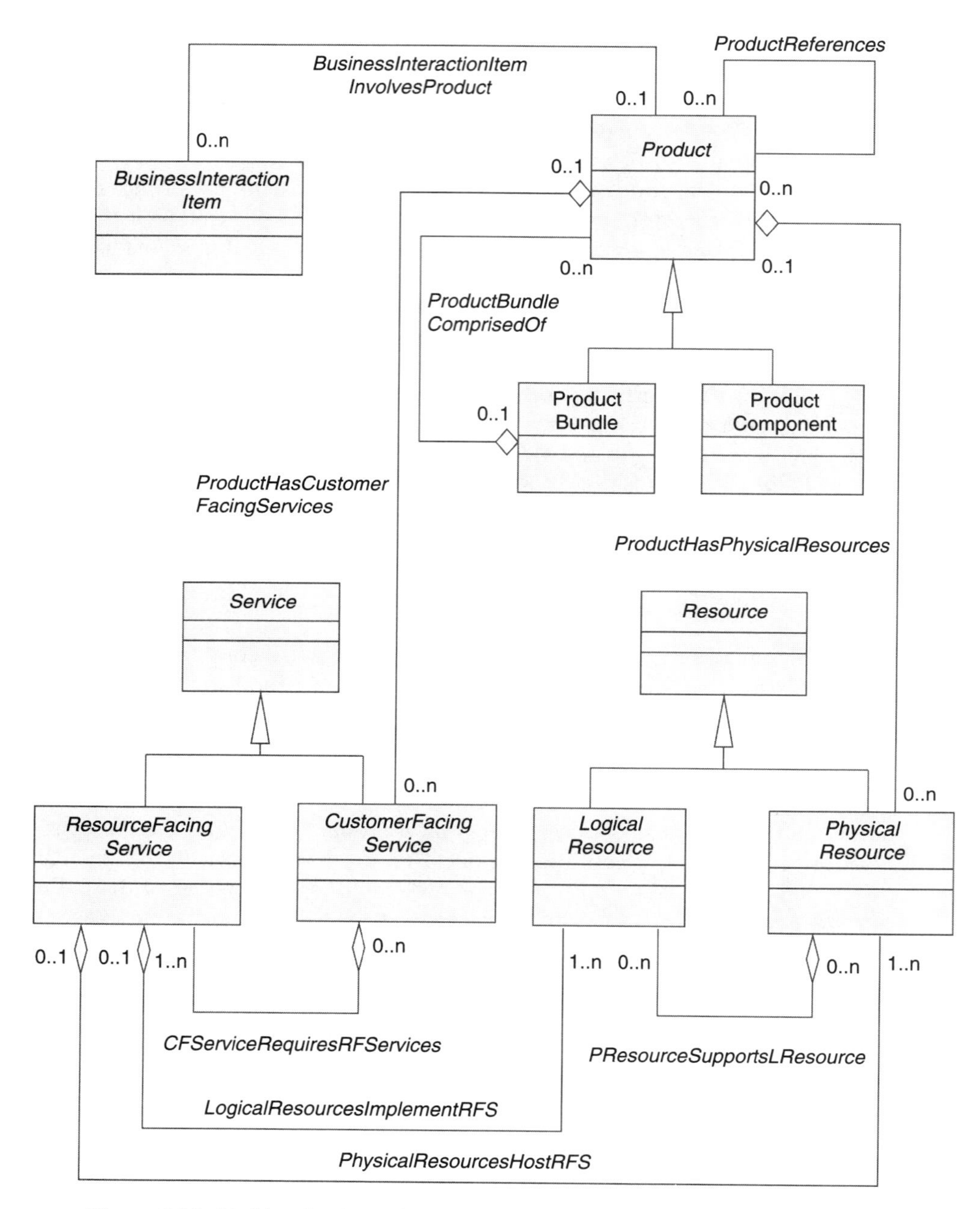

Figure 6-10 Linking business interactions to products in the DEN-ng model.

the *Service* terms and conditions. The DEN-ng model in Figure 6-10 enables each of these situations to be handled individually, while still maintaining their association with a particular *Product*.

The strength of this model is that *BusinessInteractions* associate directly with *Products*. This enables different methods and types of interaction to different *Products* (i.e., services and resources) to be customized to suit the specific needs of the *Customer*.

(Note that Figure 6-10 is a simplified model. There are many other important business entities, such as *Market* and *MarketSegment*, which are linked to this model. They have been left out to keep the model as simple as possible.)

Policies can be applied at many different points to ensure consistent and correct operation of the business processes that govern each of these entities. Figure 6-11 shows some of the more important places where policy can be applied, using the DEN-ng model as an example.

Figure 6-11 shows policy being applied as association classes. This is because the particular policy applied depends on the two entities that are being associated. For example, consider the *InteractionRolePolicy* at the top of the figure. This association class provides the semantic details for the *RoleParticipates InInteraction* association. This enables policy to be used to specify which roles are allowed to participate in a given interaction. Since a role can involve a person, *PhysicalResource*, *Logical Resource*, or *Service*, this is indeed a powerful way to control what types of entities are *allowed* to participate in a given interaction.

The *InteractionPolicy* association class is used to provide the semantics for controlling what *BusinessInteractionAuthority* is allowed to manage this particular *BusinessInteraction*. For example, this association class can provide a specific policy for defining which *PartyRoles* are enabled to initiate a given *Business Interaction* as well as different policies for approving the *BusinessInteraction* and implementing the *BusinessInteraction*.

As will be seen in Chapter 11, the Intelliden product uses these types of *InteractionPolicies* to control:

- Who is allowed to construct a configuration change
- Who must approve a configuration change
- When and where the configuration change must be implemented

The *InterationItemPolicy* is used to control how a particular *Product* is presented to the *Customer*, and is linked to terms, conditions, and pricing for that customer. This supports the earlier example, in which different policies are used in offering a hardware upgrade versus a service upgrade.

Finally, the *ResourcePolicy* and *ServicePolicy* association classes control how a particular type of *Resource* and *Service* are interacted with, respectively. This is different than the *InteractionItemPolicy*, in that *InteractionItemPolicy* gives specifics for a particular *category* of *ProductItem*. For example, a *ProductItem* specifies a customer premise *Router*—it does not specify the exact type of customer premise *Router* that is being used. Thus, *InteractionItemPolicy* can specify general rules and policies for all types of customer premise *Routers* (such as their installation instructions), and *ResourcePolicy* can be used to provide explicit instructions on how to configure a particular type of customer premise *Router*.

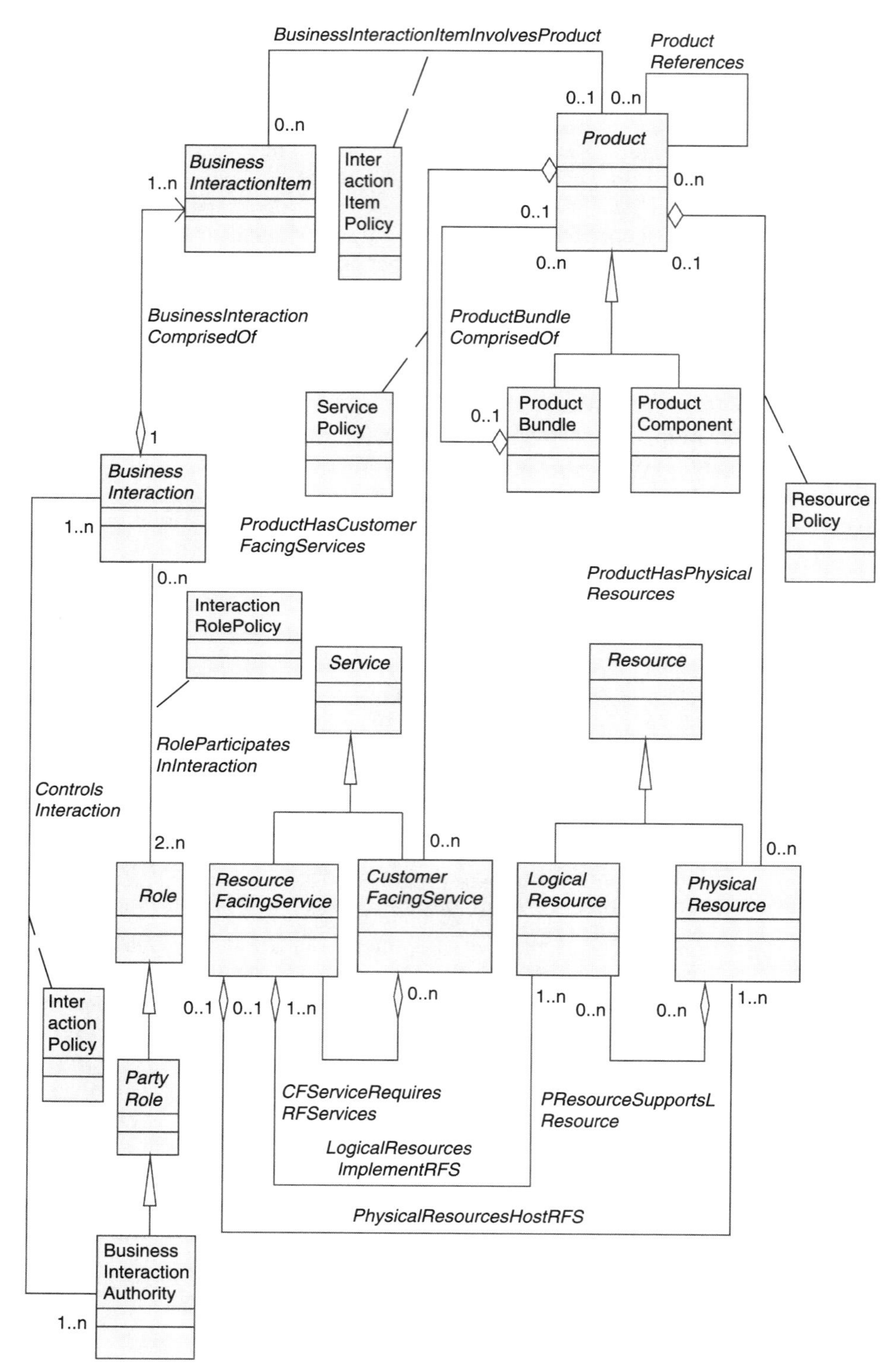

Figure 6-11 Applying policy to business interactions in the DEN-ng model.

Additional business policies can be defined. These fall into two categories: policies that are purely business in nature, and policies that cover other aspects of agreements and interactions. Examples of business policies that affect products are shown in Figure 6-12.

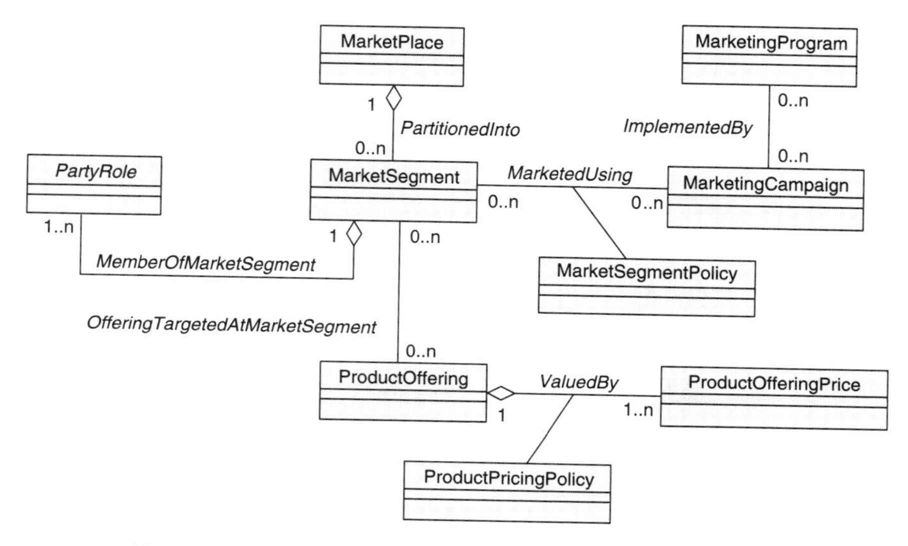

Figure 6-12 Additional DEN-ng product business policies.

Examples of additional policies that affect more granular aspects of *BusinessInteractions* are shown in Figure 6-13.

Figure 6-13 shows more detail about some additional business entities in the DEN-ng model. This figure shows two additional relationships to *Interaction RolePolicy*: one to define the policies that determine the access rights of a *PartyRole* in this *BusinessInteraction*, and one to define the policies that determine the usage rights of a *PartyRole* in this *BusinessInteraction*. By relating the *UsagePolicy* and *AccessRightsPolicy* entities to the *InteractionRolePolicy*, usage and access rights can be tailored to each role interaction.

Figure 6-13 also supplies additional details on policies that affect *AgreementTerms*. DEN-ng defines a large number of different types of agreement terms and conditions as subclasses of the *AgreementTerm* entity. Two of these, *FinancialTerms* and *GeographicUsageTerms*, are shown in Figure 6-13. The *AgreementTermPolicy* association class defines policies that control which *AgreementTerms* should be part of this agreement. The *FinancialPolicy* and *GeographicUsagePolicy* association classes define policies that determine additional semantics on the *FinanceAgreementSpecifiedBy* and *GeographicUsage AgreementSpecifiedBy* associations, respectively. Whereas the *AgreementTerm Policy* association class defines how financial terms and geographic usage terms can relate to an agreement, the *FinancialPolicy* and *GeographicUsagePolicy* association classes define the specific policies to be applied to a particular *FinancialTerm* and *GeographicUsageTerm*, respectively.

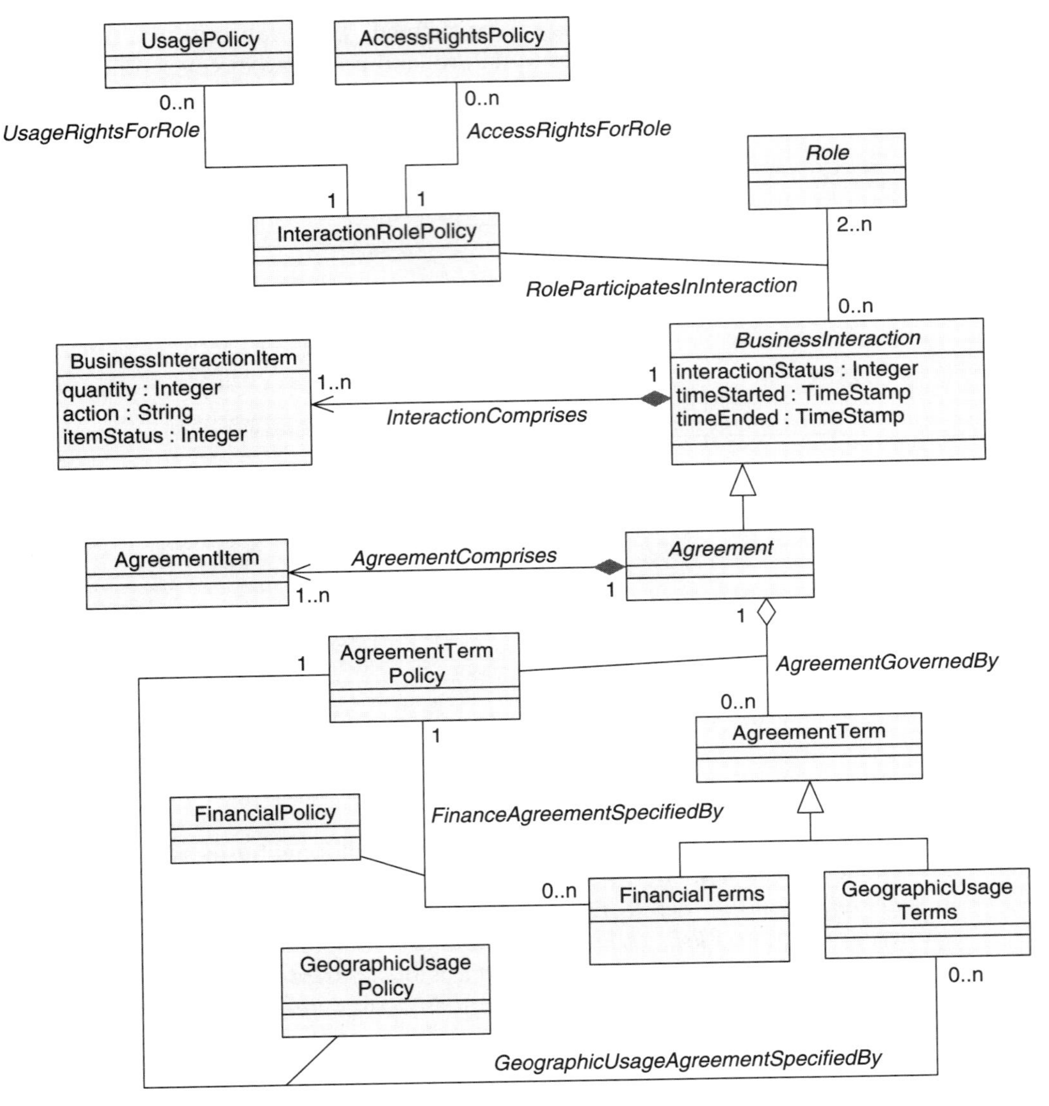

Figure 6-13 Additional DEN-ng BusinessInteraction policies.

6.4.4 Using Business Policies to Drive Configuration

The concept is indeed simple—the network should be configured for providing and supporting services that have been contracted to the customers of the organization owning the network. Thus, if the business needs of the organization change, the network should be reconfigured to support those changes. PBNM has been used to control the changing of network device configuration commands with little or no relation to the underlying business rules and processes that the organization is running. The DEN-ng policy model, along with other DEN-ng domain models, were built to address this need and to ensure that the network reflects how the business should be run. The process of using business rules and processes to (re)configure the network results in a closer association between how the business is run and the services that are offered by the network.

An example will help explain how business rules can be used to drive the configuration of network devices. Suppose a service provider offers virtual private network (VPN) connectivity as a product offering. The most popular of these is the so-called BGP-MPLS VPN.[31] This type of VPN uses the global Internet as a backbone to connect different users to different resources. This provides a virtual private network using a public IP network. The users of this type of VPN only see the VPN connections within their own group. Figure 6-14 shows a simple MPLS VPN.

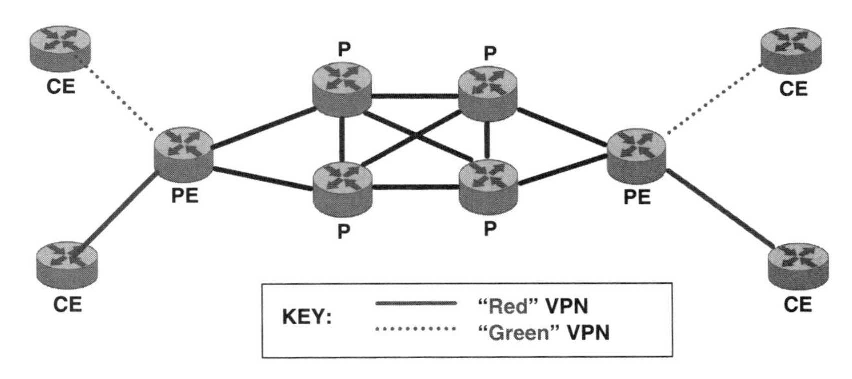

Figure 6-14 A simple MPLS VPN.

This simple MPLS VPN has a common core of provider (P) routers, arranged in a full-mesh. They are connected to provider edge (PE) routers, which designate the edge of the provider network. Two customer VPNs are supported; one is designated by the two red connections from the customer edge (CE) routers to their respective PE routers, and the other is designated by the two green connections. Thus, our common provider backbone supports two VPNs. MPLS VPNs will be covered in more detail in Chapters 9 and 10.

This type of VPN can be modeled as consisting of routers that play three different *roles*:

- CE routers are associated with customer sites and are usually managed by the customer (though they do not have to be)
- PE routers serve as the entry and exit points to and from the VPN and are managed by the provider
- P routers are any router that does not interface directly with a CE router and are managed by the provider

The use of roles will be critical in reducing the complexity of different devices with different operating systems that each have different commands and features. Roles will enable the product characteristics to be abstracted into a set of resource and service requirements, which can then be implemented using vendor-specific models.

Clearly, each of these three roles will require different configuration commands. If policies are assigned to each of these roles, each type of configuration can be governed in a standard and consistent fashion independent of the vendor-specific features and programming models used to actually configure the device.

Strassner (2002)[26] describes a set of architectural and integration problems that prevent business processes from being used to drive the configuration and management of network resources in current network management architectures, as well as the use of DEN-ng to solve this problem. This paper describes a configuration workflow process, whose general form is shown in Figure 6-15. A more thorough example of how an MPLS VPN is represented in DEN-ng will be provided in Chapters 9 and 10.

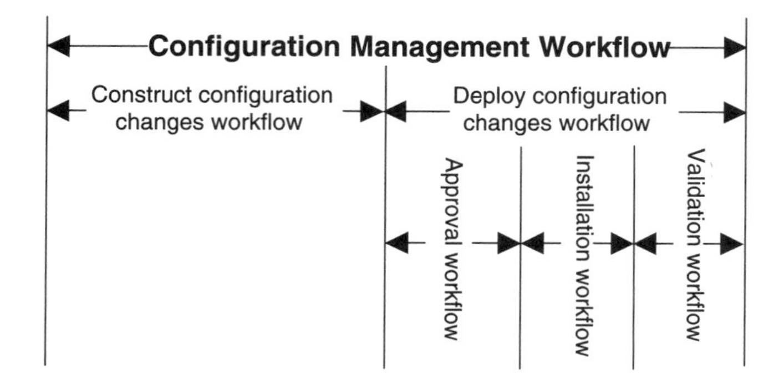

Figure 6-15 Configuration management workflow in DEN-ng.

The idea, then, is to define three different configuration management workflows, each targeted to the specific needs of one of the three roles defined for this VPN. In this way, the VPN can be managed in a consistent manner regardless of which vendor devices are being used to build the VPN. Furthermore, the VPN is defined as a product, with associated SLAs, market strategies, and so forth. Thus, as these are changed to address the needs of the market, the effect of these changes can be related to the specification of the resources and services used to define the VPN. Since all of these concepts and entities are represented in a single information model, changes to any one entity can be related to the other entities.

This forms a critical abstraction that enables the different capabilities and features of different vendor devices to be managed and controlled in a consistent fashion.

6.5 Summary

This chapter has discussed the motivation for developing a new policy model. The most important of these were: (1) building a model that mapped a detailed architecture, (2) ensuring that the policy information model was tightly integrated with other information models that described the managed objects that the policy model was managing and controlling, (3) developing interfaces between the business, system, and implementation views of how policy (and other domain information models) interact with each other, and (4) defining policy in a dynamic as well as a static sense. Part of this theme included rebuilding other domain models (such as those representing a network device) to provide more detail. This was necessary to show how policy affected, for example, the changing of device configuration files. Since the current information models did not address these concepts, DEN-ng was built.

The role of standards was then covered. Standards describing information and data models are significantly different than standards defining, for example, a protocol. The role of the TeleManagement Forum was examined and shown to be the most relevant for policy, which is why the new DEN-ng policy information model was built there.

The TMF NGOSS program was briefly described and characterized as a blueprint from which the DEN-ng policy model was built. Special emphasis was given to the catalyst programs of the TMF, since they are a perfect way for vendors to get used to the power of the SID and DEN-ng models, and hopefully will create demand from service providers and enterprises for more vendors to implement a standards-based information model.

This led to a discussion of shortcomings in previous models. These included a lack of UML compliance, a lack of specificity in representing different types of resources and services, and a lack of integration, both between models and between different expressions of policy. For example, the problem of managing devices from different vendors was analyzed. Most current PBNM solutions are built to manage only a particular set of devices. Worse, they usually manage only a subset of the functionality that they provide. The DEN-ng approach is to normalize these different offerings using the concepts of capabilities and constraints to build extensible information models based on the policy continuum. This approach normalizes the same function independent of device implementation, providing additional layers to model different implementations if required.

Even more important, however, are the needs to incorporate state and runtime information and, most importantly, to use the model to integrate the business, system, and implementation worlds. The interaction models of the TMF SID and DEN-ng were described as a means to model how different types of interactions affect people, places, services, and entities. The DEN-ng SLA model was then described, which linked agreement (which was one of the five fundamental types of interactions) to SLAs, SLOs, and SLSs. The DEN-ng product model was then introduced and linked to business interactions. Policies to manage and control these interactions as well as other business concepts were briefly described. The above culminated in an example that showed how business policies could be used to control various aspects of configuring an MPLS VPN.

The rest of the chapter concentrated on how the business and system viewpoints are connected in the DEN-ng policy model. The concept of a business interaction was defined. This enables users, products, services, and resources to interact with each other. Closely related to business interactions are the concepts of SLAs, SLOs, and SLS. The high-level DEN-ng models for these concepts were explained, as was their relationship to business interactions.

Then, DEN-ng models that showed how these different concepts were tied together were shown. This is the first time that any information model has tried to relate entities at such different conceptual levels as these; therefore it is a significant contribution to the industry. The DEN-ng models are more focused on networking and implementation details, whereas the SID is more focused on business needs. However, a stated objective of both models is to converge as fast as possible. Indeed, as will be shown in later chapters of this book, this is already happening to a large degree.

6.6 Recommended Further Reading and References

The following is a set of references for this chapter.

1. Strassner, J., *Directory Enabled Networks*, Chapter 10, Macmillan Technical Publishing.
2. Moore, B., Ellesson, E., Strassner, J., Westerinen, A., *Policy Core Information Model – Version 1 Specification*, RFC 3060, February 2001.
3. Moore, B., Rafalow, L., Ramberg, Y., Snir, Y., Westerinen, A., Chadha, R., Brunner, M., Cohen, R., Strassner, J., *Policy Core Information Model Extensions*, draft-ietf-policy-pcim-ext-06.txt, November 2001.
4. Snir, Y., Ramberg, Y., Strassner, J., Cohen, R., Moore, B., *Policy QoS Information Model*, draft-ietf-policy-qos-info-model-04.txt, November 2001.
5. Gai, S., Strassner, J., Durham, D., Herzog, S., Mahon, H., Reichmeyer, F., *QoS Policy Framework Architecture*, February 1999.
6. Current work is in the DMTF Members Only site: http://www.dmtf.org/apps/org/workgroup/policy/ The current release of CIM's policy model as of this writing is version 2.7.1, and is located at: http://www.dmtf.org/standards/standard_cim.php
7. The Workshop on Policies for Distributed Systems and Networks had its fourth meeting on June 4-6 of 2003. Information on these conferences is:
   Policy 2003: http://www.labs.agilent.com/policy2003/
   Policy 2002: http://www.policy-workshop.org/2002/
   Policy 2001: http://www-dse.doc.ic.ac.uk/events/policy-2001/
   Policy 1999: http://www-dse.doc.ic.ac.uk/events/policy-99/
8. The IP Policing Conference had a special focus and emphasis on policy. Please see: http://www.upperside.fr/ippol2001/ippol2001pro.htm or http://www.upperside.fr/ippol2000/ippol.htm
9. The NetCon conference, or Network Control and Engineering for QoS, Security and Mobility, with Focus on Policy-Based Networking, can be found at: http://net-con.utt.fr/netcon/arret_marc_7_octobre_2002/index.html
10. Please go to the following website: http://www.noms2004.org/
11. Please go to the following website: http://www.im2003.org
12. Strassner, J., Wheeler, J., *Policy, QoS, and DEN—Tutorial T248*, Network + Interop.

13. Please see the following website for detailed information about Ponder: http://www-dse.doc.ic.ac.uk/Research/policies/ponder.shtml
14. Please see the following website for an overview of the mission of the DMTF: http://www.dmtf.org/about/index.php
15. Please see the following website for an overview of the mission of the TMF: http://www.tmforum.org/ browse.asp?catID=930&sNode=930&Exp=Y
16. Please see the following website, which is the home page of the ebXML effort: http://www.ebxml.org/
17. TMF, *NGOSS Architecture Technology Neutral Specification*, TMF053, v. 3.0, April 2003.
18. Strassner, J., *NGOSS Technology Overview*, TMW Asia-Pacific Conference, August 2002.
19. The UML 1.4 specification is downloadable from the following website: http://www.rational.com/uml/resources/documentation/
20. This is a TMF member's only document, called *Mining CIM for SID*. If you are a TMF member, you can download it from: http://teamlink.tmforum.org/Information%20and%20Data%20Modeling
21. The DMTF *meta-schema* can be found at the following website: http://www.dmtf.org/standards/documents/CIM/DSP0004.pdf
22. Please see the following website about the OMG's Model Driven Architecture initiative: www.omg.org/mda
23. ITU-T, *Generic Network Information Model*, Recommendation M.3100, July 1995.
24. See the following page for comprehensive information about the Trust Policy Language (TPL): http://www.haifa.il.ibm.com/projects/software/e-Business/TrustManager/PolicyLanguage.html
25. See the following page for comprehensive information about KeyNote: http://www.cis.upenn.edu/~keynote/
26. Strassner, J., *A New Paradigm for Network Management: Business Driven Network Management*, SSGRR summer conference, L'Aquila, Italy, July 2002.
27. See the following URL, which defines the router product families for Cisco: http://www.cisco.com/en/US/products/hw/routers/index.html
28. Westerinen, A., Schnizlein, J., Strassner, J., Scherling, M., Quinn, B., Herzog, S., Huynh, A., Carlson, M., Perry, J., Waldbusser, S., *Terminology for Policy-Based Management*, RFC3198, November 2001.
29. TMF, *Shared Information/Data (SID) Model—Addendum 1BI—Common Business Entity Definitions—Business Interaction*, GB922, Version 3.0, June 2003.
30. TeleManagement Forum, *Wireless Service Measurements Handbook*, GB923, August 2002.
31. Rosen, E., Rekhter, Y., Bogovic, T., Brannon, S., Vaidyanathan, R., Carugi, M., Chase, C., Fang, L., Chung, T., De Clercq, J., Dean, E., Hitchen, P., Smith, A., Leelanivas, M., Marshall, D., Martini, L., Morrow, M., Srinivasan, V., Vedrenne, A., *BGP/MPLS VPNs*, draft-ietf-ppvpn-rfc2547bis-03.txt, October 2002.

Chapter 7

The DEN-ng Policy Model

This chapter builds on the preceding chapter, which described the motivation for creating a new policy model, by describing the new DEN-ng policy information model. This model strives to reuse established concepts (such as a PolicyRule) but will enhance the semantics of these concepts, plus introduce new formalisms, to make the DEN-ng policy model a better and more extensible information model. Particular emphasis will be placed on using constraints to guide how the model is used by different applications.

7.1 Introduction

This chapter will explore the main features of the DEN-ng policy model. This model meets the requirements set forth in the preceding chapter, and, unlike previous industry policy models, is completely UML compliant.

The first two sections will introduce the concepts of a *Policy* and a *PolicySet*, which is the superclass of *PolicyGroup* and *PolicyRule*. The next two sections discuss *PolicyGroups* and *PolicyRules*, and their differences from previous industry models. The next two sections—*Events* and *PolicyStatements*—form a departure from most previous industry models and deserve careful reading. This is followed by detailing the design of the *PolicyCondition* and *PolicyAction* hierarchies, along with advanced uses of these classes.

Reusable versus ad hoc policies are then discussed. This is followed by a discussion of how other DEN-ng models interact with the DEN-ng policy model. References 1.3 contain, more information on this design.

7.2 The DEN-ng Concept of a Policy

DEN-ng takes a different approach than most other standards-based information models for representing policy. DEN-ng represents a policy rule as a container that consists of four important components: (1) metadata, (2) an event clause, (3) a condition clause, and (4) an action clause. Here, a clause is an expression that consists of a set of terms.

Policy rules often have special semantics associated with them. These semantics govern the usage and/or operation of the policy rule and are not part of the event, condition, or action clauses. The DEN-ng model defines the *Policy* class as the base class of the policy hierarchy. Metadata is part of every policy and takes the form of components that are part of each subclass of *Policy*, as well as external components, such as relationships and constraints. For example, Figure 7-1 shows the *Policy* base class and some important relationships that help connect the *Policy* domain to other DEN-ng domains.

The *AppliesTo* association enables a particular Policy to be applied to one or more PolicyDomains. The semantics of this association are defined by the *PolicyAppliesToDetails* association class. This association class is used to explicitly define which ManagedEntities in a PolicyDomain this Policy information applies to. Applying a Policy to multiple domains is useful to distribute control via policy to multiple domains.

The *PolicyStoredIn* association defines the sets of *PolicyRepositories* in which this particular *Policy* can be stored.

The purpose of the event clause is to represent the set of events that can trigger the evaluation of a *PolicyRule*. DEN-ng mandates that all policy rules have an event clause. Even if the *PolicyRule* is set to evaluate immediately (e.g., when a user logs onto a system), there is always an underlying event that can be modeled that corresponds to this "automatic" evaluation. The representation of a *PolicyRule* is made consistent by always including an event clause.

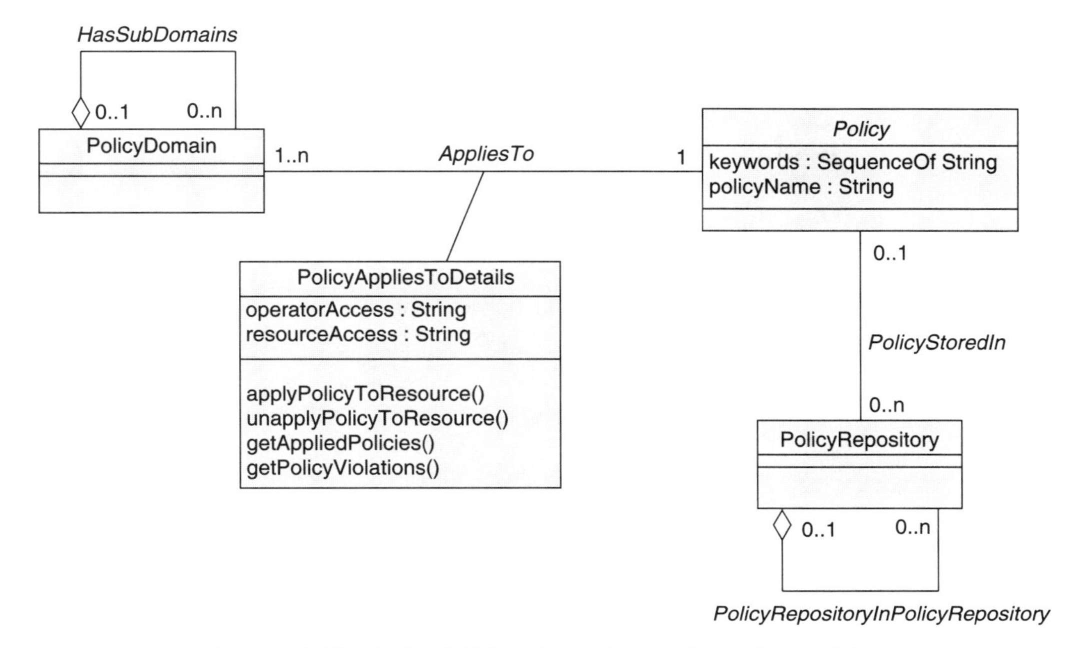

Figure 7-1 The *Policy* DEN-ng base class and its relationships.

Conditions and actions are the heart of the *PolicyRule*. The DEN-ng semantics are more powerful than previous models in this area. Conditions are again restricted to being expressed in Boolean. However, positive and negative actions clauses can now be expressed. A positive action clause is one that is executed because the condition clause evaluated to TRUE, whereas a negative action clause is one that is executed because the condition clause evaluated to FALSE. This book will only show a single action clause, for simplicity, which is triggered by the condition clause evaluating to TRUE, since experience has shown that having a positive and a negative action clause causes a large increase in complexity for a relatively small gain.

Policy rules can be grouped into sets, called *PolicyGroups*. Sometimes, however, it is important to define a nested *PolicyRule*, which is a *PolicyRule* that contains one or more *PolicyRules*. This brief summary leads to the following summary representation of the structure of a DEN-ng *PolicySet*.

Figure 7-2 points out some additional subtle differences between a DEN-ng *PolicyRule* and previous definitions of a *PolicyRule*, such as those from the IETF and the DMTF. First, a DEN-ng *PolicyRule* mandates that at least one event clause, one condition clause, and one action clause are always present. This is done through the cardinality present in the *IsTriggeredBy*, *PolicyConditionIn PolicyRule*, and *PolicyActionInPolicyRule* aggregations, respectively. That is, since the cardinality of these three aggregations is "1..n" on the aggregate side, the components being aggregated (*PolicyEventSet*, *PolicyCondition*, and *PolicyAction*, respectively) must be present. This provides a consistent structure for *PolicyRules*, which in turn simplifies implementation. The DMTF/IETF

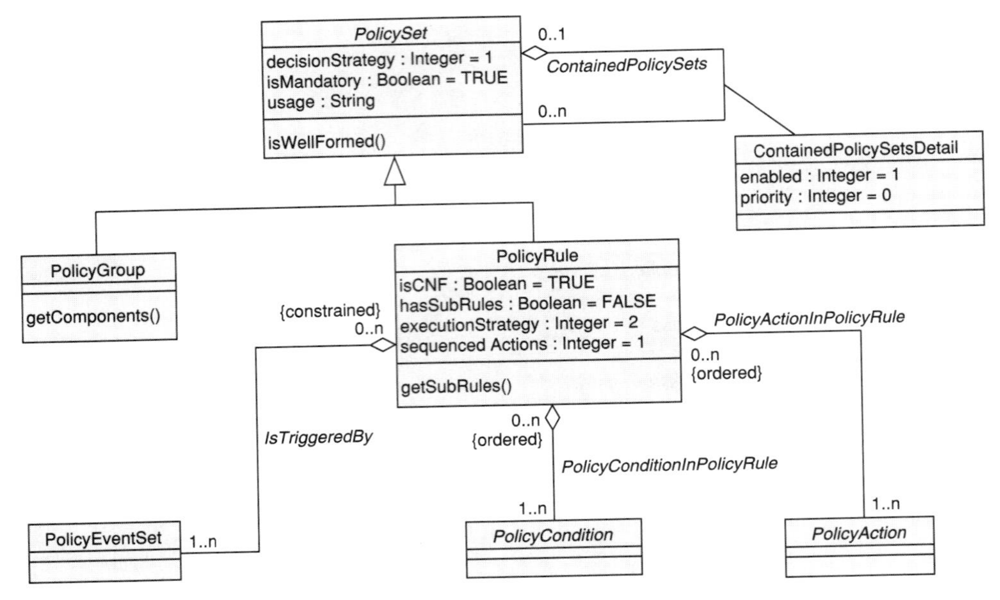

Figure 7-2 Simplified representation of a DEN-ng *PolicySet*.

approach specified this cardinality as 0..n to 0..n, which literally means that neither the PolicyRule nor the aggregated entity needs to be instantiated. This not only provides an inconsistent structure to defining and building *PolicyRules*, it enables incorrect and useless *PolicyRules* to be specified. For example, one could define a *PolicyRule* that has no conditions and no actions! Inconsistencies and lack of specificity such as this example make it hard for people to use models, which theoretically are formal, consistent, and specific.

Second, each of the clauses has a constraint (written in OCL) mandating that they are ordered. This means that if the event, condition, or action clause consists of more than one term, those terms must be defined in order. This enables semantics concerning evaluation and execution order to be specified by the designer. These features are not present in the IETF and DMTF versions of a *PolicyRule*, causing interoperability problems and complicating implementation.

7.3 The DEN-ng *PolicySet*

In Chapter 2, a *policy* was defined as follows:

"Policy is a set of rules that are used to manage and control the changing and/or maintaining of the state of one or more managed objects."

The key concept here is the phrase "set of." As will be shown, both a *PolicyGroup* and a *PolicyRule* can act as intelligent containers. Therefore, as in the IETF model defined in Moore (2001),[4-5] DEN-ng uses the notion of a *PolicySet* to define common semantics for *PolicyRules* and *PolicyGroups*. However, it has different semantics than that of the IETF *PolicySet*.

A PolicySet is an abstract class for defining two types of collection classes: PolicyRules and PolicyGroups. A PolicyRule collects PolicyEvents, Policy Conditions, and PolicyActions, whereas a PolicyGroup collects PolicyRules and PolicyGroups.

A *PolicySet* captures the common semantics (attributes, methods, relationships, and constraints) for policy collection classes. Figure 7-2 shows three common attributes, along with a common relationship and method, which are inherited by the *PolicyGroup* and *PolicyRule* subclasses.

7.3.1 Attributes of a *PolicySet*

A *PolicySet* defines three attributes that both *PolicyGroups* and *PolicyRules* inherit. The first of these is the *decisionStrategy* attribute, which is an enumerated integer that specifies the particular decision strategy to be used in this *PolicySet*. This attribute defines how *PolicyRules* in a *PolicyGroup* or a *PolicyRule* are to be evaluated. (*PolicyRules* can indeed contain *PolicyRules*—these so-called nested rules will be explained shortly.) The values of this attribute can include:

- 0: Unknown
- 1: MatchFirst
- 2: MatchAll

Unknown can be used at initialization time, or if an error occurs, as an easy way to signify that an error in this *PolicySet* has occurred, since the two legal values of this attribute currently defined are MatchFirst and MatchAll.

MatchFirst enforces the actions of the first *PolicyRule* whose condition clause evaluates to True in a *PolicySet*; policy evaluation in this *PolicySet* then "stops," meaning that all subsequent *PolicySets* are no longer evaluated. The order of evaluation of the *PolicySets* is determined by the *priority* attribute in the *ContainedPolicySetsDetail* association class, which is explained below.

In contrast, MatchAll enforces the actions of all *PolicyRules* whose condition clauses evaluate to TRUE. This is equivalent to performing two passes through the set of *PolicyRules* that are contained in the *PolicySet*. The first pass gathers all *PolicyRules* whose condition clause evaluates to TRUE and places them in a group. The second pass then executes only those *PolicyRules* whose condition clauses evaluated to TRUE. Execution of these rules is controlled through the Priority property (on the aggregation *PolicySetComponent*), which defines the order in which the *PolicyRules* execute. Thus, a *PolicySet* can be configured to either execute the first *PolicyRule* whose condition clause evaluates to TRUE, or to execute all *PolicyRules* whose condition clauses evaluate to TRUE.

The semantics of this attribute are identical for *PolicyGroups* and for *PolicyRules*. In a *PolicyGroup*, this attribute defines how to evaluate the set of *PolicyRules* that are contained in a *PolicyGroup*. In a *PolicyRule*, this attribute defines how to evaluate nested *PolicyRules* within a higher-level *PolicyRule*.

The *isMandatory* attribute is a Boolean attribute that, if TRUE, signifies that evaluation (and possibly action execution) of this PolicySet is mandatory and must be attempted. If the mandatory property value of this entity is FALSE, the evaluation of this entity is considered to be "best effort" and may be ignored. Thus, in a *PolicyGroup*, this attribute tells the system that the contents of this *PolicyGroup* must be evaluated. In a *PolicyRule*, this attribute states that the evaluation of this *PolicyRule* is mandatory. This is significant, because normally, if a nested *PolicyRule* fails, the parent *PolicyRule* will also fail. Setting this attribute to "True" overrides this rule and causes the parent *PolicyRule* to be evaluated.

Finally, the *usage* attribute is a free-form string attribute that recommends how this policy object should be used. This is an optional attribute and is meant to be an aid to the administrator.

7.3.2 Methods of a *PolicySet*

The *PolicySet* class has a single method: *isWellFormed*. The purpose of this method is to determine whether this *PolicySet* is constructed correctly or not. A *PolicySet* may evaluate if, for example, its construction is not completed and it is missing one or more of its three mandatory clauses (event, condition, and action).

A well-formed *PolicySet* is defined as follows:

- If the *PolicySet* is a *PolicyRule*, that *PolicyRule* must have a complete event, condition, and action specification, each of which is syntactically correct.
- If the PolicySet is a PolicyGroup, that PolicyGroup must consist of only PolicyRules or PolicyGroups.

The return type is an integer, to enable success or failure of the operation to be indicated. Values are:

- 0: the *PolicySet* is well formed
- 1: a *PolicyRule* in the *PolicySet* is missing an event specification
- 2: a *PolicyRule* in the *PolicySet* has an incorrect event specification
- 3: a *PolicyRule* in the *PolicySet* is missing a condition specification
- 4: a *PolicyRule* in the *PolicySet* has an incorrect condition specification
- 5: a *PolicyRule* in the *PolicySet* is missing an action specification
- 6: a *PolicyRule* in the *PolicySet* has an incorrect action specification
- 7: a *PolicyGroup* in the *PolicySet* contains illegal objects

The above detail is provided as a simple example to show that it is possible to specify sufficient detail in an information model to significantly simplify implementation. Since none of the above specifies platform- and/or protocol-specific details, none of the above is specific to any particular data model. Yet, the above information is detailed enough to help guide the developer in implementation and to help ensure interoperability is achieved.

7.3.3 Relationships of a *PolicySet*

PolicySet defines a single recursive aggregation, called *ContainedPolicySets*. This aggregation is used to gather together discrete *PolicySet* objects to form a group of *PolicySet* objects. Such a group must share the same *DecisionStrategy*.

This aggregation is implemented as a class, called *ContainedPolicySets Details*. This class represents the semantics of the *ContainedPolicySets* aggregation. It provides additional semantics that enable this grouping of *PolicySets* to be prioritized and enabled, so that they can interwork with other *PolicyRules* and *PolicyGroups*. Note that each object in this aggregation must have the exact same decision strategy. This class defines two attributes: *enabled* and *priority*.

The attribute *enabled* is an enumerated integer. Values include:

- 0: Unknown (this can be used as an error indicator)
- 1: Enabled for evaluation (this signifies that this object is enabled for evaluation)
- 2: Disabled for use (this object is disabled and cannot be used or evaluated)
- 3: Enabled for debugging (this means that this rule can be used in test or debug scenarios)

Note that this object still might end up being disabled, even if this attribute has the value of 1 ("enabled"). For example, an associated *PolicyTimePeriodCondition* might evaluate to FALSE, effectively disabling this object.

The *attribute* priority is a non-negative integer used to prioritize *PolicyRules* among each other, and/or *PolicyGroups* that are contained within a *PolicyRule*. Larger integer values indicate higher priority.

Priorities apply to *PolicySets* (e.g., *PolicyGroups* and *PolicyRules*) and are to be used at the same level of scoping. That is, if a *PolicyGroup* contains "n" *PolicyGroups* or *PolicyRules*, the priority of the outermost (or parent) *PolicyGroup* can be compared with either stand-alone *PolicyRules* or *PolicyGroups* (at that same level of nesting).

The default setting of this attribute is 0, meaning that by default, no prioritization is applied to this *PolicySet*. Prioritization among policy rules provides a basic mechanism for resolving policy conflicts. However, it should only be used when required.

7.3.4 Using *PolicySets*

The use of a *PolicySet* depends on its *decisionStrategy* and *isMandatory* attributes. If the *PolicySet* contains other *PolicySets*, it also depends on how those *PolicySets* are contained. This is determined by the *enabled* and *priority* attributes of the *ContainedPolicySetsDetail* aggregation class.

The first thing to consider is the setting of the *decisionStrategy* attribute. This attribute is used to determine whether a single or potentially a group of *PolicyRules* should be evaluated. The *isMandatory* attribute should be used with care, because it will require the evaluation of the condition clause of a *PolicyRule* and the execution of the actions of a *PolicyRule*, if the condition clause evaluates to TRUE. If the *isMandatory* attribute of a *PolicyGroup* is set, all of the *PolicySets* contained in that *PolicyGroup* must be evaluated. Thus, this attribute is most useful when coupled with setting the *decisionStrategy* to MatchAll.

The *enabled* attribute is best viewed as a switch. If the value of the *enabled* attribute is "false," that *PolicySet* cannot be evaluated, no matter what the settings of the other attributes of the *PolicySet* or the *ContainedPolicySetsDetail* classes are. Finally, the *priority* attribute is used to define the order of execution of *PolicyRules*. Ensuring that the *priority* attribute is unique helps guarantee deterministic action.

Use of the MatchAll strategy, although powerful, is also tricky. Care must be taken in the policy engine to watch for conflicts between the actions of *PolicyRules*. For example, consider two *PolicyRules*, *A* and *B*. A's action is to mark traffic with a Differentiated Services Code Point (DSCP) of 5, and B's action is to mark traffic with a DSCP of 0. Assume that the *priority* of *PolicyRule A* is set to 5, and the *priority* of *PolicyRule B* is set to 10. If the *decisionStrategy* is set to MatchAll, and both of these *PolicyRules* are enabled for execution, the DSCP will first be set to 5, but then *end up set to 0*. because the lower priority *PolicyRule* will execute after the higher-priority *PolicyRule* and (in this case) do the wrong thing. Note that if the *decisionStrategy* is set to MatchFirst, the intended action (of setting the DSCP to 5) will happen, because only *PolicyRule A* is allowed to execute.

The correct solution to this dilemma is *not* to constrain the model, since the model cannot know how specific *PolicyRules* are going to be used. Instead, the correct solution is to detect and eliminate conflicts in the policy server itself. This will be described in Chapter 8.

7.4 DEN-ng *PolicyGroup* Class

A *PolicyGroup* is one of the two subclasses of *PolicySet*. This class is a generalized aggregation container. It enables either *PolicyRules* or *PolicyGroups* to be aggregated in a single container. Note that loops, including the degenerate case of a *PolicyGroup* that contains itself, are not allowed when *PolicyGroups* contain other *PolicyGroups*. Also realize that the semantics for this class are "just" those of a container. That is, its purpose is to group *PolicySets* for evaluation; it does not of itself contain any execution semantics.

A *PolicyGroup* has a single method, called *getComponents*. This method returns the set of components (*PolicyGroups* and/or *PolicyRules*) that are contained in this higher-level *PolicyGroup*. Arguments to this method define how many levels to search as well as whether to return all components or just a specific type of component. The return type is an integer, to enable success or failure of the operation to be indicated.

A *PolicyGroup*, being a container, is a perfect way to distribute notifications of events to other components of the PolicyServer. This is done by associating a *PolicySet* with a *PolicyGroup*, (Figure 7-3).

The *ControlsExecutionOf* association is used to define the *PolicyEventSet* that controls the execution of *PolicyRules* in the selected *PolicyGroup*. It is implemented as an association class to provide additional semantics that define a filtered subset of elements in the *PolicyGroup* to be executed by receiving this set of *PolicyEvents*.

PolicyGroupExecutionDetails is an association class that defines the semantics associated with a *PolicyEventSet* being applied to a *PolicyGroup*. It has four attributes and one method.

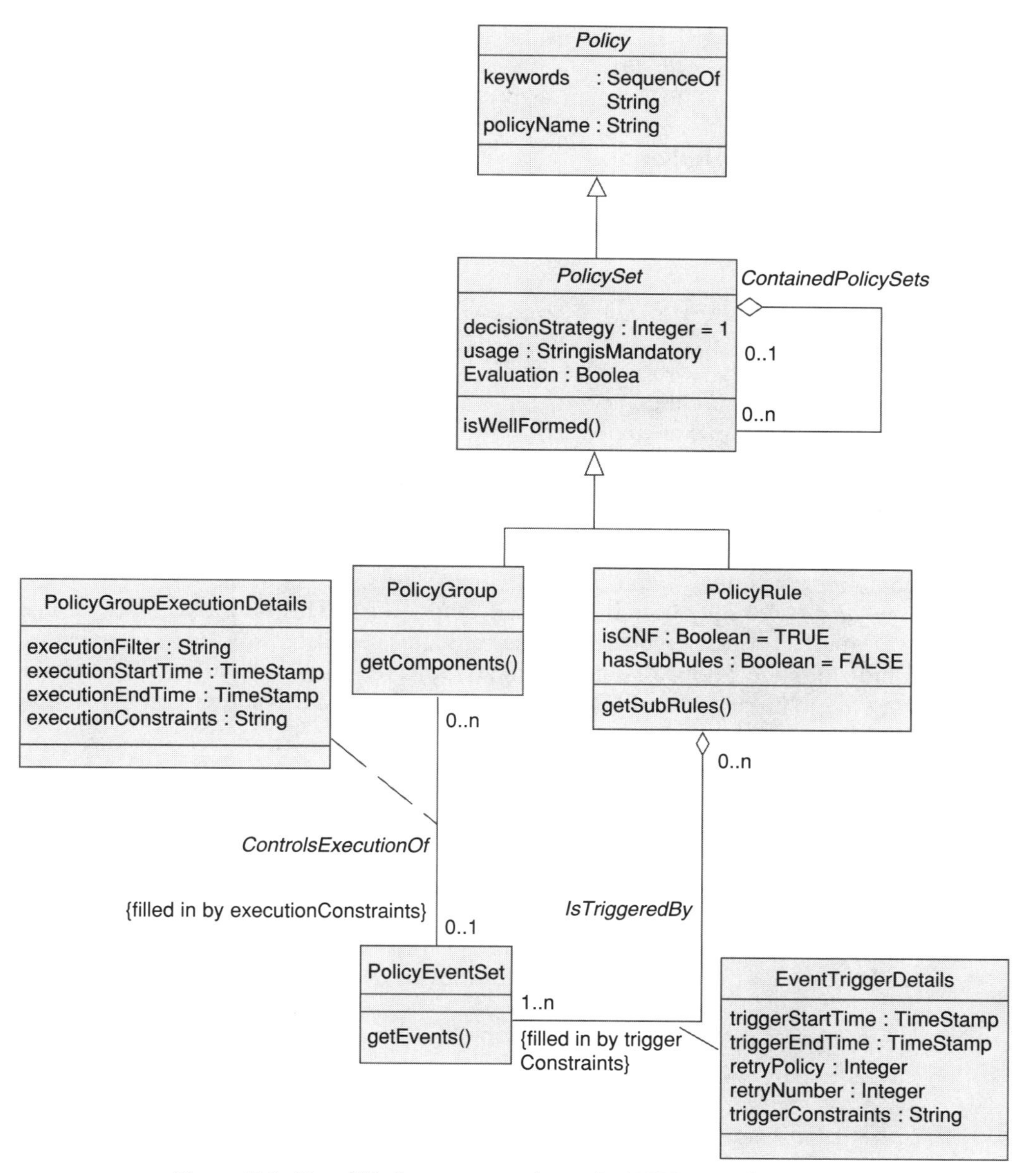

Figure 7-3 Simplified representation of a DEN-ng *PolicyGroup*.

The *ExecutionFilter* attribute is used to define which components in the *PolicyGroup* this *PolicyEventSet* will be passed for evaluation. This attribute enables an external application to explicitly control how different policy elements are contained in a particular *PolicyGroup* are to be evaluated. The *executionStartTime* and *executionEndTime* attributes define the period in which events from this *PolicyEventSet* can be used to start the evaluation of *PolicyRules* that are contained in a *PolicyGroup*. Finally, the *executionConstraints* attribute is a string that contains the specific OCL expression to constrain how a given set of *PolicyEvents* are distributed to a particular *PolicyGroup*. This enables an external application or process to specify how this *EventSet* should be distributed to one or more *PolicyGroups*.

The *executionFailure* method is a reference to a *PolicyAction* object that defines what action(s) to do in case a *PolicyEventSet* is not able to finish being distributed to all identified *PolicyGroups* within its time window.

7.5 DEN-ng *PolicyRule* Class—Basic Concepts

A *PolicyRule* is the other subclass of *PolicySet*. This is a concrete base class for realizing the "event-condition-action" semantics that form a DEN-ng *PolicyRule*. The semantics of this rule are that the rule is evaluated when an event occurs. If the condition clause is satisfied, the pass-action clause (e.g., the set of actions that are associated with the condition clause being satisfied) will be executed (otherwise, the fail-action clause will be executed). For simplicity, this book will just show the pass-action clause. The option to define pass-action and fail-action clauses, though more complex, is of critical importance to certain types of applications.

PolicyRules may be nested within *PolicyRules*. This is often needed in networking to properly represent complex sequences of actions, such as those for bandwidth allocation. Similarly, *PolicyGroups* can be nested within a *PolicyRule*.

A simplified model of the DEN-ng *PolicyRule* class is shown in Figure 7-4.

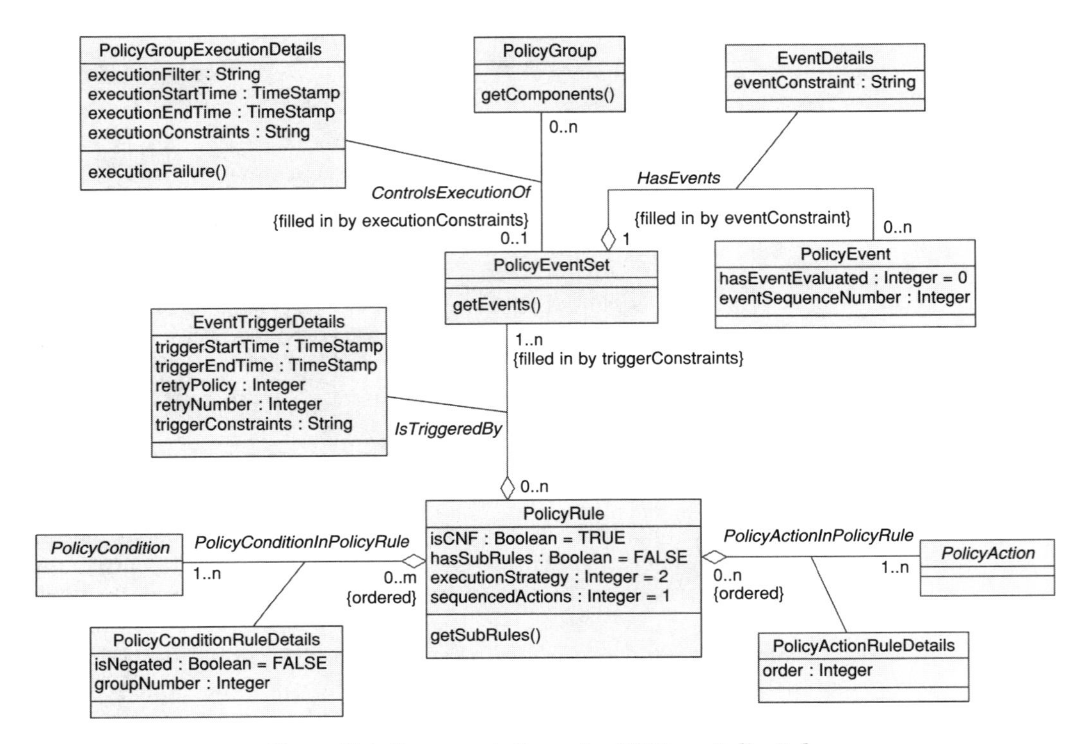

Figure 7-4 Representation of a DEN-ng *PolicyRule*.

7.5.1 Attributes of a *PolicyRule*

The attribute *isCNF* is a Boolean attribute that, if TRUE, defines the condition clause of this *PolicyRule* to be represented in conjunctive normal form (e.g., an AND of ORs). If the value of this attribute is FALSE, the condition clause will be represented in disjunctive normal form (e.g., an OR of ANDs).

There are two basic types of forms that a condition clause can take: a single term, or a collection of terms. These are represented by the *PolicyCondition Atomic* and *PolicyConditionComposite* objects (and their subclasses), respectively. The *isCNF* attribute controls the setting of how multiple condition clauses are treated in the *PolicyRule*. The *PolicyConditionAtomic* and *Policy ConditionComposite* objects will be explained in the section describing *PolicyConditions* later in this chapter.

The attribute *hasSubRules* is a Boolean attribute that, if TRUE, signifies that this *PolicyRule* has one or more sub-rules. Sub-rules are used to enforce a hierarchical nesting of rules, so that parent rules may control the execution and other semantics of sub-rules that they contain. This will be explained in the next section, when advanced concepts and applications of *PolicyRules* are discussed.

The attribute *executionStrategy* is an enumerated integer that defines the overall strategy to be used upon the sequenced actions aggregated by this *PolicyRule*. Defined execution strategies include:

- 0: Unknown (this can be used to indicate an error, or a non-initialized *PolicyRule*)
- 1: Do Until Success
- 2: Do All
- 3: Do Until Failure
- 4: Do All Without Failure or Do Nothing

"Do Until Success" means that actions should be executed according to their predefined order, until successful execution of a single action. At that point, the process should stop (even if there are additional actions that could be executed).

"Do All" means that all actions should be executed that are part of the set, according to their predefined order. This should continue even if one or more of the actions fails.

"Do Until Failure" means that all actions should be executed according to their predefined order until an action fails to execute. At that point, the entire sequence should be stopped.

Finally, "Do All Without Failure or Do Nothing" means that all actions specified should be performed. However, if a failure occurs, then *none* of the actions should be performed. This implies *either* the ability to look-ahead and ensure that the actions will be able to be performed successfully, or the ability to perform a rollback of that action and all previous actions when an error is encountered.

The final attribute in the *PolicyRule* class is called *sequencedActions*. This is an enumerated integer that defines how the ordering of the *PolicyActions* associated with this *PolicyRule* is to be interpreted. Values include:

- 0: Unknown (this can be used to indicate an error, or a non-initialized *PolicyRule*)

- 1: Mandatory (the actions must be done in the indicated order, or else not done at all)
- 2: Recommended (the actions should be done in the indicated order if possible; if this is not possible, it is permissible for them to be done in any order)
- 3: Best Effort (the actions should be done irrespective of whether they can be done in any particular order)

7.5.2 Methods of a *PolicyRule*

There is only one method defined for the *PolicyRule*. This method, called *getSubRules*, returns the set of sub-rules that are contained in this higher-level *PolicyRule*. Arguments to this method define how many levels of nested *PolicyRules* are to be searched.

The return type is an integer, to enable success or failure of the operation to be indicated.

7.5.3 Relationships of a *PolicyRule*

As can be seen from Figure 7-4, there are three aggregations defined for a *PolicyRule*. Each aggregation is implemented as a class, in order to capture the required semantics of that aggregation.

The first aggregaton, *IsTriggeredBy*, defines the set of events that are used to trigger the evaluation of the condition clause in this *PolicyRule*. A *PolicyRule* is considered malformed unless it defines at least one *Event* term. Otherwise, there is nothing to cause the evaluation of the conditions of the *PolicyRule*. This aggregation is implemented as a class in order to capture additional semantics that define how the *PolicyEventSet* is used by the policy system.

The class *EventTriggerDetails* implements the semantics needed for the *IsTriggeredBy* aggregation. This aggregation defines the set of events that can be used to trigger the evaluation of this PolicyRule. It has five attributes. The *triggerStartTime* and *triggerEndTime* attributes define the start and the end of an acceptable period in which the policies in the *PolicyEventSet* can trigger the evaluation of the condition clause in the *PolicyRule*. If the condition clause in the *PolicyRule* is not successfully triggered within this time period, the triggering activity will stop, and the *hasEventEvaluated* attribute of the *PolicyEvent* in the *PolicyEventSet* will be set to 2 (failed to evaluate).

The *retryPolicy* and *retryNumber* attributes work together. The *retryPolicy* attribute is an enumerated integer that defines the policy for retrying the sending of the *PolicyEventSet*. Values include:

- 0: Do *not* retry
- 1: Always retry (until successful acknowledgment)
- 2: Retry the number of times specified in the retryNumber attribute

The *retryNumber* attribute specifies often the *PolicyEventSet* should be resent to the specified *PolicyRule*. If the *triggerEndTime* elapses, the *PolicyEventSet* will not be resent anymore, even if the *retryNumber* indicates that additional retries should be attempted. Instead, the *hasEventEvaluated* attribute of each *PolicyEvent* in this *EventSet* must be set to 2 (failed to be evaluated).

The final attribute, *triggerConstraints*, is a string that contains the specific OCL expression to constrain how a given set of PolicyEvents are used to trigger the evaluation of the condition clause of a particular PolicyRule. Defining this as an attribute enables external applications to populate the constraint that will be applied to this PolicyRule. This is applied to the *PolicyEventSet* end of the *isTriggeredBy* aggregation in Figure 7-4.

The second aggregation to consider is *PolicyConditionInPolicyRule*. This aggregation defines the set of *PolicyConditions* that are contained in this *PolicyRule*. A *PolicyRule* is considered malformed unless it defines at least one *PolicyCondition* term. Otherwise, there is no way to tell if the actions of the *PolicyRule* should be executed or not. Note that the cardinality of this aggregation is "1..n" on the *PolicyCondition* side and 0..n on the *PolicyRule* side. This is a departure from the IETF and DMTF model, which has the cardinality for this relationship as "0..n" on both sides. The reason for this departure is to formally require the presence of a *PolicyCondition* term in a *PolicyRule* (which the DEN-ng cardinality does not the IETF and DMTF definition does not). Remember that cardinalities are read in a "reverse" order. That is, the "1..n" on the *PolicyCondition* side of the *PolicyConditionInPolicyRule* aggregation means that a *PolicyRule* requires 1 or more *PolicyConditions*. Note, however, that the cardinality on the *PolicyRule* side is 0..n. This means that a *PolicyCondition* does not require a *PolicyRule* to be placed in. This is because a *PolicyCondition* can be stored in a *PolicyRepsository* and indirectly referenced by the *PolicyRule*. It also enables *PolicyConditions* to be defined independent of *PolicyRules*. Note also that an OCL expression requiring that the *PolicyConditions* are *ordered* is defined. This is also different from the IETF and DMTF approach, which has no such expression. The result of these changes is that the DEN-ng representation is a tighter, more consistent, and more formal specification of the semantics of a *PolicyRule*. As such, it improves the ability for different implementations to be able to interoperate.

The semantics for the *PolicyConditionInPolicyRule* aggregation are implemented using the *PolicyConditionRuleDetails* class. This class has two attributes. The first, *isNegated*, is a Boolean attribute that, if TRUE, signifies that this condition is negated. The default value for this attribute is FALSE. The second, *groupNumber*, is used to define the group to which the condition is a part. It corresponds logically to the position of parentheses used to group condition terms together. Thus, if the condition clause was, for example:

IF (A and B) OR (C or D) . . .

The *groupNumber* for the "(A and B)" part of the condition would be 1, and the *groupNumber* for "(C and D)" would be 2. This will be explained in more detail in the upcoming section on *PolicyConditions*.

The final aggregation, *PolicyActionInPolicyRule*, is used to define the set of *PolicyActions* that are contained in this *PolicyRule*. A *PolicyRule* is considered malformed unless it defines at least one *PolicyAction* term. Otherwise, there is nothing for the *PolicyRule* to do. Note that the cardinality of this aggregation is "1..n" on both sides. This is another departure from the IETF and DMTF model, which has the cardinality for this relationship as "0..n" on the *PolicyAction* side and 0..n on the *PolicyRule* side. The reason for this departure is to formally require the presence of a *PolicyAction* term in a *PolicyRule* (which the DEN-ng cardinality does, the IETF and DMTF definition does not). Note also that an OCL expression requiring that the *PolicyActions* are ordered is defined. This is also different from the IETF and DMTF approach. Since this is a mirror image of what was done for specifying how a PolicyRule aggregates PolicyConditions in DEN-ng, the same benefits apply.

The semantics for the *PolicyActionInPolicyRule* aggregation are implemented using the *PolicyActionRuleDetails* class. This class defines a single attribute, *order*, which is an unsigned integer that defines the relative position of this action in relation to the overall sequence of actions that are associated with a given *PolicyRule*. Smaller integers indicate earlier positions in the sequence. The value "0" has the special semantics of "don't care".

The value of the *order* attribute is not unique. Thus, if two or more *order* attributes have the same value, they may be performed in any order, but they must all be performed at the appropriate place in the overall sequence.

7.5.4 Summary of the DEN-ng Basic *PolicyRule* Semantics

PolicyRules can be used in a stand-alone or group manner. In a stand-alone manner, it is both an intelligent container as well as a self-contained object. Its container aspect defines the event, condition, and action clauses that make up a *PolicyRule*. Events are used to trigger the evaluation of the condition clause. If the condition clause evaluates to TRUE, the action clause corresponding to a condition clause that has evaluated to TRUE is executed. Otherwise, if the condition clause evaluates to FALSE, the action clause corresponding to a condition clause that has evaluated to FALSE is executed. Note that for simplicity, the figures in this chapter have just shown the case in which action is taken when the condition evaluates to TRUE. Even though DEN-ng enables actions to be defined when the condition clause evaluates to either TRUE or FALSE, practice has shown that this is overkill. It also greatly complicates the design of the PolicyRule, and consequently the implementation of conflict detection and resolution. Practice has shown that handling "just" the case in which actions execute when the condition clause evaluates to TRUE satisfies requirements over 90% of the time. This case corresponds to defining a null action to be taken in case the condition clause evaluates to FALSE.

PolicyRules can be used as a set by grouping them either in a *PolicyGroup*, or in another *PolicyRule*. Each has very different semantics, which will be explained in the next section. Both cases use the *ContainedPolicySets* aggregation shown in Figure 7-2 to define the relationship between the "containing" *PolicyRule* and the "contained" *PolicySet*.

7.5.5 DEN-ng *PolicyRule* Class—Advanced Concepts

DEN-ng provides for two types of containment semantics for *PolicyRules*: grouping and nesting. Their differences will be explained in this section. Again there are some semantic differences between the DEN-ng policy model and the IETF/DMTF policy models in this area.[4,5]

Grouping PolicyRules

Grouping is the simpler of the two methods. It has the semantics of an assembly of *PolicyRules*. The *PolicyGroup* establishes its own hierarchy for evaluation purposes, and all immediate children (whether *PolicyGroups* or *PolicyRules*) that are *directly* contained within a given *PolicyGroup* are treated as being at the same level of containment.

This is primarily used for two reasons. The first is to ensure that different *PolicyRules* are treated as a unit. Their execution is dependent on the *decisionStrategy* attribute of their containing *PolicyRule*. For example, when a user logs in, separate authentication, authorization, and accounting policies might be placed into a common *PolicyGroup*. This helps ensure that these three related policies are loaded together and evaluated as appropriate.

The second reason is to evaluate groups of rules against other *PolicyRules* and *PolicyGroups*. Consider the following set of *PolicyRules* and *PolicyGroups* in Figure 7-5:

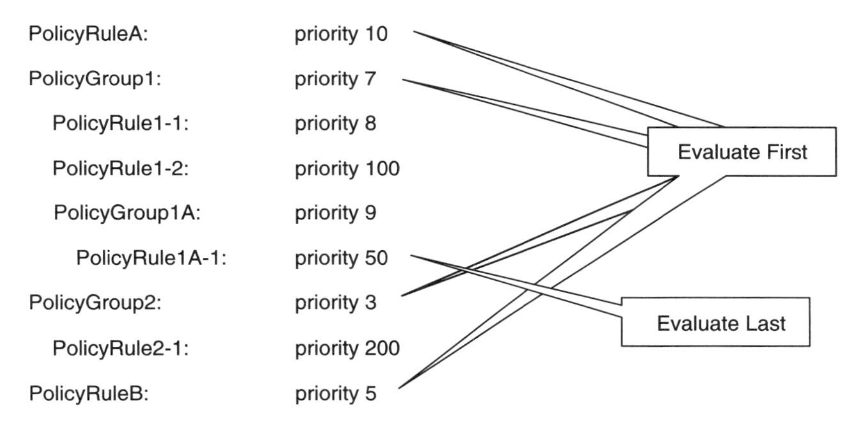

Figure 7-5 Evaluating *PolicyRules* and *PolicyGroups*.

If the above set of *PolicyRules* and *PolicyGroups* were presented as a unit for evaluation, evaluation would proceed as follows. There are three hierarchies of *PolicySets* shown in Figure 7-5. The first hierarchy consists of *PolicyRuleA*, *PolicyGroup1*, *PolicyGroup2*, and *PolicyRuleB*. The second hierarchy consists of *PolicyRule1-1*, *PolicyRule1-2*, *PolicyGroup1A*, and *PolicyRule2-1*. The third hierarchy consists of *PolicyRule1A-1*.

PolicySets are evaluated in hierarchies, with the highest priority being executed first. Thus, we do not look at the strict numerical order of the priorities—we instead look at the priorities of the rules *in the same hierarchy*. Therefore, in our first hierarchy, the execution order is *PolicyRuleA*, *PolicyGroup1*, *PolicyRuleB*, and *PolicyGroup2*, because they are all at the same hierarchy, and their priorities are 10, 7, 5, and 3, respectively.

PolicyRuleA will execute first. Next in priority is *PolicyGroup1*. Since it is a group, we must stop and look into its contents to figure out what should be executed in what order, since a *PolicyGroup* cannot be executed by itself. Thus, we see that the execution order is *PolicyRule1-2*, *PolicyGroup1A*, and *PolicyRule1-1*. Note that *PolicyRule2-1* was not considered. Although it is at the second level of the hierarchy, it is contained within a *PolicyGroup* whose priority was lower than that of the *PolicyGroup* currently being evaluated. Since *PolicyGroup1A* is a group, its contents will be executed before *PolicyRule1-1* because the priority of *PolicyGroup1A* is higher than that at *PolicyRule1.1*. This effectively moves the execution of *PolicyRule1A-1* between *PolicyRule1-2* and *PolicyRule1-1*.

Finally, *PolicyRuleB* will execute, followed by *PolicyGroup2*, which means that *PolicyRule2-1* will be the last to execute, even though it has the highest individual priority.

This example shows the following key points:

- Grouping compares the priority of *PolicySets* at the same level
- Whenever a new hierarchy is entered through a *PolicyGroup*, its *PolicySets* are compared with each other at that level of the hierarchy, before further evaluation of the previous hierarchy is continued
- The absolute priority of a *PolicySet* has nothing to do with its final execution order—it only serves to determine when it will be executed with respect to other *PolicySets* of its same hierarchy level.

Nesting PolicyRules

Nesting is the more complicated of the two methods. It has the semantics of defining one or more *PolicyRules* within a containing *PolicyRule*. This is also called a *nested PolicyRule* or a *sub-rule*. The *PolicyRule* establishes its own hierarchy for evaluation purposes, and all immediate children (whether *Policy Groups* or *PolicyRules*) that are *directly* contained by the parent *PolicyRule* are evaluated in their own context.

The parent *PolicyRule's* condition clause is used as a pre-requisite for evaluating all nested rules. If the parent *PolicyRule's* condition clause evaluates to FALSE, any sub-rules that the parent *PolicyRule* has are skipped. Alternatively, if the parent *PolicyRule's* condition evaluates to TRUE, the set of sub-rules are evaluated according to the decisionStrategy and priority attributes as discussed above. This equates to logically ANDing the condition clause of the parent *PolicyRule* to the condition clauses of each of the sub-rules.

If the parent *PolicyRule's* condition clause evaluates to TRUE, the parent *PolicyRule's* set of actions is executed before any of the actions of the sub-rules are executed. Care must be taken to avoid conflicts of actions taken by the sub-rules in this case.

Example of the Difference in Using Grouping versus Nesting of *PolicyRules*

Rule grouping provides very different semantics than rule nesting. This will be illustrated by a bandwidth allocation example.

The start of the example will be to define two *PolicyRules* that allocate the bandwidth of a device interface as follows:

- IF (protocol is UDP) THEN (guarantee 30% of available bandwidth) (1)
- IF (protocol is TCP) THEN (guarantee 40% of available bandwidth) (2)

Let's suppose we want to divide the bandwidth for UDP- and TCP-based flows by application type. For this example, assume that three UDP applications must have the following guaranteed bandwidth percentages allocated from the bandwidth assigned to all UDP flows as follows: DNS—30 %, TFTP—20%, and SNMP—10%. Furthermore, assume that two TCP applications must have the following guaranteed bandwidth percentages allocated from the bandwidth assigned to all TCP flows as follows: HTTP—40 % and FTP—20%. A flat (i.e., one that does not use nested *PolicyRules*) representation of this assignment looks as follows:

- IF (protocol is UDP AND protocol is DNS) THEN (guarantee 9% of available bandwidth) (3)
- IF (protocol is UDP AND protocol is TFTP) THEN (guarantee 6% of available bandwidth) (4)
- IF (protocol is UDP AND protocol is SNMP) THEN (guarantee 3% of available bandwidth) (5)
- IF (protocol is TCP AND protocol is HTTP) THEN (guarantee 16% of available bandwidth) (6)
- IF (protocol is TCP AND protocol is FTP) THEN (guarantee 8% of available bandwidth) (7)

Equations (3)—(7) are defined by taking the desired bandwidth guarantee of each application and multiplying it by its respective application type percentage. Thus, the DNS bandwidth guarantee is calculated by taking 30% (the desired bandwidth guarantee for DNS) of 30% (the amount of bandwidth allocated to all UDP flows), which yields 9% of the overall bandwidth of the interface.

Rule nesting enables the bandwidth allocation within a given traffic class to be more explicitly specified. Instead of the five grouped *PolicyRules* in the previous representation, two nested rules will be grouped together. This is represented as follows:

- IF (protocol is UDP) THEN (guarantee 30% of available bandwidth) (8)
    - IF (protocol is DNS) THEN (guarantee 30% of available bandwidth) (8a)
    - IF (protocol is TFTP) THEN (guarantee 20% of available bandwidth) (8b)
    - IF (protocol is SNMP) THEN (guarantee 10% of available bandwidth) (8c)
- IF (protocol is TCP) THEN (guarantee 40% of available bandwidth) (9)
    - IF (protocol is HTTP) THEN (guarantee 40% of available bandwidth) (9a)
    - IF (protocol is FTP) THEN (guarantee 20% of available bandwidth) (9b)

There is a big difference in these two representations. To see this, consider the simple case of one UDP application and one TCP application (e.g., DNS and HTTP, respectively) being applied to the interface. In the grouped representation, the bandwidth percentage allocated to these two flows is 9:16, which means that the DNS flow will consume 36% of the bandwidth, whereas the HTTP flow will consume 64% of the bandwidth. In the nested representation, the bandwidth percentage allocated to these two flows is 3:4, which means that the DNS flow will consume 43% of the bandwidth, whereas the HTTP flow will consume 57% of the bandwidth. The difference lies in the fact that the grouped representation allocated bandwidth by multiplying the bandwidth guarantee of the flow by the bandwidth percentage allocated to the type of allocation, whereas the nested representation starts with an overall limit and then takes percentages of that limit.

7.6 DEN-ng Events and EventSets

A *PolicyEvent* is a base class that represents the occurrence of an event, which is used to trigger the evaluation of the condition clause of a *PolicyRule*. Individual policy events can be supplied to a *PolicyRule*, or they can be aggregated in a *PolicyEventSet* (Figure 7-6).

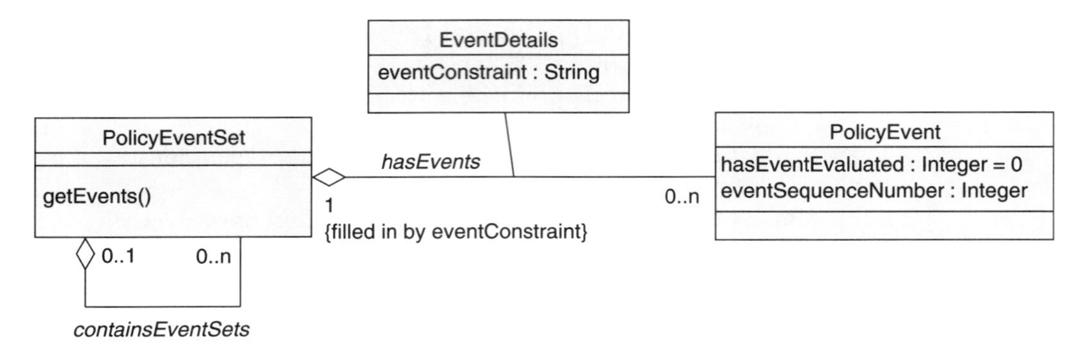

Figure 7-6 Representation of DEN-ng *PolicyEvents* and *PolicyEventSets*.

The *PolicyEvent* class has two attributes. The first, *hasEventEvaluated*, is an enumerated integer that is set by a policy application to define the state of evaluation of this event.

Values include:

- 0: Not yet evaluated
- 1: Evaluated as TRUE
- 2: Evaluated as FALSE

The value of this attribute is that it enables us to build state machines and track which state a set of managed objects are in by tracking the status of the events that were used to trigger a state transition.

The second attribute, *eventSequenceNumber*, is a non-negative integer and represents the particular sequence in which this event is parsed as part of the overall event clause. This will allow a policy engine to exert more granular control to this object as it is evaluated.

The *PolicyEventSet* class represents an aggregation of PolicyEvents, constrained according to the *eventConstraint* attribute of the *EventDetails* aggregation class. This set of *PolicyEvents* is then presented to one or more *PolicyRules* to trigger the evaluation of their condition clauses. Note that the actual constraint to be applied in grouping the *PolicyEvents* is supplied by an external process or application populating the *eventConstraint* attribute of the *EventDetails* class. The *eventConstraint* attribute contains the OCL expression that will be used by the *hasEvents* aggregation to determine what, if any, constraints are to be applied to this *PolicyEventSet*. Again, this approach enables an external application to define the appropriate constraints and supply them to the PolicyServer by populating the *eventConstraint* attribute with the appropriate OCL expression.

7.7 DEN-ng *PolicyStatements*

A *PolicyStatement* models the triplet (variable, operator, value) that is used by both the *PolicyCondition* and *PolicyAction* classes. Note that the semantics differ in how this triplet is used for a *PolicyCondition* compared with a *PolicyAction*; the difference is reflected in the types of operators that are allowed to be used in each case. For conditions, we want the semantics of "variable relates to value," where "relates to" is usually the match operator, but could also be other applicable operators (e.g., a comparison operator). For actions, we want the semantics of "set variable to value." Here, the only operator allowed is the set operator. These semantics are enforced using OCL.

Figure 7-7 shows a simplified representation of the *PolicyStatement* class and its three principal subclasses and relationships. This section will explore these classes and relationships in more detail.

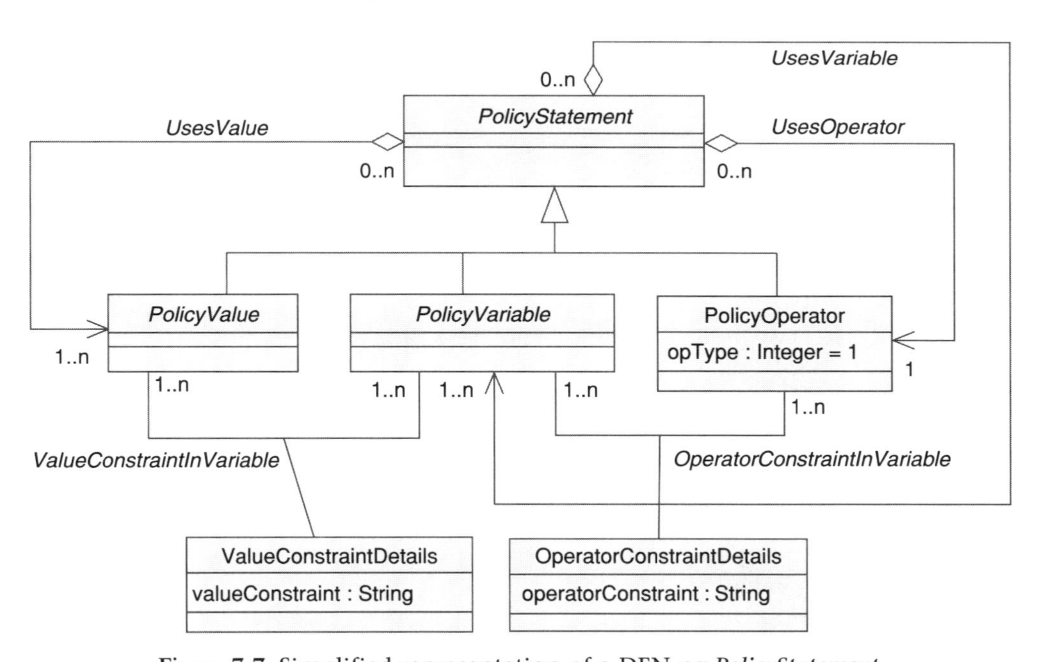

Figure 7-7 Simplified representation of a DEN-ng *PolicyStatement*.

7.7.1 Methods of a *PolicyStatement*

The *PolicyStatement* class defines a single method, *isStatementWellFormed*, which returns TRUE if this *PolicyStatement* is well formed, and FALSE if not. A *PolicyStatement* may evaluate to not well formed if its construction is not completed or if the *PolicyStatement* is not syntactically or semantically correct. For example, a well formed *PolicyStatement* must have a variable, an operator, and a value, and each must work with the other in the given context. This means that values for the *PolicyVariable* that are incompatible with the particular *PolicyValue* chosen cannot be used.

A well-formed *PolicyStatement* is defined as one that has a single variable, an operator, and a value. Furthermore, the datatype of the *PolicyVariable* object determines the datatypes of the *PolicyValue* object, as well as the type of *PolicyOperators* that are permissible to use.

The design of the *PolicyStatement* is based on a single occurrence of a *PolicyVariable*, a *PolicyOperator*, and a *PolicyValue*, though multiple Policy Values may be associated with a single variable. Although a *PolicyStatement* could be designed such that it has a variable clause (e.g., the variable portion of the *PolicyStatement* may consist of multiple terms), this complicates the design. Furthermore, as we explore the design of the *PolicyVariable* and *PolicyValue* class hierarchies, the use of a single object for the *PolicyVariable* and *PolicyValue* portions of a *PolicyStatement* enables the datatypes of each to correlate to the type of object class (e.g., subclasses of *PolicyVariable* and *PolicyValue*) that is being used. This also lends itself to a straightforward implementation of the *ValueConstraintInVariable* association multiple policy values can be accommodated, as long as they are of the same object class (i.e., datatype). This is controlled by the *ValueConstraintInVariable* association. However, if either the *PolicyVariable* object is replaced with a set of objects, or if the *PolicyValue* objects did not have to be of the same datatype, the correlation between datatype and object class, as well as the specification of the association, would be very difficult (if not impossible).

The return type of the *isStatementWellFormed* method is an integer, to enable success or failure of the operation to be indicated.

Values are:

- 0: The *PolicyStatement* is well formed
- 1: The *PolicyStatement* is missing a variable
- 2: The *PolicyStatement* is missing an operator
- 3: The *PolicyStatement* is missing a value
- 4: The object class of the *PolicyValue* does not match the object class of the *PolicyVariable*
- 5: The *opType* of the *PolicyOperator* does not match the object class of the *PolicyVariable*
- 6: The value constraint specified by the *ValueConstraintInVariable* association has not been met
- The operator constraint specified by the *OperatorConstraintInvariable* asssociation has not been met.

7.7.2 Relationships of a *PolicyStatement*

There are three relationships—all aggregations—that are defined for the *PolicyStatement* class. The first is the *UsesVariable* aggregation, which defines the specific *PolicyVariable* that this particular *PolicyStatement* uses. The cardinality is 0..n to 1..n, meaning that a particular *PolicyStatement* must consist of only one or more *PolicyVariables,* and that *PolicyVariables* can be designed and stored independently of *PolicyStatements.* This enables *PolicyVariables* to be stored in a *PolicyRepository* and then referenced indirectly by a *PolicyStatement.* The second is the *UsesOperator* aggregation, which defines the specific *PolicyOperator* that this particular *PolicyStatement* uses. Again, the cardinality is 0..n to 1, meaning that a particular *PolicyStatement* must consist of only a single *PolicyOperator*, but that *PolicyVariables* can be designed and stored independently of *PolicyStatements*. This enables *PolicyVariables* to be stored in a *PolicyRespiratory* and then referenced by a *PolicyStatement*. The third is the *UsesValue* aggregation, which defines the specific *PolicyValue* or set of *PolicyValues* that this particular *PolicyStatement* uses. The cardinality of this aggregation is 0..n to 1..n, meaning that one or more *PolicyValues* can be associated with this particular *PolicyStatement*, and that *PolicyValues* can be designed and stored independently of *PolicyStatements*. This enables *PolicyValues* to be stored in a *PolicyRepository* and then referenced indirectly by a *PolicyStatement*. As we will see, the *ValueConstraintInVariable* association is used to ensure that the object class of the set of *PolicyValues* specified is compatible with the *PolicyVariable* that is chosen.

7.7.3 *PolicyVariables*

PolicyVariable is an abstract class for modeling different types of variables that can be used to form a *PolicyStatement*. The *PolicyVariable* specifies an attribute or concept that should either be matched or in some way compared to a value when the condition is evaluated.

There are three important factors that must be taken into account in the design of *PolicyVariables*. The first is that not every combination of a *PolicyVariable* and a *PolicyValue* creates a meaningful condition that can be used to form a valid *PolicyStatement*. For example, port numbers are integers, so it is meaningless to use a *PolicyValue* that is not an integer. In addition, the value of a port number cannot be negative. These constraints can be met in a number of ways. In the DEN-ng design, the object class of the *PolicyVariable* is used to define the set of allowable *PolicyValue* object classes that can be used to form a *PolicyStatement*. This ensures that the right type of *PolicyValue* object is matched up with a particular type of *PolicyVariable*.

Second, just because a *PolicyValue* is of the correct object class (which in turn means that it is of a certain datatype) does not mean that it can be meaningfully compared to a *PolicyVariable*. To continue the above example, a port whose value is 1,000,000 is meaningless, because that far exceeds its largest legal value. This is handled in this model by specifying constraints (using OCL) that apply to a particular variable that must be obeyed by any value to which it is being compared. As shown in Figure 7-7, the *ValueConstraintInVariable* association binds a *PolicyValue* to a *PolicyVariable*. This association relates a *PolicyValue* (or set of *PolicyValues*) to a particular *PolicyVariable* and defines any constraints that exist between the *PolicyValue* (or set of *PolicyValues*) and the *PolicyVariable*. For example, when modeling a source port variable, the specific values that can be assigned to the *PortSourceVariable* are limited to the range of values that represent legal assignments for this source port.

In order to provide an extensible means to add constraints, this association is implemented by an association class. This class, *ValueConstraintDetails*, contains a single attribute called *valueConstraints*. Any constraints that are required in this binding are added to this attribute. The constraints themselves are specified using OCL, which contains the specific OCL expression that defines how a given *PolicyValue* is constrained by a particular *PolicyVariable*. The cardinality of this association is "1..n" on the *PolicyVariable* side and "1..n" on the *Policy Value* side, meaning that for this particular *PolicyVariable*, this set (one or more) of *PolicyValues* have the same constraint. Defining this as an attribute enables external applications to populate this constraint.

Another type of constraint is shown by the *OperatorConstraintInVariable* association. This association defines any constraints that exist between the type of operator in a PolicyStatement and the PolicyVariable in the PolicyStatement. For example, if the type of PolicyVariable is IPProtocolVariable, it makes no sense to have relational or fuzzy operators used. On the other hand, if the type of PolicyVariable is IPv4Variable, it may make sense to use fuzzy operators or string matching operators, since an IPv4Variable can also be a host name. The constraints themselves are specified using OCL, which contains the specific OCL expression that defines how a given *PolicyOperator* is constrained by a particular *PolicyVariable*. The cardinality of this association is "1..n" on the *PolicyVariable* side and "1..n" on the *PolicyOperator* side, meaning that for this particular *PolicyVariable*, this set (one or more) of *PolicyOperators* have the same constraint. Defining this as an attribute enables external applications to populate this constraint.

7.7.4 Standard Versus Custom *PolicyVariables*

The DEN-ng borrows this idea from Moore *et al.*[5] (where this was called implicit versus explicit), but enhances and simplifies its semantics and implementation. The name change is indicative of these semantic changes.

The idea is that a standard can and should only define a set of common variables that are largely application independent. Clearly, applications need the freedom and flexibility to create their own application-specific extensions. Therefore, the design of the DEN-ng *PolicyVariable* class hierarchy is as shown in Figure 7-8:

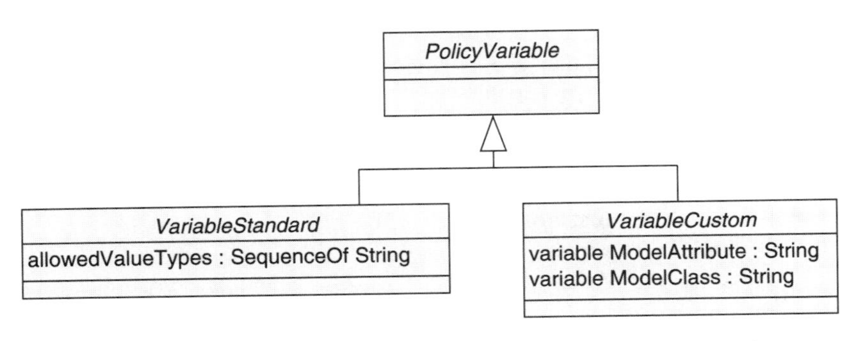

Figure 7-8 Top-level design of the *PolicyVariable* class hierarchy.

The purpose of the *PolicyVariable* class is to serve as a single point wherein relationships to other classes can be defined. This enables all of its subclasses to inherit these relationships.

VariableCustom is is an abstract base class providing two basic attributes to define custom variables that can be used in an application-specific fashion. These two attributes are called *variableModelAttribute* and *variableModelClass*. The *variableModelAttribute* is a string attribute that defines the name of the attribute within the class specified in the *variableModelClass* attribute that is to be evaluated or set as a *PolicyVariable*. The *variableModelClass* is a string attribute that defines the class name whose attribute is to be evaluated or set as a *PolicyVariable*. This combination enables new custom subclasses of *VariableCustom* to be defined that specify the class and attribute they are modeling. These new subclasses can be found by users of the current DEN-ng schema by searching for these two properties. That also enables DEN-ng users to immediately understand the purpose of new extensions.

Another use of the *VariableCustom* class is to turn the attribute of a DEN-ng class into a variable that can be tested. For example, imagine a class *SLA* that has an attribute *hasViolation*. An application may need to test this value and, if it is TRUE, change the type of queuing being used to help recover from the SLA violation. This can be easily accommodated through this class.

VariableStandard is the abstract base class for defining a set of standardized *PolicyVariables*. This set will be added to over time and represents a set of common variables that are useful in a variety of PBNM applications. This class defines a single attribute, *allowedValueTypes*, which is an array of string attributes that define the set of class names (which correspond to a set of datatypes) that are allowed to be used in conjunction with this PolicyVariable. Each string provides a class name, which corresponds to a particular subclass of PolicyValue. Each allowed subclass defines an object with a compatible datatype and semantics. Thus, this variable establishes a pre-defined link between which set of *PolicyValues* can be used with a particular *PolicyVariable*. This set of definitions can be preloaded by an application and changed between applications. Thus, it provides a set of changeable constraints that control the data integrity of a *PolicyStatement*.

The subclasses of *VariableStandard* are a set of classes that define the semantics of commonly occurring variables that occur in PBNM applications. These subclasses are optimized and more powerful when compared to Moore *et al.*[6] The classes cover the following common concepts:

- Source and destination IPv4 and IPv6 addresses
- IP version and protocol information
- IPv6 flow ID information
- Source and destination port numbers
- Source and destination MAC address numbers
- DSCP, ToS, 802.1p, and 802.1q information
- VLAN information
- Time periods
- General information encoded as strings, bit strings, or distinguished names
- Vendor-specific extensions

Figure 7-9 shows these subclasses:

Many of the classes shown in Figure 7-9 above do not have any attributes. To see why, consider the role of the *DSCPVariable* object. It will form a condition term as follows:

IF DSCP == Value . . .

In this example, we are simply testing to see if the value of the DSCP is a certain number. Note that if we change this to a constant, as in:

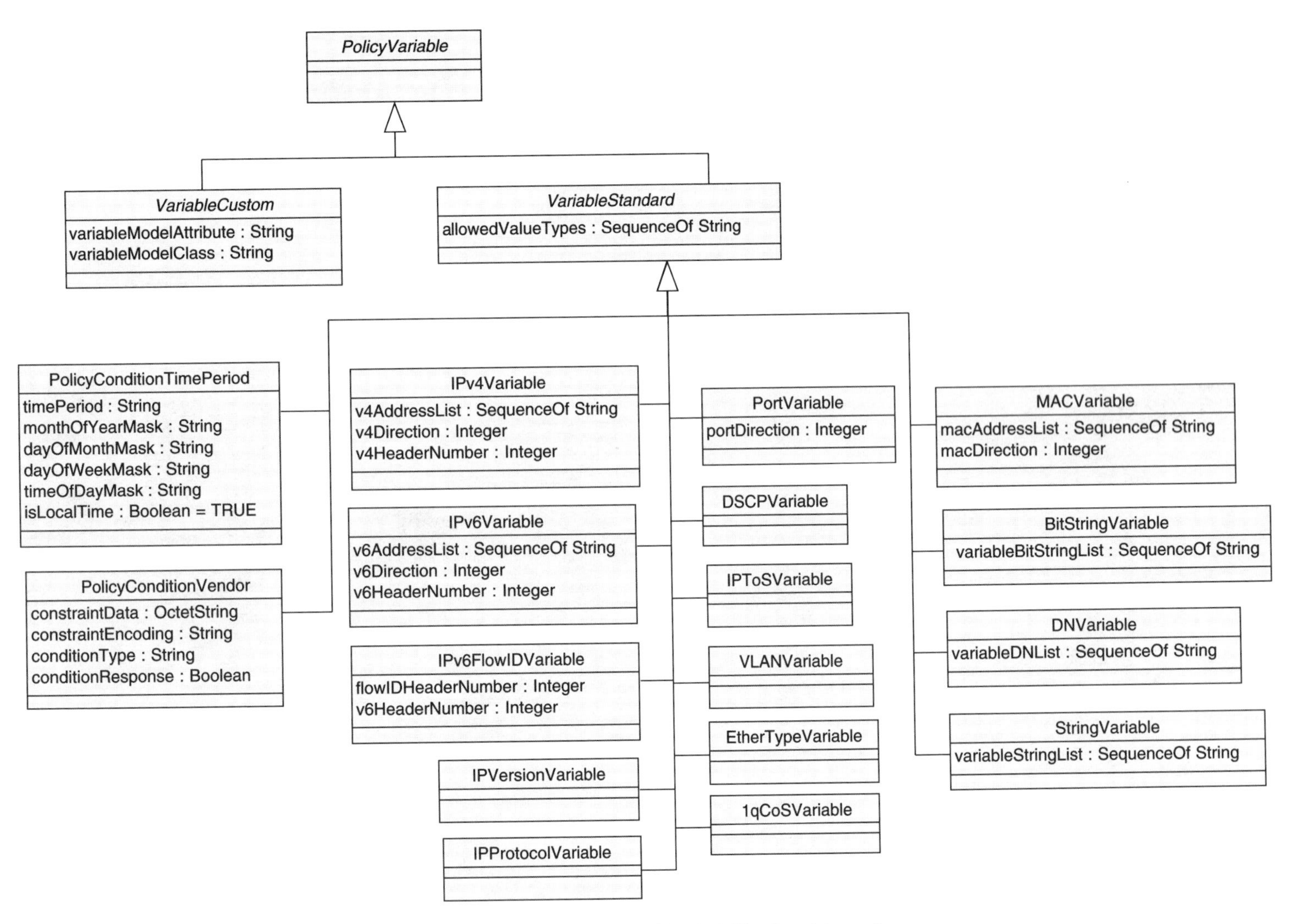

Figure 7-9 The DEN-ng *PolicyVariable* class hierarchy.

IF DSCP == SilverService . . .

The above still holds; we just need to map the constant *SilverService* to a *PolicyValue*.

Other classes do have attributes. In all cases, these attributes are used to specify more detail about the type of entity that we are testing. For example, consider *IPv4Variable*. First, there is a significant difference between different types of addresses. Figure 7-9 shows two such classes—one for IPv4 and another for IPv6. This is because the basic format of an IPv4 address is significantly different from that of an IPv6 address. Similarly, the format of an IPX address is different, (this, and other types of addresses, are not shown in Figure 7-9 in order to simplify it.) Thus, each of these significantly different formats translates into a different class.

There are many ways that an IPv4 address can be specified. Plus, details concerning the type of address (source or destination) and which address (since many can be present in a flow) need to be specified. Therefore, three attributes have been added to and are described as follows.

The *addressList* attribute is an unordered list of strings, each specifying a single IPv4 address, a hostname, or a range of IPv4 addresses. A single IPv4 address can be specified in dotted decimal with mask notation (e.g., 121.1.1.2 255.255.255.252) or in CIDR format (178.21.4.10/12). Hostnames are specified following RFC 1035.

The *direction* attribute is an enumerated integer that defines if this variable is used to represent source addresses, destination addresses, or both.

Finally, the *v4HeaderNumber* attribute is an enumerated integer that defines the particular packet header that contains the IPv4 address information. Specifying the packet header is necessary in case the address is tunneled. The syntax of this attribute assigns an integer to represent each header, starting at 0 for the outermost and increasing monotonically for each inner header.

Most of the PolicyVariables are self-explanatory, the exception being Policy TimePeriodCondition, PolicyConditionVendor, BitStringVariable, DNVariable, and StringVariable.

PolicyTimePeriodCondition is a concrete class that provides the capability of enabling or disabling a *PolicyCondition* according to a pre-determined time schedule. This class can then effectively enable or disable evaluation of the entire condition clause. The simplest way to do this is to specify a value for the *TimePeriod* class attribute; of course, this may be modified by other class attributes as necessary. These six attributes are defined as follows:

- The *timePeriod* attribute is a string that defines an overall range of calendar dates and times over which a *PolicyRule* is valid. As such, this is the most basic way to use date and/or time to enable or disable a given *PolicyRule*. This attribute is based on time as defined in RFC 2445. Briefly, this defines time as consisting of a starting and an ending date and time, separated by a forward slash ("/"). The second date and time must be later than the first.

  Dates and times are expressed as substrings of the form: "yyyymmdd Thhmmss." For example:
  20000101T080000/20000131T120000

is interpreted as:
January 1, 2000, 0800 through January 31, 2000, noon

- Two special strings are also recognized:
  - If the first date/time is replaced with the string "THISANDPRIOR," this indicates that a *PolicyRule* is valid, from now until the date/time that appears after the "/."
  - If the second date/time is replaced with the string "THISANDFUTURE," this indicates that a *PolicyRule* becomes valid on the date/time that appears before the "/", and remains valid from that point on.
- The *monthOfYearMask* attribute is a string that refines the definition of the valid time period that is defined by the *timePeriod* attribute. This is done by explicitly specifying the months when the policy is valid. Thus, the *timePeriod* attribute is used to specify the overall period during which the policy might be valid, and the *monthOfYearMask* attribute is used to pick out the specific months within that period when the *PolicyRule* is valid. This attribute consists of a fixed string of 16 bits. The first 12 bits correspond to the 12 months of the year, beginning with January and ending with December. The last four bits are always set to "0." For each month, the value "1" indicates that the policy is valid for that month, and the value "0" indicates that it is not valid. If this attribute is not specified, the *PolicyRule* is treated as valid for all 12 months.
- The *dayOfMonthMask* attribute is a string that refines the definition of the valid time period that is defined by the *timePeriod* attribute. This is done by explicitly specifying the days of the month when the policy is valid. It works in a similar fashion to the *monthOfYearMask* attribute, except that the *dayOfMonthMask* attribute consists of a fixed string of 64 bits. The first 31 bits correspond to the days of the month counting from the beginning, followed by 31 more bits identifying the days of the month counting from the end, followed by 2 bits that are always set to "0." For months with fewer than 31 days, the digits corresponding to days that the months do not have (counting in both directions) are ignored. The encoding of the 62 significant bits in the octet string matches that used for the schedDay object in the DISMAN-SCHEDULE-MIB.
- The *dayOfWeekMask* attribute is a string that refines the definition of the valid time period that is defined by the *timePeriod* attribute. This is done by explicitly specifying the days of the week when the policy is valid. It works in a similar fashion to the *monthOfYearMask* attribute, except that the *dayOfWeekMask* attribute consists of a fixed string of 8 bits. The first 7 bits identify the 7 days of the week, beginning with Sunday and ending with Saturday, followed by 1 bit that is always set to "0." For each day of the week, the value "1" indicates that the PolicyRule is valid for that day, and the value "0" indicates that it is not valid.
- The *timeOfDay* attribute is a string that refines the definition of the valid time period that is defined by the *timePeriod* attribute. This is done by explicitly specifying the time of day when the policy is valid. It works in a similar fashion to the *monthOfYearMask* attribute, except that the *timeOfDay* attribute consists of a fixed string formatted in the style of RFC 2445: a time string beginning with the character "T," followed by a

forward slash character ("/"), followed by a second time string. Each time period is expressed as a substring in the form "Thhmmss."
- The *isLocalTime* attribute is a Boolean attribute that, if TRUE, signifies that the times represented in the *timePeriod* attribute and in the various mask attributes represent local times. If the value of this attribute is FALSE, the times represented in the *timePeriod* attribute and in the various mask attributes represent UTC times.

The *PolicyTimePeriodCondition* attribute was originally defined as a subclass of *PolicyCondition* in the IETF and DMTF versions. The DEN-ng model changes this to a subclass of *PolicyVariable* because the effect of the attributes in the *PolicyTimePeriodCondition* class are to define a variable that either can be tested (when used in a *PolicyCondition*) or must be set or cleared (when used in a *PolicyAction*). The IETF/DMTF design could not use their version of the *PolicyTimePeriodCondition* class in a *PolicyAction* because it was defined as a subclass of the *PolicyCondition* class. Clearly, actions can be valid only for a particular time period, so this was a flaw. Furthermore, it was impossible in the IETF/DMTF design to isolate the effect of testing versus setting or clearing one or more fields of the time period, since it was defined as an atomic condition term. Note also that making this class a subclass of *PolicyVariable* makes for a more consistent design.

The *PolicyConditionVendor* class was also originally defined as a subclass of *PolicyCondition* in the IETF and DMTF models; the DEN-ng design moves this to a subclass of *PolicyVariable* for the same reasons as *PolicyTimePeriod Condition* was moved to be a subclass of *PolicyVariable*. In addition, it is undesirable to mask the entire condition clause (e.g., the variable, operator, and value): this makes for a system that is inherently non-reusable.

In DEN-ng, *PolicyConditionVendor* is a concrete base class that provides a general extension mechanism for representing the variable term of a condition clause that have not been modeled with the attributes specified in this model. This class uses two properties (*constraintData* and *constraintEncoding*) for defining the content and format of a vendor-specific condition, and a third property (*conditionType*) that can be used to define the type of condition that this instance is checking. Its fourth attribute, *conditionResponse*, is a Boolean attribute that is used to allow a "standard" definition of a TRUE/FALSE response to be made by an external system attached to the model at this point. This will allow integration with a generalized policy processing engine. Note that these two attributes are not defined by the IETF/DMTF model. Not only does this preclude easy and standardized integration with external policy components, it also prevents treating this object as a type of *PolicyCondition*. These two attributes enable custom PolicyConditions to be encoded and evaluated, and given a standard response value. The *constraintData* and *constraintEncoding* attributes are defined as follows:

- The *constraintData* attribute is an OctetString that provides a general extension mechanism for representing *PolicyConditions* that have not been modeled with specific attributes (as specified in this model). The format of this attribute is determined by the OID value stored in the *constraintEncoding* attribute of this class. The *constraintEncoding* attribute provides a convenient and simple mechanism by which a policy server can determine whether it supports the values stored in an instance of *constraintData*. If it does not recognize the OID value in the *constraintEncoding* attribute, it will not know how to interpret the data in the corresponding *constraintData* attribute. The action for the policy decision point to take in case it does not recognize the format of this data could itself be modeled as a *PolicyRule*, governing the behavior of the policy server.

- The *constraintEncoding* attribute is a string used to represent an OID that in turn identifies the encoding and semantics of the *constraintData* attribute in this instance.

BitStringVariable, *DNVariable*, and *StringVariable* are used to represent a single or set of bit string, distinguished name, and string variables, respectively. Therefore, only *BitStringValue*, *DNValue*, and *StringValue* classes can be used in the value portion of the condition expression with these *PolicyVariables*.

7.7.5 *PolicyOperators*

Most conditions have the common form:

PolicyVariable MATCHES PolicyValue

The DEN-ng model formalizes the above by introducing the notion of a *PolicyOperator* class (Figure 7-10):

Figure 7-10 The DEN-ng *PolicyOperator* class.

The *opType* attribute is defined in order to simplify the overloading of the semantics of the match operator. To see why this is necessary, consider a simple and common condition term:

DestinationPort match 80

In this case, the match operator is replaced by the equal operator. However, consider the following cases:

- *DestinationPort match* {80, 389, 8080}
- *DestinationPort match* {1..99, 300..1000}
- *SourceIPAddress match* acme.com

The first example is not a simple equality operator; rather, this is an example of the "IN" operator, which matches the value of the destination port to one of the values in the set {80, 389, 8080}. The second example matches the value of the destination port to one of the values in either of the ranges {1..99} or {300..1000}. The third example first looks up the IP address in DNS, and then uses the equality operator to compare it to the value of *SourceIPAddress*.

In addition, there is cause to use relational operators, as in this example:

IF DSCP is greater than or equal to SilverService . . .

In the above example, the value of *DSCP* is greater than or equal to the value of the constant *SilverService*. This means that an action will be taken if the marking for this traffic is for a class of service of Silver or higher (such as Gold).

These examples illustrate the context-dependant nature of the *match* operator. The actual operator that should be used depends on the *PolicyVariable* and *PolicyValue* instances that make up the term. Although the above examples have used a condition term, actions also require an operator. The two most common operators for actions are SET and CLEAR. Thus, the *opType* attribute is defined as an enumerated integer that specifies the various types of operators allowed for this *PolicyStatement*. Values include:

- 1: Match
- 2: Greater than
- 3: Greater than or equals
- 4: Less than
- 5: Less than or equals
- 6: Equals
- 7: Not equals
- 8: IN
- 9: NOT IN
- 10: SET
- 11: CLEAR

7.7.6 *PolicyValues*

The *PolicyValue* class is an abstract base class for modeling different types of values that occur in a *PolicyStatement*. Remember that a *PolicyStatement* can be used to form a *PolicyCondition* or a *PolicyAction*; this will be discussed shortly. The *PolicyValue* specifies an attribute that should either be set or cleared (if used in a *PolicyAction*) or matched or compared in some way to a value of the *PolicyVariable* in a *PolicyCondition*.

As with the design of the *PolicyVariable* class hierarchy, the *PolicyValue* class defines two immediate children. These two subclasses—*ValueCustom* and *ValueStandard*—are both abstract base classes. They are used to define two hierarchies: one for enabling developers to define application-specific extensions, and the other for providing a set of standard extensions that are common to a wide variety of PBNM applications.

ValueCustom is an abstract base class that provides two basic attributes to define custom value objects that can be used in an application-specific fashion. These two attributes are called *valueModelAttribute* and *valueModelClass*. The *valueModelAttribute* is a string attribute that defines the name of the attribute within the class specified in the *valueModelClass* attribute that is to be evaluated or set as a *PolicyValue*. The *valueModelClass* is a string attribute that defines the class name whose attribute is to be evaluated or set as a *PolicyValue*. This combination enables new custom subclasses of *ValueCustom* to be defined that specify the class and attribute that they are modeling. Users of the current DEN-ng schema can find these new subclasses by searching for these two properties. That also enables DEN-ng users to immediately understand the purpose of new extensions.

ValueStandard is the abstract base class for defining a set of standardized *PolicyValues*. This set of *PolicyValue*s will be added to over time and represents a set of common values that are useful in a variety of PBNM applications. The subclasses of *ValueStandard* are a set of classes that define the semantics of commonly occurring variables that occur in PBNM applications. These subclasses are optimized and more powerful when compared to the design defined in Moore et. al.

BitStringValue, *BooleanValue*, *DNValue*, and *StringValue* are used to represent bit string, Boolean, distinguished name, and string values, respectively. The other two classes, *IPAddressValue* and *MACAddressValue*, are used to represent the values of IP addresses and MAC addresses respectively.

The *IPAddressValue* class is used to provide a list of IPv4 addresses, IPv6 addresses, ranges of IPv4 addresses, ranges of IPv6 addresses, and hostname values to be matched against in a policy condition. Hostnames are matched against another valid IP address by first converting the hostname to an IP address, and then matching the two IP addresses together. The three attributes of this class are defined as follows:

- The *valueAddressFormat* attribute is an enumerated integer that contains one of five different representations for an address or address range (a single address in dot notation, an address range specified by an address in CIDR notation, an address range specified by a starting address in dot notation and an ending address in dot notation, separated by "-", an address range specified by an address and mask (the address and mask are represented in dot notation, separated by a comma), and a single hostname, as specified in RFC 1035
- The *typeOfAddress* attribute is an enumerated integer used to specify the type of IP address that this instance represents (IPv4, IPv6, or hostname)
- The *addressValueList* attribute is an array of string values. Each string value contains a particular value on which to filter. Note that all values must have the format specified in the format attribute of this class.

The *MACAddressValue* class is used to define a list of MAC addresses and MAC address range values as a list of strings, each specifying a MAC address or a range of MAC addresses according to the canonical 802 MAC address format. Each string entry is either a single MAC address or a range of MAC addresses specified by an address and mask. The three attributes of this class are defined as follows:

- The *macFormat* attribute is an enumerated integer that defines how the MAC Address is represented (either as a single address, or as an address with mask)
- The *macValueList* attribute is an array of strings, each of which contains a MAC Address. Each string must be of the same format, which is specified in the macFormat attribute of this object.

Figure 7-11 shows these subclasses:

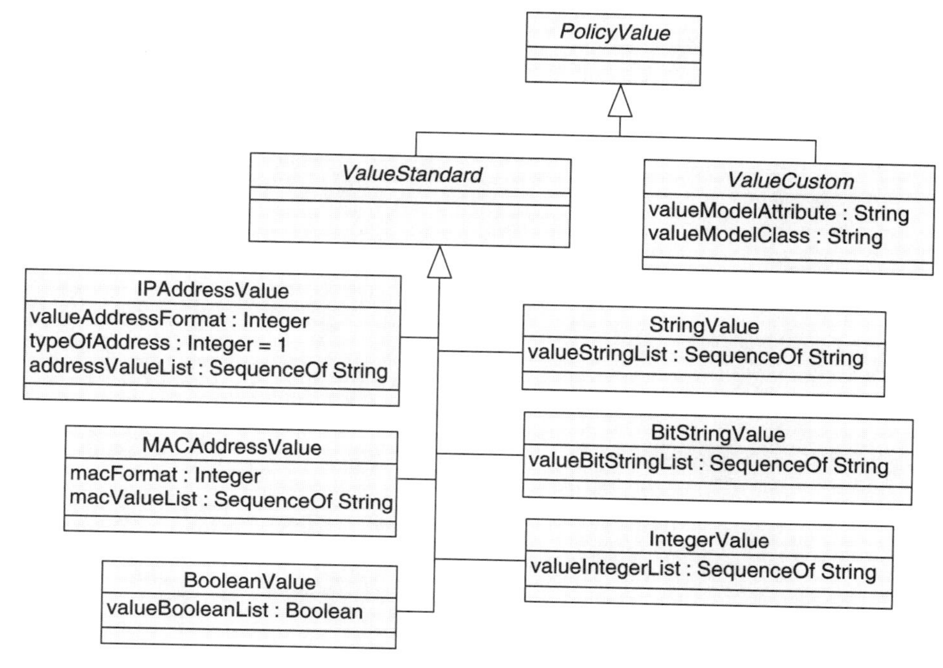

Figure 7-11 The DEN-ng *PolicyValue* class hierarchy.

7.8 DEN-ng *PolicyConditions*

The DEN-ng *PolicyCondition* class is an abstract base class that represents how to form the condition clause of a *PolicyRule*. This class can be used to represent ad hoc or reusable policy conditions (these will be covered later in this chapter). Basic *PolicyConditions* are of the form:

{variable, operator, value}

This gives the semantics of "IF the condition is TRUE (or FALSE)."
In DEN-ng, there are two types of *PolicyCondition* subclasses, called *PolicyConditionAtomic* and *PolicyConditionComposite*. The DEN-ng design of *PolicyConditions* is based around the use of a *PolicyConditionAtomic* object to represent the semantics of a condition term, and the *PolicyConditionComposite* object to aggregate *PolicyConditionAtomic* objects to form more complex condition clauses. This is shown in Figure 7-12:

There are several differences between the above DEN-ng model in Figure 7-12 and the IETF and DMTF approaches. The ones shown in Figure 7-12 are:

- The composite pattern is used in DEN-ng; this is not present in the IETF and DMTF approaches. The importance of using this pattern is the inherent extensibility that it provides, which leads to a simpler and easier implementation.
- The above causes a list of differences, not just in the structure of the *PolicyConditionComposite* (versus the *CompoundPolicyCondition* class

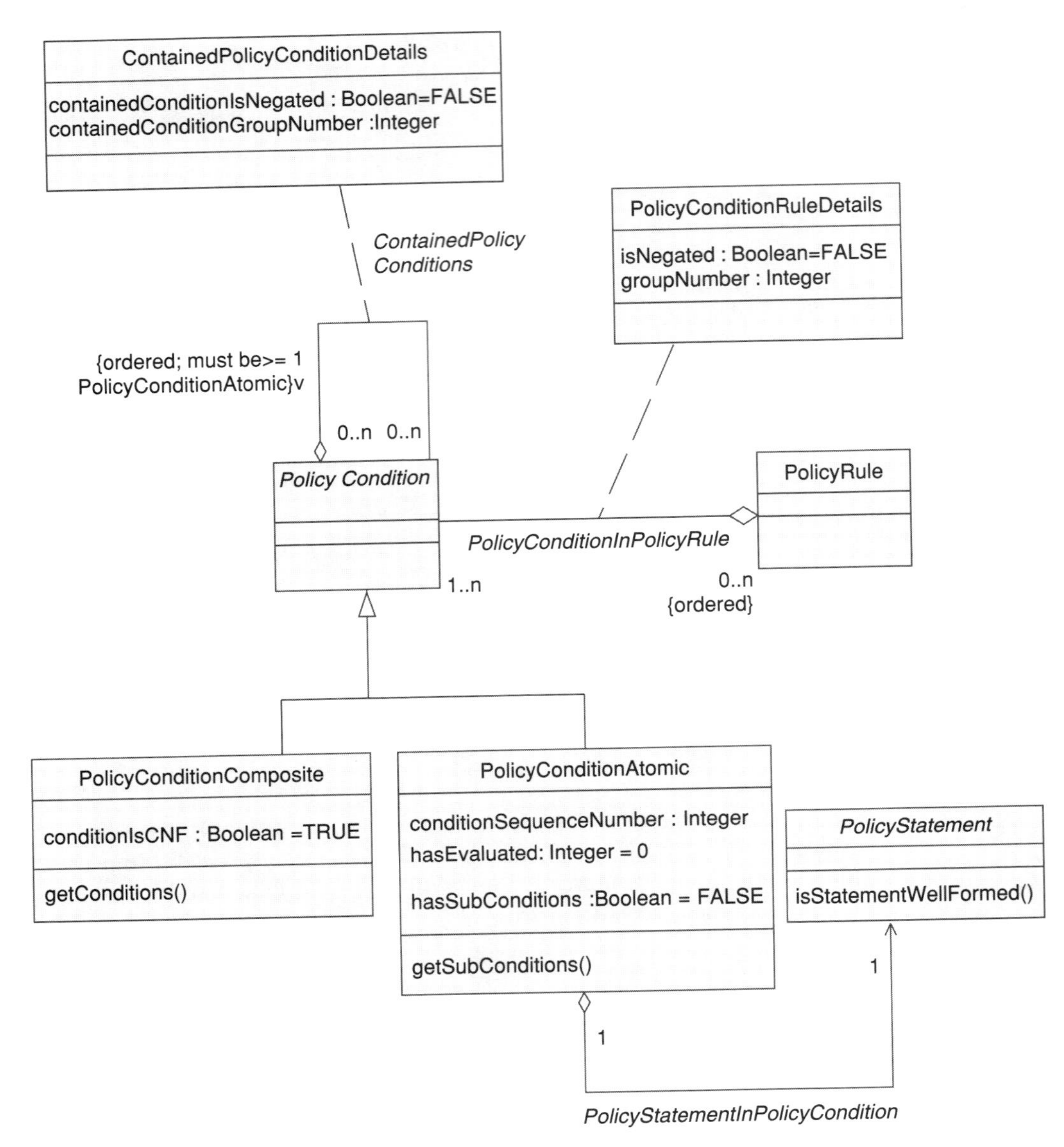

Figure 7-12 The DEN-ng *PolicyCondition* Class Hierarchy.

of the IETF/DMTF approach), but in the attributes and relationships in which each subclass of *PolicyCondition* participates.

- The structure of the *PolicyConditionAtomic* class is changed. Note that in the DEN-ng model, the *PolicyStatement* is used to provide the content of the condition—this provides an object-oriented view of the contents of the condition, improves reusability, and differentiates the semantics of the *PolicyConditionAtomic* object from all other *PolicyConditions*. In the IETF/DMTF model, there is no dependence on a *PolicyStatement* object. This misses the chance to make the contents of a *PolicyCondition* reusable.

- The recursive aggregation on *PolicyCondition* is a simple way to support simple and compound policy conditions (rather than having multiple classes and relationships as in the IETF and DMTF approach).
- The aggregation class on *PolicyCondition*, called *ContainedPolicy ConditionDetails*, provides a consistent way to define semantics for this relationship (versus a set of simple aggregation classes in the IETF/DMTF models).
- No subclasses of *PolicyConditionAtomic* are defined, whereas the IETF and DMTF approach defines *SimplePolicyCondition*, *PolicyCondition TimePeriod*, *CompoundPolicyCondition*, and *PolicyConditionVendor* as subclasses of *PolicyCondition*. Remember that two of these, *Policy ConditionTimePeriod* and *PolicyConditionVendor*, have already been analyzed and shown to be better implemented as *PolicyVariables*.
- In DEN-ng, the concept of a *PolicyStatement* has been developed to serve as the standard way for populating *PolicyConditionAtomic* (and as we will see shortly, *PolicyActionAtomic*) classes with a single condition (or action) term. This symmetry is indicative of a good, consistent model and will also ease implementation. This is not present in the IETF and DMTF approaches.

More complicated *PolicyConditions* enable one or more *PolicyConditions* to be nested inside of a higher-level *PolicyCondition*. This is termed a compound *PolicyCondition* and will be discussed later in this section.

The *ContainedPolicyConditions* aggregation is used for two distinct purposes. First, it is used to gather together discrete *PolicyConditions* to form a *PolicyCondition* clause. In this usage, the set of individual *PolicyConditionAtomic* objects are aggregated by a *PolicyConditionComposite* object. The second use is to form compound *PolicyConditions*. In this case, the *PolicyConditionAtomic* object aggregates *PolicyConditionAtomic* and/or *PolicyConditionComposite* objects. The former provides a simple nested *PolicyCondition*, whereas the latter provides structure for grouping the sub-conditions. Note that the aggregation specifies that it is optional, that the contained *PolicyConditions* are ordered, and that at least one *PolicyConditionAtomic* object must be present if it is instantiated.

This aggregation is implemented using the *ContainedPolicyCondition Details* aggregation class. This aggregation defines two attributes. The *isNegated* attribute is a Boolean attribute that, if TRUE, signifies that this condition term is negated. The default value for this attribute is FALSE. The *groupNumber* attribute is used to define the group to which the condition is a part of. It corresponds logical to the position of parentheses used to group condition terms together.

7.8.1 The *PolicyConditionComposite* Class

Since the composite pattern is being used, the *PolicyConditionComposite* class is designed to be an intelligent container. The *PolicyConditionComposite* class is the base class for all complex policy conditions. A complex *PolicyCondition* consists of an aggregation of *PolicyConditionAtomic* and/or *PolicyCondition Composite* objects, which in turn form a complex Boolean statement. Such an object still evaluates to a single Boolean TRUE or FALSE value.

The *PolicyConditionComposite* class defines one attribute and one method. The *conditionIsCNF* attribute is a Boolean attribute that, if TRUE, defines the condition clause of this *PolicyConditionComposite* object as being represented in conjunctive normal form (e.g., an AND of ORs). If the value of this attribute is FALSE, the condition clause will be represented in disjunctive normal form (e.g., an OR of ANDs). This attribute controls the setting of how multiple condition clauses (as represented by individual *PolicyConditionAtomic* objects) are treated in the *PolicyConditionComposite* object. As such, it is different from the attribute *PolicyRule.IsCNF*, which controls how multiple *PolicyCondition Composite* objects are treated in a *PolicyRule* object.

The *getConditions* method returns the set of *PolicyConditions* that are contained in this higher-level *PolicyConditionComposite* object. This enables primitive conditions to be reused, so that they can be formed into more complex conditions. Arguments to this method define how many levels to search, as well as the ability to search for a specific type of *PolicyCondition* object. The return type is an integer, to enable success or failure of the operation to be indicated.

7.8.2 The *PolicyConditionAtomic* Class

As stated earlier, the DEN-ng design of *PolicyConditions* (Figure 7-12) is based around the use of a *PolicyConditionAtomic* object to represent the semantics of a condition term, and the *PolicyConditionComposite* object to aggregate *PolicyConditionAtomic* objects to form more complex condition clauses. While this holds true for simple *PolicyConditions*, it is modified slightly to accommodate compound (also called nested) *PolicyConditions*.

The DEN-ng design revolves around the concept of using a *PolicyStatement* as the means to define a single condition term. This is being stressed, because otherwise each condition term could have a different structure. This impedes interoperability and makes it harder to automate.

The *PolicyConditionAtomic* class is treated as an intelligent container that defines common semantics for all simple policy conditions. The actual condition definition is applied using a *PolicyStatement*. It is this fact—that a *PolicyConditionAtomic* object is also a form of container—that enables this structure to also support nested *PolicyConditions*.

The *PolicyConditionAtomic* class defines three attributes. The *condition SequenceNumber* attribute is a non-negative integer and represents the particular sequence in which this element is parsed as part of the overall policy structure. This will allow a policy engine to reference externally supplied arguments to this object as it is evaluated. The *hasEvaluated* attribute is an enumerated integer that is set by a policy application to define the state of evaluation of this *policyCondition* after it has been evaluted. Values include:

- 0: Not yet evaluated
- 1: Evaluated as TRUE
- 2: Evaluated as FALSE

The *hasSubConditions* attribute is a Boolean attribute that, if TRUE, signifies that this *PolicyCondition* has one or more sub-conditions. Sub-conditions are used to enforce a hierarchical nesting of conditions, so that parent conditions may control the evaluation and other semantics of sub-conditions that they contain. However, sub-conditions are handled differently than regular conditions. Therefore, this attribute serves as a convenient "flag" to alert processing applications that a sub-condition is going to be encountered.

The *PolicyConditionInPolicyRule* aggregation was defined earlier in this chapter. The last remaining relationship to describe is the *PolicyStatementIn PolicyCondition* aggregation, which is used to associate a *PolicyStatement* with a *PolicyConditionAtomic* class.

The *PolicyStatementInPolicyCondition* aggregation models the binding of a *PolicyStatement* object to a *PolicyConditionAtomic* object. It is implemented as an aggregation (instead of an association) to emphasize the whole-part nature of this relationship. It is not implemented as a composition in order to enable *PolicyStatements* to be built independently and exist independently of *PolicyConditions* (and *PolicyActions*).

A *PolicyStatement* is used both in conditions as well as in actions. The difference in usage (in a condition versus an action) is based on the values that the operator can take on, which is represented by the *opType* attribute of the *PolicyOperator* object that is part of the *PolicyStatement*. This consistency is indicative of a good design and simplifies the resulting implementation.

The *PolicyStatement* serves as the main body of the *PolicyConditionAtomic* object. This means that the design of the *PolicyConditionAtomic* object is simplified; the *PolicyConditionAtomic* object becomes an intelligent container that adds semantics to the *PolicyStatement*. Note that a *PolicyStatement* is used to form the condition (or action) body of the *PolicyConditionAtomic* (or PolicyActionAtomic) objects ONLY. Other types of *PolicyConditions* (or *Policy Actions*) do not use a *PolicyStatement*. Again, this increases consistency, and makes the design easy to debug and automate.

7.8.3 Advanced Concepts—Compound *PolicyConditions*

A compound *PolicyCondition* is one that contains a Boolean combination of other *PolicyConditions*. The conditions being combined may be *PolicyConditionAtomic* and/or *PolicyConditionComposite* objects, though at least one *PolicyConditionAtomic* object must be present.

In the IETF and DMTF approach, a special class (*CompoundPolicy Condition*) and special relationships (*PolicyConditionInPolicyCondition*, which actually necessitated other changes) are required to define this concept. In the DEN-ng model, this concept is defined inherently by using the composite pattern. *No special classes or relationships are required.*

A compound *PolicyCondition* is a convenient construct to represent a sequence of conditions that must be tested yet treated as a single atomic condition within a *PolicyRule*. This is conceptually equivalent to using a set of simple conditions. However, from an implementation point-of-view, it enables the DEN-ng information model to better match the environment that it is modeling, resulting in simpler and more efficient implementations. This also simplifies the mapping of the DEN-ng information model to different types of data models. Finally, it enhances the manageability and reusability of complex conditions.

Consider, for example, the concept of a filter. Most filters consist of multiple terms. The problem, then, is that if the only construct is a *PolicyCondition Atomic* object, we cannot refer to a filter as an atomic condition, because we will need to combine multiple instances of the *PolicyConditionAtomic* object to construct the filter.

The DEN-ng use of the composite pattern solves this problem nicely. Remember that the *containedPolicyConditions* aggregation, (Figure 7-12) is a recursive aggregation between a *PolicyCondition* and a *PolicyCondition*. This means that both the *PolicyConditionComposite* as well as the *Policy ConditionAtomic* classes inherit this relationship. Therefore, this one relationship supports:

- *PolicyConditionAtomic* objects by themselves (i.e., with no encapsulating *PolicyConditionComposite* object needed) because the cardinality of the *containedPolicyConditions* aggregation is "0..1" to "0..n"—the 0s on each side of the relationship define this aggregation as optional
- *PolicyConditionComposite* objects can contain *PolicyCondition Composite* as well as *PolicyConditionAtomic* objects
- *PolicyConditionAtomic* objects can contain *PolicyConditionComposite* as well as *PolicyConditionAtomic* objects

Compound conditions should be used when the definition of a set of terms that should be treated atomically (e.g., as a single condition) is required. One such example is the common case of filtering on a five- or six-tuple (e.g., the source and destination address and ports, protocol, and optionally, the DSCP or ToS bytes) that occurs when systems use packet markings to signify which packets should receive better than best-effort traffic conditioning. This type of filter can be modeled as a container that holds one or more basic *PolicyConditions* that are grouped together into a single entity. This grouping maintains the correspondence to the underlying implementation, which is going to test this condition as an atomic entity. Grouping also enables individual terms, such as testing the value of a DSCP, to be reused as individual conditions. This will be explained more under the "Reusable versus Ad Hoc Policy Elements" section later in this chapter.

For example, consider the modeling of the previous compound condition (for simplicity, only two terms will be defined, though this is clearly extensible to six). The logic to be expressed is:

SourceIPAddress = 178.20.10.4/12 AND DSCP = 5

This can be conceptually implemented as shown in Figure 7-13:

The design shown in Figure 7-13 is based on building reusable policy condition clauses. PolicyStatement C models the IPv4 clause, and PolicyStatement D models the DSCP clause. Each of these can be reused by other PolicyRules, or they can be embedded as is shown in Figure 7-13 in a PolicyConditionAtomic object. Reusing the PolicyStatement enables the clause itself to be reused; reusing the PolicyConditionAtomic object also reuses the associated semantics of the

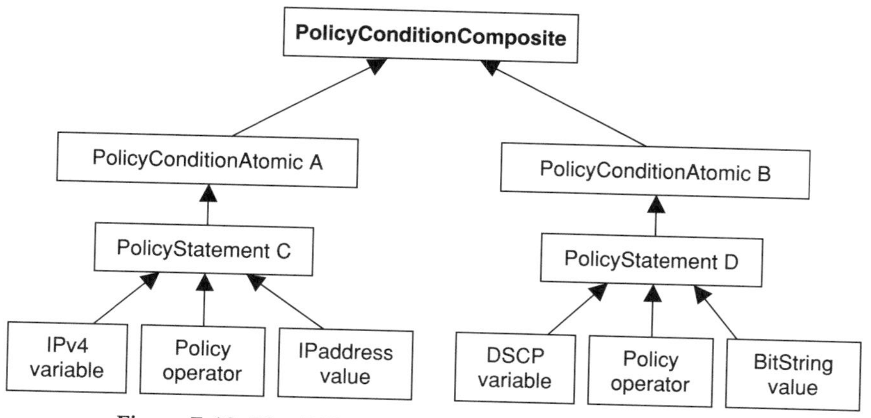

Figure 7-13 The DEN-ng *PolicyCondition* Class Hierarchy.

PolicyCondition. That is, if the PolicyStatement object is reused, other PolicyRules can reuse the statement (i.e., if DSCP == 5). However, if the PolicyConditionAtomic object is reused, the embedded PolicyStatement is reused along with its associated semantics (e.g., conditionSequenceNumber).

Note that there are several ways to build this condition—the example shown in Figure 7-13 is one of those ways.

To show that the information model contains enough detail to help guide implementations, simply match the preceding discussion about the various DEN-ng elements to the boxes shown in Figure 7-13. For example,

- The *PolicyConditionAtomic* object A contains the following *Policy Statement C* objects:
  - *IPv4Variable*
    - *v4AddressList* = 178.20.10.4/12
    - *v4Direction* = 0 (specifies source address)
    - *v4HeaderNumber* = 0 (outermost header)
  - *PolicyOperator.opType* = 6 (defines the equals operator)
  - *IPAddressValue*
    - *valueAddressFormat* = 1 (defines CIDR format)
    - *typeOfAddress* = 1 (specifies IPv4)
    - *addressValueList* = 178.20.10.4/12
- The *PolicyConditionAtomic* object B contains the following *Policy Statement D* objects:
  - *DSCPVariable*
  - *PolicyOperator.opType* = 6 (defines the equals operator)
  - *BitString Value*
    - *valueIntegerList* = 5

In this example, note that the *PolicyConditionAtomic* objects are containers that hold the two *PolicyStatements*. The *PolicyConditionComposite* object serves to group the two condition terms together, forming a compound condition.

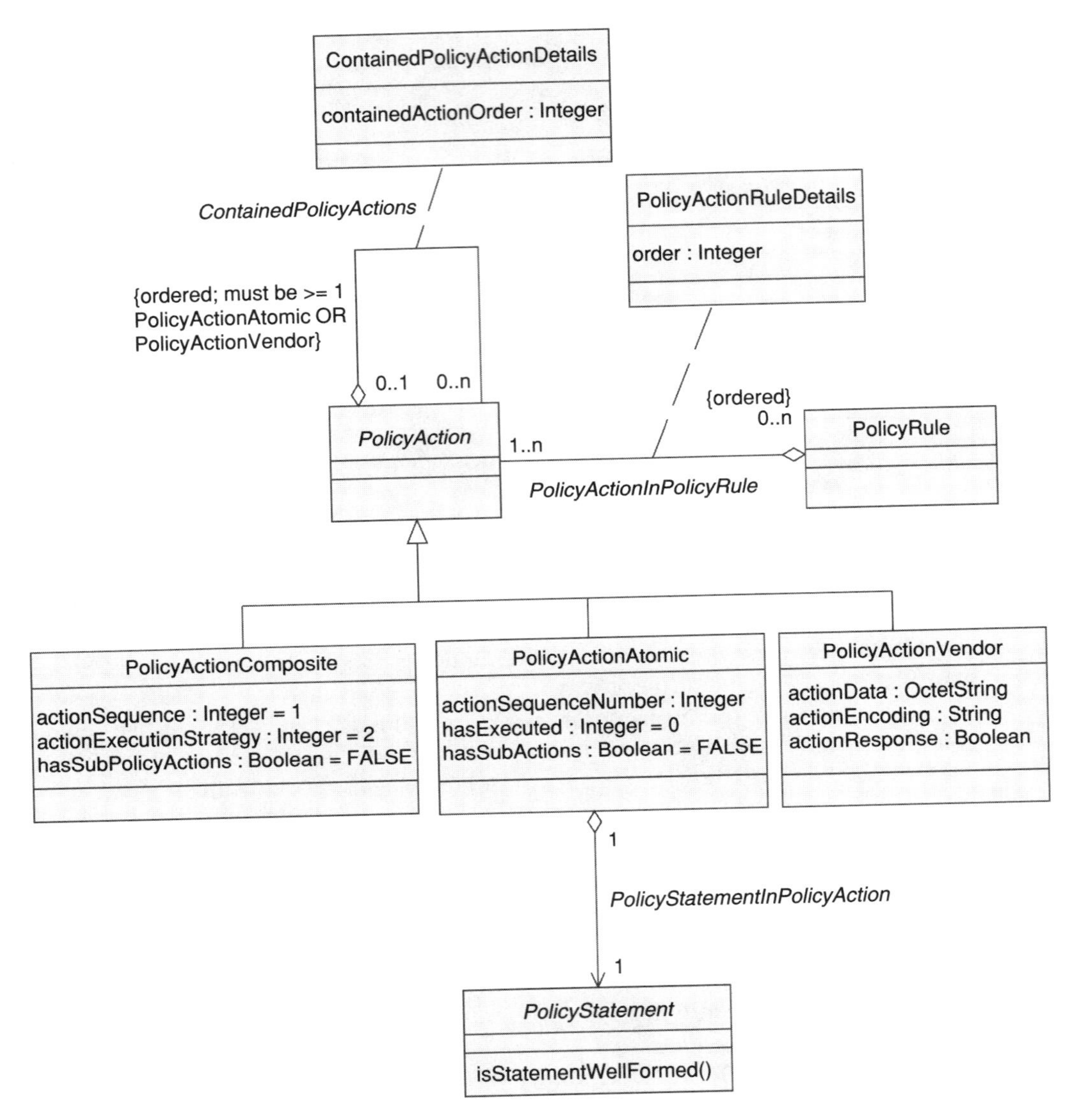

Figure 7-14 The DEN-ng *PolicyAction* Class Hierarchy.

7.9 DEN-ng *PolicyActions*

The DEN-ng *PolicyAction* class is an abstract base class that represents how to form the action clause of a *PolicyRule*. This consists of a single occurrence of a *PolicyStatement*, which is of the form:

{SET | CLEAR} PolicyVariable to PolicyValue

This is the same three-tuple of variable, operator, and value that we saw for the PolicyCondition side. The difference is that the PolicyOperator is first and is limited (for now) to the two choices of SET or CLEAR.

There are two types of actions:

- Pass actions are invoked if the condition clause was TRUE
- Fail actions are invoked if the condition clause was FALSE

For the purpose of this chapter, we will just show the "pass action" clause for simplicity, as the "fail action" clause is structured in an identical fashion.

The *PolicyAction* class hierarchy is shown in Figure 7-14:

There are several differences between the above DEN-ng model and the IETF and DMTF approaches. The ones shown in Figure 7-14 include:

- The composite pattern is used in DEN-ng; this is not present in the IETF and DMTF approaches. The use of this pattern provides inherent extensibility and makes for a simpler and easier-to-implement model.
- The previous statement causes a list of differences, not just in the structure of the *PolicyActionComposite* (versus the *CompoundPolicyAction* class of the IETF/DMTF), but in the attributes and relationships in which each subclass of *PolicyAction* participates.
- The structure of the *PolicyActionAtomic* class is changed. Note that in the DEN-ng model, the *PolicyStatement* is used to provide the content of the condition; this provides an object-oriented view of the contents of the condition, improves reusability, and differentiates the semantics of the *PolicyConditionAtomic* object from all other *PolicyActions*. In the IETF/DMTF model, there is no dependence on the *PolicyStatement*. This misses the chance to make the contents of a *PolicyAction* reusable.
- The recursive aggregation on *PolicyAction* is a simple way to support simple and compound policy actions (rather than having multiple classes and relationships as in the IETF and DMTF approach).
- The aggregation class on *PolicyAction*, called *ContainedPolicyAction Details*, provides a consistent way to define semantics for this relationship (versus a set of simple aggregation classes in the IETF and DMTF models).
- The IETF and DMTF approach defines *SimplePolicyAction*, *Compound PolicyAction*, and *PolicyActionVendor* as subclasses of *PolicyAction*, whereas the DEN-ng approach separates *PolicyActionAtomic* (which uses a *PolicyStatement*) from *PolicyActionVendor* (which does not use a *PolicyStatement*). In addition, there is no need to build a special *CompoundPolicyAction* class.
- In DEN-ng, the concept of a *PolicyStatement* has been developed to serve as the standard way for populating *PolicyActionAtomic* (and as seen earlier, *PolicyConditionAtomic*) classes with a single action (or condition) earlier, term. This symmetry is indicative of a good, consistent model, and will also ease implementation. This is not present in the IETF and DMTF approaches.

More complicated PolicyActions enable one or more PolicyActions to be nested inside of a higher-level PolicyAction. This is termed a compound PolicyAction and will be discussed later in this section.

The *containedPolicyActions* aggregation is used for two distinct purposes. First, it is used to gather together discrete *PolicyActions* to form a *PolicyAction* clause. In this usage, the set of individual *PolicyActionAtomic* objects are aggregated by a *PolicyActionComposite* object.

The second use is to form compound *PolicyActions*. In this case, the *PolicyActionAtomic* object aggregates *PolicyActionAtomic* and/or *PolicyAction Composite* objects. The former provides a simple nested *PolicyAction*, whereas the latter provides structure for grouping the sub-actions that are dependent on a higher-level action.

Note that the cardinality of this aggregation is "0..1" to "0..n," meaning that it is optional. This enables PolicyAction objects to be instantiated without this aggregation. In addition, notice the two OCL statements. If this aggregation is instantiated, not only do the PolicyAction objects have to be ordered, but at least one PolicyActionAtomic object or one PolicyActionVendor object must be present (otherwise, only containers would be present, and containers do not have actions—they *contain* actions). This level of specificity is not present in the IETF/DMTF model.

The *ContainedPolicyActionDetails* aggregation is implemented by an aggregation class, *ContainedPolicyActionDetails*. This class defines a single attribute, *Contained ActionOrder*, which is an unsigned integer attribute that defines the relative position of this action in relation to the overall sequence of actions that are associated with a given *PolicyAction*. Smaller integers indicate earlier positions in the sequence. The value "0" has the special semantics of "don't care."

This value is not unique. Thus, if two or more attributes have the same value, they may be performed in any order, but they must all be performed at the appropriate place in the overall sequence.

7.9.1 The *PolicyActionAtomic* Class

The DEN-ng design of *PolicyActions* shown in Figure 7-14 is based around the use of a *PolicyActionAtomic* object to represent the semantics of an action term, and the *PolicyActionComposite* object to aggregate *PolicyActionAtomic* objects to form more complex action clauses.

The DEN-ng design revolves around the concept of using a *PolicyStatement* as the means to define a single action term for standard usage. This is being stressed, because otherwise each action term could have a different structure. This impedes interoperability and makes it harder to automate.

The *PolicyActionAtomic* class is treated as an intelligent container that defines common semantics for all simple policy actions. The actual action definition is applied using a *PolicyStatement*. This fact—that a *PolicyAction Atomic* object is also a form of container—enables this structure to also support compound *PolicyActions*.

PolicyActionAtomic objects can also be used to form more complex action structures. A *PolicyActionComposite* object can be used to contain a group of *PolicyActionAtomic* objects—this grouping enables multiple *PolicyAction Atomic* objects to be executed as a group. Alternatively, a *PolicyActionAtomic* object can contain one or more *PolicyActionAtomic* objects (and also *PolicyActionComposite* groups if desired) to provide the semantics of a compound *PolicyAction*. In either case, the aggregation is done using the *Contained PolicyActions* aggregation.

This class defines three attributes and one method. The *actionSequence Number* attribute is a non-negative integer and represents the particular sequence in which this element is executed as part of the overall policy structure. This will allow a policy engine to reference externally supplied arguments to this object as it is executed.

The second attribute, *hasExecuted*, is an enumerated integer that is set by a policy application to define the state of execution of this *PolicyAction*. Values include:

- 0: Not yet executed
- 1: Executed with no errors
- 2: Executed with errors but successfully rolled back
- 3: Executed with errors and did not roll back
- 4: Did not complete execution but successfully rolled back
- 5: Did not complete execution and did not roll back

The final attribute, *hasSubActions*, is a Boolean attribute that, if TRUE, signifies that this *PolicyAction* has one or more sub-actions. Sub-actions are used to enforce a hierarchical nesting of actions, so that parent actions may control the evaluation and other semantics of the sub-actions that they contain. This attribute serves as a flag to alert the processing engine that special procedures must be invoked, since sub-actions are present.

The *getSubActions* method returns the set of sub-actions that are contained in this higher-level *PolicyAction*. This corresponds to a "compound action," which is an action formed by a set of actions. This enables primitive actions to be reused, so that they can be formed into more complex actions. Arguments to this method define how many levels to search, and whether to search for just PolicyActionAtomic or other types of entities. The return type is an integer to enable success or failure of the operation to be indicated.

7.9.2 The *PolicyActionVendor* Class

This is a concrete base class that provides a general extension mechanism for representing *PolicyActions* that cannot be modeled with the attributes specified in this model. This class uses its two properties (*actionData* and *actionEncoding*) for defining the content and format of a vendor-specific condition. Standardized extensions are not expected to use this class.

This class defines two attributes. The *actionData* attribute provides a general extension mechanism for representing policy actions that have not been modeled with the specific properties contained in this model. The format of this attribute is determined by the OID value stored in the *actionEncoding* attribute. A policy decision point can readily determine whether it supports the values stored in an instance of *actionData* by checking the OID value from *actionEncoding* against the set of OIDs it recognizes. The action for the policy decision point to take in case it does not recognize the format of this data could itself be modeled as a *PolicyRule*, governing the behavior of the policy decision point.

The second attribute, *actionEncoding*, is a string attribute. It is used to represent an OID that in turn identifies the encoding and semantics of the *actionData* attribute value array in this instance. The value of this property is a single string, representing a single OID.

The final attribute, *actionResponse*, is a Boolean attribute that is used to allow a "standard" definition of a TRUE/FALSE response to be made by an external system attached to the model at this point. This will allow integration with a generalized policy processing engine.

7.9.3 The *PolicyActionComposite* Class

The *PolicyActionComposite* class is the base class for serving as a generic container in which to place *PolicyActionAtomic*, *PolicyActionVendor*, or *PolicyActionComposite* classes. The first two provide actions that this container groups, whereas the latter establishes a hierarchy in which to order the execution of *PolicyActions*.

Both simple as well as complex *PolicyActions* can be placed in this container.

Each *PolicyActionAtomic* and *PolicyActionVendor* object is linked to this object using the *containedPolicyActions* association.

This class defines three attributes and one method. The *actionSequence* attribute is an enumerated integer that defines how the ordering of the *PolicyActions* associated with this *PolicyActionComposite* object is to be interpreted. Values include:

- 0: Unknown
- 1: Mandatory
- 2: Recommended
- 3: Best Effort

The value of 1 (mandatory) means that the actions must be done in the indicated order or not at all. The value of 2 (recommended) means that the actions should be done in the indicated order if possible; if this is not possible, it is permissible for them to be done in another order. The value of 3 (best effort) means that the actions should be done regardless of whether they can be done in the specified order or not.

The second attribute, *actionExecutionStrategy*, is an enumerated integer that defines the execution strategy to be used upon the sequenced actions aggregated by this *PolicyActionComposite* object. Defined execution strategies include:

- 0: Unknown
- 1: Do Until Success
- 2: Do All
- 3: Do Until Failure
- 4: Do All Without Failure or Do Nothing

"Do Until Success" means that actions should be executed according to their predefined order, until successful execution of a single action. At that point, the process should stop (even if there are additional actions that could be executed). "Do All" means that all actions should be executed that are part of the set, according to their predefined order. This should continue, even if one or more of the actions fails. "Do Until Failure" means that actions should be executed according to their predefined order until an action fails to execute. At that point, the entire sequence should be stopped. Finally, "Do All Without Failure or Do Nothing" means that *all* actions specified should be performed. However, if a failure occurs, then *none* of the actions should be performed. This implies *either* the ability to look ahead and ensure that the actions can be performed successfully, or the ability to perform a rollback of that action and all previous actions when an error is encountered.

The final attribute, *hasSubActions*, is a Boolean attribute that, if TRUE, signifies that this *PolicyAction* has one or more sub-actions. Sub-actions are used to enforce a hierarchical nesting of actions, so that parent actions may control the evaluation and other semantics of sub-actions that they contain.

The *getActions* method returns the set of *PolicyActions* that are contained in this *PolicyActionComposite* object. This will retrieve all types of *PolicyActions*—single and compound—unless it is desired to limit the type of *PolicyActions* that are sought. This, as well as limiting how many levels to search for, are controlled by arguments supplied to this method. The return type is an integer, to enable success or failure of the operation to be indicated.

7.9.4 Advanced Concepts—Compound *PolicyActions*

A compound *PolicyAction* is a convenient means to represent a sequence of *PolicyActions* to be applied as a single atomic *PolicyAction* within a *PolicyRule*. In many cases, *PolicyActions* are related to each other and should be looked upon as sub-actions of one higher-level action. For example, consider the common act of first rate limiting the input stream, and then marking the packets with a particular ToS or DSCP value. Although this can be thought of as a single action, it is in reality two very different actions, and must be performed in a well-defined order and as a complete set. This is best implemented as a compound *PolicyAction*, as opposed to having two separate *PolicyActionAtomic* objects grouped in a *PolicyActionComposite* object.

In the IETF and DMTF approach, a special class (*CompoundPolicyAction*) and special relationships (*PolicyActionInPolicyAction*, which actually necessitated other changes) are required to define this concept. In the DEN-ng, this concept is defined inherently by using the composite pattern. *No special classes or relationships are required.*

7.10 Reusable versus Ad Hoc Policy Elements

There is a fundamental difference between how a particular policy element (i.e., a *PolicyCondition*, *PolicyAction*, *PolicyStatement*, *PolicyRule*, or *Policy Group*, or any of their subclasses) is intended to be used. A *reusable* policy element is designed to be used by multiple applications (or even multiple other policy elements). In contrast, an ad hoc policy element is not designed to be reusable.

The different class hierarchies reviewed up to this point are all inherently suitable as either reusable or ad hoc policy elements. The big difference, then, is *how* they are accessed and used.

Reusable policy elements need to be generally accessible by multiple applications. This is subject, of course, to access control and other security requirements. Therefore, it makes sense to define the concept of a reusable policy container—a common repository that can be used to contain policy elements that are intended to be reusable.

It should be emphasized that this does *not* change the design of *any* of the previously described class hierarchies. Reusability is in the eye of the developer, and the only thing that changes is the intent of the developer to treat this policy element as a reusable policy element.

The fact that a policy element is reusable means that it will be accessed from a common repository. This means that a reusable PolicyRule whose elements are all reusable may take multiple accesses from this repository in order to retrieve all of its reusable policy elements so that it can then be processed. This is compared with an ad hoc PolicyRule, which could be stored and retrieved as a single atomic object. The principal advantages and disadvantages of this approach are shown in Figure 7-15.

Feature	Reusable Policy Elements	Ad Hoc Policy Elements
Access speed	Slow: One access per reusable element	Fast: Can be accessed all at once
Reuse	High: Any element can be reused	Nonexistent: Built for one use
Storage location	In a central repository, or even in multiple separate repositories	In application-specific repositories
Can be thought of as	Elements linked together	Elements embedded in one element
Functionality	Identical	Identical

Figure 7-15 Comparison of reusable versus ad hoc policy elements.

7.11 The DEN-ng Policy Model—Interfacing to Other DEN-ng Models

This part of the chapter will focus on how the policy model interfaces to other DEN-ng models. There is an opportunity to use PBNM in a wide variety of situations. In order to prove this, the intent of this section is to provide three different examples of where policy can be used to manage and control other domain information models.

7.11.1 Interface to the DEN-ng Resource Model

The purpose of the DEN-ng Resource model is to define the physical and logical aspects of managed objects. However, in order to understand the top-level DEN-ng resource class hierarchy, its superclasses (*ManagedEntity* and *Entity*, respectively) must first be explained. Entity is an abstract base class used to represent entities that cannot be managed directly (e.g., a hub). *ManagedEntity* is an abstract base class that is used to represent entities in a managed environment that have the following semantics in common: (1) a *ManagedBusinessEntity* owns or is otherwise responsible for them, (2) management of the entity is critical for providing a service and/or maintaining the environment, and (3) the entity is "important" from a management point-of-view. As such, it is a point at which semantically important relationships can be created; this avoids duplicating these relationships on multiple subclasses, and can be used to define appropriate *ManagedObjects* to support these relationships. The attributes of *ManagedEntity* are used to specify how the *ManagedEntity* is currently being managed, as well as the management methods that it supports.

In DEN-ng, a *Resource* is defined as the abstract base class for all entities that are inherently manageable and make up a *Product*. Examples that are not manageable include legacy hubs that do not support any type of management protocol; rather, they must be managed manually. Entities like *Policy* are of course manageable, but do not make up or are found packaged inside of a *Product*. Rather, *Policy* is used to manage and control the behavior of a *Product*, or the use of a *Product*.

Resource defines a single attribute, *usageState*. This is an enumerated integer that defines the current usage of the object.

Some *Resources* can be very complex and difficult to model. DEN-ng uses the principle of abstraction and models the physical and logical aspects of a *Resource* using the *PhysicalResource* and *LogicalResource* subclasses. A *PhysicalResource* is an abstract base class for describing different types of hardware that constitute a *Product*. It has two main purposes: (1) to collect common attributes and relationships for all hardware, and (2) to provide a convenient, single point where relationships with other managed objects can be defined. Its two main subclasses are *PhysicalDevice* and *Hardware*. *PhysicalDevice* is the base class for representing the device as a whole, whereas *Hardware* is the base class for representing the different components of a *PhysicalDevice*. Both of these classes will be discussed in subsequent chapters.

A *LogicalResource* is an abstract base class for describing different logical aspects of devices (e.g., services) that constitute a *Product*. It has two main purposes: (1) to collect common attributes and relationships for all logical entities, and (2) to provide a convenient, single point where relationships with other managed objects can be defined.

A *LogicalDevice* is an abstract base class for representing logical concepts and services that can be managed, which are associated with the *LogicalDevice* as a whole. This class represents a convenient aggregation point for combining different aspects of a *LogicalDevice* (e.g., software contained in the device, protocols that the devices runs, the set of services that it offers). It also enables the *LogicalDevice* itself to have a single logical manifestation. To show the usefulness of this concept, relations to *Protocol*, *DeviceInterface* (which is used to model the programming of the device), and *Software* (which models the different software features of the device) are shown in Figure 7-16.

A *ManagedTransmissionEntity* is an abstract base class for describing different types of logical entities that help to form connections that transmit and/or receive information. This is a base class that contains the important concepts of *Connection*, *Trail*, and various types of *TerminationPoint* from the ITU. The DEN-ng model provides additional structure so that additional IETF and IEEE concepts, like routes, can be defined in this class hierarchy. This is not present in the IETF and DMTF models.

A simplified view of the top-level resource class hierarchy is shown in Figure 7-16.

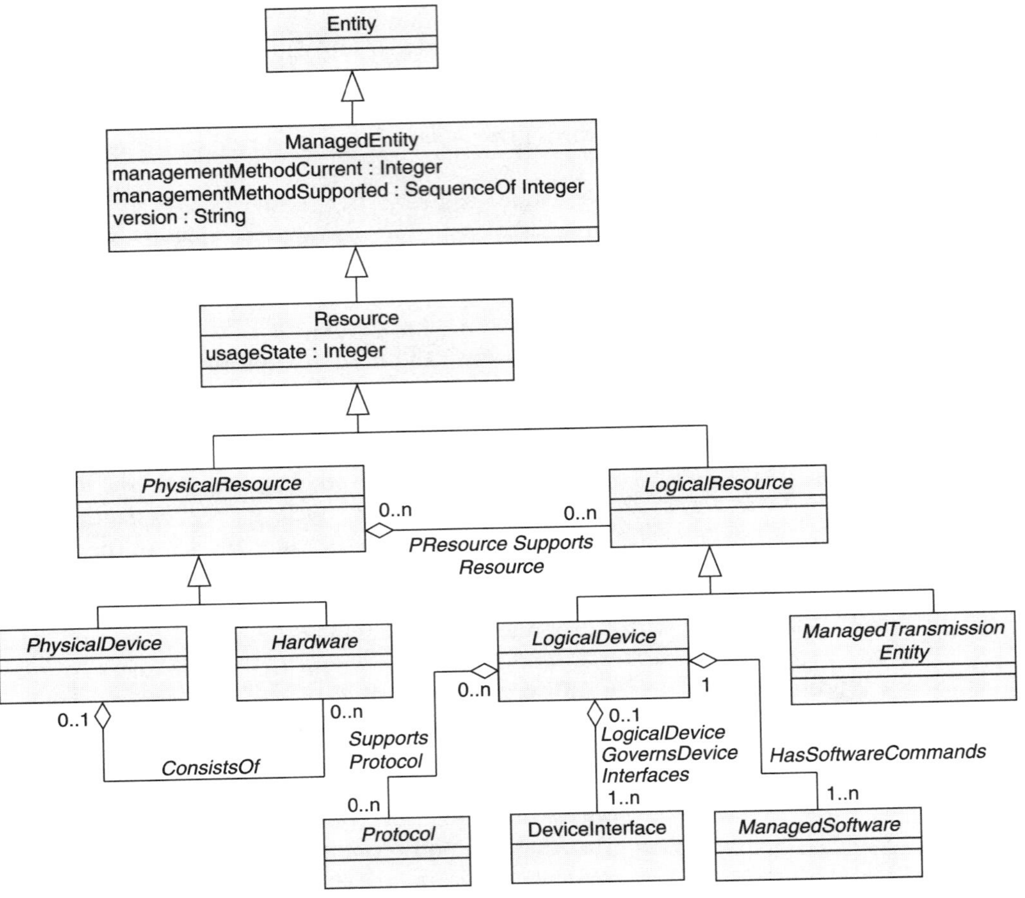

Figure 7-16 Top-level view of the DEN-ng resource class hierarchy.

In DEN-ng, *Policy* is used to manage individual objects (e.g., a *PhysicalDevice*) as well as relationships between different managed objects. For example, in Figure 7-16, policy can be used to:

- Control which features of which protocols are used by a device
- Control when, how, and by whom a device interface is programmed
- Control the enabling and disabling of different software features that are available
- Define how routes are built between two devices
- Control physical access to *PhysicalDevices*
- Policies that define installation, power-up, and power-down instructions for physical entities
- Policies that define maintenance and service instructions

Clearly, there are many more types of policies that can be defined. The next few sections will provide brief glimpses into other DEN-ng domain models in which policy can be used. These will set the stage for Chapters 9 and 10, which will discuss examples of how policy is used.

The DEN-ng resource model is documented in references 6 through 8.

7.11.2 Interface to the DEN-ng Service Model

Chapters 9 and 10 will provide specific examples of how services are controlled by policy. Figure 7-17 shows a simplified view of how the DEN-ng Service model is constructed.

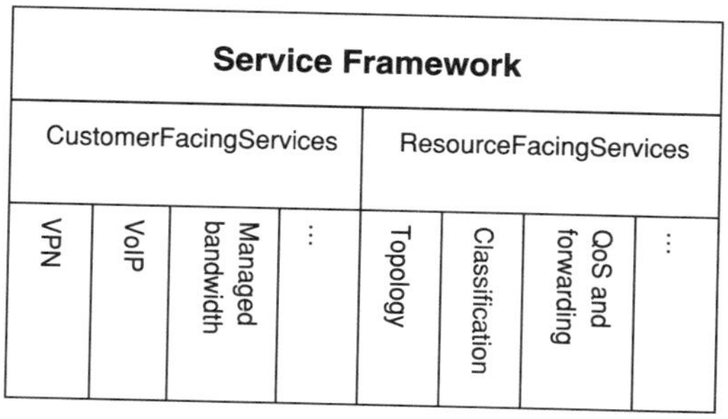

Service Framework

CustomerFacingServices

ResourceFacingServices

VPN

VoIP

Managed bandwidth

...

Topology

Classification

QoS and forwarding

...

Figure 7-17 Top-level view of the DEN-ng Service class hierarchy.

The top of the DEN-ng *Service* class hierarchy, denoted by service framework in Figure 7-17, consists of infrastructure that enables additional *Services* to be added over time. In DEN-ng, there are two types of *Services*. A *Customer FacingService* is an abstraction that defines the characteristics and behavior of a particular *Service* as seen by the *Customer*. This means that a *Customer* purchases and/or is directly aware of this type of *Service*, and is in direct contrast to *ResourceFacingServices*, which support *CustomerFacingServices* but are *not* seen or purchased directly by the *Customer*. A VPN is an example of a *Customer FacingService*, whereas the sub-services that perform different types of routing between network devices making up the VPN are examples of *Resource FacingServices*. The rest of the DEN-ng service framework consists of *Service Roles*, *ServiceSpecifications*, and additional classes that support these concepts.

Policy is used in many different ways in this hierarchy. Examples of its use include controlling and managing:

- Logical maintenance operations such as installation, power-up, and power-down instructions
- Logical administrative functions, such as authentication, authorization, accounting, auditing, and negotiation
- Logical network functions, such as negotiation, traffic conditioning, and providing a particular level of quality of service (QoS)
- How a particular protocol is run (e.g., how a dynamically assigned IP address should be chosen)
- How a service is constructed and programmed

- Which *ResourceFacingServices* are made available for which *CustomerFacingServices*
- Which *CustomerFacingServices* are made available to which *Customers*

The DEN-ng Service model is documented in references 9 and 10.

7.11.3 Interface to the DEN-ng Party Model

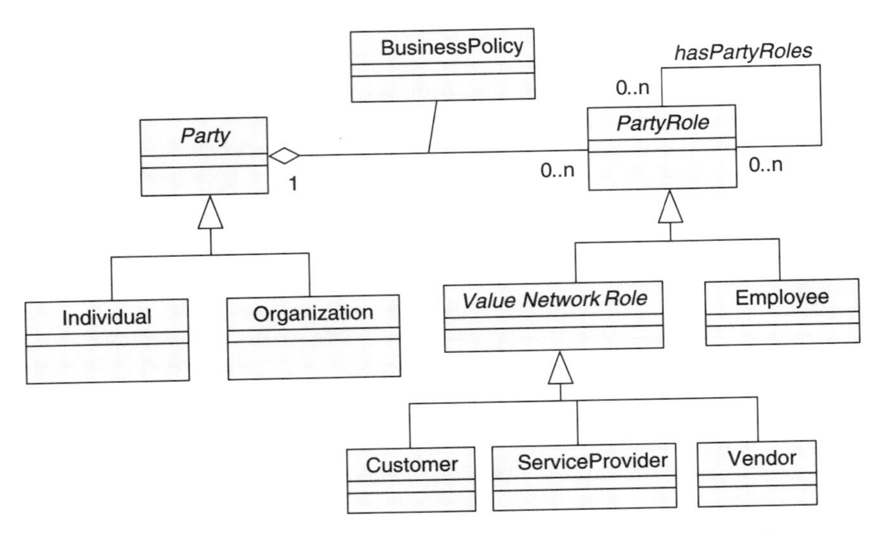

Figure 7-18 Top-level view of the DEN-ng Party class hierarchy.

Policy plays an integral role in determining things like access control, authentication, authorization, accounting, and auditing. These have been covered in many other references. DEN-ng includes these but adds to them in innovative ways. For example, Figure 7-18 shows a business policy that determines which roles an individual or an organization is allowed to assume.

There are numerous additional examples of how policy can be used in this model, but they are beyond the scope of this book.

The SID party model is documented in reference. The DEN-ng extensions to this are in the process of being documented.

7.12 Summary

This chapter has discussed the DEN-ng policy model in depth. This model is significantly different in structure and functionality compared with previous policy models. Although the same concepts (such as a PolicyRule) were used, some new additional concepts, plus a repackaging of existing concepts, were also used in order to build a more powerful and simpler policy information model. The key points of this redesign include the following:

- A *PolicyRule* is redefined to consist of an event clause, a condition clause, and two action clauses—one to be used if the condition clause evaluated to TRUE, and the other if FALSE
- Constraints are used to control how different clauses are evaluated
- Semantics are more fully specified than previous models, often with the use of association classes; this is important, since it significantly simplifies implementation and helps interoperability
- Mechanisms are put in place to enable external applications to define many of the constraints, thereby easing implementation
- Events are used for both triggering the evaluation of the condition clause of a *PolicyRule* as well as for distributing policies to *PolicyGroups*
- The notion of a *PolicyStatement* was defined; this is used to formalize the definition of a triplet consisting of a variable, an operator, and a value
  - The design of the *PolicyStatement* was generic enough to be used by both *PolicyConditionsAtomic* and *PolicyActionsAtomic*
  - A *PolicyStatement* is used to differentiate policies that can be modeled in a standard way from application-specific policies by treating a *PolicyConditionAtomic* as well as a *PolicyActionAtomic* as containers to hold *PolicyStatements*
  - A *PolicyStatement* specifies constraints that provide restrictions on semantics; these constraints enable the generic *PolicyStatement* to be applicable to a wide variety of different applications
  - The concept of standard and custom *PolicyVariables* and *Policy Values*, as well as different types of *PolicyOperators*, were introduced; this enables a class hierarchy to be developed to model each aspect of the *PolicyStatement*
- Artificial concepts, like the *CompoundPolicyCondition* class discussed in the literature (and their attendant additional relationships), are replaced by a more elegant and efficient information model that supports their functionality without having to create these artificial classes; this is done through the use of the composite pattern

The model described defines a set of extensible policy elements. Each of these elements can be used in an ad hoc or a reusable fashion. The difference in this is solely in the purpose of the administrator; the actual functionality is the same. The *effects* of using a reusable versus an ad-hoc policy element are a combination of differences in access speed, how the elements are stored, and of course whether multiple applications can use them.

Finally, three examples of how the DEN-ng policy model would interface to other DEN-ng domain models were given. These examples were the DEN-ng resource, service, and party models. In each case, concepts were defined that lent themselves to being managed and controlled by policy. Chapters 9 and 10 will give further examples of many of these.

7.13 Recommended Further Reading and References

The following is a set of references for this chapter.

1. TMF, *NGOSS Architecture Technology Neutral Specification—Behavior and Control Services*, Annex TMF053C, Version 1.0, February 2003.
2. TMF, *NGOSS Architecture Technology Neutral Specification—Policy Management*, Annex TMF053P, to be published.
3. TMF, *Shared Information/Data (SID) Model—Addendum 1POL—Common Business Entity Definitions—Policy*, GB922, July 2003.
4. Moore, B., Ellesson, E., Strassner, J., Westerinen, A., *Policy Core Information Model—Version 1 Specification*, RFC 3060, February 2001.
5. Moore, B., Rafalow, L., Ramberg, Y., Snir, Y., Westerinen, A., Chadha, R., Brunner, M., Cohen, R., Strassner, J., *Policy Core Information Model Extensions*, draft-ietf-policy-pcim-ext-06.txt, November 2001.
6. TMF, *Shared Information/Data (SID) Model—Addendum 5PR—PhysicalResource Business Entity Definitions*, GB922, Version 3.0, June 2003.
7. TMF, *Shared Information/Data (SID) Model—Addendum 5RO—Resource Overview Business Entity Definitions*, GB922, to be published.
8. TMF, *Shared Information/Data (SID) Model—Addendum 5LR—LogicalResource Business Entity Definitions*, GB922, Version 1.0, June 2003.
9. TMF, *Shared Information/Data (SID) Model—Addendum 4SO—Service Overview Business Entity Definitions*, GB922, Version 2.3, June 2003..
10. TMF, *Shared Information/Data (SID) Model—Addendum 4VPN—MPLS VPN Service Business Entity Definitions*, GB922, to be published
11. TMF, *Shared Information/Data (SID) Model—Addendum 1P—Common Business Entity Definitions—Party*, GB922, Version 3.0, June 2003.

Chapter 8

Components of a PBNM System

This chapter will describe the recommended architecture for a PBNM system. This architecture will be derived from analyzing the behavior of some example policies. Through this analysis, conclusions can be derived that affect how the architecture will be built. The proposed architectural principles are then explained, with an emphasis on how the novel use of information modeling and a language, coupled by using a data dictionary, support a distributed implementation of a PBNM system.

8.1 Introduction

The purpose of this chapter is to build on the previous chapters and define the detailed architecture of a PBNM system. The public literature talks mostly about the simplistic IETF model at a very high level. This chapter develops a more realistic PBNM architecture by taking several example policies and tracing their execution through the system. This motivates the need for additional components and architectural changes to be made to the IETF model.

This chapter discusses system behavior and how to manage it through policy. This is done by showing how our three building blocks—the information and data models, the languages, and the data dictionary—are used by the various components of the system to express and control behavior. An architectural blueprint is then developed that unites these three components. Last, it provides the basis for understanding the features and limitations of commercial PBNM systems, which will be covered in Chapters 9–11.

8.2 Requirements of Different Policies on the PBNM System

Different policies place different requirements on the architecture of the PBNM system because different types of policies are constructed using different levels of abstraction and are intended to be understood and executed by different sets of people. Therefore, by understanding the policies that are going to be applied to the system, we can develop a greater appreciation for the subtleties of executing policies, and a greater understanding for constructing a PBNM architecture.

8.2.1 Interoperability

First and foremost, our architecture must promote a scalable framework for interoperable policy systems. There are several levels of interoperability. For example, two different systems should be able to:

- Exchange policy definitions with each other
- Exchange policy objects and information
- Execute policies on behalf of each other

The first bullet is easily satisfied by the two systems using the same language or data model. Note that sharing either a language or a data model is sufficient, as this first requirement does not place any component-level restrictions on how the architecture is built.

The second bullet point does, however, start to place component-level restrictions on the architecture. In order for two systems to be able to exchange policy information with each other, they either must have a common definition of how the policy information is structured and represented or must use adaptation and/or mediation software to translate one system's policy objects to the other system's representation and structure.

The third bullet point mandates that a tight architectural relationship exist between the two systems. For one system to be able to execute policies on behalf on another system, it must understand the representation of the policy used by the other system as well as how policies are executed in that system. For example, if one system executes a policy to determine the order in which a set of processes are carried out, the other system must perform the exact same sequence of steps.

Clearly, the goal of our architecture should be to provide maximum interoperability between two systems. However, based on the complexity implied in satisfying the third bullet, an information model alone is insufficient. This is because we have defined policy as a continuum. As we proceed toward low-level implementation details, policies become dependent on device-specific implementations. We concluded in Chapter 3 that in such cases, we needed a combination of a set of models and a set of languages, each oriented at a specific level of abstraction in the policy continuum, to properly represent the semantics of policies at different levels of abstraction. Therefore, assuming that two systems do not use the same languages and models, our architecture must facilitate the mapping between information and data models as well as languages between the systems.

Another aspect of interoperability is the ability to monitor the application and result of a policy. Without such an ability, there is no way to assess either the compliance of the PBNM implementation or the effect that applying the policy has had on the device in particular and the system in general. Unfortunately, this aspect is often ignored in vendor products. Furthermore, although a vendor *may* provide the ability to monitor the application of an individual policy, it is even more important to monitor the *system* to which the policy has been applied.

One way to ensure that monitoring is taken into account is if the PBNM solution has employed a shared information model. If so, it should be simple and straightforward to relate the status of a managed object to a policy. DEN-ng goes one step farther and uses a finite state machine. This enables policies to be represented as managed entities that force a state transition. Since the DEN-ng information model is used to define a managed object as well as a policy, it is possible to relate the state of a managed object to a policy at a very granular level.

8.2.2 Managing Networks, Not Just Device Interfaces

Much of the existing work on PBNM systems has focused on "individual device management." Here, the goal is to use policy to determine when settings in a device configuration file should change. Although valuable, this is not the way to manage a network. First, networks have too many interfaces to manage individually. Second, network services span multiple types of interfaces, such as ATM and IP. Most interface types have semantics that cannot be easily managed by logging into a single interface. Finally, configuring individual device interfaces is very different in nature than defining a policy that is applied across multiple interfaces of different types of devices. This latter is an example of a network policy, whose objective is to coordinate the changes to different types of device interfaces (e.g., hosts, firewalls, routers, and servers) in order to provide a common service.

8.2.3 Use of Business Rules to Drive Network Configuration

Business rules are fundamental to defining the policies and processes used by an organization to manage its assets (whether network devices or other types of assets). The ability to define business rules to drive the configuration of network devices and services is an extremely powerful concept, since it enables the network to be managed using the business rules and processes of an organization. More importantly, this means that network services are now being controlled by business rules. Unless this is done, the network remains a cost center, not a profit center. Strassner[1] discussed this subject extensively.

Ideally, business rules will define management processes as a function of who is implementing the change (via the role that the person plays in managing the network), the type of change that is being implemented, and the role that the device (or device interface) plays in the network.

This presents the first step in the evolution of network intelligence. If business rules are used to drive the configuration of network services, the services that the network offers can be dynamically adjusted to support the changing needs of its users, applications, and environment.

8.2.4 Unification of Policy-Based Management and Process Management

The unification of the heretofore disparate worlds of policy-based management and process management presents a very powerful solution. In this approach, policies are used to control which processes are used to implement a given change. The result of these processes is fed back to the system, so that the state of the network can be appropriately adjusted to reflect the changes made. These changes then affect which policies are next used. This forms a simple yet effective closed loop system based on a finite state machine.

Intuitively, *behavior* defines a way of acting, or a way of describing how an entity responds to a certain set of conditions, and *control* defines a set of mechanisms to exercise restraint and/or directing how a particular behavior is to be managed or executed. The TMF has formed a sub-team, called the Behavior and Control team, whose purpose is to holistically integrate policy-based management and process management.[2,3]

8.2.5 The Importance of Sharing and Reusing Information

Operational support systems (OSSs) are commonly built using best-of-breed products that provide solutions for different functions needed by the OSS. For example, an OSS may consist of one application for fault and performance measurement, another application for billing, and yet another application for configuration management and service activation. Each of these applications has specific requirements. Thus, common practice has been to purchase best-of-breed products for each niche application and then integrate them.

The problem, of course, is that the task of integrating diverse applications is difficult and time-consuming. The two main reasons for this are the lack of a common definition and representation of common information, and the lack of a standard means for different components to communicate with each other.

Using a single common model provides two important benefits. First, it lets multiple applications share and reuse the same data. For example, the output of a discovery application can be used to feed an inventory application to dynamically catalog the devices that are currently operational. Second, it enables processes to be built to do a function once, which is applied in different applications under different contexts. Continuing the above example, other applications (e.g., configuration and service activation) could also use the output of the discovery application as input for different processes. Thus, the requirement is for *a set of common information to be used to represent common functions, enabling different applications to share and reuse common information*.

However, the problem is more complicated, because different applications have different views of the same information, and not every application needs to share all of the information that is presented to it. Think of two applications: a human resources (HR) management application and a billing application. The common concept of a *user* is present in both of these systems, but the attributes of the user are very different for each application. The HR application does not need to know anything about the credit history of the user, what applications cost how much, or what applications the user can run. Similarly, the billing application does not need to know anything about the employees that work for the user, or the particular employment details of the user.

Now add a provisioning application to the mix. Although the provisioning application may know about the user, the network elements that it provisions certainly do not. All they can know are network representations of the user, such as an IP address. Not only do we now need to correlate between possibly different user names (e.g., *johns* versus *john.strassner*), we now also need to match a user name (or equivalent human concept, such as an employee id) to a non-human concept, such as an IP address. The same goes for other concepts, such as services and service qualities.

The result is that unless there is a common information model that can serve as a data dictionary to correlate these different names and concepts, applications are unable to share and reuse data.

The solution is to define and use a common information model that defines the representation, semantics, and structure of data that is shared and reused between OSS components.

The TMF has taken a detailed look at the second problem and defined the notion of a contract to be used as the standard means for different components in an NGOSS system to be able to exchange information with each other.[4] The contract is the fundamental unit of interoperability in the NGOSS architecture. Specifically, in an NGOSS system:

- A service is the functionality made available through a contract-defined interface to another component.
- A contract is a technology neutral representation of an interface to a service—it expresses the behavior required for a component to use a particular service.

- An interface is a named set of operations that characterize the behavior of a service.
- The formal arguments of an interface are defined using shared information to ensure interoperability.

Note that an NGOSS contract depends on the use of NGOSS shared information (which is supplied by DEN-ng and the SID).

8.2.6 Communication

It is imperative that different OSS components be able to communicate effectively and efficiently with each other. Two fundamental requirements are the amount of communications and feedback between components. Clearly, it is not desirable to have "chatty" components that broadcast their data to the entire world. It is much better to either direct or somehow filter such communications, so that only essential data are transmitted between components.

Likewise, quick feedback is important, so that the system can tell as rapidly as possible whether the policy is having the desired effect or not as well as to keep up to date on changes. For example, consider a router. Routers have complicated configuration files that consist of many commands, many of which affect other commands or which affect the same interface. Suppose that after the policy server issues a command, someone logs into the router and changes one or more commands in the configuration file without telling the policy server. Then suppose that subsequently, the policy server logs into the same router and makes another change. This change should be fine as far as the policy server is concerned, because it computed the change based on the last known configuration file for that router. However, the configuration file has been changed, so this seemingly innocent change made by the policy server could in fact be disastrous. Note that this test is usually *not* done by most current commercial PBNM products.

Chapter 4 provided examples of two different types of communication mechanisms that could be used to accomplish both of these goals—a publish-subscribe event messaging system and a Jini-based system. In the publish-subscribe approach, filtering is done by the message server, to ensure that components only receive events containing information that they have subscribed to. Feedback is provided by ensuring that each component that needs to communicate with other components is a publisher and/or a subscriber. In the Jini/Javaspace approach, services are made that control which objects enter and leave the Javaspace. This ebb and flow of object into and out of the Javaspace is, in fact, controlled by policy in a DEN-ng (or NGOSS) system.

8.3 The Life of a Policy

In order to better understand how to build a PBNM system, this section provides two example use cases of how policies are used in a PBNM system. These scenarios are provided from two different points of view—one from the network manager installing and acting on a policy, and one from the policy server receiving a new policy.

It is important to first look at both viewpoints to ensure that the architecture will match the needs of the person using the system. These two specific use cases were selected so that two important design principles for our PBNM system could be emphasized. These principles are:

- How information is organized and flows through the system, and
- How the architecture must be built in order to support distributed processing and control.

Each of these design principles are covered in subsequent sections in this chapter. This section will conclude with examining the system-wide effects of the "life and death" aspects of a policy.

All sub-sections in this section are based on the PBNM architecture that has been alluded to in the previous chapters. This architecture is *not* the simplified architecture proposed by the IETF/DMTF; rather, it is the more robust DEN-ng architecture that is being adopted in the TMF. For convenience, it is redrawn in simplified form as Figure 8-1.

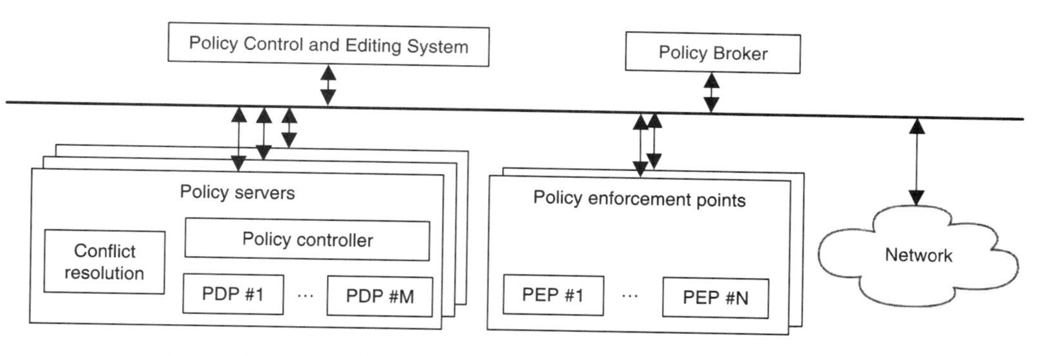

Figure 8-1 Simplified DEN-ng architecture of a PBNM system.

8.3.1 First Things First

The first thing that the PBNM system must do is validate that the user has, in fact, the permission to perform the desired operation on the policy. One of the characteristics that has aggravated many network administrators is the tendency of current management software and tools to assign a complete set of permissions to any user that is assigned the role "administrator." This ignores the fact that most networks are designed as a set of domains, each of which is owned and run by different people, different organizations, or both. Therefore, simply assigning the same privileges to each administrator enables that administrator to manage devices that he or she might not own.

Instead, the PBNM system proposed in this book takes a different approach. By intersecting the *role* of *a user* with the *role* that *a device* plays, a precise definition of what permissions the user is actually allowed to have, as a function of the device that the user is trying to manage, can be implemented. This can be supplemented with external information, such as the time of day, that may affect either the intended operation (as in no, the operation cannot be performed right now) and/or the deployment of that operation (as in no, the policy cannot be installed now and must wait until this evening). This conceptual approach is shown in Figure 8-2.

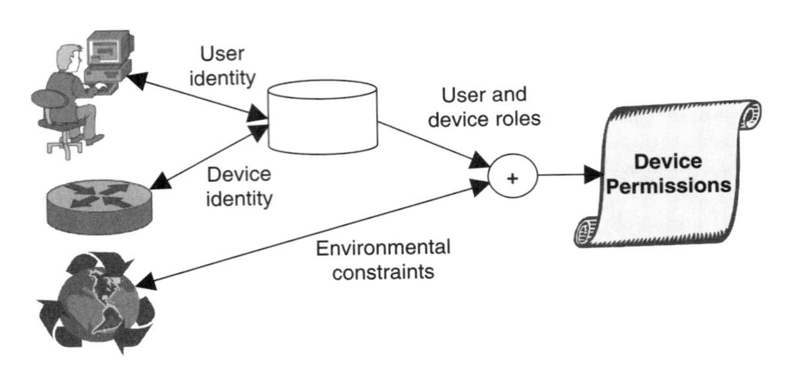

Figure 8-2 How permissions should be defined in a PBNM system.

Figure 8.2 drives home an important point: This approach works because the PBNM system is *in control* of the session. Should the PBNM system accept policies written outside of its control (e.g., by a simple text editor)? This has a surprisingly difficult answer. If external files are not allowed to be submitted to the system, then users will be unhappy. If they are, the system must ensure that sufficient information exists for the user and device identities and roles to be determined. Furthermore, the system must ensure that the submitted policy contains the proper information corresponding to the device permissions that the system will compute. Although it is easy to ignore extraneous information (or alternatively, to deny its use) and send a message to that effect, it is in general impossible to manufacture missing information that was not provided.

To allow as flexible a system as possible, the recommended solution is to allow only external policies that use templates generated by the PBNM system. This dramatically simplifies the template validation problem. Note that external constraints can still be filled in by the system if they are provided in the template. Finally, the presence of the DEN-ng model and associated languages enables the necessary information for the policy template to be easily supplied, verified, and updated.

Figure 8-3 shows a simplified use case diagram identifying the main components needed to establish the identity of the user. This same procedure can be used to establish the identity of a resource as well.

Figure 8-3 has been generalized to illustrate some of the options available in designing a PBNM system. Note for example that two repositories have been identified: a policy repository and a data repository. The former is dedicated to the storage and retrieval of policy rules and other policy information, whereas the latter is devoted to the storage and retrieval of user and resource information. Clearly, these two repositories could be combined or further partitioned; the reason they are shown as two separate repositories is that fundamentally,

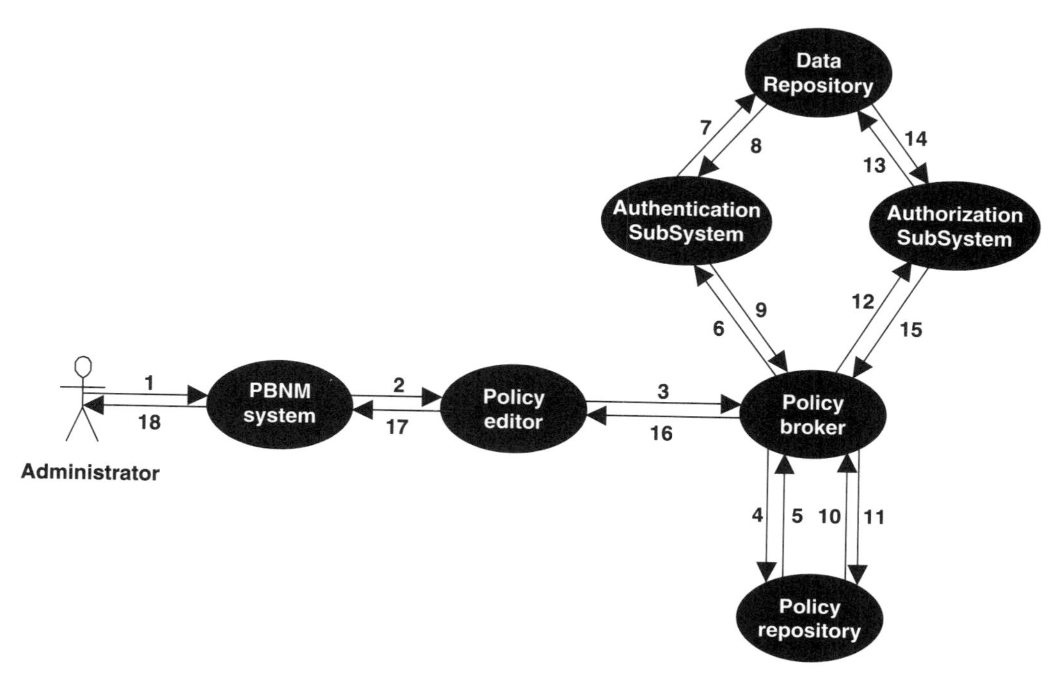

Figure 8-3 Authentication and authorization of a user in a PBNM system.

the management characteristics of policy information are different than the management characteristics of user and resource data. Furthermore, these two types of data are usually owned by different parts of the organization.

The sequence of steps is defined as follows. Note that some of these steps may be combined or further subdivided as necessary to suit the specific nature of the implementation. The main point of this use case is to show the complexity of determining the appropriate access rights and privileges for a given user, and that this computation should be hidden from the user.

1. The administrator requests access to the Policy Editor in order to create, read, update, or delete a policy.
2. The PBNM system invokes the Policy Editor, passing the user credentials to the Policy Editor.
3. The Policy Editor passes this request to the Policy Broker, who is the overall controller of the PBNM system. The Policy Broker is responsible for coordinating the actions of multiple Policy Servers. Thus, the editing session may be denied, depending on the overall state of the system.
4. The Policy Broker requests the specific authentication policy and roles for this user.
5. The Policy Repository returns the requested information.
6. The Policy Broker passes this information to the authentication subsystem. This step is necessary because the user may be required to use different authentication methods based on how the user is connecting to

the system. Furthermore, different operations may be permitted depending on how this connection is made.

7. The Authentication Subsystem selects an appropriate authentication method (this is, of course, a *policy*) and requests the appropriate type of credential information for this user when using this method.
8. The Data Repository returns the requested information.
9. The Authentication Subsystem computes the permissions for this user for this session (note that these permissions may change from session to session) and passes them back to the Policy Broker.
10. The Policy Broker examines the result. If the user has been authenticated, the Policy Broker may ask the Policy Repository for the appropriate authorization policy for this user for this session. Note that this enables different authorization schemes, as well as additional checking, (e.g., a second level of authentification) to be used if desired.
11. The Policy Repository returns the requested information back to the Policy Broker.
12. The Policy Broker passes this information to the Authorization Subsystem.
13. The Authorization Subsystem selects the appropriate authorization mechanism (which is, of course, another policy) and requests additional data from the Data Repository as necessary.
14. The Data Repository returns the requested information to the Authorization Subsystem.
15. The Authorization Subsystem returns the set of operations that this user is authorized to perform for this session to the Policy Broker.
16. The Policy Broker passes this information back to the Policy Editor.
17. The Policy Editor informs the PBNM System that it can or cannot be launched.
18. The PBNM system communicates this fact to the administrator and, if the Policy Editor can be launched, passes control to the Policy Editor. The editing session now starts.

8.3.2 From the Network Manager's Point-of-View

This section describes a "day in a life" of a policy from the perspective of the network manager installing a new policy, modifying an existing policy, or deleting an existing policy. Figure 8-4 picks up the flow after the user has been authenticated and the operations that the user can perform have been determined (as described in Figure 8-3).

The network manager should not require any explicit awareness of the functionality of the PBNM system, or how it operates. Rather, the network manager should be able to think of this as a *service* that can be invoked. For example, rather than providing a list of what the user *cannot* do because that user does not have sufficient permissions to perform that operation, the PBNM system should hide this information from the user. This makes for a better end-user experience, since now the user can concentrate on the task at hand instead of wondering why he or she cannot access a particular function or operation.

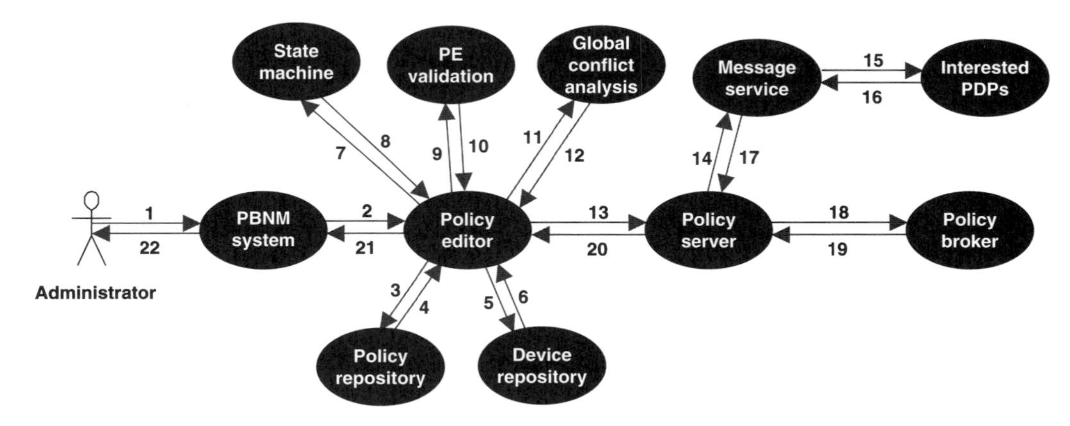

Figure 8-4 How the PBNM system processes a request to modify an existing policy.

This scenario starts after the processes conceptually described in Figure 8-3 have already been performed. Note that the user did not have to be aware of any of these steps explicitly; other than supplying the appropriate credentials when asked, all the user had to do was enter his or her name. In a single sign-on system, the user may not have even had to enter additional credentials besides those he or she used to log into the system.

The following is a sequence of steps that is followed when the user tries to modify a previously entered policy. For simplicity, assume that the GUI uses a Web-based form to guide the user's input. Furthermore, just the positive (i.e., no errors) path of processing as shown:

1. The PBNM system connects the administrator with the Policy Editor.
2. The administrator searches for a previously stored policy in order to modify it.
3. The Policy Editor contacts the Policy Repository to retrieve the policy.
4. The Policy Repository returns the policy to the Policy Editor (or a suitable error if the policy was not found).
5. The administrator selects a Policy Domain, and one or more network devices with that Policy Domain. The Policy Editor asks the Device Repository for information concerning the selected Policy Domain and network devices. (Note that this is shown as a separate repository for maximum flexibility. This is not a requirement of this architecture.)
6. The Device Repository returns the requested information to the Policy Editor.
7. The Policy Editor establishes a state machine to track the state of the policy being changed.
8. When the state machine has been established, the Policy Editor starts the editing session.
9. When the edits are finished and submitted, the Policy Editor does a final validation of the policy. (This is not meant to exclude the ability for the Policy Editor to perform per-field validation during the editing process;

rather, even if such validation is performed, the Policy Editor will have to validate the entire policy at the end of the session.) For maximum flexibility, this is shown as a separate module. This enables multiple, physically distributed Policy Editors to use the same validation mechanism, which in turn ensures editing consistency.

10. The result of the validation operation is passed back to the Policy Editor. If the validation fails, it returns the policy to the administrator for further correction. The state machine is updated as appropriate.
11. If the validation succeeds, the Policy Editor passes the validated policy to the Global Conflict Analysis module. This module is responsible for checking to ensure that common global data and information are correct. It also checks for conflicts with other deployed policies at the global level (e.g., John gets Gold service and John gets Bronze service, at the same time), not at a specific device or interface level. That is done later in the Policy Server.
12. The Global Conflict Analysis module passes the results back to the Policy Editor.
13. If there were conflicts, the Policy Editor passes the policy back to the administrator with a message stating what the conflict was. If there were no conflicts, then as far as the Policy Editor is concerned, this policy is okay. The user can choose either to stop and simply store the policy in the Policy Repository or to perform further validation that this policy is able to be installed by checking with other interested components of the PBNM system, such as the set of PDPs that will deploy this policy. (Assume that the user has chosen to install the policy, which means further processing is necessary). In this event, the Policy Editor sends the changed policy to the Policy Server. The state machine for that device is updated.
14. The Policy Server is responsible for sending the changed policy on to all PDPs registered for this type of policy, using the Message Passing Service, that are resident in this Policy Server.
15. Each participating PDP performs appropriate checking on the policy. This may include local conflict checking (e.g., against a specific interface) of this policy against other policies this PDP is handling as well as satisfiability and feasibility checking (see Chapter 4 for definitions of these terms). Each PDP is also responsible for establishing a state machine to control any devices that are affected by this operation as well as supplying information to update the state machine of the original device.
16. Each PDP evaluates the policy. These replies state whether the policy can or cannot be implemented (and why). Note that a policy may not be able to be implemented, not just because it conflicts with another policy, but also because that policy is not relevant to a particular set of devices for a given PDP. The state machines for that device, as well as for the original device that the policy applied to, are updated.
17. The Message Passing Service routes the set of PDP messages back to the Policy Server.

18. The Policy Server evaluates the PDP responses and then tells the Policy Broker that the policy is either acceptable unconditionally, acceptable under a set of conditions, or not acceptable (and why, of course). This enables the Policy Broker to communicate to other Policy Servers to coordinate the implementation of this policy, if desired.
19. The Policy Broker communicates with all other Policy Servers that have registered interest in this policy (this step is not shown for simplicity, but basically consists of repeating the appropriate steps from this use case). The Policy Broker gathers the set of replies from each of the interested Policy Servers. If one or more replies state that the policy is not acceptable (because it conflicts with the actions of another Policy Server), the Policy Broker relays this message to the Policy Server, which relays this message to the Policy Editor, which in turn informs the user. (Clearly, a real-life system could choose to return additional information, which could then be used to [for example] request a variation of the policy that would be more acceptable to the component. However, this is beyond the scope of this book.) Otherwise, the Policy Broker instructs the Policy Server that the policy is ready to be installed in one or more additional Policy Servers, and will govern that application. The state machine for each device affected in each Policy Server, along with the device of this Policy Server, is updated.
20. This is communicated back to the Policy Editor.
21. This is communicated back to the PBNM system, which then asks if the user wants to install this policy now or at a later time. The state machine for that device is updated.
22. The user is now free to commence installing the policy.

8.3.3 From the Network Element's Point-of-View

The following section describes a "day in a life" of a policy from the perspective of the PBNM system. For simplicity, we will describe installing a new policy, modifying an existing policy, or deleting an existing policy to a Network Element. This is described using Figure 8-5.

The following describes the sequence of steps that is followed when the device tries to install a policy. This process takes place after the policy has already been validated, as described in the previous section. For simplicity, only the actions of a single PDP in a Policy Server will be considered. Furthermore, just the positive (i.e., no errors) path of processing will be shown. Lastly, it will be assumed that each PDP in a Policy Server has already been associated with its set of devices (and vice-versa).

1. A PDP determines the Policy Targets a given policy applies.
2. The PDP asks the Policy Repository for information about each network device that is targeted by the policy.
3. The Policy Repository returns the requested information.
4. The PDP connects to the set of devices to which it wants to send a policy. It then starts a state machine to keep track of changes to each device.

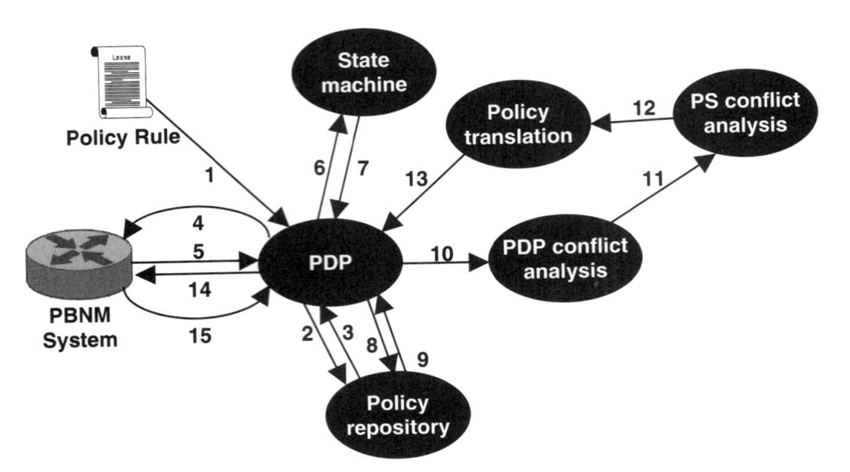

Figure 8-5 How the network element responds to a request to install a policy.

5. For each device, the PDP uploads the current configuration, including role definitions. This ensures that the PDP has the most current information on each Policy Target (i.e., another management station might have modified that device's configuration without telling the PDP). The state machine for that device is updated.
6. For each device, the PDP compares the current configuration to its last saved configuration. If there are differences, it updates itself and the state machine for each device.
7. The PDP computes the set of other policies that apply to that device (for all devices in which the policy is going to be installed).
8. For each device, the PDP requests the complete set of policies that are currently relevant to that device from the Policy Repository.
9. The Policy Repository returns the requested information to the PDP.
10. The PDP sends the set of policies to its own local conflict analysis module and updates the state machine for that device.
11. If the PDP's conflict analysis module finds a conflict, it informs the PDP, which informs the user. Otherwise, the PDP conflict analysis module then forwards its results to the Policy Server's conflict analysis module. In either case, the state machine for that device is updated.
12. The Policy Server's conflict analysis module forwards its results to the PDP.
13. If the Policy Server's conflict analysis module found a conflict, the PDP informs the user. Otherwise, the PDP conflict analysis module then forwards its results to the policy translation module, which then translates the policy into a vendor-specific form to install on the device. The state machine for that device is updated with the new state of the policy.
14. The PDP provides a set of instructions, in vendor-specific form, to each device.
15. The device informs the PDP of its success (or failure) in installing the policy. Alternatively, the PDP must monitor the device to determine if

the installation of the policy succeeded or failed. The state machine for that device is updated with the new state of the policy.

In the previous (simplified) process flow, the PDP may have needed to talk to a policy proxy if the PDP was not able to communicate directly with the device.

8.3.4 Additional Life Cycle Aspects of a Policy

The life and death of a policy can have a significant effect on the managed system. Introducing a new policy to the system can have effects that range from solving a problem to destabilizing the system. This is also true when a policy is removed from a system. The system is protected by the robustness of its conflict detection and resolution component. However, this is more complex than it first appears. This complexity is mainly because the introduction or removal of a policy fundamentally changes the states of all managed objects that it affects. At a minimum, two tasks must be done in order to introduce a new policy to the system:

- Perform conflict detection and resolution between the new policy and existing policies that are already deployed in the system; this helps ensure that introducing the new policy will not cause an unanticipated state transition.
- Ensure that all side effects of introducing a new policy are determined.

Similarly, when a policy is removed, the following two tasks at a minimum must be checked:

- Account for any and all dependencies that other managed objects have on a particular policy, and ensure that no harm will result to those managed objects if the policy is removed.
- Ensure that all side effects of removing an existing policy are determined.

8.4 The Building Blocks of a PBNM System

This section will discuss the three building blocks of PBNM systems—an information model, a policy language, and a data dictionary—and how they work together to bind the theory of PBNM systems to implementation. This subject was introduced in Chapter 3 and elaborated on in Chapter 5. The conclusion was that a combination of a set of languages and a set of models must be used, each bound to a particular layer of the policy continuum. Models enable new users to quickly understand the content and help guide the design and implementation of the PBNM information through its rigorous structure. Models define the high-level features of the PBNM system and provide the overall guidance for understanding the business as well as system features of the entities that the PBNM system is managing. Languages provide conciseness and the ability to directly express semantics that enable each layer of the policy continuum to translate its information and concepts to other layers of the policy continuum that need that information. More importantly, languages are machine readable and lend themselves to sophisticated uses, such as conflict detection. Finally, the data dictionary ensures that different models and languages operating at different levels of abstraction have a common source of understanding. The data dictionary is used to ensure that all of the metadata and semantics of each entity are appropriately captured and named. It can also be used to translate between different users of the same or different level of the policy continuum. This is because the data dictionary defines a set of synonyms and aliases that a given entity is known by, so that different users know that the same entity is being referenced even though it has a different name. This is crucial for building a complex system such as a PBNM system that is trying to address multiple user constituencies.

The following sections elaborate on these concepts with respect to how each influences the architecture of a PBNM system.

8.4.1 Information Model

An *information model* is an abstraction and representation of the entities in a managed environment. This includes definition of their attributes, operations, and relationships. It is independent of any specific type of repository, software usage, or access protocol. A *data model* is a concrete implementation of an information model in terms appropriate to a specific type of repository that uses a specific access protocol or protocols. It includes data structures, operations, constraints, and rules that define how the data are stored, accessed, and manipulated.

The information model is designed to be more than just a standard representation of data—it also defines semantics and behavior of, and interaction between, managed entities. This set of information is provided in a standard representation using standard data types to enable different components to share and exchange information. Using the information model to describe different types of managed entities (not just routers and switches, but also users, services, and much more) achieves a very important result: *Relationships can be established between different entities at different levels of abstraction.*

The importance of this statement cannot be overemphasized. Consider a service provider that offers a set of different classes of service to its customers. Although the concept of "class of service" can be used in many ways, fundamentally it is a way of being able to differentiate traffic when the network becomes congested. That is, if the network is not congested, by definition there is not a problem, and each application can happily transmit as much traffic as desired. However, if the network is congested, by definition the traffic for one or more applications will be adversely impacted.

In this example, class of service is defined to provide two distinct benefits: grouping different applications together to provide additional functionality, and treating a group of applications collectively better than other groups of applications. This concept is illustrated in Figure 8-6.

In Figure 8-6, we see an immediate difference between Gold, Silver, and Bronze service: functionality. That is, while Data and Web services are present in each, Silver adds VoIP and Gold adds SAP. The advantage of encapsulating these different applications in Gold, Silver, or Bronze service is that the network can

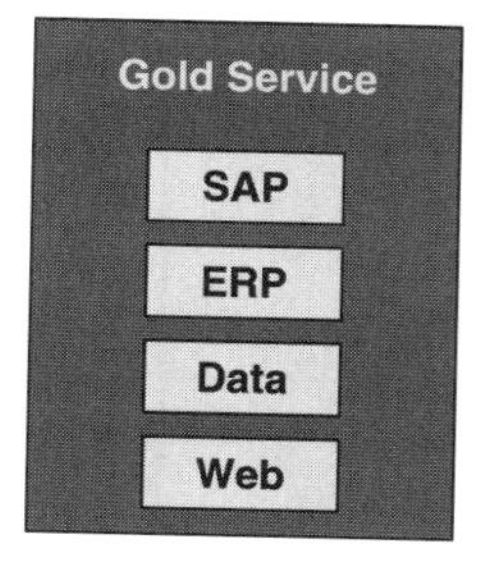

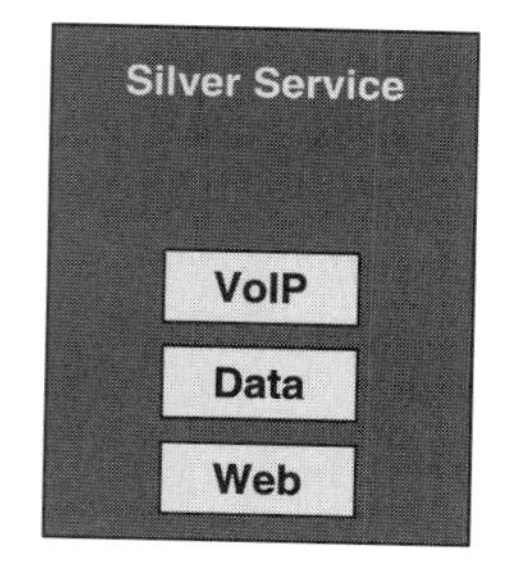

Figure 8-6 How the network element responds to a request to install a policy.

now be optimized to give better service *collectively* to that *group* of applications. This enables the service provider to also add *quality* to its offerings. For example, Data traffic is provided in Gold, Silver, and Bronze traffic, but the quality of service assigned to Data traffic in Gold Service is "better" than Data traffic in Silver service (which of course has a "better" quality of service than Data traffic in Bronze service).

But how do we relate users to services to devices to device configurations? This is where the power of the information model can help. Clearly, if we build a model that describes the characteristics of a given entity, we can describe its salient features and define how that entity can be managed. Now, suppose we build a framework that can accommodate all of the different entities that exist in our system. If we then define a model that contains these different types of entities and treats them all as *managed entities*, we can simply relate one entity to another using standard UML constructs. This is wonderful, because this establishes important relationships between these entities. These relationships can be of two types. This is shown conceptually in Figure 8-7.

Even though Figure 8-7 has been simplified considerably, note that there are many different types of managed entities at many different levels of abstraction (e.g., a PhysicalPort versus a Service versus a Customer). The power of the information model is its ability to treat any managed entity as an object, thereby enabling us to relate different entities at different levels of abstraction to each other.

DEN-ng, as well as the SID, has been constructed as a federated set of models. Each model in the federation covers one or more management domains, because, if a single model was used to represent the full diversity of management information that was needed, the resulting model would be much too large to be understood. In addition, two other important disadvantages would surface. First, this monolithic packaging would mean that any time one element changed, the entire model may have to change. Second, it enables applications to view the federated model as a set of extensible building blocks. The applications do not need to use the entire model (which will be quite large); rather, they use only those portions that are applicable to their application.

Perhaps the most important benefit of using an information model is one of abstraction. The diversity of our management data means that we will not be able to use a single type of data repository. The advantage of using a single information model is that it can serve as the single "source of truth" for defining

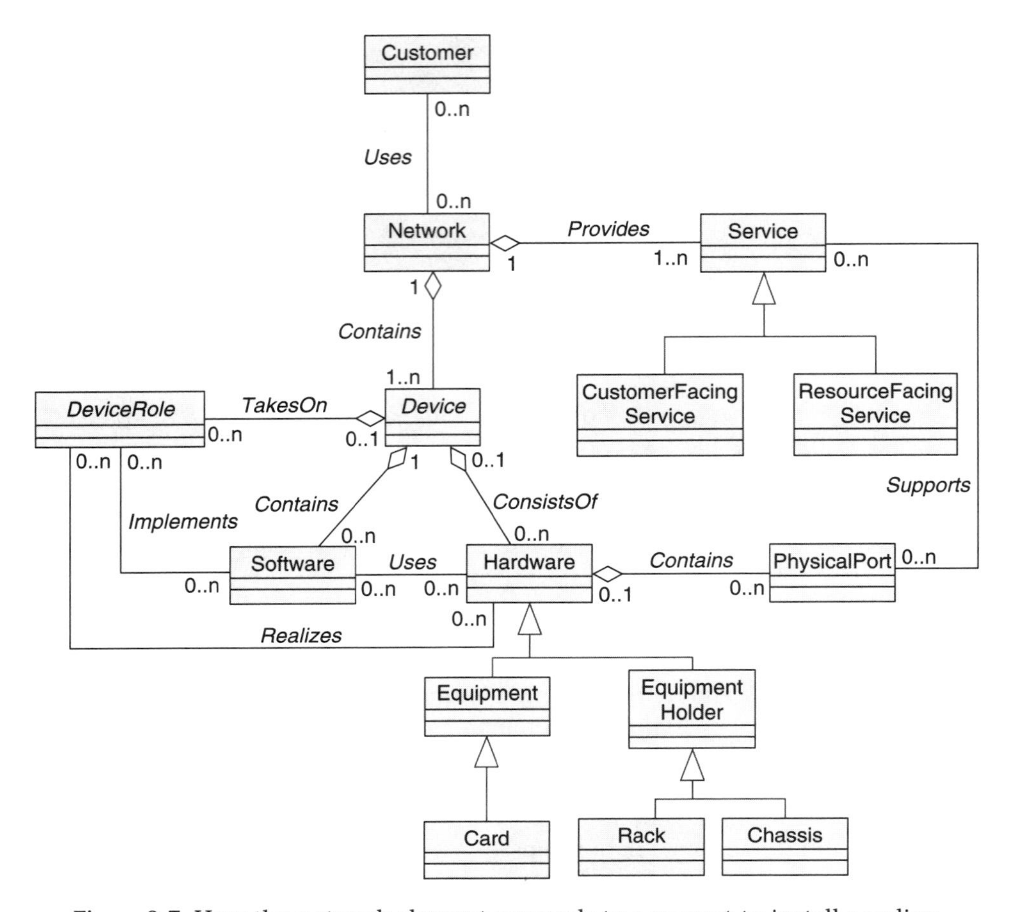

Figure 8-7 How the network element responds to a request to install a policy.

objects in which we are interested. This single information model is then translated to a set of data models that each application needs. Each data model is optimized for a specific combination of repository and access protocol that is being used to steward particular types of data. This enables applications to use only the particular set of data models that they require. If there were multiple information models, data coherency of a given entity could not be maintained across different implementations in different repositories, because one would not know which information model to use to build the translation. Furthermore, one would then have to keep the set of information models synchronized. If different mappings were built off of different information models, it rapidly becomes impossible to guarantee that the correct changes to the information models have been used to define the data models.

8.4.2 Policy Language

As pointed out in chapter 3, the information model is very important. However, it does not obviate the need for a language. Friedman et al.[5] provided good overview of the usefulness of programming languages. Three important uses of a policy language are: (1) to provide a specification for implementing the behavior described in the models, (2) to express denotational semantics in a concrete way, and (3) to bind together the different abstractions from which different policies are built when different levels of the policy continuum are used. Although the first use is obvious, the other uses deserve further elaboration.

One of the important uses of a language is to be able to express denotational semantics (e.g., the literal semantics, instead of connotational semantics, or the meaning associated with the semantics) in a concrete way. For example, a particular policy can have many different implications. Although these are important, the literal meaning of the policy is critical if a system is going to be able to understand the actions and side effects that deploying this policy has.

This is critical in PBNM systems. While models are able to specify the high-level relationships between different managed entities, the actual *behavior* of these entities must be more precisely specified than what can be conveyed in a model. Denotational semantics are well suited for such specification. Furthermore, the need to define explicitly and carefully literal semantics is especially critical when entities at different levels of abstraction are related to each other. For example, suppose an entity representing a business policy is defined. Assume that an association is defined between this entity and another entity. It is critically important to represent precisely the semantics of that business policy in a form that apply directly to the entity to which it relates. A policy language is best suited for this task, since the language itself contains the ability to represent the semantics of how these two entities are associated.

Another important use of a language is to bind together the different abstractions that different policies are built from, using different levels of the policy continuum. Since these are different abstractions, a means must be provided to translate the semantics from one level of abstraction to another. Such translations require precise semantics for transforming one representation to a different representation. An analogy is a multi-phase compiler, in which the output of one phase of the compiler is in an intermediate form that can be used as the input to the next phase of the compiler. This is a powerful analogy, since it makes explicit the implied association between different levels of the policy continuum. This is in reality the most difficult part of the policy language: ensuring that the different languages used for each level of the policy continuum can be united into a single "compound" language that can be understood by the PBNM system. This in turn ensures that each user at each level of the policy continuum can communicate with every other user at other levels of the policy continuum.

8.4.3 Data Dictionary

The reason why a data dictionary was proposed for DEN-ng was twofold: to provide a set of aliases, synonyms, and cross-references between different models: and to capture additional information to supplement the models.

First, different types of users (business users, system designers, and implementers) were targeted as users of DEN-ng. This was a problem, because just like in the policy continuum, each of these different users had their own grammar and terminology. Without a means of translating between different names, terminologies, and so forth, it would be impossible to build a system that supported the policy continuum.

It should be noted that the TMF came to a similar conclusion in the SID. The GB922 guidebook[6] for the TMF Shared Information and Data model (which contains many of the DEN-ng models) mandates the use of a data dictionary. This is because the TMF SID team is building a combination of business and system models. The use of the data dictionary enables information contained in business views to be related to information contained in system views. In the DEN-ng policy model, the same approach applies, but there are more than just the business and system views defined in the TMF.

The second reason was to capture additional information pertinent to the understanding of the models. The motivation here was to enable the information captured in the policy to feed into the policy language, and vice-versa. Development of the DEN-ng models was started before the DEN-ng policy language; hence, a means for associating these two efforts was needed. The beauty of the data dictionary is that it can be used to feed information back into the DEN-ng models in a standard, structured manner.

8.4.4 Putting It All Together: Model Driven Architectures

The information model by itself is just a tool. The problem to be solved is the mapping of business rules and procedures to network and system configuration, so that system and network resources are allocated according to the business rules that are being executed. This mapping is done by representing business rules as part of the information model, and then using policies to control the allocation and use of managed resources. As stated earlier, if business rules as well as managed entities can be represented in the same object-oriented framework, each can be related to the other by building relationships to relate them together.

This is in reality a specialized example of what the OMG calls a Model Driven Architecture (MDA).[7] The thesis of the MDA is that there will never be complete agreement on hardware platforms, operating systems, protocols, and programming languages. Therefore, the only way for different system components to interoperate will be through a standard set of interfaces. MDA is a prescription for writing specifications and developing applications that are based on a platform-independent model (PIM). The object is to insulate systems from continually having to be rebuilt to accommodate the benefits of new technology. An MDA specification consists of a single platform-independent UML model, plus one or more platform-specific UML models that describe how the platform-independent model is implemented on different platforms.

Fundamentally, UML is the key enabling technology for the MDA. Modeling using UML enables the creation of applications that are portable, and interoperate naturally, across a wide variety of systems. This is done using the following three steps:

- Start with a detailed PIM, including all necessary pre- and post-conditions (using OCL) and semantics in the UML Action Language: this captures the business functionality and behavior without being influenced by technical implementations

- Use an OMG MDA tool to map from a PIM to a supported PSM (e.g., CORBA, Java EJB, SOAP)
- Use OMG MDA tool to generate application interface code for the set of Platform Specific Models that are being using (this is a combination of automated code and hand-crafted code that is added by the developer)

This approach facilitates the building of mediation software to enable different technologies (e.g., CORBA and Java) to interoperate. The MDA can also be used to reverse-engineer a legacy application into a model. Once the model is derived, the MDA tools can then be applied to integrate it into other MDA applications.

The DEN-ng approach builds on the original DEN specification,[8] which, though less in scope, is similar to the MDA approach. DEN and DEN-ng both prescribe the use of a single information model to define the structure and behavior of managed entities, followed by a set of mappings to translate this single model into a set of data models. However, DEN and DEN-ng recommended that an additional mapping layer be inserted, as shown in Figure 8-8.

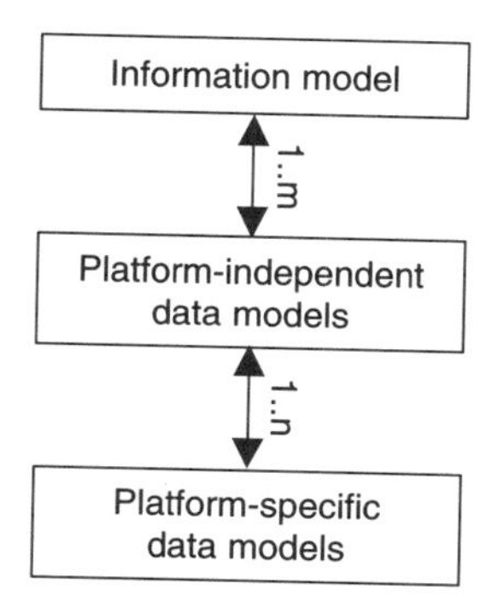

Figure 8-8 DEN and DEN-ng model mappings.

Figure 8-8 shows two levels of mapping. The first transforms a repository- and protocol-independent information model into a form that is optimized for a particular type of repository. This form is based on standards as much as possible. Sadly, however, a fact of life is that vendor implementations vary from the standard. Thus, a second level of mapping is often needed. For example, such a second level of mapping is often needed when using directories, since there is quite a bit of variance in how the directory standards are implemented by different vendors.[8]

DEN-ng is a natural outgrowth of the original DEN approach and is heavily influenced by the MDA approach. The PBNM approach described in this book is similarly influenced by the MDA approach. The difference is the inclusion of a formal policy language not just to guide implementation, but also to define the semantics of how policy is applied. This necessitated the use of a data dictionary to help bind the model and the language together. This should not be seen as contradictory to the MDA approach. Rather, it builds on the MDA approach.

8.5 Architecture of a PBNM System

The architecture of a PBNM system is driven by both the types of operations that it provides and the types of entities on which it will perform those operations. To a lesser extent, it is driven by the type of application that is using the system. This can be categorized broadly into applications that require quantitative results from policy versus applications that want qualitative results. Applications that want quantitative results, such as VoIP, need a specific quantity, amount, or other discrete measure of one or more attributes that are being controlled by policy. Qualitative applications, such as a "mission-critical" application, are typified by demanding an approximate behavior (which may relate to an unspecified amount of one or more attributes, such as bandwidth) that gives these applications the desired character or "quality" that distinguishes it from other applications. As an example, characterizing some applications as "mission-critical" and others as not "mission-critical" can be used to define different classes of service. The class of service used for "mission-critical" traffic can have different traffic conditioning, or even different routing, to give that traffic "better" service than other types of traffic. Here, "better" service means that one or more metrics, such as bandwidth, jitter, and delay, are given preferential treatment compared with those same metrics for other classes of traffic.

This section first discusses the effect of different types of policies that have different types of characteristics and uses on the design of the PBNM system as a whole. Then, it describes the difference between policy for quantitative versus qualitative applications. Finally, it relates the different types of policies to the needs of quantitative and qualitative applications and sets the stage for defining the architecture of our PBNM System.

8.5.1 Overview of the Operation of the PBNM System

This section starts by providing a quick overview of how the PBNM system works. Then, the different facets of a PBNM architecture are examined in greater detail. This way, the detailed descriptions of how particular components work can be related to the overall operation of the PBNM system.

There are several types of policies that can be defined for network devices: high-level policies that control the types of behavior specified by business rules to control a managed object, intermediate-level policies that control how different managed objects operate in the overall system, and low-level policies that control the function that the device performs. In general, the set of policies that can be defined for a network corresponds to the different types of policies that make up the policy continuum.

The PBNM architecture must provide a closed-loop system that consists of monitoring the managed devices to ensure that their current state is known, providing commands that transition one or more managed devices to a new state as dictated by the appropriate finite state machine, and then check the result of the commands by monitoring not just the changed managed devices, but the rest of the managed devices as well. This ensures that changing the state of a managed device did not produce any undesirable side effects.

The use of a finite state machine in this architecture is a characteristic of the DEN and DEN-ng approaches and is recommended for use in PBNM solutions. It is used to model the state of the system as well as the managed entities in the system. The power of the finite state machine is that it serves as a blueprint for tracking which policy-selected processes have been applied as well as the success of their application. This is especially important for the DEN-ng approach, whose policy model allows separate policies to be specified in a single policy rule for conditions that are met as well as those that are not. Conceptually, this takes the form of specifying behavior in response to expected versus unexpected events. The finite state machine enables each event to be tracked through using different state transitions.

Policy-Defined Views of the Managed System

In real life, the policy management is complicated by several factors. The most basic of these is that not every administrator is allowed to see every device in the network. This can be represented by authentication and authorization policies.

Authentication policies are used to prove the user's identity. Currently, many systems define this as an exceedingly simplistic process that is made up of simple discrete actions, such as asking for a password or certificate. This approach is rejected in the DEN-ng PBNM approach. Instead, policies are used to determine what type of authentication (or authorization) method should be used. This is because the environment may change underneath the policy system, or because a different user may have different avatars that require different sets of authentication mechanisms, or even because the organization may have a different set of business policies (which of course are used to drive the authentication method in the first place!) that are used as a function of the environment and/or how the user is logging onto the network. By making this a policy (instead of a static check of a system), the same network can accommodate different uses without having to be redesigned for each of these use cases. Instead, the only thing that needs to be changed is the control mechanism, which has been abstracted through policy.

Authorization policies define what a given user is allowed to do as a function of the current state of the managed environment. As in the case for authentication policies, authorization policies are used because a business rule says that they should be used. This enables business rules to limit what a given user is allowed to see (beyond what he or she is normally allowed to see) under different circumstances. For example, a business may have a policy that prohibits anyone from downloading source code over the public Internet. It is irrelevant if the user provides the correct authentication credentials: the business policy overrules the fact that the user has given the correct authentication credentials and prohibits the user from accessing the code server, because it knows that the user is currently logged in over the public Internet.

These are two examples of how policy can be used to define how users access a set of services. The use of policy provides two different abstractions—one for the user, and one for the network administrator. Each are provided a set of views of the network. The user is given a set of service-oriented views that enables the user to think of the network as an intelligent provider of various services. The administrator is given a set of views describing the set of managed devices that can be seen and operated. These administrator-oriented views define the set of operations that a particular administrator is permitted to perform on a given device. This is shown conceptually in Figure 8-9.

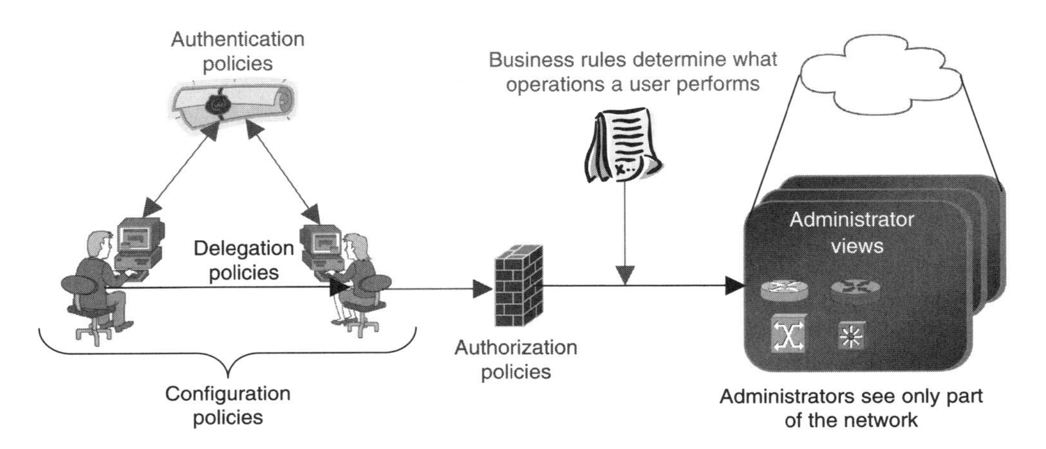

Figure 8-9 Policies define views of the network per administrator.

Furthermore, if a change to a device is required, a proper accounting of the change as well as an audit of the change must take place. (This, as we will see later in this chapter, requires the use of processes that are called by policies.) Since these policies work together, they are referred to as AAAA (Authentication, Authorization, Accounting, and Auditing) policies in this book. Note that most other systems do not discuss auditing, and many also don't discuss accounting.

It is important to emphasize the role that the business rules play in Figure 8-9. Many current systems take a much too simplistic view of the world and enable a privileged user, such as an "administrator" to perform any function on any device. This is extremely dangerous and likely violates fundamental business rules that specify that only people with appropriate skills perform certain operations. For example, although almost anyone can change simple commands (e.g., SNMP community strings, or NTP addresses), changing BGP routing policies is quite complex and should only be done by someone with the appropriate technical skill and background.

There is a tendency in the design of current policy systems to be able to abstract devices to facilitate their mass or group configuration. This can be done easily through the use of roles. However, not every device should be treated the same. This is especially true for lower levels of granularity, such as device interfaces. This does not mean that roles should not be used; rather, it means that they should be used intelligently. That is, just because a particular user has been assigned the role of administrator should not necessarily mean that the user has free access and rights for all devices in the network! This makes the very bad assumption that all such devices are equal. Not only do the devices serve different roles, but they are usually owned by different groups. Hence, we need a finer level of control exercised in order to avoid compromising the system.

Therefore, in an ideal world, the PBNM system should manage which users can perform what types of operations on which devices. *This is in itself a policy and should be recognized as such.*

Using Business Rules to Define Which Operations Can Be Performed By Which Operator

Note that an operation can be as simple as viewing a device, or as complicated as changing parts of its configuration to build a new service or to modify an existing service supported by the device. As stated previously, it is critical to use business rules to determine which operations a given user can perform on which devices. This is illustrated in Figure 8-10.

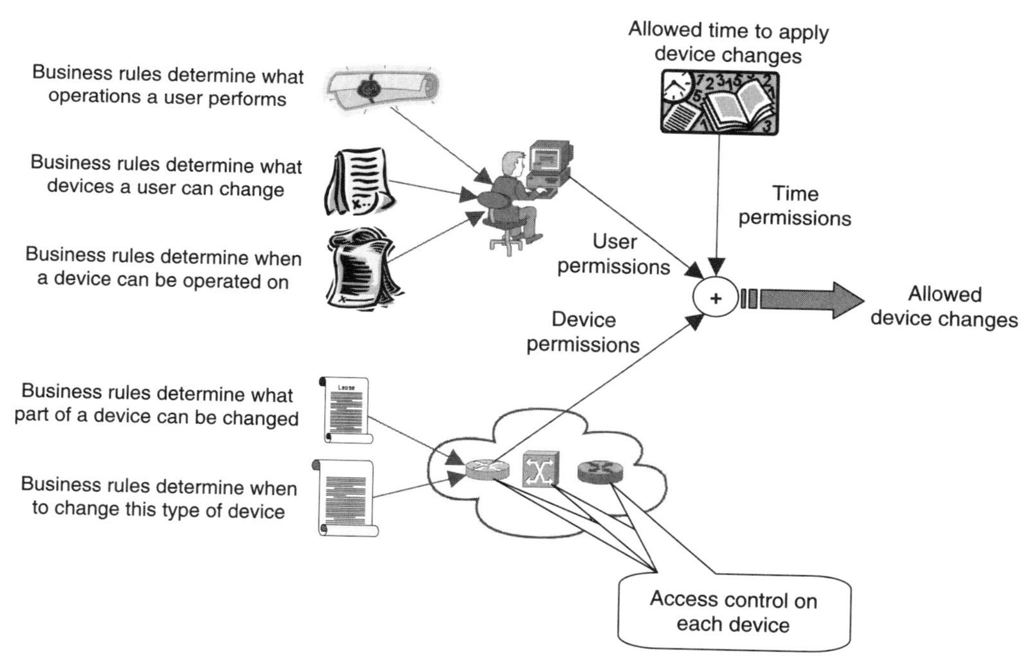

Figure 8-10 The different policies used to control device changes in a PBNM system.

Figure 8-10 shows the resources controlled by the PBNM system are managed using three main types of policies. User-oriented policies determine what operations which users can perform on what devices. Note that a good PBNM system will provide sufficient detail to enable a user to not necessarily be able to:

- Do the same operations on the exact same device if it belongs to a different organization, plays a different role, is located in a different part of the network, or otherwise differs

- Perform all possible operations on a particular device
- Even see a given device

This is represented in Figure 8-10 by different business rules that result in a consolidated set of user permissions.

It is equally important to perform the same analyses on devices. Otherwise, a system is created in which specific users are bound to specific devices. Although it may indeed be a business policy to only allow a particular user to change a given device, this is more likely the exception than the rule. Therefore, what we want to do is to abstract the concept of user and device permissions with the concept of *roles* that are used (at least in part) to determine those permissions. This will help provide as fine a level of granularity as required (e.g., by subclassing a role so that only a specific person or set of people belong to that role). However, the main advantage of using roles is that it provides a convenient abstraction mechanism to assign different people, groups, or even organizations to perform a task. Similarly, devices can have roles, which enable us to select one or more devices that play a given role.

Although this seems to be enough, in reality it is not. For example, consider a user that has the permission of editing access control lists. For certain cases, it is sometimes important to be able to restrict which devices that user can access. Thus, the privilege of editing access control lists should be logically ANDed with the permission to operate on a device. The result of this logical AND then determines if the user can perform the specified operation on the given device. Note that this is a trivial example: real systems will need to logically AND many more variables. However, the principle is the same.

Businesses often use the notion of *time* to determine whether or not a given business rule should be executed. Thus, the conceptual diagram of Figure 8-10 can be simplified to the following diagram in Figure 8-11.

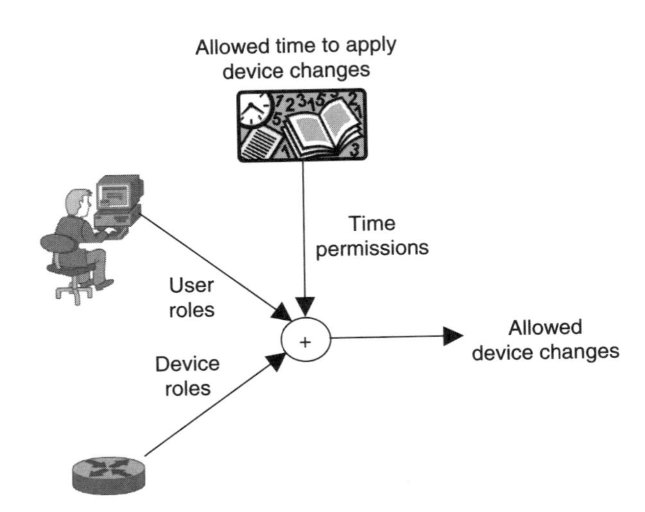

Figure 8-11 How roles determine the use of policies to control device changes in a PBNM system.

Roles enable the right user to perform the appropriate operations on the appropriate device. However, it is also important for a PBNM system to ensure that an operation happens at the correct time. This requires more than the use of time as a simple "gating" operation that says when an operation can be performed. Instead, we need to introduce the concept of a workflow-driven system, operating under policy control, to ensure that the correct processes execute at the correct time. This is more than scheduling an operation—it also includes the sequencing of operations as well as the assurance that any dependencies between these operations are observed. Workflow is a good start, but a full-fledged business process automation system[9] is really needed to exercise enough control.

Integrating Policy Management and Process Management

Many people have mistakenly believed that policy management and process management are incompatible. Nothing could be further from the truth. Instead, we need to be able to holistically combine them. If we do this, then we get two benefits:

- Policy can be used to select the appropriate process to execute
- The results of executing a process can be fed back to the policy system to fine-tune its control

Policy is **not** just the changing of a command in a configuration file. Instead, policy is the selection of a process that is used to perform an operation. Such operations are not limited to simply changing configurations. Depending on the system, policy may dictate that any operation on a device, even looking at its configuration or pinging it, should be audited. **This combined use of policy and process is an essential feature of the PBNM architecture described in this book.**

The PBNM architecture must facilitate the integration and coordination of different policies to perform these operations by authorized users only. This in turn means that an integral part of policy management is accounting for changes made to a device and auditing those changes. If something goes wrong, the organization can better understand what happened through the accounting and auditing information associated with changing a device. This is facilitated by the system described in Figure 8-11. Roles can be used to identify users as well as devices to facilitate determining which user is allowed to access what device.

However, from an operational point-of-view, we need more than just roles. We need the ability to control the execution of each sub-task that forms a higher-level task, in order to ensure complete control over the task at hand. For example, the high-level task of changing the configuration of a device can be separated into a number of sub-tasks. The most important of these are approving the configuration, installing the configuration, and verifying that the configuration changes were accepted.

Note that this latter can be broken in turn into two distinct sub-tasks: verifying that the new configuration changes were accepted in the device in which they were installed, and verifying that those changes did not cause any other errors in any other devices.

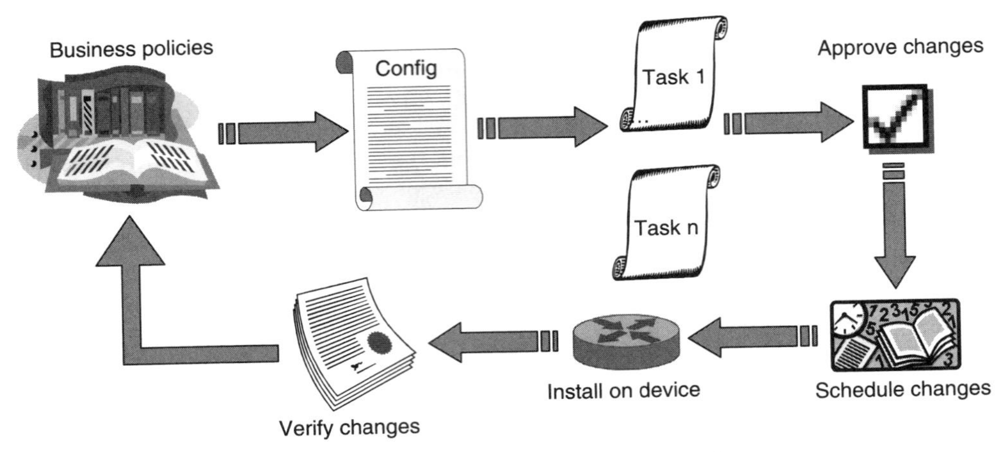

Figure 8-12 The configuration process in a PBNM system.

These are some of the important sub-tasks in changing the configuration of a device. The overall configuration process is shown in conceptual form in Figure 8-12, and defines the following major tasks:

- Business policies are invoked to start the overall configuration change process. These select who is going to work on which device, who must approve it, what the proper procedures are for making the changes, and so forth.
- The next major step is to construct the configuration. Clearly, this could be its own complex workflow, with different people responsible for different portions of the configuration. So . . .
- . . . It is shown conceptually as a set of *n* tasks. A good PBNM system should enable a task to be separated into a set of tasks; the granularity of this separation should be configurable. Note that the use of a finite state machine greatly facilitates this.
- The next major task is to approve the configuration work to be done.
- Once approved, the task is scheduled for installation on the target device or set of devices.
- After installation, the configuration needs to be verified. As stated above, this could include checking other devices to ensure that they were not adversely affected by this configuration change.
- Finally, the results need to be fed back to the PBNM system, so that it can ensure that the device as well as the system are in the correct state.

Filtering Actions

In general, AAAA policies impose a set of constraints that define which actions may be taken for a given device. This is a function of who is operating on the device, the role of the device, the nature of the changes, time of day, and many other factors. Furthermore, the set of actions that are scheduled to be performed may not be able to be executed immediately on the target device. This could be because of a number of factors, such as the action conflicting with an already executing action or the device being in a particular state that does not allow that action to be performed.

Therefore, the decisions must be filtered to allow only those decisions that can be performed at any given time to proceed and turn into actions. This should be implemented as a set of policies, because the state of the managed environment will change over time, and the set of permitted decisions will consequently change over time. Policies provide the flexibility to capture the set of decisions that are possible at any given time. Filtering therefore provides us a way to adjust the output to suit the specific needs of the application.

Feedback Within a PBNM System

These factors lead to Figure 8-13, which shows how feedback is performed in a PBNM system. The set of managed objects shown to an administrator is the set of managed objects that a given administrator has permission to see. Figure 8-13 indicates this concept by showing a set of devices in each administrator view. AAAA policies restrict administrators to performing a set of operations on one

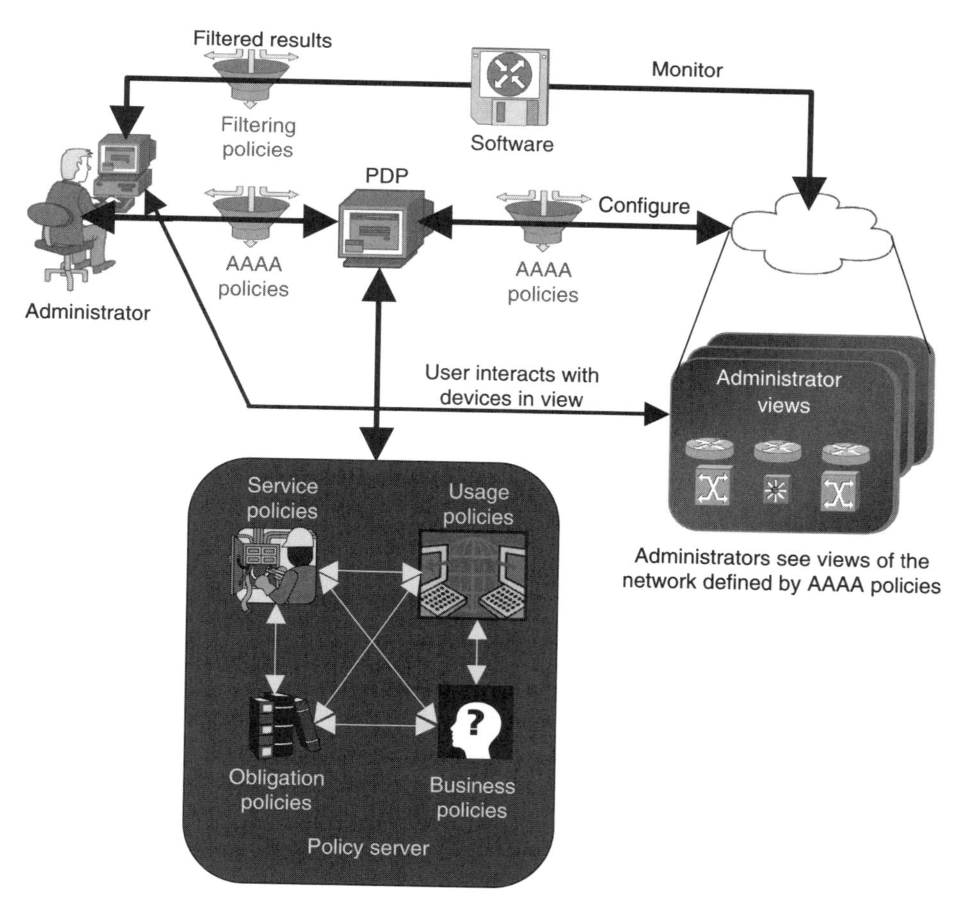

Figure 8-13 Feedback within a PBNM system.

or more subsets of devices based on geographic connectivity, administrative roles, or other factors.

Figure 8-13 assumes that the administrator has first chosen one or more policy domains of interest. This is represented by an administrator-defined view of the network devices in one or more networks. The selection of these policy domains causes the PBNM system to execute a set of obligation policies to decide what work needs to be done in what order. These different policies combine in guiding the administrator through a set of tasks to complete. These are, of course, invoked as a function of who the administrator is, into which administrative domain the administrator is logging, and what the state of the managed device and the managed environment is at that particular time. The devices in the administrator-defined view will be shown on the policy console, as indicated by the bi-directional arrow connecting the administrator-defined view with the policy console.

Obligation policies are responsible for correlating the work that can be done with the "allowed device changes" to each device being operated on in the given policy domain, as defined in Figure 8-11. It is the responsibility of the PBNM user interface to filter this information so that the PBNM user knows what he or she can and cannot do. Although it is beyond the scope of this book to prescribe an appropriate user interface, the user's tasks to be performed are simpler to describe and understand by filtering the data. This is a basic principle of human-computer interaction and is called *progressive disclosure*. The key is to avoid information overload and build an interface that allows, or even encourages, further exploration by the user.

Figure 8-13 therefore starts with the concept of the user performing an operation on one or more devices within an administrator-defined view, such as a configuration operation. This figure illustrates the feedback necessary for the PBNM system to ensure that the requested operation is performed correctly and has the desired effect on the managed device and/or system. This is indicated in the DEN-ng model by treating the new configuration as a state transition. The PBNM system can therefore verify that the operation was successful if the end state of the system matches the "good" end state in the state machine.

The state transition is modelled in Figure 8-13 as a sequence of the following steps:

- Administrator sends completed set of operations to the PDP (AAAA policies are used to govern which operations can be sent to which devices as part of this process).
- The PDP then translates these operations either to the form required by the devices (e.g., vendor-specific CLI) or to an intermediate form that another entity can use to perform the required translation. This is shown in the "Configure" arrow. Note that separate AAAA policies govern what the PDP can do.
- These are implemented on the selected devices.
- Separate software monitors the results of the operations and presents filtered results back to the administrator. Optionally, the PBNM system could determine if the results match the expected end state and alert the administrator if this is the case.

Depending on the type of operation and the overall state of the managed environment, one or more policies that invoke one or more operations may be executed by the policy server to help make a decision on what to do. These decisions are filtered to restrict those decisions that are not applicable to the target-managed objects before they are passed to control software that processes them and produces actions to send to the target-managed devices. The need for filtering arises from the desire to *loosely couple* different components of the PBNM solution in order to not adversely affect the scalability of the system. This means that no two components need to know all of the implementation details about each other. Hence, one policy server could have computed a decision that is no longer applicable because the target object has just changed. The PBNM system must be aware of and track such changes to the system, and how those changes correlate to the overall finite state machine that represents the current and desired states of the system being managed. It is therefore able to discard non-applicable decisions and forward only those decisions that are applicable to the target-managed object at the current time.

One interesting example of the principles of loose coupling for OSS systems is the TMF NGOSS architecture.[9,10] The *contract* is the central concept of interoperability in the NGOSS architecture. Contracts define public-shared services in a technology-neutral fashion. This enables technology-specific implementations of the contract to be developed. A real-life example of this was implemented in the Fine Grain Catalyst project.[11]

For any operation that changes the state of a managed entity, it is crucial to monitor the results to ensure that the entity has transitioned to the appropriate state. Monitoring software is used to provide the current state of individual or groups of managed objects as instructed by the administrator. Note that in general, the information being monitored that the administrator can see is a subset of the information being monitored by the entire system. This is important, because enabling the administrator to see information that he or she is not supposed to see could compromise the system. Of course, the flip side is that if the administrator is not given enough information, the validity of the decision that the administrator is trying to make is jeopardized.

The state of the monitored objects will be communicated to interested parties and clients using some type of distributed communications mechanism. For example, a publish-subscribe system or a Jini system, as described in Chapter 4, could be used. It is up to the PBNM system to ensure that these data are sent only to clients with the appropriate permissions.

For simplicity, the process of applying actions to a managed object is shown as a single (and probably too simple!) entity. In reality, the decision as to *which* actions to perform, in which circumstances, and to which managed objects changes dynamically according to changes in the device, the managed environment as a whole, and the needs of the users of the PBNM system. For maximum flexibility, the information that is monitored should be controlled by an appropriate policy. This is discussed in more detail in later sections of this chapter.

Actions enable a managed object either to maintain its current state or to transition to a new state. However, without proper feedback, the management system (as well as the administrator) cannot tell if the actions that were applied to the managed object took effect and resulted in the behavior that was expected. Therefore, monitoring actions is essential in order to provide feedback to ensure that the managed objects are in the proper state after the policy has been applied.

8.5.2 Policies Are Static, Environments Are Dynamic

People tend to think of policies as being "static" or "dynamic." This is incorrect. A "policy" is always a static entity. It may be parameterized, but it is still static because the policy must be invariant in order to produce predictable behavior. Clearly, if the policy is dynamic, it is impossible to predict what will happen when it is applied. This in turn means that it would be impossible to use policy to predictively govern the state of the system. This would defeat one of the main purposes in using policy.

On the other hand, management environments are inherently dynamic. This is in fact one important reason why policies are of interest in the first place. If the environment never changed, it would be trivial to ensure that its managed objects were in the correct states, and policy would not be needed. It is exactly because the managed environment can change frequently that it is desired to use policy to ensure that a standard response can be provided.

However, users and administrators of the system, as well as their needs, can also change. This means that different policies, serving different purposes, are needed by the same system. This leads to the realization that different policies that do not inherently conflict when they are designed, but do conflict when they are executed, need to be stored in the same policy repository.

If this is done, different policies can be used for different purposes by the same system. The policies themselves are differentiated by specifying different events to trigger their evaluation and/or different conditions to determine when a particular policy rule is applicable or not. In this way, the use of different events and conditions has the effect of enabling or disabling what would be conflicting actions in different policy rules.

8.5.3 Scalability

There are three types of scalability: (1) the ability to add capacity to existing hardware components via additional resources, such as memory and processors, (2) the ability to add additional hardware components, such as servers, to increase the overall processing power of the system, and (3) the ability to add logical processing power, such as increasing the number of threads that are working on a single job. Each of these types of scalability can be used to scale the entire policy system.

Adding Additional Capacity to an Existing Hardware Component

Ideally, the architecture should support the notion of capacity, which is the ability to define the minimum and maximum requirements, limits, or other variable features of a component. For various reasons, it may be preferable to gradually increase the capacity of various physical components as the system itself grows.

It is important to ensure that such additions do not cause any disruption in offered services. One way of ensuring this is to use an object-oriented information model, such as DEN-ng, to model capacity for certain managed entities. This enables policies to be defined that control when capacity is added to a particular set of system components.

Adding Additional Hardware Components. The architecture should also be able to support the ability to add additional hardware components to increase capacity, processing power, storage, or other variable features of the system.

The system must be able to tell when a new hardware component is installed. A PBNM architecture should use an active registry, such as that specified in the TMF NGOSS architecture,[10] to enable the recording of new hardware and software components. The term active means that the registry should be able to interact with other essential framework services, so that its contents can be kept current with changes in the PBNM system.

The most powerful type of registry is one that is logically centralized, yet physically distributed. This enables the registry to be located near components that need to use it. It also provides failover and additional capacity, but of course requires additional processes to keep it updated. Note that JavaSpaces are a "natural" type of distributed repository/registry, because they can grow to accommodate new parts of a system that are added dynamically. This is done using the standard Jini discovery mechanism. See references 12 and 13 for more information on the JavaSpace specification and a commercial implementation of a JavaSpace, respectively.

The purpose of the registry is to enable new components to advertise their existence, as well as to record important system information, including:

- The location and state of all managed nodes in the managed environment
- The location and state of all management nodes in the environment (e.g., nodes that enable services and devices to be managed)
- The location and state of all management entities in the environment (e.g., people and/or processes that can manage certain devices and services)
- The location and state of "knowledge information" (e.g., an information model, or a state machine), which are externalized descriptions of content and/or behavior
- The location and state of all policies and policy information (so that it may be easily accessible by entities that need to refer to, use, or enforce policies)

The NGOSS Registry supports all of the above features, along with the following:

- A Trading Service, which enables objects to be accessed, selected, and located without knowing the precise name of the object (i.e., the attributes are used as trading parameters)
- One or more Naming Services that enable logical names and/or aliases to be assigned to a manageable entity that renders a service

- Information that supports the use of the Contracts, such as the location and state of all Contracts that are implemented by software entities in the system, or the set of user and role information that is used by different Policies. This information is defined by the NGOSS SID.[6]

Contracts are a very powerful way of controlling how components interact with each other, and they will be discussed below.

The existence of a single logical registry is assumed, regardless of whether it is physically implemented as a single or a distributed system.

Adding Logical Processing Power

There are many ways to add additional computing power. Two simple examples are to apply additional process threads to a given job, or to split up the job into multiple jobs that can be executed in parallel. However, in almost all cases, the components themselves must have been architected to enable such functionality. In other words, it is usually impossible to simply add a new component and have its processing power magically increase existing processing power.

In order for a system to be fully distributed, the intelligence of that system must be distributed among all of its components. It is important to note that this definition include not just hardware and software components, but human resources participating in the generation and delivery of services as well. Contracts can help control how additional logic resources are invoked and executed.

Accommodating Multiple Policy Domains

The system described in this chapter defines a logically centralized management of policy, with the ability to provide distributed execution.

Recall the following definition of a policy domain (from Chapter 2):

A policy domain is a collection of entities and services that are administered in a coordinated fashion using a set of policies. The policies are used to control the set of services and entities according to a common methodology, such as a finite state machine.

In general, a PBNM system is usually used to control a set of policy domains. As we will see, a policy domain may consist of multiple Policy Servers, in which each Policy Server is responsible for managing a particular set of services. However, a policy domain may contain many different types of network devices, each with their own set of capabilities and programming models. This makes it difficult to support a particular service, let alone multiple services, within a given policy domain.

The solution to this dilemma is to treat the Policy Server as a fundamental unit of distribution. This means that the Policy Server must contain a set of entities that are responsible for controlling one or more services for one or more devices. Here, "controlling" consists of two phases: responding to requests that want to invoke policy decisions, and enforcing those decisions. These entities are referred to as the Policy Decision Point (PDP) and the Policy Enforcement Point (PEP), and will be described in detail later in this chapter. (Note that for the sake of generality, these elements are described as components that are distinct from the network devices themselves. Nothing precludes a manufacturer from embedding these functions into a network device.)

A Policy Server therefore may consist of multiple PDPs and PEPs, each responsible for controlling a different type of device, function of a device, and/or technology. Furthermore, multiple instances of the same policy server may need to be deployed because of administrative, political, and/or geographical considerations. The system must therefore scale through deploying multiple Policy Servers, each of which has a set of PDPs and PEPs that control a set of services provided by that domain. Decision logic is provided to coordinate three different levels of abstraction:

- Coordinating the operations of multiple Policy Servers
- Coordinating the actions among the different PDPs and PEPs resident in a given Policy Server
- Coordinating the configuration of devices that support a particular type of service

This distributed approach enables the policy system to scale to meet an increased number of services by increasing the number of components that control an ever-increasing network. It also enables more advanced services to be supported by enabling the complexity of policy decisions that need to be taken. As we will see, by separating the knowledge used to configure a device from the device configuration itself, new devices that have different functional capabilities can be introduced into the same network without having to re-engineer the previously specified policies.

8.5.4 Principles of Distributed Computing

An exposition of the principles of distributed computing is beyond the scope of this book. See for example the many references in reference 14. However, there are a number of important principles of distributed computing that are required to be implemented in a PBNM system that deserve explicit mentioning. These include:

- Entities must be able to run independently of one another
- The failure of one component does not directly impact other components (e.g., by intelligently switching to a new component)
- Upgrading a component doesn't necessitate reprogramming other components, so long as its interface remains the same (or at least backwards compatible)

In order to realize these and other distributed computing goals, it is important to realize that the resulting architecture must move away from the construction of monolithic applications to a component-based model.

There is a series of benefits that results from building such an architecture. Arguably, the most important of these is that business logic can now be externalized. In other words, it is no longer necessary to hard-code business logic into a monolithic component. Rather, the business logic can be distributed among a number of components, and its semantics can be controlled using a combination of policy-based management and process management. Policy-based management is used to select and coordinate the appropriate processes to implement the business logic, and each process is used to control interaction among the appropriate components comprising a business process.

Componentized architectures enable finer levels of granularity to be achieved. The overall logic flow is externalized from any one component and is instead spread among multiple components. Depending on the specific technologies used to implement the PBNM system, a range of possibilities (from parallelized to spontaneous computing if Jini is used) exists. For example, with Jini, its network services architecture promotes and enables a network environment in which devices, upon connecting to a network, spontaneously become participants in the network as providers and/or consumers of services. This automatic network awareness happens without manual administrative intervention to configure the system to accommodate the new device as it launches. As mentioned in Chapter 4, the Fine Grain NGOSS architecture[11,15] has demonstrated that Jini principles apply equally well to software as well as hardware.[16,17] These two devices mentioned in the Fine Grain projects are "ahead of the pack" in terms of delivering Jini-based services. This should not matter, as the Jini Surrogate Architecture can instead be used. The Jini Surrogate Architecture provides a framework for allowing an external surrogate process that is fully Jini enabled to act as a stand-in on behalf of the device, permitting the non-Jini device to participate within the Jini community and appear to that community as a full-fledged Jini player. The Jini Surrogate Architecture provides a good interim mechanism for supporting hardware elements that are not ready to run Java natively. In fact, it was used successfully in the second phase of the Fine Grain project to connect legacy SONET equipment to a new version of the Fine Grain system.

8.5.5 Contracts

The NGOSS Contract[4] is used to define interoperability in an NGOSS Architecture. A contract is a binding agreement to provide information and/or perform an operation. It is similar in concept to a Java interface but much more powerful. This is because an NGOSS contract specification defines the semantics and behavior of the service being offered. All shared public services are described and defined using contracts.

Contract specifications have a technology-neutral as well as a technology-specific part. The technology-neutral portion ensures that common terminology, structure, and high-level semantics are preserved between different implementations. Similar in concept to how a single technology-independent information model can drive the development of multiple technology-specific data models, the technology-neutral portion of a contract is used to enable different technology-specific implementations to be interoperable.

There are many ways to implement a contract. The TMF is currently leaning toward using an XML document for specifying the contents of the contract in a human-readable form. However, it is likely that a simple XML document by itself will have insufficient structure to include a detailed specification of all parts of the contract. For example, the pre- and post-conditions, exceptions, and some additional metadata do not lend themselves well to being placed in a "flat" description. Thus, it is likely that these and other parts of the contract will instead be encapsulated in an XSD (XML Schema Definition), if XML continues to be used. This enables additional facilities, such as the ability to use XML namespaces, to be incorporated in the XSD.

At the time of this writing, there is a debate as to whether contracts are required for all interactions between all components. Certainly, it is difficult if not impossible to use contracts when the system is in bootstrap mode, as critical framework services such as the registry may not yet be initialized. However, this book recommends that contracts be used wherever possible. This enables a standard set of definitions to be used to invoke different services. Without this standard set of definitions, it is almost impossible to implement a component-based architecture, since each component will have the freedom to implement its own interface. If there is not a standard way to invoke services, each component must have explicit knowledge of every other component with which it wants to invoke or exchange data. This simply cannot scale. Plus, it is a very *brittle* architecture, in that it prohibits new components from being introduced, since they do not have a standard interface.

Furthermore, this approach explicitly defines key elements of the expected behavior of invoking a contract (e.g., its pre- and post-conditions, exceptions, and other features). This makes the important point that it is not enough in a PBNM system to simply specify an interface—the *behavior* of that interface, as well as the components using that interface, must be specified.

8.5.6 Input Tools

In general, a PBNM system can be interacted with using a GUI, an API, or both. This is shown conceptually in Figure 8-14.

The Policy Console and Editing System (PCES) is the component used by the network administrator to input, modify, store, view, search, and delete policies and related information. It provides a logically centralized interface to the

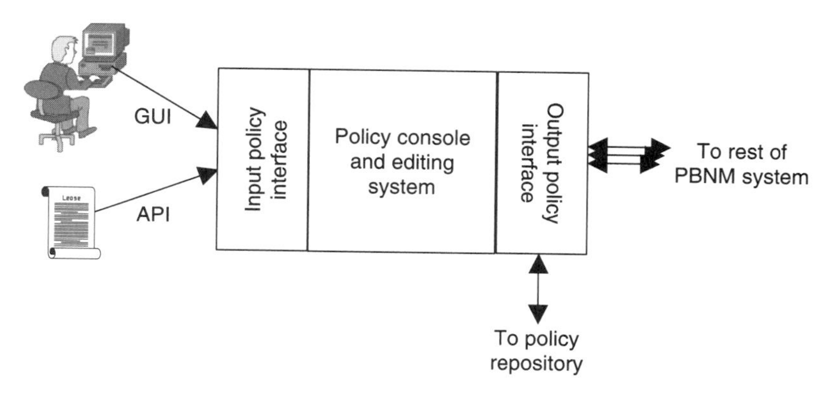

Figure 8-14 The input and output interfaces of a PBNM system.

outside world for storing, managing, and retrieving policies and policy information. Although sometimes clients need to communicate directly with another component in the PBNM system (e.g., a user that needs to interact directly with a PDP), the entity doing so should work directly with the PCES anytime that policy information needs to be stored, edited, retrieved, or administered. The PCES also presents a logically centralized interface for all PBNM components to communicate with when policy information needs to be modified in some way and stored.

The PCES presents a mechanism for entering, viewing, and editing policy rules in the Policy Repository. As a minimum, the PCES consists of an input interface, an output interface, an editor, and a means to communicate with the Policy Repository. There are two separate interfaces shown for the PCES: a dedicated interface to communicate with the Policy Repository, and a set of interfaces to communicate with the rest of the PBNM system components. In this way, external clients of the PBNM system can be directed to go through a single, standardized interface (as opposed to each having to figure out how to interface to a set of repositories, where each repository could use a different protocol). This simplifies the design of the PCES, since supporting one standard interface is much easier than supporting multiple different interfaces.

Similarly, the output of the PCES (not to the Policy Repository, but to the rest of the PBNM system components) can be a single system (e.g., a bus or a Jini interface) or a set of systems. This, being internal to the PBNM system, is a function of the design criteria of the PBNM system itself, though clearly supporting a single or small number of interfaces is greatly preferred.

Along the lines of interfacing to external components, the PCES may be used to facilitate the monitoring of policies, as well as to manage the life cycle of policy rules and policy information (whether they have been deployed or not, or just added to or removed from a system). Both of these activities will use a state machine to ensure that the component either is behaving as expected or is in the correct state. However, in general, the software required to monitor policy rules and policy information is different than the software required to track the life cycle of selected policy rules and policy information. This is because monitoring software is interested in comparing the current values of policy targets to the expected or desired values of those policy targets. (Remember from Chapter 2 that a policy target is a set of entities to which a set of policies will be applied.) This is significantly different than software used to manage the life cycle of policy information. Such software will not be *measuring* parameters of these managed entities, but rather will be tracking the *state* (e.g., whether the component is in a running or failed state) of a component.

The interface to the Policy Repository will be discussed in the next section.

In general, the PCES must be aware of the *type* of component that is to be stored, edited, or retrieved. There is a difference between working on an entire policy rule and working on a component of a policy rule, such as a policy condition or action. In particular, some of the validation and conflict detection logic must be disabled for policy components, because they lack sufficient context to be properly evaluated. For example, the structure of a policy condition can certainly be evaluated independent of any containing policy rule. However, once this policy condition is inserted into a policy rule, the context of the policy condition is defined, and now this same policy condition must be evaluated against the other policy conditions that are active in this policy rule. (Remember that a policy rule is a triplet, consisting of event, condition, and action clauses. Therefore, it is possible that a given condition may not even be called based on the event that is triggering evaluation of the policy rule.)

Once a policy rule has been entered into the PCES, simple validation of that policy rule should usually be performed. This enables malformed policy rules to be found and corrected before they are stored in the Policy Repository (or worse, enter other components of the PBNM system, which could cause needless additional work). This should be able to be overridden, of course, for efficiency and ease of use. If this facility is not bypassed, the PCES should provide feedback to the administrator of the validation results. At the simplest level, this could result in a "Valid/Invalid Policy" message. More useful, however, would be for the Policy Editor to specify what portions of the policy rule were in error. This is likely to require a combination of an information model as well as a language. Although the information model can prescribe what is possible, it will not be able to define specifically how to compare different instances of different policy subjects and targets with each other.

Part of the validation process can also be used to check for simple global conflicts. As described in Chapter 4, this function should be shared between the Policy Broker and the PCES. The PCES can have a set of simple, globally valid checks that can be used to validate policy rules and policy rule components. These will be driven in large part by the language, though the information model will also help at a high level. Examples of these simple checks include checking the syntax of different components of a policy rule, whether the policy rule is correctly formed or not (e.g., does it contain event, condition, and action clauses), and whether two identical conditions result in calling two different actions. For example, suppose that policy rule A, which says "John gets Gold Service," is already registered in the system, and a user tries to add a new policy rule B, which says "John gets Bronze Service." Clearly, this is a conflict and will fail, if the event clause is the same. The objective of these simple checks is to detect obvious policy rules that cannot be evaluated correctly, or policy rules that are not properly structured, in order to save future processing time in the system. As we will see, in order for the system to be properly distributed, a great amount of parallel processing could be performed (e.g., by different PDPs in a Policy Server). It is therefore much more preferable for the system to detect that a particular policy rule cannot be properly executed (or at the minimum, to identify which policy rules it obviously conflicts) in order to cut down additional processing that would not change the result. For example, this saves running this same policy rule (which cannot work) against all of the different policies supported by all of the PEPs in each Policy Server in the system.

The Policy Broker controls how different policy servers interact. As part of this process, the Policy Broker has knowledge of different policies that are active in the system. The PCES cannot have such knowledge, because it does not have a global view of the system or its components. Rather, the PCES is simply focused on providing the ability to store, edit, and retrieve policy rules and policies out of the Policy Repository.

Thus, we end up with the more detailed picture of the PCES shown in Figure 8-15.

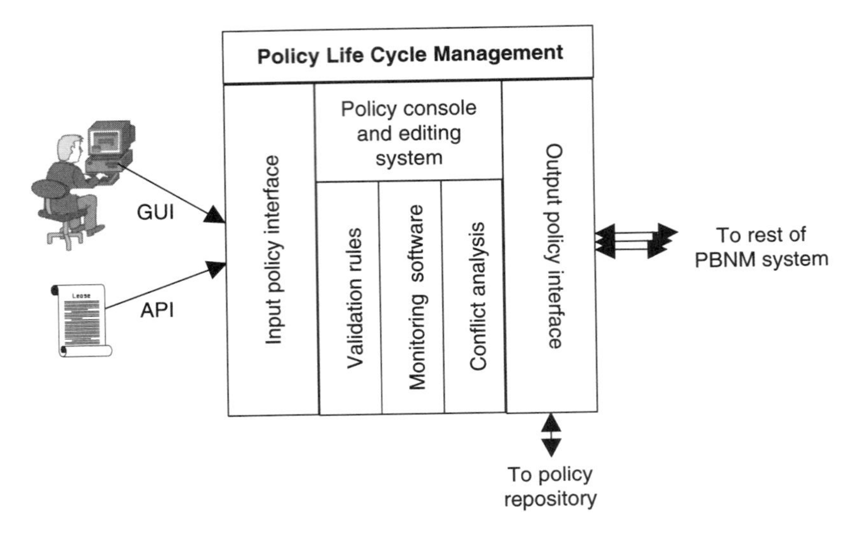

Figure 8-15 A more detailed view of the PCES.

8.5.7 Policy Repository

The Policy Repository is generally assumed to be a *passive* store. In other words, the Policy Repository is used to simply store data—it is not used to process and act on data.

Once a policy rule has been translated and verified, its storage in the Policy Repository is required. This may be done before or after the component(s) interested in using the policy rule have started the processing of the policy rule. (Some early implementations recommended storing the policy first, before any valida-

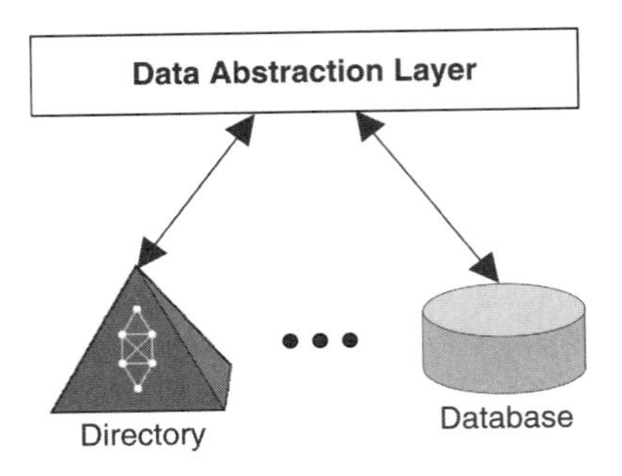

Figure 8-16 The policy repository.

tion or processing of it was done. Although this certainly can be done, it is somewhat inefficient.)

The PBNM system places a variety of demands on the Policy Repository. These come in two forms; however, each leads to the structure shown in Figure 8-16.

Figure 8-16 emphasizes the principle of abstraction. The simple fact is that different applications make use of different management information. Each type of management information has its own storage, access, and other requirements. It is impossible to design a single repository that can simultaneously meet all of these different needs.

This problem is solved by using a combination of a common information model and a set of standard model mappings. The role of the common information model is to define the information in a technology-neutral fashion *once*, so that different model mappings can be used to translate these data according to the specific needs of a particular repository in an optimal fashion. Thus, the common information model and the set of model mappings are used together to implement the conceptual Policy Repository shown in Figure 8-16.

This approach places two demands on the PBNM system. The first demand is the most obvious—the simple fact that the wide variety of management data that are of interest to a PBNM system requires multiple types of data stores. This is because no one data store has all of the necessary features to facilitate the storage and retrieval of these varied types of data.

The second demand is more subtle. If policy rules are used to maintain or change a system/device state, a wide variety of different policy information must be used. We also need access to the underlying state machine(s) that are being used. These different types of information require not only different types of data stores, but also different ways of communicating and interacting with the Policy Repository.

Without a common data abstraction layer, the n different systems that need to access data in the Policy Repository must access that data from potentially m different data stores. This is undesirable because each of the n data processing systems must support up to m different protocols and interfaces. A much better, and more scalable, solution is to provide a common data abstraction layer, whose purpose is to determine where to get the requested data. This data abstraction layer may also need to reformulate the original query posed by the data processing component in order for that query to be executed in a more efficient manner. Thus, we centralize this problem and require just the data abstraction layer of the Policy Repository to shoulder the burden of interfacing with the plethora of different data processing components that comprise the rest of the PBNM system.

8.5.8 Policy Broker

The Policy Broker controls how different policy servers interact with each other. In this regard, it has two different functions, as described in Chapter 4.

The first function is to ensure that conflicts between different policy rules do not exist when different Policy Servers are asked to work together. The second is to coordinate the application of different policies in different Policy Servers.

In a distributed system, one or more components must be aware of what the other components are doing in order to coordinate their work. This is the purpose of the Policy Broker. It enables different Policy Servers to communicate and work together. It is cognizant of all changes in the Policy Servers that it represents. Indeed, it is cognizant of the state of the Policy Servers that it controls. The Policy Broker acts as an overall commander and coordinator of multiple Policy Servers. It is therefore required in order to provide a predictable policy change process across multiple independent components. This is shown in Figure 8-17.

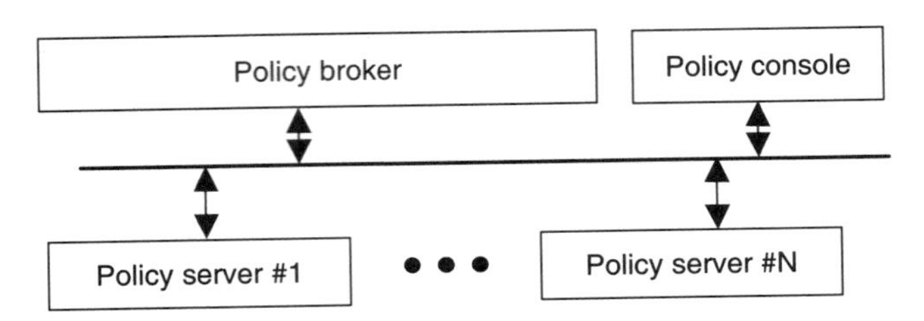

Figure 8-17 The policy broker.

The Policy Broker coordinates the efforts of a set of Policy Servers. Each Policy Server consists of a number of different components, such as PDPs as well as conflict detection and resolution components. If we examine just a single operation—the act of changing an existing policy that has been already deployed by one or more PDPs—we see that a component is needed to coordinate the efforts of the different PDPs. Specifically:

- PDPs must be informed of proposed policy changes that are applicable to them. Another entity must be responsible for this decision process, because otherwise each PDP must know the entire state of the system, which cannot scale.
- Each PDP must validate the proposed policy changes and perform satisfiability and feasibility checking on them. Thus, the PDPs need a single component to which they can report the results of validating proposed policy changes.
- GUI and API clients require a consolidated response to their submitted policy change requests.

The most notable example of a Policy Broker is the so-called bandwidth broker.[18] However, this is beyond the scope of this book.

8.5.9 Policy Server

The Policy Server is a set of components that make decisions and perform actions that affect one or more network elements. The Policy Server consists of two main components: decision logic and knowledge interface. The decision logic is in turn separated into a number of sub-components. These sub-components first translate the received policy into the appropriate level of abstraction (remember, it could have come from any constituent from the policy continuum, and therefore been in any number of forms). Then, the policy must be validated to ensure that it is of the proper form. Next, conflict detection analysis must be performed. This analysis must be performed twice—first, against policies that are already deployed, and second, against policies that could be deployed. The first check ensures that applying this policy will not adversely affect previously deployed policies. The second check forms a decision tree that defines which policies conflict with each other and why. This enables the Policy Server to understand if a future set of policy rules will conflict with each other or not.

The other component is an extensible infrastructure that supports interfacing to different policy-specific components that provide specific rules and knowledge about specific devices and/or technologies. This is necessary to enable the Policy Server to be able to control different types of network devices. For example, the same router that runs a different version of that vendor's operating system may have a different set of capabilities, even though the two routers are physically identical. This new capability translates into one or more new features that in turn require one or more new control mechanisms. Therefore, the Policy Server must include components that provide detailed knowledge about the capabilities of the network devices that are being controlled by the Policy Server, so that higher-level policies can be correctly translated into device-specific policies.

The Policy Server must also include logic to translate environmental constraints imposed on the system to a form that can be used to determine what functionality must be restricted. This restriction will have an associated time-to-live, varying from as small as this one particular application of policy to as long as the system is running. Thus, we see that logic must also be provided to map these environmental restrictions to a policy that controls when this restriction must be applied.

Policies can make fundamental changes to the system. Therefore, it is prudent for the Policy Server to first check to ensure that the policies are safe to deploy. This means:

- They will not adversely affect any deployed policies in the system
- They will not cause irreparable harm to the target object(s) to which they are being applied
- The goal(s) of the policies can be satisfied
- They are feasible to deploy

In general, these all require separate logic.

The first two safety checks can be accommodated by using a finite state machine. The state machine can define which policies are supposed to be active at any given time in the system. As long as the policies have been correctly designed, no harm should come (or else the state machine is wrong).

The concepts of satisfiability and feasibility are essential to policy-based management. A policy can be satisfied if the resources needed by that policy, in isolation from all other local policies, are available in the target objects to which this policy applies. For example, suppose that a policy requires that a certain amount of bandwidth is to be reserved for a given application. If there is not enough bandwidth at the time of application of the policy, the needs of the policy are not satisfied. Thus, the policy itself is not satisfied, implying that this policy cannot be implemented in this system at the present time.

Feasibility checking compares the available services of the network with respect to the full set of policies that want to use those services. Feasibility checking will most likely require post-policy deployment checking that is sensitized to

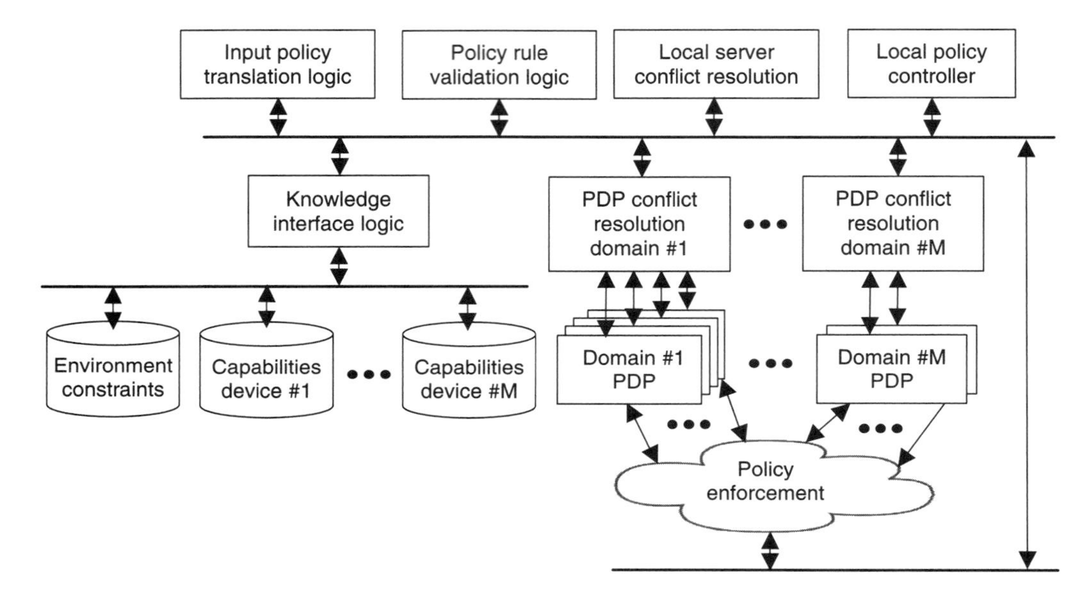

Figure 8-18 The policy server.

the particular network elements involved as well as the nature and effects of the deployed policies. This is beyond the scope of this book.

Therefore, Figure 8-18 shows the basic elements of the Policy Server. The elements of a Policy Server are discussed in the following sections.

Local Policy Controller

The Local Policy Controller (LPC) is responsible for coordinating the actions of this Policy Server. Its purpose is to coordinate the actions of the different components that make up this Policy Server. In this way, the Policy Server can be treated as an atomic building block. This enables multiple Policy Servers to be deployed. In such a distributed deployment, a Policy Broker will control the function of each Policy Server by communicating with the LPC of each Policy Server.

One of the critical functions in the Policy Server is to coordinate the decisions of each of the PDPs. Each PDP decision directly affects one or more network elements, is enforced by one or more PEPs, and is controlled by an individual state machine constructed by the LPC. This is because in general, policies can have different effects on the network. Therefore, by constructing an individual state machine for each decision request, the particular operational semantics of each policy rule can be managed. However, depending on the nature of the policy decision, other network elements can be indirectly affected that are not governed by this Policy Server. In this event, the LPC will communicate with other Policy Servers through the Policy Broker.

The state machine controls the sequencing of steps necessary to translate this policy into a form that can be implemented against the set of devices that is controlled by this Policy Server, validate this policy rule, and perform conflict detection and resolution analysis. (Each of these functional blocks will be explained in more detail in the following sections.) Finally, the LPC coordinates the communication of policy decisions to the set of PEPs that this Policy Controller manages.

Input Policy Translation Logic

Policy decision requests can enter the Policy Server from a variety of sources. Requests from network elements will be at a low, vendor-specific level of abstraction, whereas requests from an operator may be at a high, vendor-independent level of abstraction. In order for the Policy Server to coordinate multiple decision requests from multiple sources, the set of requests that it receives must be converted into a normalized form to enable their comparison. For example, consider the following two decision requests:

- Set "Gold Service" for user John
- Guarantee bandwidth for a flow

These two requests are at different levels of abstraction and therefore cannot be directly compared (though we could easily imagine that "Gold Service" translates to a particular type of conditioning for traffic). The purpose of the Input Policy Translation Logic is to translate each policy decision request to a normalized form so that they can be compared to ensure that a new policy request does not conflict with an existing policy rule that has already been deployed.

Such conflicts are not necessarily bad. For example, a new request for additional bandwidth to be allocated to a flow is going to conflict with the existing state of the network, because this request conflicts with the existing state of the network. Similarly, the upgrading of the service provided to a user will conflict with the service definition that the user previously had. The point is that each new request conforms to a new state in the state machine formed by the LPC.

Once the policy request is translated to a normalized form, it is handed off to the validation logic to ensure that the request is valid.

Policy Rule Validation Logic

The Policy Rule Validation Logic is a set of rules that is used by this particular Policy Server to validate the structure and semantics of all policy requests sent to this Policy Server.

Although the structure of the policy rule is generic and therefore easily validated, the particular contents of the policy rule conditions and actions vary. This is because the structure and semantics of policy conditions and policy actions are directly related to the set of features that is being controlled by the set of PDPs in this Policy Server. For example, the structure and semantics of a QoS policy rule are generally different than the structure and semantics of a security policy rule. This in turn implies that if we were to demand that this module be able to validate each policy rule for all of the different PDPs that make up this Policy Server, the design and implementation of this module would be exceedingly complicated. Therefore, this module only validates the structure of incoming policy requests; validation of the contents and semantics of an individual policy rule are performed with each PDP.

Knowledge Interface Logic

The purpose of the Knowledge Interface Logic (KIL) component is to assist each individual PDP in validating the structure, syntax, and semantics of each policy decision received, as well as to aid in deciding whether this request should be accepted or denied.

The KIL provides a set of detailed information describing the capabilities of each network device that it manages as well as the set of constraints imposed on the network by the operating environment. These concepts are taken directly from the capabilities and constraints model of DEN-ng. This in effect normalizes the different functionalities of each network device that is controlled by this policy server. It also ensures that any constraints imposed by the environment are enforced. Such constraints can be modeled by restricting or prohibiting the use of one or more features of the set of network devices that are controlled by this Policy Server. This can be modeled by *refrain* policies, as was discussed in Chapter 5. The result is the set of allowed features for each network device controlled by the Policy Server.

It is important to design the KIL in such a way that its structure can be extended to accommodate new devices (or new features of existing devices). The same analogous function regarding the addition (or removal) of environmental constraints must also be accommodated by the KIL. The DEN-ng concept of capabilities and constraints, as explained in Chapter 6, enables both device capabilities as well as environmental constraints to be inherently extensible. From an implementation point-of-view, the KIL contains a set of device capability data stores and an environmental constraint data store. By abstracting the different features of each device as a set of common capabilities, the set of constraints can be applied in a normalized fashion against the capabilities of the devices managed by this Policy Server.

Policy Decision Points (PDPs)

PDPs use policies to configure or answer queries from policy-capable network elements or from an operator of the PBNM system. An example of a query is an RSVP admission request—a decision must be made to either accept or reject the request by the receiving network element. Since the network element does not have any idea about the overall state of the network, it makes a request of the PDP, asking the PDP if the RSVP request should be allowed. Thus, the PDP serves as the interface between the network and higher level processes.

The PDP configures policy-capable network elements by using a policy protocol such as COPS.[19] This may include using policies to control specific traffic conditioning mechanisms, such as traffic shaping and queuing. A Policy Proxy (see next section) may also be used to map COPS commands into vendor-specific configuration commands for devices that are not policy capable (e.g., to CLI or SNMP). This enables the PDP to manage those elements that are not policy capable but are otherwise capable of providing services that should be managed by the PDP.

This architecture is explicitly designed to accommodate multiple PDPs, because most vendors are not only building their own PDPs—but they are building different PDPs to govern specific functionality. Thus, in a single Policy Domain, it is possible to have multiple PDPs (whether from different vendors or not) to manage different devices and/or different functions of the same device. In either case, some entity other than the PDPs must coordinate policies delivered by the multiple different PDPs to a single device.

Another example requiring the use of multiple PDPs is due to geographic distribution. Imagine a large enterprise or Service Provider, which contains multiple PDPs. Each PDP has local knowledge to control a set of local network elements. This local knowledge could be due to physical, geographical, or technical constraints. Therefore, since each PDP only controls a portion of the devices in the network, multiple PDPs are required.

Each PDP consists of logic that compares the capabilities of each network device that it manages to the policy request. The knowledge data stores of the KIL are used to represent these capabilities and to enable translation between the low-level device-specific implementation of these features and the higher-level abstraction of these features. This enables the different features of each device to be coordinated. For example, different devices may implement different queuing algorithms, whereas other devices may not have this capability. The PDP can use the high-level abstraction of queuing to separate the devices that have this capability from the devices that do not. Then, the PDP can use lower-level abstractions of queuing to assign the appropriate relative queuing to each device. This ensures that each device does its relative part in queuing traffic. Once these capabilities are defined, the PDP can use the device capability data stores to translate these abstractions into the appropriate set of commands for each specific type of device.

Policy Proxies

A Policy Proxy is a device that acts as a proxy for objects controlled by the policy system. It is needed when the target object (i.e., the network device that is supposed to implement the policy) cannot understand or directly communicate with the PDP. This may be because the network device and the PDP use different protocols, or because the network device does not understand the commands given to it by the PDP. The former is a problem, because the use of different protocols means that the two components cannot communicate with each other and thus require a proxy service. The latter is a problem, because the PDP may be general in nature and may not know the specific CLI spoken by the device.

Even if there was a standard policy language, it is doubtful that all vendors would implement support for it in their devices. First, this means that the vendors must now perform extra work. Second, it means that the vendors are enabling other entities to control their devices. Third, most vendors provide their own management interface, and only that interface, because of otherwise adverse effects management functions can have on a device. For example, consider SNMP. Even though the third version of this protocol has been introduced, many vendors still do not support SNMP functions for monitoring, let alone configuring, their devices. A policy language is at a much higher level of abstraction than SNMP, and therefore will meet with much more resistance by vendors for implementation.

In addition, such implementation requires extra work as well as extra resources that must be resident on the network element itself. For example, many network elements are memory-space constrained, either by cost or device size. Furthermore, even if a policy language was standardized, its implementation will differ based on the functionality of the device that is implementing it.

Clearly, the PDP must be able to communicate with the network element using the language that a particular network element expects. The question, therefore, is whether this communication should be embedded within a PDP or externalized using a Policy Proxy.

There is no one right answer to this question. Often, the design of the PDP will be based on using one or more device-centric languages (e.g., CLI) to enable the PDP to communicate directly with the network elements that it is managing. However, this proposition gets complicated. Many devices support multiple languages (e.g., CLI and SNMP). Unfortunately, most devices do not have a common information model, and so it is difficult (if not impossible) to correlate monitoring commands using one language with configuration commands using another language. Furthermore, this design often precludes the ability for a PDP to talk with older versions of the same network element as well as new devices.

These problems can be solved with the use of the Policy Proxy. Architecturally, the inclusion of the Policy Proxy ensures that a given PDP will be able to communicate with the set of network devices that it is managing. Therefore, it is recommended to be included as a fundamental building block of a Policy Server.

Conflict Detection and Resolution

The subject of conflict detection and resolution is an open area of research. This section will present the basic issues in this area and justify the inclusion of architectural elements to support this feature in the Policy Server.

One way to detect conflicts is to represent them in a language and exhaustively simulate the operation of all policies to catch all instances where there is a conflict. This is computationally very expensive, because it in effect amounts to analyzing all permutations of a set of policy rules to see if they conflict or not. Furthermore, it is unclear whether this can work in the most general sense, because different devices have different capabilities and different functions can affect the same resource. For example, a seemingly innocent policy may direct the administrator to perform SNMP operations concurrent with delivering a guaranteed bandwidth service. The administrator may think that this is safe, because the SNMP operation is monitoring a variable on one interface while the guaranteed bandwidth is being provided on a completely separate interface. However, if the hardware only has a single processor, this could be a problem, since both the SNMP and the guaranteed bandwidth processes require processing resources. The SNMP operation may therefore unknowingly adversely affect the guaranteed bandwidth function because it consumes valuable CPU and memory resources that the guaranteed bandwidth function needs.

The problem in this example is not that the policies directly conflict with each other. In fact, they do not. However, each policy requires the usage of shared resources. When these resources are exhausted, one of the policies in effect will cause irreparable harm for a process that it is not controlling. Hence, what appears to be the safe application of two dissimilar policies is actually a conflict.

This analysis of policies directly affects the architecture of our PBNM system. Specifically:

- Policies must be able to be applied at the right time and place. (*This can be done through the use of AAAA policies that govern the application and usage of other policies.*)
- Policies must be able to be categorized by one or more mechanisms that are independent of the conditions and actions that characterize each policy. (*We will use the notions of role, policy target, and a "processing tuple" that defines what resources and components are affected by the execution of the policy.*)
- Seemingly conflicting policies must be able to be stored together, in the same repository, because the environment can change to resolve the conflict. (*We will use roles and metadata to differentiate the policies.*)
- The satisfiability and feasibility (refer back to Chapter 4 for complete definitions of these properties) of a policy must be computed and stored, so that a policy execution entity knows the effect of applying a policy. Otherwise, different policies can conflict with each other, even though their conditions and actions appear to be independent of each other. (*We will compute and store these values and attach them to the policies; they can be used as metadata to determine which policies can be executed together.*)

Local Server Conflict Detection and Resolution

The Local Server Conflict Detection and Resolution (LSCDR) component of the Policy Server is used when a policy change is submitted to the Policy Server. Here, a change is defined as being a change to an existing policy, the removal of an existing policy, or the addition of a new policy. Part of the general verification process of ensuring that the policy change request is acceptable is determining if the new policy globally conflicts with any other policy in the Policy Domain.

The LSCDR is responsible for detecting and attempting to resolve conflicts that are global to its Policy Server (e.g., inter-PDP conflicts). Note that the LSCDR defines conflicts based on the properties of the policy, not on the specific devices (or their interfaces) to which the policy might apply.

Two policies conflict with each other at the Policy Server level when the following conditions are met:

- They are triggered by the same set of events

- The condition clause of each policy rule is satisfied (or not satisfied), causing the policies to be evaluated at the same time
- The execution time of each policy is either the same or overlaps
- One or more of the actions of one policy conflict with one or more of the actions of another policy.

The first condition is required because events can be used to resolve policy conflicts. The second condition is worded carefully to emphasize that the evaluation of each policy rule of the entire condition clause must be evaluated at the same time in order for a conflict to occur. Remember, the DEN-ng policy model differs from that of the IETF and the DMTF, in that it also defines actions when the condition clause evaluates to false. Thus, one may have a conflict between two policy rules in three different cases: when their condition clauses are both satisfied, when their condition clauses are both not satisfied, and even if one condition clause is satisfied but the condition clause of the other policy rule is not satisfied. This is because the actions of a policy rule for a condition clause that succeeds may in fact conflict with the actions of another policy rule when that (other) policy rule's condition clause fails. Although this may seem overly complex, the DEN-ng specification does not *mandate* the use of actions for both evaluations of the condition clause of a policy rule. Furthermore, for certain conditions, the use of such rules makes the system much more simple and easier to maintain.

The third condition is required because time can often be used to resolve policy conflicts. For example, consider a network that supports a financial trading house that uses the same network to support its internal needs as well as its external clients. Clearly, if the stock market crashes, the clients will demand premium service to place their orders. This will cause an increase in the traffic generated by their applications and may consequently overload the current capacity of the network. Thus, the prudent network administrator may plan for this event by designing a set of configuration changes that adjust the traffic conditioning to enable this new demand to be serviced. This "configuration policy" by definition conflicts with the "configuration policy" used to (re)establish normal network operating conditions. Yet, these two configuration policies in reality do not conflict, because (1) they are triggered by different events, and (2) they can be implemented at different times.

The final condition is the crux of the conflict. The event, condition clause, and time conditions have simply defined the possibility of a conflict occurring—they have not yet defined that a conflict exists. A conflict only exists if the actions of two policies will cause different operations to be applied to the same managed entity and/or the managed entity to be in two different states.

PDP Conflict Detection and Resolution Component

The PDP Conflict Detection and Resolution (PCDR) component is an integral part of the PDP. Whereas the LSCDR checks for policy conflicts that do not apply to any specific network device, the PCDR checks for policy conflicts that apply to all network devices controlled by a particular PDP or Policy Proxy.

The PCDR detects conflicts between different instances of the same PDP. It also checks for satisfiability and feasibility of a policy (new, changed, or deleted) in which this PDP has interest. This means that each PDP must have access to the KIL, which defines the capabilities of the devices that the PDP controls, as well as the environmental constraints that are currently imposed.

This architecture recommends that an explicit separation be made between the capabilities of a device and its configuration. If this is not done, the PDP is forced to try and compare a set of devices that may have different programming models, different features, and other differences that will likely make this impossible. If this recommendation is accommodated, however, it must not be done lightly, since it has two important consequences. First, adopting this recommendation means that a robust and detailed object-oriented information model must be used to "normalize" the different capabilities of each of the devices that it controls. This task means that different versions of the same network operating system must be accommodated. A network operating system can change for many reasons, such as bug fixes, or to introduce new hardware and/or software functionality.

Second, this means that more work needs to be done as part of the conflict detection and resolution process. Specifically:

- The features of each device must first be abstracted into a common set of capabilities
- These capabilities are checked for conflicts (along with checking to ensure that any environmental constraints have not been violated)
- If a conflict is identified, the capabilities that are in conflict must be translated back to vendor-specific features

However, the performance of these tasks ensures that vendor-specific implementation differences will be accommodated. Furthermore, this set of abstractions more accurately ensures that conflicts are indeed detected and resolved.

Managing the Policy Enforcement Point (PEP)

The Policy Enforcement Point (PEP) is the point in the network that enforces the policies through whatever hardware and software means are appropriate.

The user-programmable features of the PEP are represented by capabilities. This abstraction enables a comparison of different types of network elements that use different features to perform the same type of operation. For example, different types of network elements support different types of QoS mechanisms. These mechanisms can be available either in software or in hardware. Therefore, policies must be abstract enough to allow for those differences while remaining specific enough to provide useful direction to the PEP.

The problem to be overcome is that different network devices often use different programming models (even if they are manufactured by the same vendor). Here, "programming model" means the combination of protocol and command to implement a function.

However, controlling a PEP is much harder than it first appears. Many user-programmable features are often directly related to the environment on which they are used. For example, a line card that is able to perform advanced traffic conditioning functions at a low traffic rate may not be able to perform those same functions at a higher traffic rate. This type of conflict cannot be detected using "just" the vendor-specific programming model of the device, since it will not appear as a user-programmable function.

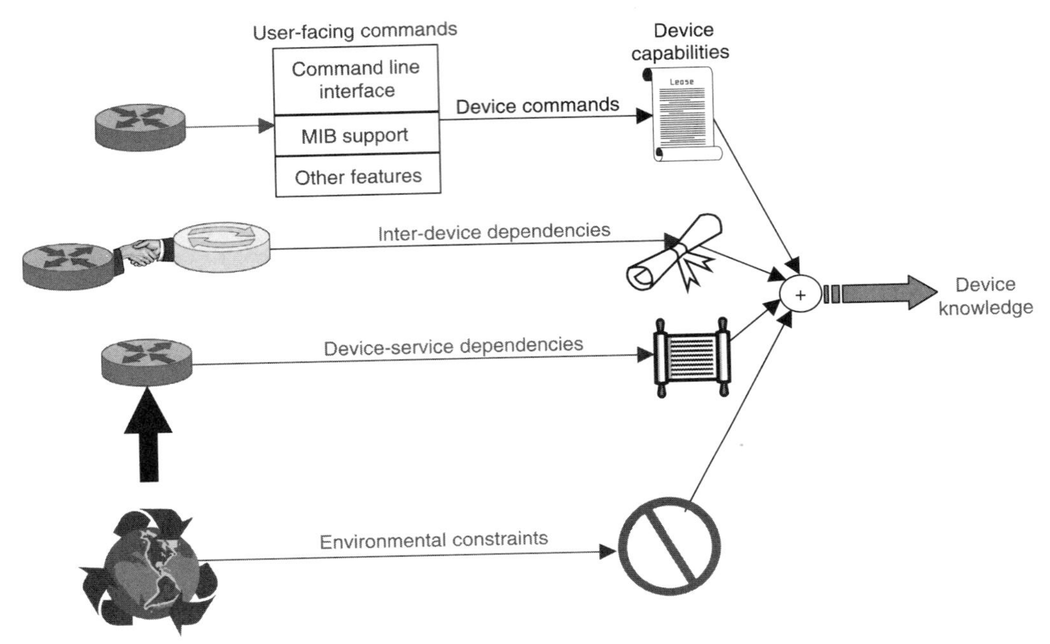

Figure 8-19 Sources for device knowledge in controlling a PEP.

If the above example is not bad enough, consider the case in which two user-visible features require the same resources of the device. Although it may be permissible to use these two features separately, it may not be to use them at the same time. Worse, it may be permissible to use them at the same time only if a set of other conditions are satisfied. Again, this type of conflict cannot be detected using "just" the vendor-specific programming model of the device, since it will not appear as a user-programmable function.

This is conceptually represented in Figure 8-19.

The two examples mentioned previously indicate the need to use an object-oriented information model, such as DEN-ng. The use of such a model enables different commands to be related to each other as well as different combinations of hardware, software, and operating conditions. In effect, the DEN-ng model *guides the use and application of vendor-specific features, functions, and devices.*

8.5.10 Messaging Subsystem

Both the different components of the Policy Server as well as the administrator require feedback to ensure that the policy change was either accepted or not accepted. This communication is provided by the messaging subsystem, which is indicated by the bus and arrows in Figure 8-18. It is used for incremental updates to policies as well as to signal the coordination of the creation, modification, validation, administration, management, and installation of policies. The messaging subsystem is recommended to be a publish-subscribe system, a service-oriented subsystem (e.g., Jini-based), or a combination of both. If it is a service-oriented subsystem, a JavaSpace-based implementation should be considered. Both of these approaches were covered in Chapter 4.

The messaging system (whether Jini or a more traditional publish-subscribe system) is used to provide incremental updates of policy information, as well as the signaling of information to coordinate the creation, modification, deletion, validation, and installation of policies. In a publish-subscribe system, messages are used to perform various operations, including the enabling, disabling, and changing of policy rules and information. In a Jini-based system, services are used to perform the various operations.

8.5.11 Operational Considerations for a PDP

A number of operational considerations must be taken into account when a PDP is asked to provide a decision on a policy request.

The Effect of Implementing and Changing Policy on a Network Device

Modification of policies may end up changing the configuration of some network elements. The component of the Policy Server that is best able to determine which devices are affected by a particular policy is the PDP because the PDP is the entity that best understands what the device is capable of doing and how the device operates. This means that the PDP must first understand what policies are currently installed in the devices that it controls.

The recommended approach is that the PDP create a state machine to define the current and desired state of the devices that it is controlling. This state machine is affected every time a new policy is installed and deployed in the system. The state machine enables the PDP to understand the effect of different policies on the device.

Protecting the Integrity of Policy Decisions

A PBNM system will only work correctly if the entities that it is managing remain under complete control of the Policy Server. However, it is a fact of life that simpler control mechanisms, such as telnet/CLI and SNMP, will continue to be used. The problem is that the Policy Server bases its command decisions on its knowledge of the current configuration of the devices that it controls. If some other entity changes these configurations and does not tell the Policy Server about those changes, the Policy Server may make a change that it thinks is correct, but that could have catastrophic consequences on the network devices that it controls.

Therefore, it is recommended that such mechanisms be disallowed (or at least temporarily blocked) for the duration that the Policy Server is communicating with and managing a network device. Once a policy is terminated (either because the flow is completed or the policy's time to live expires), the Policy Server may relinquish control of the network device and allow other mechanisms to control it. However, if this is done, other mechanisms that change the configuration of the network device must inform the Policy Server of the change. Otherwise, the Policy Server has no choice but to poll each and every network device that it manages, obtain their current configurations, compare them with the configurations that it had, and upodate its information. For a large network, this could be very time-consuming and resource intensive.

8.6 Summary

This chapter started by discussing the requirements that different policies place on a PBNM architecture. The architecture of a PBNM system is determined by several factors, such as the types of policies that are implemented and the granularity of control. The difficulty is in achieving interoperability—not just among the different policies that are used in a system, but also to ensure that data can be shared and reused between the components of a PBNM system as well as to other components of an OSS. This led to the integration of information models and policy languages (each oriented toward one or more levels of the policy continuum) through the use of a data dictionary.

A novel feature of this PBNM architecture is the use of policy-based management and process management. Policy-based management is used to select the appropriate processes to execute, and feedback from those processes is used to adjust which policies are then used, according to a finite state machine.

The reference architecture of a PBNM system was developed. To better understand the complexities of this architecture, four simple use cases were examined. The first use case discussed the procedure for the PBNM system to determine what functions a particular user was allowed to perform on a given device. A novel use of role-based access control was explained, in which the roles of the user and the devices that the user is managing are logically ANDed. These were then restricted by any applicable environmental constraints, with the result being the set of operations that could be performed on a given device. This provides great flexibility, enabling the system to enforce define a unique set of operations that can be performed by the same user on a per-device basis.

The second use case discussed how the system authenticated a user and then authorized privileges for that user. Separate policies to control the authentication and authorization processes were explained.

The third and fourth use cases described a policy from two different, yet related, viewpoints. The third use case concentrated on understanding what operations are performed on a policy from the network manager's point-of-view, and the fourth use case described the operations that the PBNM system must do to support the changing of a policy.

Given this understanding of requirements, the interaction among an information model, a policy language, and a data dictionary was examined. It was shown that this approach builds on that of the original DEN approach and is similar to the new MDA approach from the OMG. In fact, the MDA approach codifies the need for a standard set of tools that can be used to implement these mappings.

The architecture was then discussed in detail. A defining part of the architecture is the use of a finite state machine that coordinates the application of different policies and processes, each of which uses data from the DEN-ng models, to implement behavioral control over the managed entities controlled by the PBNM system. The process starts with defining the linkage between business rules and the authentication and authorization policies that are used to determine the set of operations that a particular user can execute on a given set of devices. Particular emphasis was made on the novel combination of policy-based and process management for controlling the behavior of the system. DEN-ng does this by using a finite state machine, though other mechanisms are possible.

Important elements of the architecture include:

- The Policy Console and Editing System (used as a logically centralized interface between the outside world and the PBNM system)
- The Policy Repository (used to *passively* store and retrieve policies and policy information)
- The Policy Broker (used to coordinate the actions of multiple Policy Servers)
- The Policy Server (our fundamental processing building block, consisting of a set of PDPs and PEPs that make and enforce policy decisions on network elements, along with appropriate processing components to determine which policy should be applied to what device when and where)

Particular attention was paid to the design of the Policy Server and the many different functions that it performs. This is one of the key differentiators (the other being the unique use of models and languages) in the PBNM system defined in this book. The design of the Policy Server is fundamentally based on the assumption that network devices can be viewed as intelligent containers of aggregated functions. Each function or set of functions can be controlled by policy. Thus, PBNM systems are as much about coordinating the application of multiple policies as they are executing a given policy to control a particular function.

Finally, operational considerations for deploying policies were briefly covered. It is important for the administrator to understand the effect of changing a policy not just on the network device to which the policy is being applied, but to the network as a whole. PBNM systems in general facilitate this. Without such systems, the administrator is forced to query individual device interfaces and "piece together" a picture of the system. In addition, it is important to protect the integrity of the PBNM system itself. Any PBNM system is only going to succeed if it remains in control of the system. If another management system alters the configurations of devices that the PBNM system is controlling, the PBNM system will no longer be synchronized with the network devices and will not be able to determine whether applying a policy could harm a device or a service. Workflow systems can help solve this problem by ensuring that the PBNM system always has the correct version of the device configuration that it is managing.

8.7 Recommended Further Reading and References

The following is a set of references for this chapter.

1. Strassner, J., *A New Paradigm for Network Management: Business Driven Device Management*, proceedings of the SSGRR summer conference, July 2002.
2. Strassner, J., *NGOSS Behavior and Control Project Plan*, 1 July 2002.
3. TMF, *NGOSS Architecture Technology Neutral Specification—Behavior and Control Services*, Annex TMF053C, Version 1.0, February 2003.
4. TMF, *NGOSS Architecture Technology Neutral Specification—Contracts*, Annex TMF053B, Version 3.0, April 2003.
5. Friedman, D. P., Wand, M., Haynes, S., *Essentials of Programming Languages*, MIT Press, second edition, ISBN 0-262-06217-8.
6. TMF, *Shared Information and Data model*, GB922, Version 3.0, July 2003. Note that this is a members-only working group of the TMF, with the following members-only site: http://teamlink.tmforum.org/Information%20 and%20Data%20Modeling
7. This is the Model Driven Architecture work of the Object Management Group. Please see: http://www.omg.org/mda
8. Strassner, J., *Directory Enabled Networks*, Macmillan Technical Publishing, September 1999.
9. TMF, *NGOSS Architecture Technology Neutral Specification—Process Management*, Annex TMF053M, Version 1.0, to be published.
10. TMF, *NGOSS Architecture Technology Neutral Specification*, Version 3.0, April 2003.
11. This is a TMF member's-only document that describes the Interface Implementation Specification of the Fine Grain NGOSS Catalyst Project. It is specified in the following document: TMF839v1.5 and in the following (members-only) Web site: www.tmforum.org/sdata/documents/TMFC1379 %20TMFC1000%20TMF839v1[1].5.pdf
12. Please see the following for the JavaSpaces Service Specification: wwws.sun.com/software/jini/specs/jini1.2html/js-spec.html
13. This is a very interesting example of a commercial implementation of a JavaSpace. It features some notable extensions, the most important of which is the concept of *evolution*. Please see: www.intamission.com
14. The following are some of the many distributed computing references:
    http://www.gridcomputing.com/
    http://www.gridcomputing.com/
    http://eu-datagrid.web.cern.ch/eu-datagrid/
    http://www.globus.org/
    http://www.dstc.edu.au/
    http://www-dsg.stanford.edu/
15. A public presentation at the 2002 JavaOne conference on Fine Grain was given in session T3269.
16. This is an example of a next-generation multi-service optical device. It merges the delivery of SONET, Fast Ethernet, and Gigabit Ethernet. More importantly, it uses a Java-based management system, and for the Fine

Grain project, built a Jini interface to their management software. Please see www.turinnetworks.com

17. This is an example of another next-generation device. It is aimed at delivering *last-mile* Ethernet services over existing copper and fiber infrastructure. It used Jini-based management software in Fine Grain. Please see www.hatterasnetworks.com
18. The following are some of the many bandwidth broker references:
    http://www.merit.edu/working.groups/i2-qbone-bb/
    http://www.mnlab.cs.depaul.edu/mmns2001/proceeding/22160160.pdf
    http://www.caip.rutgers.edu/~manishm/bandwidthbroker/CABB.htm
    ftp://ftp.isi.edu/in-notes/rfc2638.txt
19. Please see work in the RAP (Resource Allocation Protocol) working group, at: www.ietf.org/html.charters/rap-charter.html

Examples of Using the Policy Continuum

This chapter will discuss how the business, system, and implementation views can be coordinated using the DEN-ng information model. The conceptual framework for doing this is the policy continuum, which defines the scope of the different models that are used to represent the needs of the business, system, and implementers that are defined in the DEN-ng information model. This will be illustrated using a Multi-Protocol Label Switching Virtual Private Network (MPLS VPN) to illustrate the different types of managed objects that are required of the information model by the business, system, and implementation views.

9.1 Introduction

The purpose of this chapter is to examine a detailed example of how the various levels of the policy continuum work together to implement policies and processes. The DEN-ng approach used in this book centers on the use of a shared information model that has different viewpoints, corresponding to the needs of the different levels of the policy continuum. This enables each constituency of the policy continuum to use entities that they are familiar with and understand. The entities and the relationships between the entities enable one to map between the different levels of the policy continuum.

To show the power of this approach, we will use a single example—a Multi-Protocol Label Switching (MPLS) VPN—and trace its development through various levels of the policy continuum.

9.2 Preliminaries

This section will define the business view of our MPLS VPN. It will use the DEN-ng[1] [DEN-ng] and SID (Shared Information and Data)[2] models as examples of the shared information model that underlies this MPLS VPN. There is a close correlation between these two models. DEN-ng is specifically focused on modeling network elements and services. Unlike the original DEN specification, however, DEN-ng has a very strong business focus. DEN-ng is one of the inputs to the SID.

9.2.1 The Relationship Between DEN-ng and the SID

The SID, like DEN-ng, is a layered object-oriented information model. This means that both models consist of a set of knowledge domains. Each domain is focused on providing shared information and data models for specific management domains. However, the focus is different. DEN-ng views policy as a means to coordinate, manage, and control other managed entities. Hence, DEN-ng is focused on modeling network elements and services according to the policy continuum. Since the policy continuum ranges from business to system to implementation views, it necessarily includes entities from other domains. For this, it relies on the SID.

The SID is meant to be much broader in scope than DEN-ng. It is aimed at addressing the information and communication service industry's need for shared information/data definitions and models. Specifically, the SID is meant to support the NGOSS architecture defined by the TMF.[3] This places certain requirements on the design of the SID, such as defining which domains must be included. TMF Membership determines the order of the work in completing the modeling of a given domain. This is captured by the SID Framework, which is used to organize the SID, and is explained in GB922.[2] An example of the top-level of the Service domain of the SID Framework is shown in Figure 9-1.

DEN-ng will not develop the entities to model these different business uses (e.g., service outage), which are in reality related to the business processes

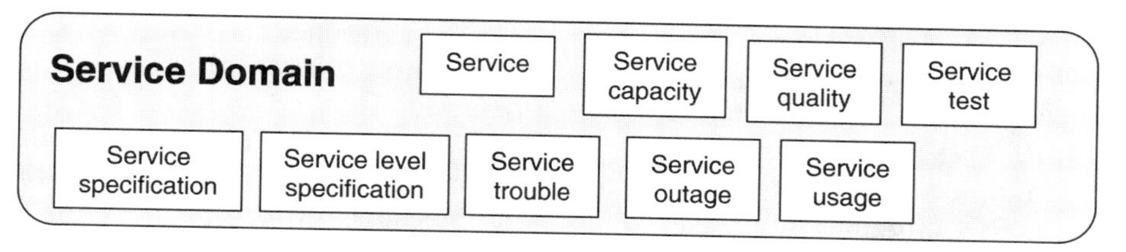

Figure 9-1 The service SID framework domain.

defined in the eTOM.[4] Rather, DEN-ng will concentrate on building a framework, then building detailed examples of different parts of the framework to ensure that the framework does not need to be changed.

DEN-ng is one of the primary inputs to the SID. Others are provided by MetaSolv, British Telecom, and Telstra. This shows that both DEN-ng and the SID are being embraced by companies, who are willing to devote some of their own intellectual property into this growing effort.

Thus, we see that the two modeling efforts overlap. This is a "good thing," because it enables modeling engineers and analysts to contribute to whichever effort in which they are most interested.

9.2.2 The Types of Entities Provided by DEN-ng and the SID

The difference in focus of the DEN-ng and SID models has a direct influence on the types of entities that are built in each model. The SID is currently focused on defining a complete set of business entities, whereas DEN-ng is focused on building a solid framework and then a small set of detailed examples. Since DEN-ng is built in direct support of the policy continuum, it contains a set of business, system, and implementation entities.

What is the difference between these types of entities? A business entity is a thing of interest to the business, and its attributes are facts that further describe that entity in business terms. Together, the definitions provide a business-oriented perspective of the information and data. Usually, business entities do not have methods shown, because methods are not needed to specify business concepts. Methods are more of the domain of system modeling. Business models will contain relationships that link business concepts together. Association classes are usually not used, because this represents a level of detail that is often inappropriate for business users.

System models add to the information represented in the business view. They address design-oriented needs and seek to specify all phases of the system. This often requires the use of other UML tools in addition to class diagrams, such as sequence diagrams and collaboration diagrams.

The focus of the business and system views must be device independent, or else the system will be locked into the use of a particular set of devices. This is not true for the implementation view. Implementation models add device- and instance-specific information, often as subclasses, of the more generic entities and concepts present in the business and system views.

DEN-ng uses this set of entities to link together the business, system, and implementation views of a project. For example, suppose that an organization wants to offer a new service as a Product. The Product may require specialized hardware to run the Service, and will most likely offer the Service under a particular set of SLAs. Just this simple definition makes it obvious that for this Product to be successful, the service rules and contracts defined for its offerings must be related to the implementation of the Service. The policy continuum comes into play immediately, since the business people designing the set of Service offerings are most likely not networking experts and do not know how to change the network to support these new Services.

The advantage of the DEN-ng integrated modeling approach is to *ensure* that this integration takes place through the use of shared entities. This is conceptually depicted in Figure 9-2.

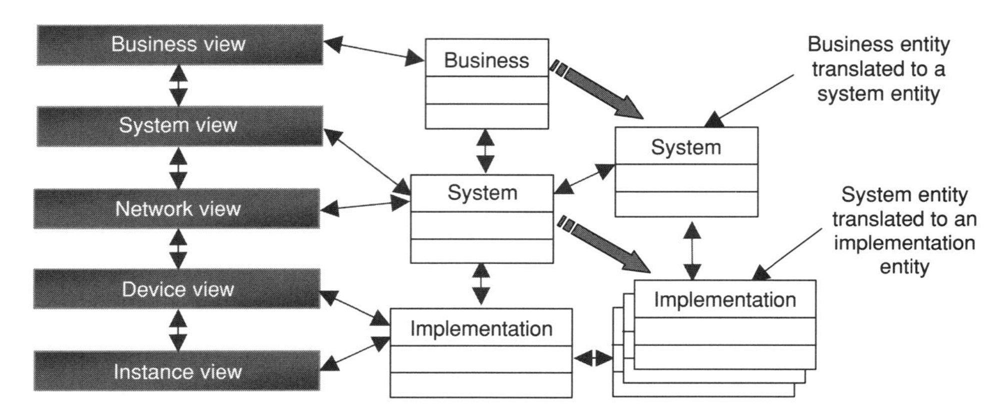

Figure 9-2 Each new view expands upon previous views.

Here, we start with a simple business view, and create an entity to represent it. When we consider how this entity acts in the system, we may need to add attributes, define methods, and even define new entities related to it. This process repeats for each level of the policy continuum (though only three are shown in Figure 9-2).

9.2.3 Policy and Process Management

One of the novel features of the DEN-ng approach is that it considers policy management and process management to be "two ends of a pendulum." That is, policy management provides a declarative approach to choosing which processes need to be executed to accomplish a certain goal, whereas process management is an imperative approach that defines how to best accomplish a particular set of tasks. Each feeds the other—a given set of policies can be used to choose which set of processes to use to accomplish a goal, and feedback from the execution of those policies can be used to change the set of policies that is in force at any particular time.

For example, consider the process of making a change to the configuration of a network device. This seemingly simple process can be a complex web of policies and processes that all work together to accomplish this goal. This is illustrated in Figure 9-3.

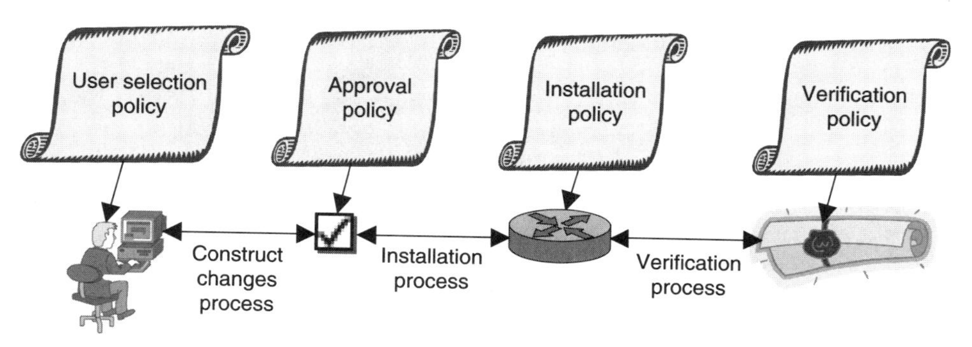

Figure 9-3 Policy and process work together to accomplish tasks.

The configuration change process can be a lot more complicated—the high-level aspects of Intelliden's approach will be described in Chapter 11. However, the above process is sufficient for understanding the problem.

No two configuration changes are alike. In general, the following variables need to be taken into account when making configuration changes:

- The nature of the change, which will determine
  - What commands are affected
  - Which operators have appropriate knowledge of those commands
  - What particular business rules are affected by this change
  - Whether a given change requires other people to approve that change
- The current state of the device
- Whether additional devices are affected, and how
- The current state of the environment

Clearly, policies can be invoked even before anyone lifts a finger to type in the first change (e.g., to choose a set of people to work on the configuration change). These policies can also invoke certain procedures. For example, a particular Service Provider has defined various sets of commands to perform certain tasks in a consistent, predictable way. A policy may be in force to use such changes where they exist. Such policies are very common, since they can be used by the Service Provider as an "audit trail" to show that a particular problem was indeed fixed at a given time. This may trigger additional policies that must be evaluated depending on the nature of the change. For example, it may make no sense to change one device if another device cannot be changed at this particular time. Thus, processes may need to be invoked to ensure that a given set of pre-conditions can be satisfied.

The installation process is a good example of a nested set of processes, often implemented using a workflow engine. One of the important truths of PBNM systems is that the policy server must always be in charge of making changes to a device. If this does not happen, the PBNM system can get "out of sync" and authorize potentially harmful changes. For example, think of what might happen when an engineer fixes a problem without using or informing the policy server. The change could be vitally important, but this could change something that the policy server then changes without knowing that an update has occurred. The policy server could then innocently overwrite this change, and a Service Provider may then be liable for paying SLA violation fees.

A better way to proceed is for the policy server to verify that the current device configuration is unchanged. This can be done by comparing the current device configuration to the device configuration before the change was made. If they are the same, the policy server will know that it is safe to go ahead and install the change. Intelliden uses an elegant variation of this approach, which will be described in Chapter 11.

The verification process is also important. Network devices are very complicated entities, and experience has shown that it is important to be "extra sure" that a change really did what the installer thought it would. Depending on the change, it may be appropriate:

- To check the device while the change is being installed (e.g., using SysLog messages) and/or
- To do another comparison operation (this time, to log into the device and ensure that its running configuration is now identical to what the policy server thought it should be)
- To issue some simple commands that prove that the change is working the way it was intended

Thus, we see that what appeared on the surface to be a simple process change could be a set of complex interactions between different policies and different processes. Chapter 10 will provide some examples of how policy management, process management, and their combination are being used in today's PBNM environments.

9.2.4 Beginnings—What IS a VPN?

There are many different definitions of VPNs.[5,6] For this book, we will use a simple definition of a VPN:

A VPN is a private network constructed within a larger public network.

A VPN is "virtual" in the sense that the private network is a virtual creation–there is no corresponding physical counterpart to it in the network in which it is created. In other words, the network is constructed by using a logical partitioning of some underlying common shared resource, rather than by using a set of discrete and dedicated physical resources. A VPN is a "private" network through its enforcement of the following three essential concepts. First, communications between devices in a VPN is kept confidential, such that devices that are not a part of the VPN are completely unaware of its existence and hence cannot see the content of the communications. Second, privacy means that access is restricted to a defined set of entities. Finally, within a VPN, the addressing and routing used is separate from that of the underlying shared network as well as from that of other VPNs.

There are many types of VPNs in existence today. The major types can be classified as network-layer VPNs (e.g., using different network layer protocols to control how routing information is conveyed from one point in the network to another), tunneling, encryption, link-layer, and hybrid.

Network-layer VPNs can be divided into "peer" and "overlay" VPN models. Peer VPN models forward traffic on a hop-by-hop basis. This class of VPN is represented by traditional routed networks. Overlay VPN models forward traffic by using the intermediate link layer network as a shortcut to another edge node. Examples of overlay VPN models are ATM, Frame Relay, and tunneling implementations.

Tunneling examples include Generic Routing Encapsulation (GRE), router-to-router and host-to-host tunneling protocols (e.g., Layer Two Tunneling Protocol and Point-to-Point Tunneling Protocol, respectively), and others that are more specific in nature (e.g., Distance Vector Multicast Routing Protocol–based tunnels, which are used to connect to the MBone). Tunnels work across a common host network, where each point of attachment between the tunnel and the common network is configured as a physical link that uses addressing and routing from the common host network. Each tunnel endpoint logically links this point of attachment to other remote points from the same VPN. This enables the tunnel egress addresses to be defined within the address space of the common host network, whereas the packets carried within the tunnel use the address space of the VPN. This isolates the routing for the VPN from the routing of the common host network.

Encryption-based VPNs use network-layer encryption technologies such as IPsec, which focus on providing security. This can be implemented either end-to-end, between participating hosts, or only between intermediate devices (routers). Encryption by itself does not provide the entire functionality required by a VPN.

Link-layer VPNs, such as those that use Frame Relay and ATM, attempt to provide the functionality of a VPN by using a common switched public network. This type of network is hoped to reduce cost while providing good scalability. A set of VPNs may share the same infrastructure but maintains their privacy.

The best known hybrid approach is an MPLS VPN. The most popular of these is the so-called BGP-MPLS VPN, defined by reference 7. This is shown in Figure 9-4.

Our simple MPLS VPN has a common core of Provider (P) routers, arranged in a full-mesh. They are connected to Provider Edge (PE) routers, which designate the edge of the Provider network. Two customer VPNs are supported; one is designated by the two red connections from the Customer Edge (CE) routers to their respective PE routers, and the other is designated by the two green connections. Thus, our common Provider backbone supports two VPNs.

This type of MPLS VPN uses MPLS for connectivity; route distribution is based on the Border Gateway Protocol (BGP). A set of administrative policies is used to determine how connectivity is achieved among the sites and what additional functionality (such as QoS and security) is provided. Connectivity is achieved using interior gateway protocols such as OSPF, EIGRP, or IS-IS. Label Switched Paths (LSPs) that represent pre-configured paths between different destination end-points that have appropriate service characteristics, such as QoS and

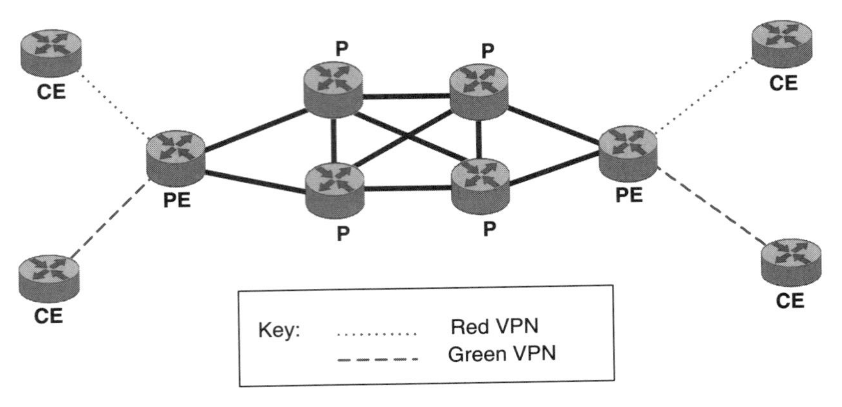

Figure 9-4 A simple MPLS VPN. CE, Customer Edge; PE, Provider Edge; P, Provider.

bandwidth, are created. The addressing used by the routers connected to the customers of the VPN are separate from the addressing used by the routers that provide the infrastructure supporting the VPN. A sample LSP is shown in Figure 9-5.

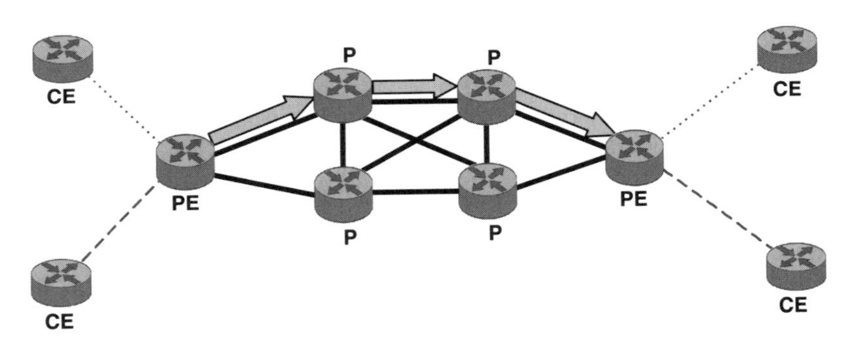

Figure 9-5 An example LSP for the red VPN. See Figure 9-4 for key and abbreviations.

This type of VPN uses routers that play three different *roles*:

- CE routers are associated with customer sites and are usually managed by the customer (though they do not have to be)
- PE routers serve as the entry and exit points to and from the VPN and are managed by the Provider
- P routers are any router that does not interface directly with a CE router and are managed by the Provider

Looking at these three types of roles, we see that each has a particular network function:

- CE routers *advertise* customer (i.e., local) routes to the Provider and *learn* remote VPN routes from their PE router

- PE routers provide most of the networking functionality in this solution. They connect the customer through the Provider core to the desired destination. Since they only need to maintain VPN routes for those VPNs to which it is directly attached, the overall scalability of the solution is enhanced. A PE router functions as the ingress or egress Label Switched Router in any given path. PE routers use an extended form of the Border Gateway Protocol to exchange routing information and to populate each VPN's routing tables. This means that the P routers only have to perform forwarding for the MPLS LSPs that are carrying traffic between the PE routers.
- P routers are primarily concerned with forwarding VPN traffic. Similar to the DiffServ approach, this places the bulk of the decision-making on the PE routers.

Consequently, each type of router is going to route traffic differently, and use a different set of protocols and commands, to accomplish its tasks. This is a "poster-child" application for PBNM for two reasons. First, the functionality for these three types of routers is different from each other, but roughly the same among each type of router. Thus, we can apply policy to ensure that each router playing a particular *role* is configured in a like manner. Second, although the design of this VPN is conceptually simple, there is in reality a relatively large number of configuration steps to be performed to define the specific routing and forwarding components of each VPN, and to relate those with the appropriate MPLS components.

The reader should pay note to the important role that policies play in this type of VPN. This will be a recurring theme used in this example in this chapter.

9.3 Brief Guide to DEN-ng

This section will describe portions of the business view for our fictional MPLS VPN product. It will begin with describing the product and then develop business entities to represent key concepts for our product offering. These classes will form the basis for the system and implementation views that will follow. In addition, the DEN-ng and SID models will be compared to other industry models in this section.

9.3.1 Product and Service

The Product domain is essential to the overall SID and DEN-ng models. Both the SID and the DEN-ng approaches associate the business world with other domains. Users of the model need to be able to express their needs in plain English (as opposed to esoteric networking terminology), yet be assured that these can be translated into the appropriate implementation. Therefore, a Service Provider or even an Enterprise will define one or more product offerings in business terms, which must then be matched to technical specifications and implemented to realize the product offerings. Thus, we can make the following simple definitions:

- A *ProductOffering* represents what is externally presented to the market for the customer's use.
- A *ProductSpecification* is a reusable grouping of invariant characteristics and behavior that typify a *ProductOffering*.
- A *Product* represents a particular instantiation of a given *Product Offering*.

This results in the simple model shown in Figure 9-6.

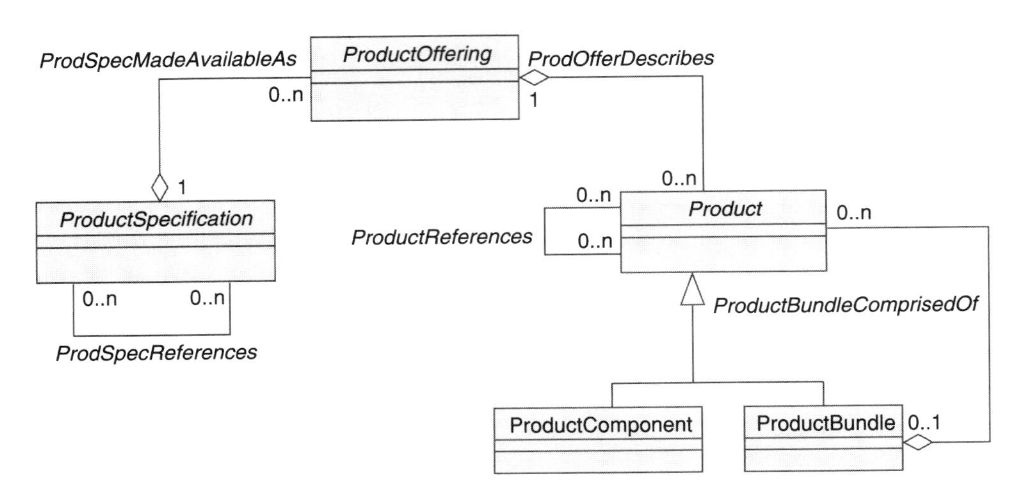

Figure 9-6 A simple model of *Product*.

This model is described more completely in reference 8.

This also leads to an obvious but *very* important observation:

Services are bound to Products.

Think about it for a moment—Services do not "magically" exist in the ether! Even for an Enterprise, services that are offered, such as an intranet, are rooted in the building of a Product. An organization's private intranet is a product of its organization and offers Services to its employees. Therefore, both SID and DEN-ng model Services are bound to, or scoped by, a Product. This is an important differentiation between the DEN-ng and SID models and other industry models, such as the DMTF CIM, which represent a Service as simply another type of logical element.

The reader may wonder about non-commercial Services, or Services provided for internal use within an organization. This is because most people think of a "Product" as something that needs to be *purchased*. The model in Figure 9-6 can still be used, since the relationship defined between Product and Service does not require either to be purchased. For example, corporate e-mail can be provided free of charge to its employees.

Services need physical and logical resources to support them. For example, our MPLS VPN is going to run on a particular set of routers, which are modeled as PhysicalResources. The VPN itself is formed by using different Logical

Resources, such as device interfaces, protocols, and so forth. This can be represented in the following excerpt[11]

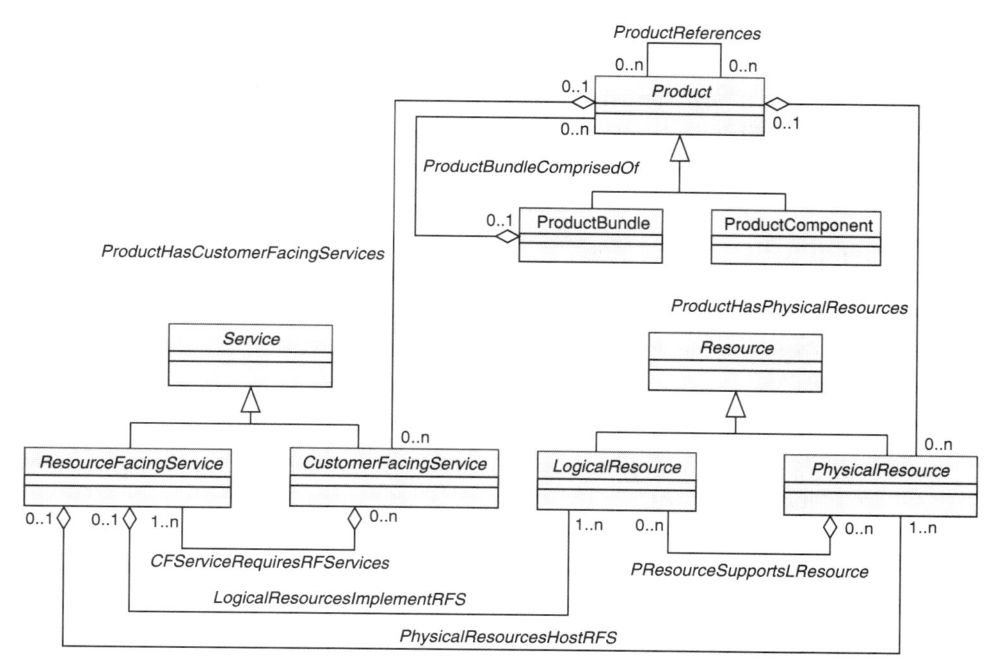

Figure 9-7 Associating *ManagedResources* with *Products* and *Services*.

Figure 9-7 shows the interplay between the physical and logical aspects of a *Resource* (such as a router), a *Service*, and a *Product*. For example, in order to support our VPN, we need a router with certain physical characteristics (e.g., the ability to support the desired physical media, enough memory and processing power to support routing and forwarding [which includes traffic conditioning functions and so forth]. That same router also needs to have a particular set of logical characteristics (e.g., the ability to run a particular protocol, or the ability to recognize special types of addresses like a VPN-IPv4 Address Family[10]). Physical and logical relationships are modeled by the three bottom associations. The *PResourceSupportsLResource* association represents the binding between a *PhysicalResource* and a *LogicalResource* (in the system view, the semantics of this binding are represented by the *LogicalPhysicalResource* association class). In other words, the *PResourceSupportsLResource* association defines the set of *LogicalResources* that a particular *PhysicalResource* supports. For example, a given *DeviceInterface*, which is of course a *logical* entity, is supported by a particular *PhysicalPort*. The two associations, *PhysicalResourceHostsRFS* and *LogicalResourcesImplementRFS*, define the set of *PhysicalResources* and *LogicalResources*, respectively, that a particular *ResourceFacingService* depends on in order for that *ResourceFacingService* to be realized. Again, any specific semantics required by this association are implemented in the *ServicePRDependency* association class (not shown in Figure 9-7 for the sake of simplicity).

Product has two important relationships. The *ProductHasCustomerFacing Services* aggregation is used to define the semantics of defining services within a *Product*. It defines the set of Services that is required to realize a particular *Product*. The *ResourceRealizedAs* association defines the set of *Resources* (again, physical and/or logical) that a particular *Product* uses in order for it to be realized (again, any specific semantics required by this association are implemented in the *ServiceLRDependency* [which is not shown in Figure 9-7 for the sake of simplicity] association class). This reinforces the original semantics of a *Product* being the means to deliver *CustomerFacingServices* as well as other types of *Resources* (e.g., a physical device, connectivity, and so forth).

Note in particular the binding of *Resource* to *ProductItem*. This reinforces the original semantics of a *Product* being the means to deliver *CustomerFacing Services* as well as other types of *Resources* (e.g., a physical device, connectivity, and so forth). Remember that *Resource* is the superclass for both *Physical Resource* as well as *LogicalResource*. This is a very important semantic, it more precisely models how Enterprises and Service Providers build actual *Products* and *Services*.

Finally, the *CFRequiresRFServices* defines the set of ResourceFacingServices that are required for a particular CustomerFacingService to operate correctly. This works in conjunction with the bottom three relationships to ensure that the correct semantics of using Resources (physical as well as the *ResourceFacingServices* that they provide) are attached to the appropriate *Customer-FacingService*.

Note also that the semantics integrating the business and system views are missing from most other models, because most other models are not focused on integrating these viewpoints. Rather, most other models focus specifically on one domain (usually the system view) almost to the exclusion of the other domain. DEN-ng, as well as the SID, are exactly the opposite, in that they both are geared to integrate the business, system, and implementation viewpoints. As mentioned in Chapter 1, this is critical if the functions of the network are to be driven by business needs. Only in this way can the network transition from a cost center to a revenue center.

Thus:

- *PhysicalResources*, such as a *Customer's* router, can be tied to a *Product* through the *ResourceRealizedAs* aggregation
- *LogicalResources*, such as the programming of a particular DeviceInterface, are represented as *ResourceFacingServices*; hence, the PResource Supports *Resource* aggregation relates these *LogicalResources* to a *ResourceFacing Service*
- Abstract items like "*GoldService*" are *CustomerFacingServices* and are represented using the *ServiceRealizedAs* aggregation
- *GoldService* is supported by particular types of *ResourceFacingServices* through the required aggregation (it is an aggregation because it may take several *ResourceFacingServices* to build a single *CustomerFacingService*).

Now, think about how we would sell our VPN Product. Some customers may already have a suitable router, so they would not need to buy one. However, others may need to buy a CE router. The *ResourceRealizedAs* aggregation can be used to define the physical (and logical) Resources that are required in order for this Product to be usable. Note that we are talking about product *instances* right now—we will revisit this concept when we talk about *Product specifications*.

Similarly, a *Product* may or may not offer a set of *Services*. Ours is going to offer a *VPNService* as well as different usage options. These can be represented by the *ServiceRealizedAs* aggregation.

DEN-ng defines a *Service* as an entity that represents logical functionality packaged as part of a *Product*. The purpose of this class is twofold. First, it is used to define attributes, methods, and relationships that are common to all *Services*. Second, it provides a convenient point to define how *Services* interact with other parts of the DEN-ng model.

A *CustomerFacingService* is an abstraction that defines the characteristics and behavior of a particular *Service* as seen by the *Customer*. This means that a *Customer* purchases and/or is directly aware of this type of *Service*; this is in direct contrast to *ResourceFacingServices*, which support *CustomerFacing-Services* but are not seen or purchased directly by the *Customer*. For example, a VPN is an example of a CustomerFacingService, whereas the sub-services that perform different types of routing between network devices making up the VPN are examples of *ResourceFacingServices*.

A *ResourceFacingService* is an abstraction that defines the characteristics and behavior of a particular *Service* that is not directly seen or purchased by the *Customer*. *ResourceFacingServices* are "internal" *Services* that are required to support a *CustomerFacingService*. The *Customer* purchases *Customer-FacingServices* and is not aware of the *ResourceFacingServices* that support the *CustomerFacingService(s)* being purchased directly by the *Customer*. For example, a VPN is an example of a *CustomerFacingService*. This particular type of VPN may require *BGP* to support it. *Customers* do not purchase *BGP*, and hopefully are not even aware that *BGP* is running. Therefore, *BGP* is an example of a *ResourceFacingService*.

In short, DEN-ng and the SID are the only industry models that seek to integrate entities in the business and system views across multiple domains.[11]

9.3.2 Product and Service Specifications

Figure 9-6 introduced the concept of a *ProductSpecification*. Similarly, there are *ServiceSpecifications* and *ResourceSpecifications* defined in DEN-ng and the SID.

The reasoning for this is simple. *Product*, *Service*, and *Resource* are all related, as Figure 9-7 illustrates. VPNs may be offered as a stand-alone *Product* or incorporated into other *ProductOfferings*. Suppose we build a set of slightly different VPNs, all of which use the same IGP for connecting the CE router to the PE router. Now assume that for whatever reason, we change that IGP. Most of the VPN configuration files will stay the same—all we need to do is to change the definition of how a CE talks to a PE, and vice-versa. It would be a shame to try and find all VPN instances and update them individually.

More to the point, no matter what type of VPN it is, the Customer must connect to a Provider network. In addition, certain features, such as ensuring that different Customer's routes are not leaked through the Provider backbone to other

Customers, stay constant. This argues for the concept of a core set of invariant features from which application-specific VPNs can be built. This is implemented in SID and DEN-ng through the concept of a specification, which is shown in Figure 9-8.

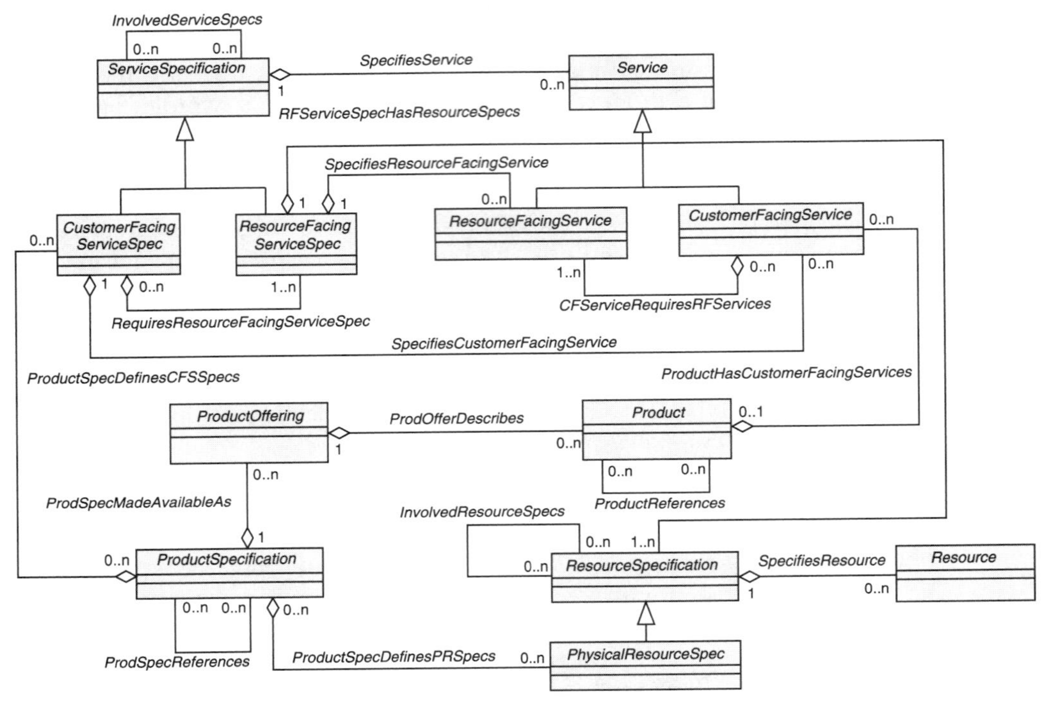

Figure 9-8 Associating *ProductSpecifications* and *Service Specifications*.

In both the SID as well as DEN-ng, a *Product* is the "instance" that a *Customer* purchases. A *Product* is made available as part of a *ProductOffering*. *Products* and *ProductOfferings* have both an invariant as well as changeable parts of their definitions. A *ProductSpecification* characterizes these invariant parts.

A *Specification* (which is not shown in Figure 9-8 for the sake of simplicity) is an abstract base class that, in the DEN-ng model, serves as a common root for all Specification classes for each domain.

A *ProductSpecification* can specify tangible or intangible objects. Tangible objects are entities that you can pick up and hold, such as a CE router. Intangible objects are entities that exist but cannot be picked up and held. For example, a VPN service consists of packets. It exists, because you can communicate using it—however, you cannot pick up an IP packet. A *ProductOffering* exists to collect and present one or more *ProductSpecifications* to the marketplace for sale, rental, or lease. A *ProductOffering* may target one or more *MarketSegments*, may be included in one or more *ProductCatalogs*, may be presented in support of one or more *ProductStrategies*, and may be made available in one or more *Locations*. (Please see reference 8 for more detail about these entities.)

A *ProductOffering* may represent a simple offering of a single or set of *ProductSpecifications*, or could represent a bundling of one or more other *ProductOfferings*.

A *Product* is made available to Customers as part of a *ProductOffering* to *Customers* or other *Parties* playing a *PartyRole*. A *ProductSpecification* may consist of other *ProductSpecifications* supplied together as a collection. *ProductSpecifications* may also exist within groupings, such as *Product Categories*, *ProductLines*, and *ProductTypes*.

A *ServiceSpecification* is an abstract base class for defining the *Service Specification* class hierarchy. All *Services* are characterized as being either directly visible and usable by a *Customer* or not. This gives rise to the two subclasses of *Service*: *CustomerFacingService* and *ResourceFacingService*. However, each instance of a *Service* is made up of changeable as well as invariant attributes, methods, relationships, and constraints. A *ServiceSpecification* defines the invariant characteristics of a *Service*, which can be thought of as a template from which different *Service* instances can be instantiated. Each of these *Service* instances will have the same invariant characteristics. However, the other characteristics of the instantiated *Service* will be specific to each instance.

A *ServiceSpecification* may consist of other *ServiceSpecifications* supplied together as a collection. Members of the collection may be offered individually or collectively. *ServiceSpecifications*, like *Services*, have *CustomerFacingService Specs* and *ResourceFacingServiceSpecs*.

There is no direct relationship between *ServiceSpecification* and *Product-Specification*, just as there is no direct relationship between *Service* and *Product*. Instead, there is the concept of a *Product*, which can specify one or more (since a *CustomerFacingService* cannot exist without at least one *ResourceFacingService* to implement it) *CustomerFacingServices*. A *CustomerFacingService* can in turn require zero or more *ResourceFacingServices*. (The *SpecifiesService* aggregation is defined between *ServiceSpecification* and *Service*, and is specialized by the

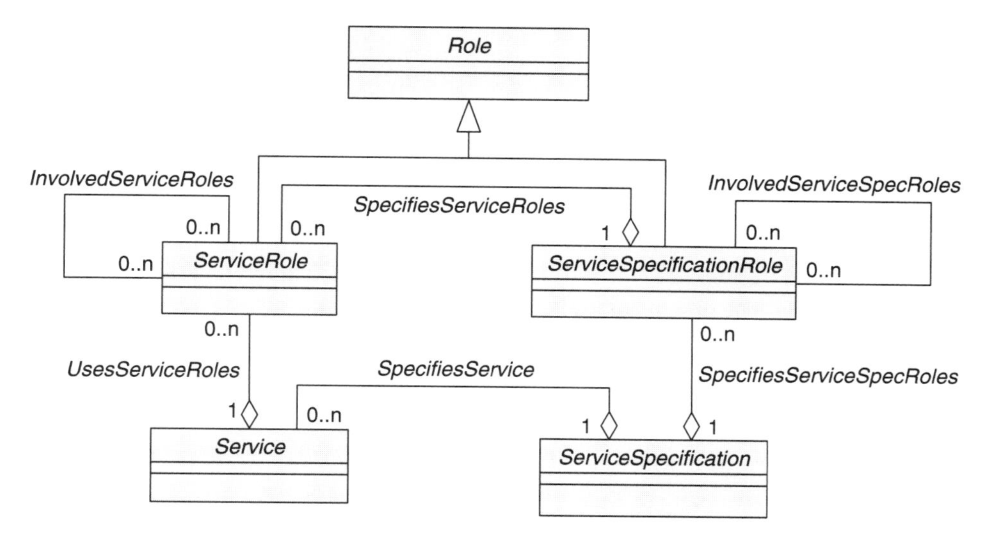

Figure 9-9 Simplified DEN-ng *Service* model.

SpecifiesCustomerFacingService, CustomerServiceSpecs, and ResourceService Specs aggregations.)

Figure 9-9 introduces two additional powerful concepts: *ServiceRoles* and *ServiceSpecificationRoles*.

A *ServiceSpecificationRole* is an abstract base class that defines a *ServiceSpecification* in terms of a set of roles. The roles are then used to characterize the invariant functionality of the *Service*, regardless of whether it is a *ResourceFacingService* or a *CustomerFacingService*. The full DEN-ng model defines subclasses of each of the entities shown in Figure 9-9.

ServiceSpecificationRoles represent the invariant functionality of a *Service*. Representing a *Service* in terms of *ServiceSpecificationRoles* enables the functionality of the *Service* to be defined independently of *BusinessActors*, *PhysicalResources*, *LogicalResources*, or other *Services*.

A *ServiceRole* is an abstract base class that defines a *Service* in terms of a set of roles. The roles are then used to characterize the invariant functionality of the *Service*, regardless of whether it is a *ResourceFacingService* or a *CustomerFacingService*.

ServiceRoles represent the functionality of a *Service*, and as such are a mix of the invariant and changeable characteristics and behavior of a *Service*. Representing a *Service* in terms of *ServiceRoles* enables the functionality of the *Service* to be defined independently of *BusinessActors*, *PhysicalResources*, *LogicalResources*, or other *Services*.

Specifications form a crucial part of the overall model, since they enable the invariant behavior and characteristics of a managed entity to be specified once as well as to enable those specifications to drive the common definition of different entities that each possess one or more new, instance-specific characteristics. This feature is unique to DEN-ng, SID, and OSS/J.

9.3.3 Resources and *ResourceSpecifications*

In DEN-ng and the SID, a Resource is defined as an abstract base class for all entities that are inherently manageable and make up a Product. Again, note the strong correlation between a *Product* and entities that make up a *Product*. Also, not all entities in a managed environment are inherently manageable, such as legacy Hubs that do not support any type of management protocol; rather, they must be manually managed. Entities like *Policy* are of course manageable but do not make up a *Product*, nor are they found packaged inside of a *Product*. Hence, we have the following snippet of the DEN-ng model that defines *Resources* and how they interact with other entities in a managed environment. This enables a complex entity, like a router, to be split into its physical and logical components. This split is important, because it enables the modeling of each to proceed in parallel. Otherwise, the entire model would have to change each time a new feature was added. This could not scale.

Figure 9-10 shows the top-level of the DEN-ng Resource model, emphasizing relationships between *Resources*, *ResourceSpecifications*, and *ResourceRoles*. Note the symmetry in the diagram, as well as the pattern of relationships that are defined between *PhysicalResources* and *LogicalResources*.

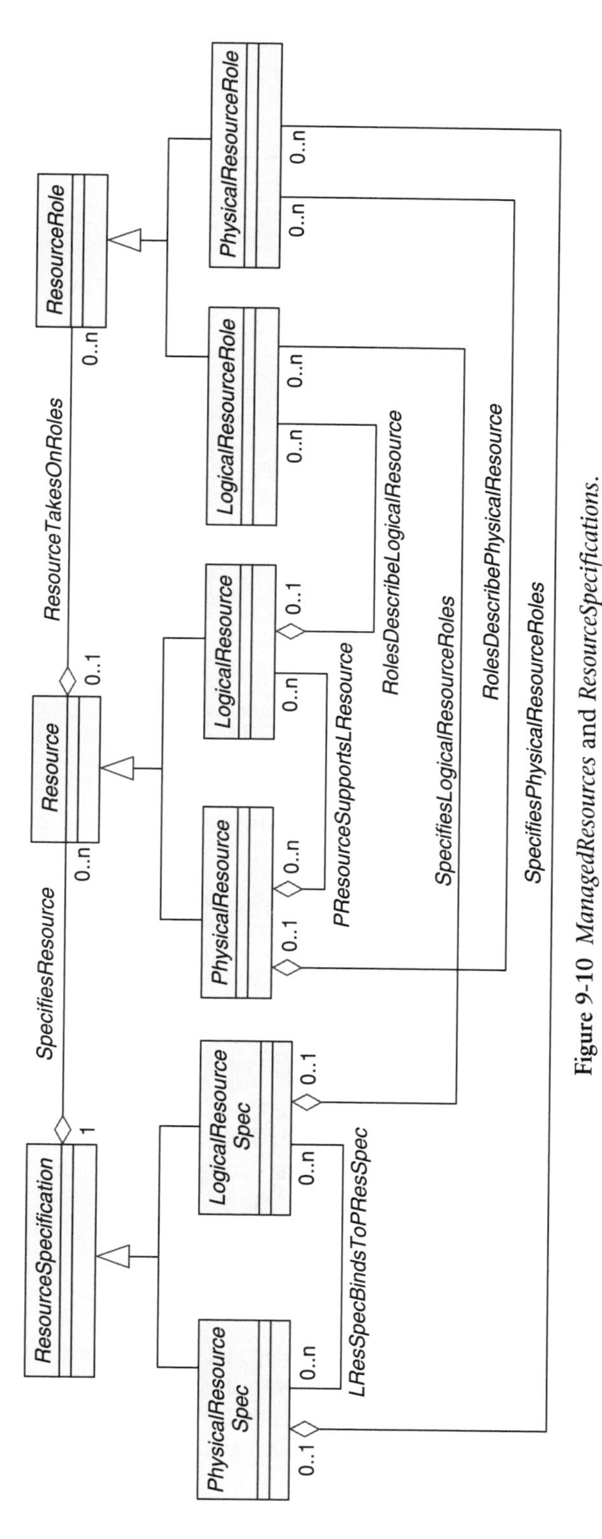

Figure 9-10 *ManagedResources* and *ResourceSpecifications.*

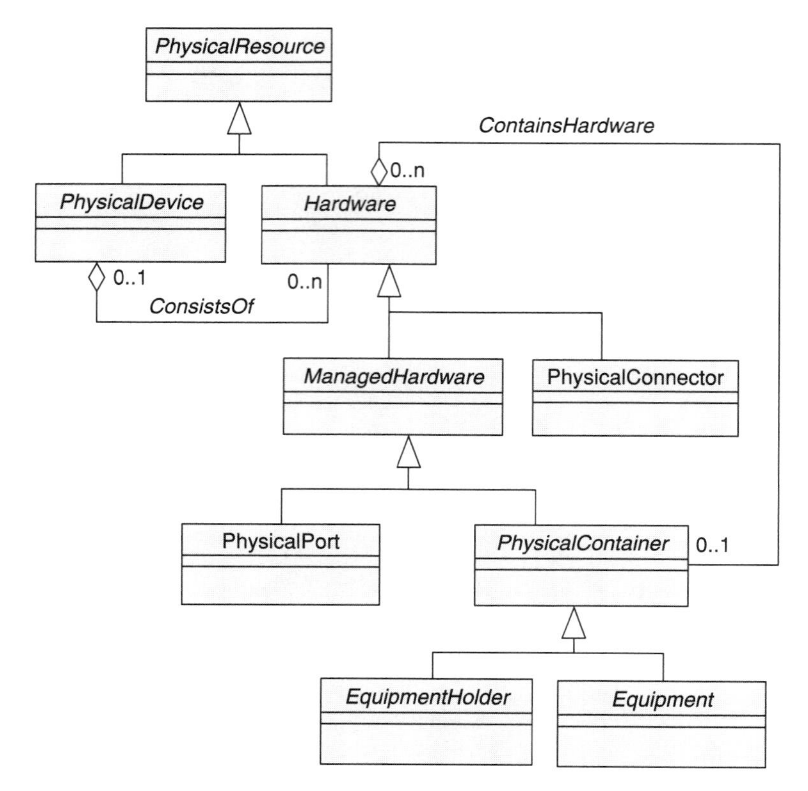

Figure 9-11 Simplified DEN-ng *PhysicalResource* model.

3.3.1 The PhysicalResource Model

The DEN-ng *PhysicalResource* model is identical to that of the SID at the business layer. Figure 9-11 shows a snippet of this model in simplified form. This diagram has been simplified to present the concepts. Reference 13 describes this model.

The two main subclasses of the *PhysicalResource* class are *PhysicalDevice* and *Hardware*. *PhysicalDevice* is an abstract base class used to represent hardware devices that can be managed. This class represents a convenient aggregation point for combining different aspects of a device (e.g., its physical composition as well as the set of services that it offers). It also enables the device itself to have a physical manifestation. Examples of this class include routers and switches, computers, and other end-devices that are managed.

In contrast, *Hardware* is an abstract base class that represents any type of hardware unit that exists as an atomic unit but that is not a *PhysicalLink* or *PhysicalConnector*. *Hardware* is defined as any component that has a distinct physical identity and can be a component of a *PhysicalDevice*. An object has a physical identity if it has a physical manifestation that enables it to be held and have a label attached to it. Thus, software, files, protocols, and policies are not physical objects.

If a router is to be managed in DEN-ng, the router itself will be represented as an instance of the *PhysicalDevice* class. However, if managing different physical components of the router (such as a networking card) is desired, the components that are to be managed will be represented as subclasses of the *Hardware* class. The components of the router are related back to the router itself by instantiating the *ConsistsOf* aggregation.

Since not all *Hardware* is inherently manageable, a subclass called *ManagedHardware* is defined. *ManagedHardware* is an abstract base class that adds the semantics of being manageable. For example, three basic concepts are what the usage and operational states of the *Hardware* are, and whether a fault exists at the physical connectivity layer. As an example of entities that are not directly manageable, *PhysicalConnector* is a peer of *ManagedHardware*, since there is no fault and other "operational status" information available for the *PhysicalConnector*.

ManagedHardware defines two subclasses: *PhysicalPort* and *Physical Container*. *PhysicalPort* represents an actual or potential end point of a topological (physical) link and corresponds directly to a physical port on a topology map. Thus, it is a concrete class (instead of an abstract class) so that it can be instantiated. *PhysicalPorts* are always contained by another physical object—they cannot exist by themselves. The two most common examples are *Physical Ports* on a *Card* and on a *Chassis*. *PhysicalContainer* is an abstract class used to add additional semantics to the *ManagedHardware* class. Its three attributes define whether a *ManagedHardware* object can be removed and/or replaced, and whether this action requires power to be removed or not when the action is performed. This is implemented as a subclass because not all *ManagedHardware* has containment capabilities. For example, these attributes do not apply to a *PhysicalPort*.

Equipment is an abstract base class based on the ITU's M.3100 specification.[14] It represents a class of managed objects that represents physical components of a managed element, including replaceable components. An instance of this object class is present in a single geographic location. An *Equipment* may be nested within another *Equipment*, thereby creating a containment relationship. For example, a *PortAdapterCard* is attached to a *LineCard*. A *Card* is one example of an *Equipment*.

Items such as *Cooling Devices*, *PowerSupplies*, and other items are *not* subclassed from *Equipment*; rather, they are subclassed from a class (not shown in Figure 9-11) called *AuxiliaryComponent*. *AuxiliaryComponents* are managed entities, subclassed from *PhysicalContainer*, such as *PowerSupplies* and *CoolingDevices*. *AuxiliaryComponents* are required for proper operation of the *PhysicalDevice* but have a function that is different than the primary end-user function(s) of the Device. For example, a router will not run well without power, but a router is characterized by its ability to route and forward traffic, not by its power supplies! In fact, to emphasize this point, there is a *ResourceCharacteristic* (and *ResourceCharacteristicSpec*) class that signifies the salient features of an entity, but this is beyond the scope of this discussion. See reference 13 for a description of *Characteristic* classes and how they are used.

EquipmentHolder is an abstract base class based on the ITU's M.3100 specification. It represents physical objects that are both manageable as well as able to host, hold, or contain other physical objects. Examples of physical objects that can be represented by instances of this object class are *Racks*, *Chassis*, *Shelfs*, *Cards*, and *Slots*.

DEN-ng and the SID define different types of *PhysicalResourceRoles*. The difference between an *EquipmentHolder* and a *Hardware* having the role of *Holder* is that an *EquipmentHolder* is *dedicated* to functioning as a holder. In contrast, a managed entity (like a *Card*) sometimes needs to hold other managed entities. The *Holder* class adds this capability as a role that can be played by a managed object. Although not strictly necessary, the *Holder* role greatly simplifies the model (removing the need to instantiate a *lot* more classes and relationships) and provides a better conceptual understanding of the multiple nature of the physical equipment. Roles will be discussed more in the section after next.

The DEN-ng and SID *PhysicalResource* models are identical at the top levels—the DEN-ng model has additional detail that is currently out of scope for the SID effort. The features that separate the DEN-ng and SID models from other models include their use of specifications, roles, and different types of patterns. This enables the DEN-ng and SID models to be inherently more extensible than other industry models while accommodating a larger variety of features and functionality.

There is more to a *PhysicalResource* than what was shown in Figure 9-11. For example, the real *PhysicalResource* model includes relationships to subclasses of *PhysicalResourceSpecification*, *PhysicalResourceRole*, and other important entities. These have been omitted to keep Figure 9-11 simple.

3.3.2 The DEN-ng LogicalResource Model

The DEN-ng *LogicalResource* model is identical to that of the SID at the business layer. Figure 9-12 shows a snippet of this model in simplified form.

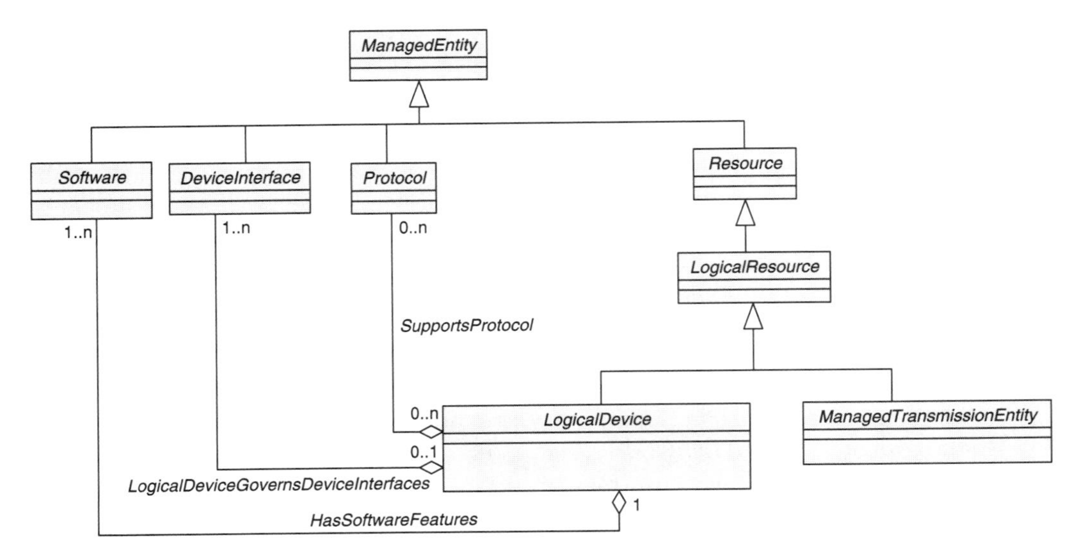

Figure 9-12 Simplified DEN-ng *LogicalResource* model.

This model represents a simplified view of a *LogicalResource*. In particular, the different relationships and subclasses of the aforementioned classes have been removed to keep the diagram simple.

LogicalResource is an abstract base class for all logical resources that are inherently manageable and make up a *Product*. It is used to describe different types of logical features and *Services* that constitute a *Product*. It has two main purposes: (1) to collect common attributes and relationships for all logical resources, and (2) to provide a convenient, single point at which relationships with other managed objects can be defined. Its subclasses are described as follows:

- *Protocol* is an abstract base class for representing protocols that can be managed. This class represents a convenient aggregation point for defining how protocols are managed and used.
- *LogicalDevice* is an abstract base class for representing logical concepts and *Services* that can be managed, but which are associated with the device as a whole. This class represents a convenient aggregation point for combining different aspects of a device (e.g., software contained in the device, protocols that the devices runs, the set of *Services* that it offers, and so forth). It also enables the device itself to have a single logical manifestation.
- *DeviceInterface* is a concrete class that can be used to model a network connection for a device; in DEN-ng, this is used to translate the model into appropriate Command Line Interface (CLI), Simple Network Management Protocol (SNMP), or Transaction Language One (TL1) commands. CLI and SNMP are widely used for network devices such as routers and switches, whereas TL1 is used primarily for optical network devices. The DEN-ng model provides a framework that can represent different types of *DeviceInterfaces*, such as *EthernetInterface* or *Sonet Interface*, using a command class hierarchy. Furthermore, this framework supports the use of CLI, SNMP, TL1, and other (including proprietary) programming models for configuring and/or monitoring such interfaces.
- *ManagedTransmissionEntity* is a superclass used for binding together ITU and IETF concepts of logical entities that form or help to form connections that transmit and/or receive information. This includes the concepts of connections and trails, as well as IP-specific concepts.
- *Software* is an abstract base class for representing software. The *hasSoftware* association is used to define software that is associated in some way with a *LogicalDevice*, such as programs and operating systems. Since this *Software* can be associated with either devices or device components, this association is defined between the roots of the two classes. Software may be nested within other software, thereby creating a containment relationship.

The DEN-ng and SID LogicalResource models are identical at the top levels—the DEN-ng model has additional detail that is currently out of scope for the SID effort. The features that separate the DEN-ng and SID models from other models include their use of specifications, roles, and different types of patterns, as well as the depth of the modeling that is done. The primary use case of the

DEN-ng *LogicalResource* model is to support the modeling of configuration and provisioning. This can be seen from the discussion of the *DeviceInterface* class just mentioned, and its ability to be matched to different management protocols. Other models did not consider this as a use case and consequently do not have this detail. This also sets the stage for building new, unique models that integrate heretofore disparate models. For example, it is common practice for an OSS to integrate different best-of-breed components, such as a configuration system, a fault correlation system, and a billing system. This integration is very difficult, because the lack of shared information makes it difficult to relate a fault to a service to a customer to a bill. A goal of the DEN-ng model is to provide a high-level view of these different activities and provide links between them.

There is more to a *LogicalResource* than that was shown in Figure 9-12. For example, the real *LogicalResource* model includes relationships to subclasses of *LogicalResourceSpecification*, *LogicalResourceRole*, and other important entities. These have been omitted to keep Figure 9-12 simple.

9.3.4 Resource Roles

Abstraction is very important and enables us to not only simplify the resulting model, but ensure that it can accommodate a wide variety of different needs. One of the principal ways in which DEN-ng (and the SID) do this is through the concept of roles.

Roles enable characteristics and behavior to be grouped together, based on a particular context or participation in a certain interaction. These groupings of characteristics and/or behavior will change over time, and this will cause problems if they are modeled using inheritance and specialization. Furthermore, if an entity can play more than one role at a time (e.g., a person can be a *Customer and* a *Vendor and* an *Employee*), the model becomes ponderous and unmanageable without the use of roles.

An innovative use of roles in DEN-ng (and in the SID) is to define a *pattern*, based on the original Role pattern[15], for not just people, but also physical and logical characteristics and behavior. In fact, DEN-ng has extended this to the Policy and Service domains as well.

So far, we have referred to two types of roles: *PhysicalResourceRoles* and *LogicalDeviceRoles*. *PhysicalResourceRoles* represent the different types of roles that various physical resources can have. This enables a single object, such as a *Card*, to have additional functions. For example, a *Card* may also serve as a motherboard or hosting board for another *Card*. In this situation, there is not a separate *EquipmentHolder*—rather, the *Card* acts as a holder in addition to providing its normal functions. *LogicalDeviceRoles* are used to define required logical features to implement the different roles played by different *LogicalDevices* that are used in a *Product* or *Service*.

Assume we want to model a (layer 2) *Switch* (which forwards traffic) and a *Router* (which routes and forwards traffic) as types of *Devices*. This could be simplistically modeled as shown in Figure 9-13.

But what about the so-called "Layer 3" switches, which are *Switches* that have routing capability? One is tempted to invent a new subclass, called *Layer3Switch*,

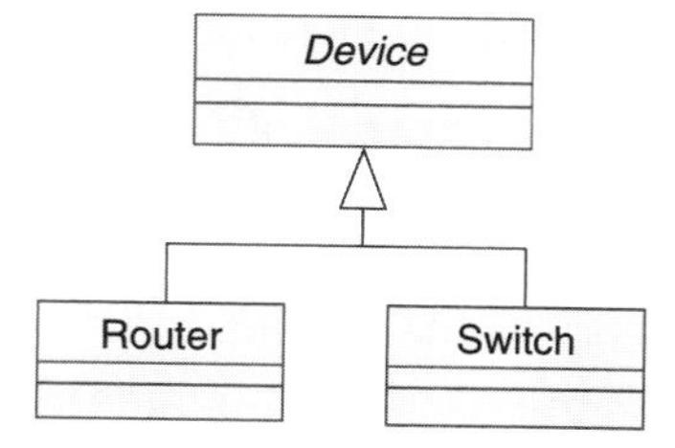

Figure 9-13 Simplified approach to modeling different types of devices.

which subclasses a *Switch* and adds routing capabilities. This is a poor solution, because now every time routing changes, the *Router* and the Layer3Switch class must be updated. Besides, this implies that a *Layer3Switch* can do everything that a full-blown *Router* can do, which is almost never the case. One may also be tempted to use multiple inheritance. If we ignore the fact that multiple inheritance is not always supported when the system is implemented, the problem then becomes one of extensibility. What if there is a "*Layer4Switch*"? (Unfortunately, some vendors do define such an animal!) What if we want to differentiate between the type of routing done in a *Router* versus. the type of routing done in the *Layer3Switch* versus. the type of routing done in the *Layer4Switch*? What if there is a *Router* that has firewalling capabilities? The list is endless.

Instead, a much more elegant and extensible solution is available: we can use the notion of roles. This simplifies our approach tremendously. Instead of trying to either define many subclasses or introduce multiple inheritance, we can instead define a set of roles that the device is meant to play. (This is another reason why the concept of a managed device is a good one—now, we can define a base concept of a managed device and model its functionality by attaching one or more roles to it as appropriate.) This solves the mess of having the same generic function (such as routing) assigned to two different types of devices that implement that same generic function in different ways, producing different subsets of functionality. More importantly, it greatly simplifies the process of generating CLI, SNMP, or TL1 from a model to provision or monitor a device.

Thus, by modeling *DeviceRole* as a separate concept from *Device*, we can represent these complex behaviors in an extensible manner. *LogicalDevice* roles are prominently used in DEN-ng, and a subset of them is shown in Figure 9-14.

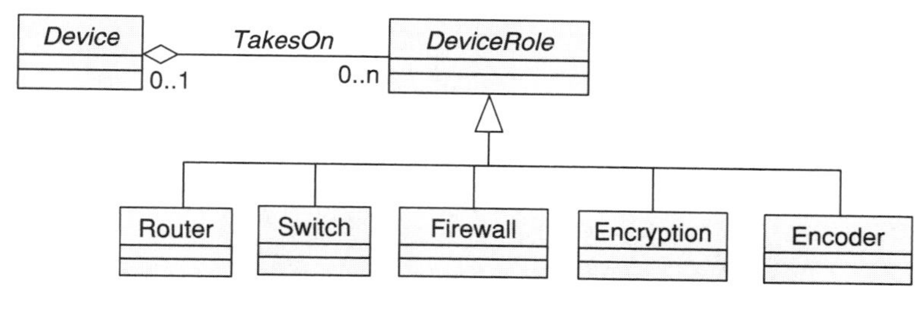

Figure 9-14 *ManagedResources* and *ResourceSpecifications*.

This is very important, because as we will see, building a model of VPNs will depend on and will use *LogicalDeviceRoles* (and other types of roles).

This has been a simple example of how physical as well as logical roles are used in the DEN-ng (and SID) *PhysicalResource* and *LogicalResource* models, respectively. As stated previously, this use of roles is unique to DEN-ng, SID, and, to a lesser extent, OSS/J.

9.3.5 Representing People and Organizations

Both SID and the DEN-ng models define the concept of a *Party*, which represents an individual, organization, or organizational unit. It should be used in places where the business says an organization, organizational unit, or an individual can be used, without having to specify the exact type of entity that is being identified (Figure 9-15).

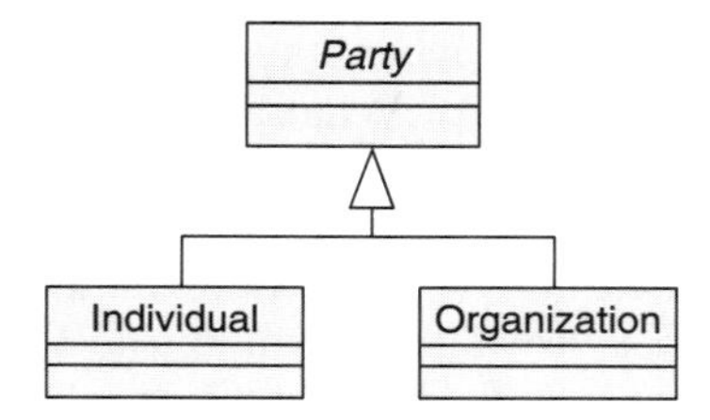

Figure 9-15 The concept of a *Party*.

Usually, people perform multiple functions. Roles can be used to effectively and efficiently capture this. For example, an Individual can be an *Employee*, a member of the *CustomerSupportGroup*, and a *Manager*. The TMF's eTOM does an excellent job of defining different roles. A fragment of the DEN-ng model is shown in Figure 9-16.

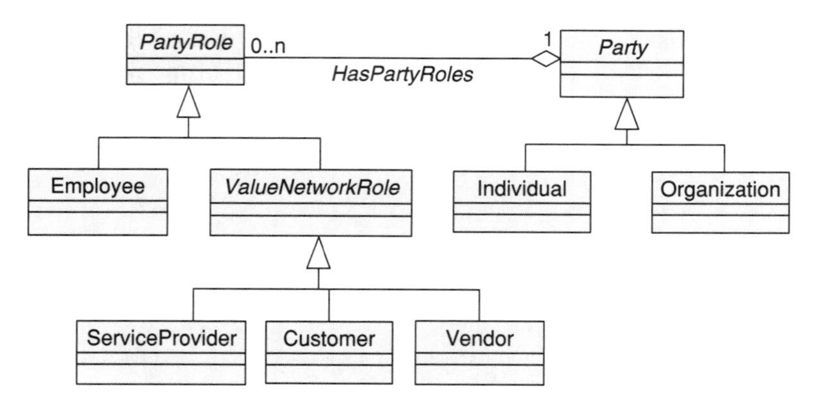

Figure 9-16 The concept of *PartyRoles*.

With these preliminaries, we are now ready to build a simple model of our MPLS VPN.

9.4 The Business View of an MPLS VPN

The Business View primarily addresses the needs of the business process planner and analyst. The business view is primarily concerned with producing a compelling *ProductOffering* that provides generic *Services* that many different *Customers* will use. There are many such VPN offerings (see reference 16 as one example). Thus, one of the common *ProductOfferings* in such *Services* are to offer different Classes of Service (CoS) that can be used to provide different levels of service for different applications. The AT&T VPN Service[16] provides four such classes of services, described as follows:

- CoS1: Designed solely for high-speed voice and video transmissions
- CoS2: Designed to carry important business communications that are data-centric (as opposed to voice-or video-centric); these applications are characterized by requiring strict guarantees for end-to-end delay and packet delivery
- CoS3: Designed to carry standard data-centric business communications; these applications are characterized by requiring strict guarantees for delay and packet delivery (*not* end-to-end)
- CoS4: Designed to carry generic data-centric communications; these applications do not require strict guarantees for delay and packet delivery

There are other basic business requirements:

- Must be able to connect to a wide variety of customer networks
- Must be able to offer additional services to the basic four classes of service without having to redesign the network
- Must be able to use customer CE routers or supply a customer an approved CE router as part of the ProductOffering
- Must guarantee service availability and state in writing when the service will not be available
- Must specify a technical contact person from the Customer's organization with whom to work
- Must use an agreed-upon SLA

9.4.1 Developing MPLS VPN Product Model Extension

From the requirements mentioned previously, we can start to build a business model of our VPN. First, we will start with modeling our VPN Products. This simplified model will be represented using classes that are derived from the DEN-ng model (Figure 9-17).

Figure 9-17 shows four subclasses defined from the standard DEN-ng model, focused on defining the business needs of our VPN. The *MPLSVPN ProductSpecification* is the base class for defining invariant characteristics and behavior of all MPLS VPNs. This is where we define essential characteristics of an MPLS VPN that are independent of any specific *Product* or *ProductOffering*.

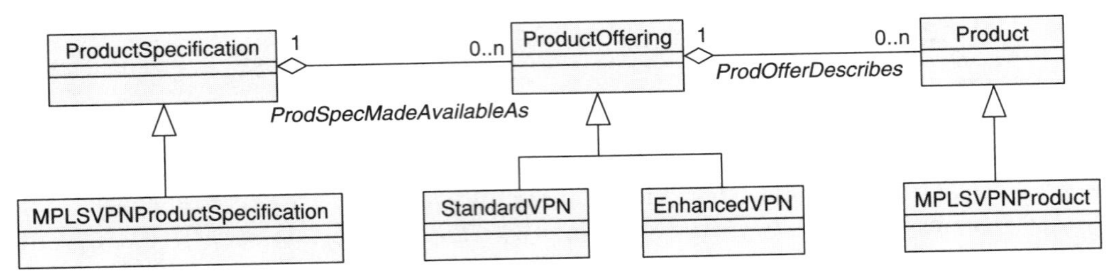

Figure 9-17 Deriving a high-level *VPNProduct* from the DEN-ng *Product* model.

For example, all VPNs will have to define the following:

- Maintenance intervals, in which Customer access to the VPN is denied
- Routing of best-effort application traffic (the number of better-than-best-effort traffic is one of the things that can provide different *Product Offerings*)
- Specification and use of a common set of P routers
- Specification and use of a common set of PE routers

The maintenance interval is going to be implemented as an association to one or more *MaintenancePlans*. These entities describe the specific time periods when *Customer* access to the VPN product is denied, and in turn have associations to *PartyRoles* that implement the maintenance. This is not shown in Figure 9-17 for simplicity.

The routing of best-effort application traffic has a customer portion and an internal portion. Figure 9-8 shows that a *Service* has two subclasses: a *CustomerFacingService* and a *ResourceFacingService*. The former will be used to capture the specification of how customer traffic is handled. The latter will be used to specify how customer services are provisioned internally in the Provider's network.

The last two requirements can be handled through DEN-ng's *PhysicalResoucesRoles* and *LogicalResourcesRoles*. The three different types of routers in our MPLS VPN—P, PE, and CE routers—each have common sets of characteristics that distinguish them from the other two types of routers. These common characteristics can be categorized as physical and logical *DeviceRoles*. *PhysicalDeviceRoles* characterize various physical requirements that the device must meet. For example, a *PhysicalDeviceRole* may specify that a router must have certain *LineCards* that can carry specific types of traffic. *LogicalDeviceRoles* may specify that certain routing protocols are used, or that traffic is forwarded a particular way. Figure 9-18 shows only *some* of the physical roles. More logical roles exist, but their design is beyond the scope of this book.

The three subclasses of *PhysicalDeviceRole* are shown as having dotted lines to denote that several intermediate subclasses have been removed to simplify the diagram. This is also true for *ServicePackage* and *ServiceBundle*, which package sets of *Services* requiring QoS and sets of specifications for providing CoSs for QoS, respectively.

We take these four characteristics and develop two different *Product Offerings*. Our basic VPN *ProductOffering* will come in two flavors. Both will supply the MPLS VPN Service; the difference will be whether the *Customer* also purchases a CE router, and what type of CE router. Our *StandardVPN*

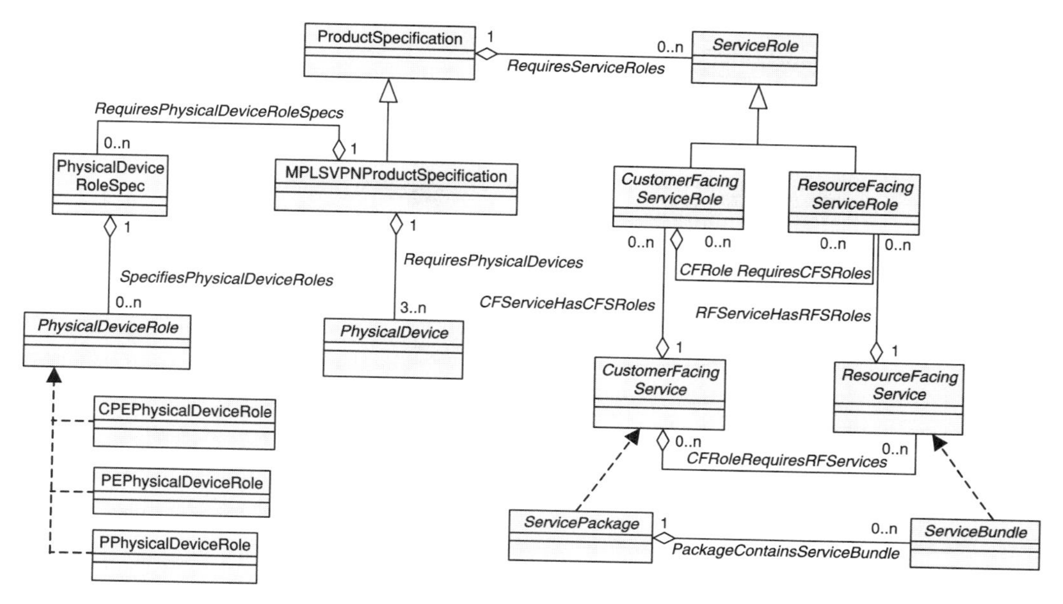

Figure 9-18 Defining more detailed *VPNProduct* subclasses from the DEN-ng *Product* model.

ProductOffering will add two additional classes of service, so we have built up CoS2, CoS3, and CoS4. Our *EnhancedVPNProductOffering* will add CoS1—the ability to handle high-speed voice and video traffic.

Since a particular *Customer* can have different specifications of *Service* (whether it is through a CoS or not), it is convenient to package them together. In DEN-ng, this takes the form of the *ServicePackage* class. The *ServicePackage* class defines the concept of bundling a set of different *CustomerFacingService Specifications* so that a *Customer* can purchase a single *Product*, which uses one or more services. Treating this set of services as a single object is very important for building complex services, such as a VPN. This enables a single *Product* to be offered to the customer, even though in reality the *Product* consists of a set of different *Services* that must work together to provide the functionality that the *Customer* needs. The *ServicePackage* class is shown in Figure 9-19.

The *ServicePackage* class uses the composite pattern for extensibility. The *ServicePackageAtomic* class is used to model different service packages as a set of different instances of individual services. This is fundamentally different than *ServicePackageComposite*, which models one *ServicePackage* as the combination of other existing *ServicePackages* (as well as providing its own extensions).

DEN-ng provides a set of five packages as examples of how to use this feature (Figure 9-19). Each *ServicePackage* is used to identify a set of applications that are to receive the same QoS, access permissions, and/or other services. This chapter will concentrate on the specification of QoS. This is done in Figure 9-19 by linking one or more *ServiceBundles* to the appropriate *ServicePackage* through the *ServicePackageSpecUsesServiceBundleSpecs* aggregation. This enables each "package" to specify its set of CoS specifications.

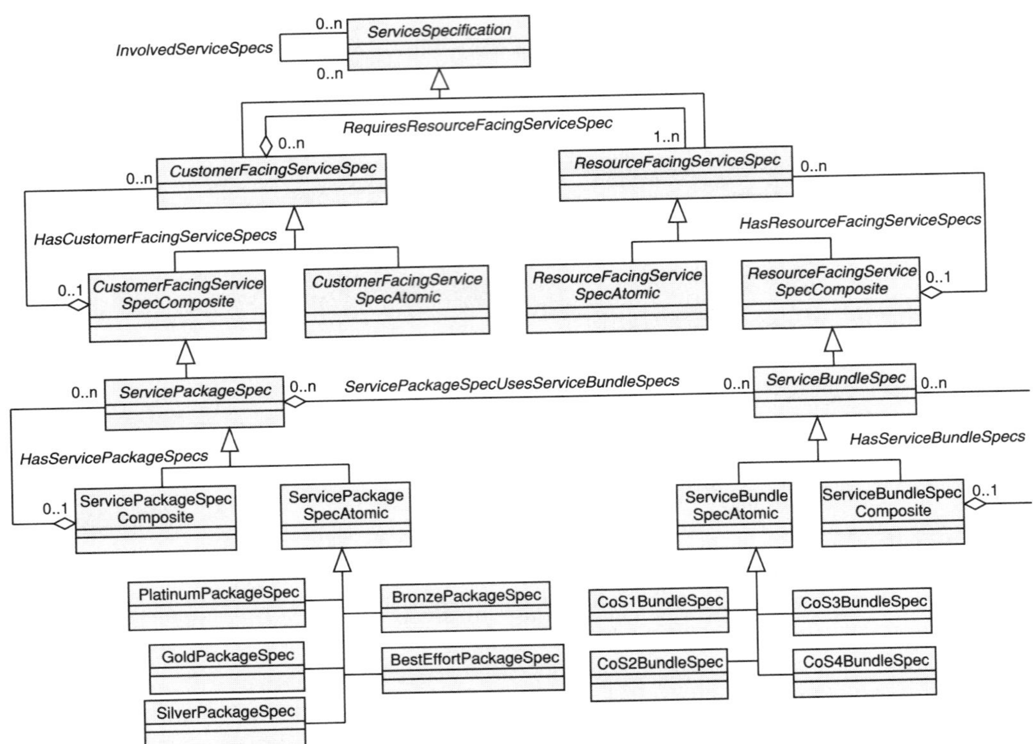

Figure 9-19 Defining basic *VPNProductOfferings* from the DEN-ng *Product* model.

A *ServiceBundle* is a base class from the DEN-ng *Service* model. It is used to define the different Classes of Service to which a *Customer* can subscribe. It is thought of as a "bundle" to enable the needs of different sets of *Services* to be grouped together. The "bundle" conveys the concept of grouped *Services* that are related, enabling a *Customer* to upgrade to a higher-level Service (or downgrade to a lower-level Service) as desired. In our example, the *StandardVPN* uses CoS2, CoS3, and CoS4, whereas the *EnhancedVPN* adds Cos1 to these Service Levels. (Note that these are classes, because each one of these definitions is going to be related to multiple managed objects as our example progresses. This also enables us to use the power of object-orientation and to develop the concept of a Class of Service as an aggregate entity that is related to physical and logical resources.) The *ServiceBundle* also uses the composite pattern; the logic is analogous to the *ServicePackage* classes.

9.4.2 The Concept of a Service Bundle

Figure 9-19 defines our four CoSs as four classes of *ServiceBundle*. Figure 9-20 shows how *ServiceBundles* are related to different levels of QoS. As Figure 9-20 shows, *ServiceBundle* and *QoSService* are both subclasses of *Resource-FacingServiceAtomic*. This means that both can be used to build a *ResourceFacingService*.

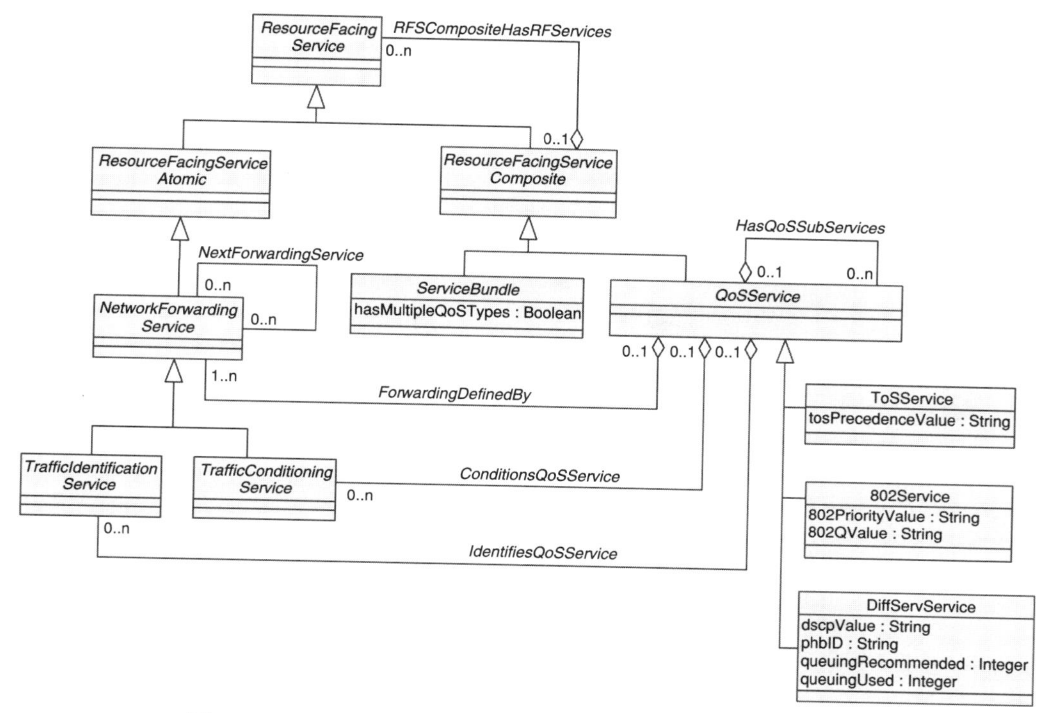

Figure 9-20 Defining different QoS levels per *ServicePackage*.

QoSService is a base class that is used to relate different QoS technologies, such as Differentiated Services, to a particular CoS. The implementation of QoSService is a system issue and will be covered later in this chapter.

The subclasses of *QoSService* can specify their own *NetworkForwarding Services*. This means that each type of technology (e.g., switches using 802.1Q and routers using Differentiated Services) can define how traffic is conditioned using its specific technology. Since there is a single common definition of QoS, these different technology-specific definitions can be linked together to provide an end-to-end definition of QoS.

BGP is not shown in Figure 9-20 for the sake of simplicity. BGP is a *Protocol*, and the *BGPService* uses that *Protocol* to provide different functionality, such as route advertisement. Thus, *BGPService* would be a subclass of *ResourceFacing Service* that has a "uses" association with the *BGPProtocol* class. However, the relationships are more complicated than this and consequently are not shown in Figure 9-20.

9.4.3 Device Extensions for VPNs

In Figure 9-4, we saw that there are three different types of devices used in this particular type of MPLS VPN: a CE router, a PE router, and a P router. This can be efficiently modeled through extensions to DEN-ng's PhysicalResource and LogicalResource models.

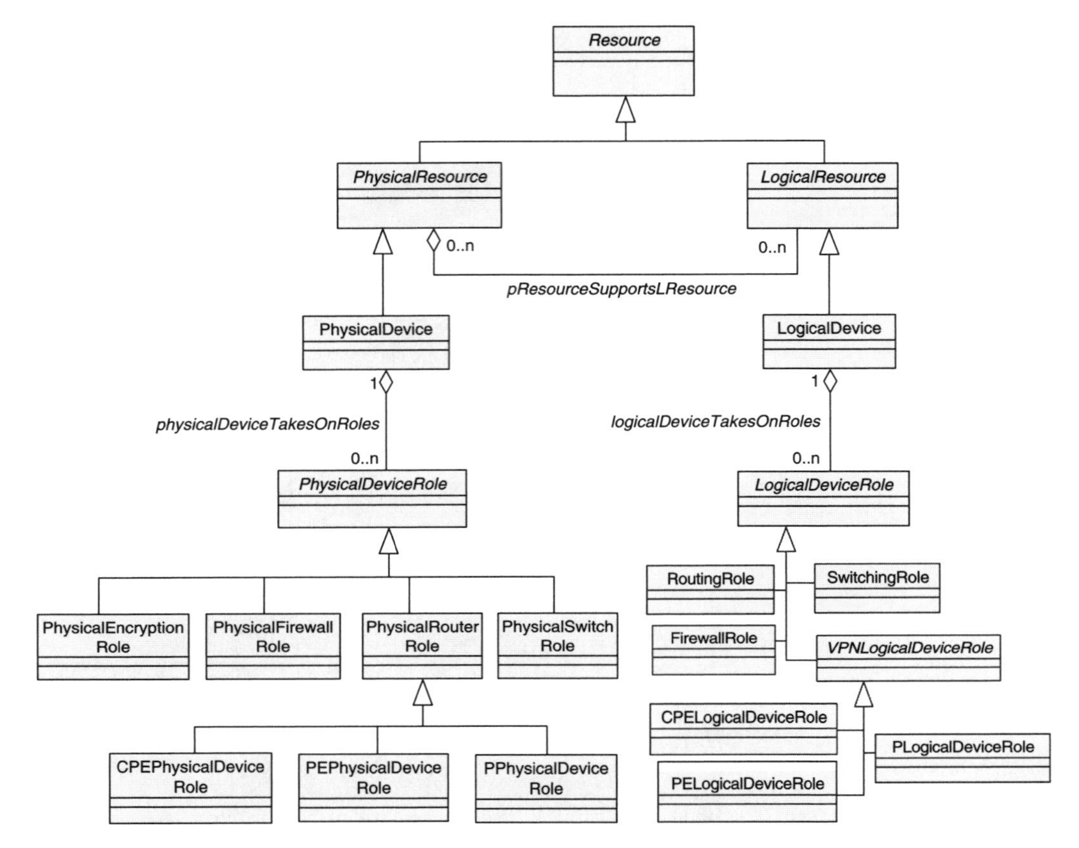

Figure 9-21 Device extensions for VPNs.

We have already seen that a device, such as a router, is represented as having both physical and logical characteristics. The most basic extension necessary to model VPNs is to define a set of physical as well as logical roles that capture the basic functions that our CE, PE, and P routers have in the VPN (Figure 9-21). The various Specification classes (such as *ServiceSpecification*, *ManagedResource Specification*, and role specification classes) are not shown in Figure 9-21 for the sake of simplicity.

This background provides a solid foundation on which to build our MPLS VPN model. First, it is important to recognize that the same router can be used for multiple *roles* in an MPLS VPN. The only thing that makes a particular router a PE router as opposed to a P router is the function that it plays in making up the MPLS VPN. Thus, the correct way to model the different routers that make up an MPLS VPN is by using *roles*.

The simplest way to model this in an extensible fashion is in two steps. First, we will think of the device as having a physical and a logical part. This is shown in simplified form in Figure 9-22.

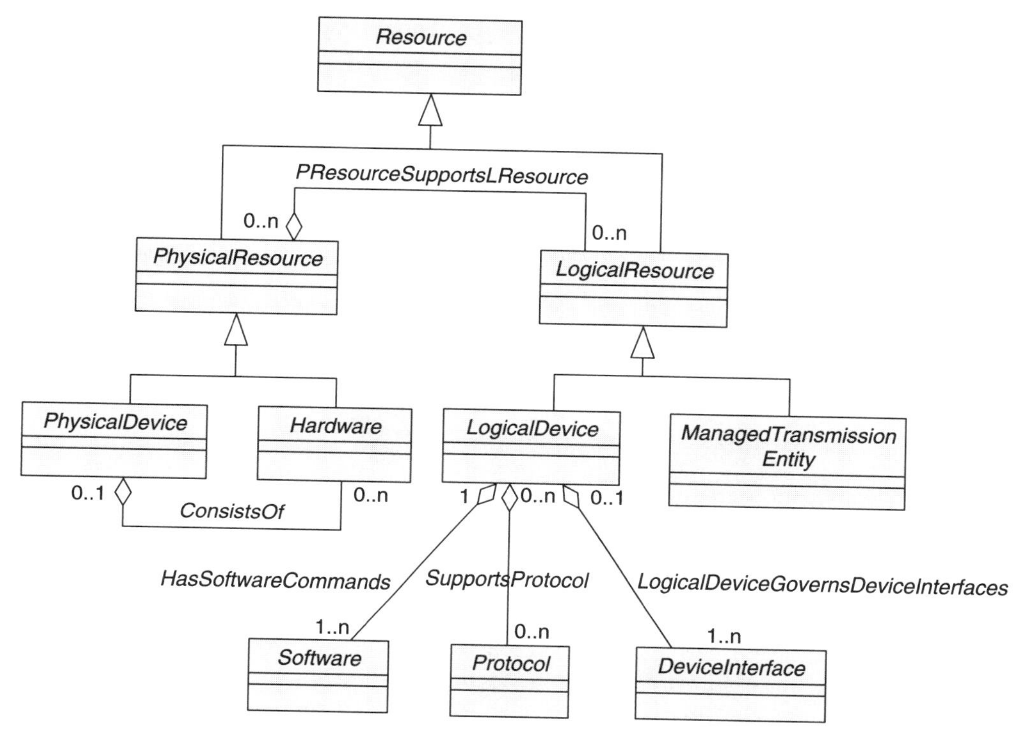

Figure 9-22 Physical and logical resources and their specifications.

This can be represented by the *PhysicalDevice* and *LogicalDevice* classes of the DEN-ng model. The *PhysicalDevice* class is an abstract base class for representing hardware devices that can be managed (such as routers). This class represents a convenient aggregation point for combining different aspects of a device (e.g., its physical composition as well as the physical resources that it provides, such as a *PhysicalPort*, to support logical functions). It also enables the device itself to have a single physical manifestation. Examples of this class include routers and switches, computers, and other end-devices that are managed. The *Logical-Device* class is an abstract base class that defines required logical features to implement the different roles played by different *LogicalDevices* used in a *Product* or *Service*. Some of the more important entities include the *Software* that the *LogicalDevice* runs; the *DeviceInterfaces* that are part of the *Device*, and the set of *Protocols* that are supported by this *Device*.

The second step is to associate different roles with the *PhysicalDevice* and *LogicalDevice* classes. This is done by defining two aggregations, *Specifies-PhysicalDeviceRoles* and *SpecifiesLogicalDeviceRoles*, which enable a *Physical-Device* and a *LogicalDevice* to aggregate zero or more *PhysicalDeviceRoles* and *LogicalDeviceRoles*, respectively.

The relationship between *PhysicalResourceRole* and *PhysicalResourceRole Spec* is shown in Figure 9-23.

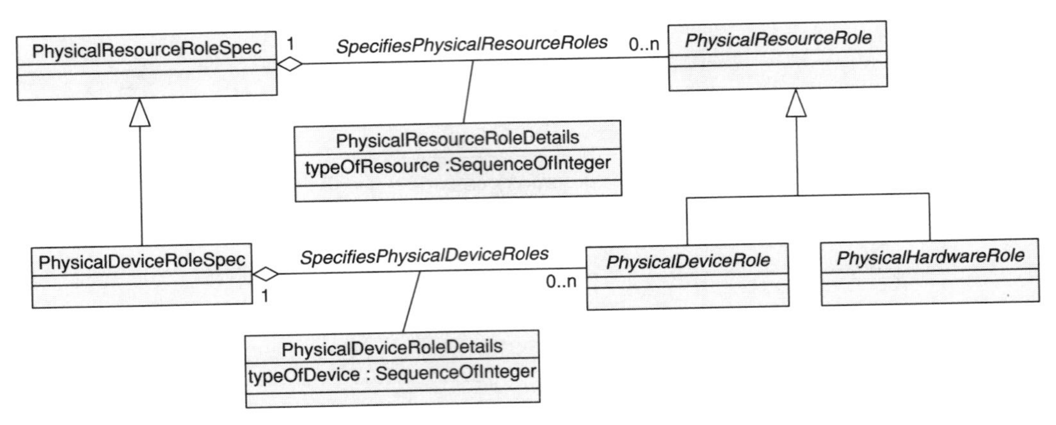

Figure 9-23 *PhysicalResourceRoles* and their specifications.

Figure 9-23 shows that specifications can be built at several different levels of abstraction. The *PhysicalResourceRoleSpec* class is a concrete base class for all *PhysicalResourceRoleSpecification* subclasses and is used to define relationships between it and other classes in the DEN-ng core model. This helps prevent relationship explosion. The *PhysicalDeviceRoleSpec* class is a subclass of the *PhysicalResourceRoleSpec* class. The difference in these specifications is that the latter is more specific than the former. This causes the *PhysicalDeviceRole Details* association class to have more granular semantics than its analogous *PhysicalResourceRoleDetails* association class.

PhysicalResourceRoles are aggregated by *PhysicalResources*, as shown in Figure 9-24.

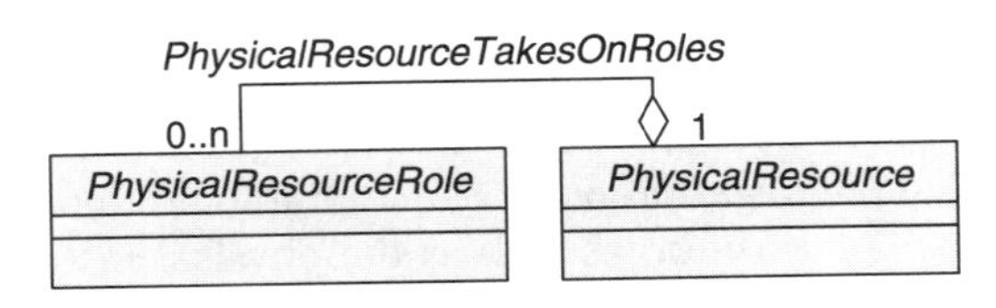

Figure 9-24 *PhysicalResources* aggregate *PhysicalResourceRoles*.

The power of the *PhysicalResourceTakesOnRoles* aggregation is its ability to define the set of physical roles that a particular *PhysicalDevice* can play. This enables functionality to be related to specific physical components (e.g., if a particular type of *LineCard* is installed in a given *PhysicalResource*, that *LineCard* could support a particular *Protocol* that enables it to play a specific role, such as a *CERouter*.

Given this infrastructure, a more complete set of *PhysicalResourceRoles* can be defined (Figure 9-25).

Each of these aggregations is focused on representing the characteristics and behavior of CPE, PE, and P routers from a physical as well as a logical point-of-view. There are separate roles for defining the physical and logical characteristics of CPE, PE, and P routers. The three *PhysicalDeviceRoles* focus on defining

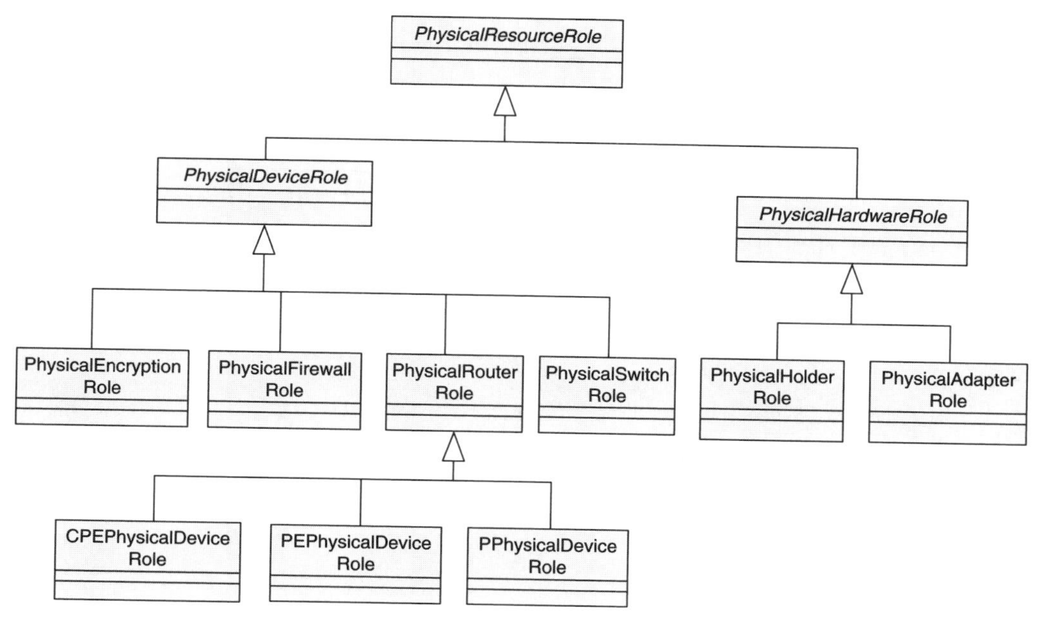

Figure 9-25 More of the DEN-ng *PhysicalRole* hierarchy.

the required physical characteristics that a particular type of router must have. For example, a CPE router may be connecting to a PE router using a particular type of media.

The *LogicalResourceRoles* follow a similar design as the *PhysicalResource-Roles* and *PhysicalDeviceRoles*. The VPN design will use two main subclasses of *LogicalResourceRoles*: *LogicalDeviceRole* and *DeviceInterfaceRole*. A *Logical-DeviceRole* defines required logical features to implement the different roles played by different *LogicalDevices* that are used in a *Product* or *Service*. This type of role affects the overall logical working of the device as a whole. As such, it is differentiated from *DeviceInterfaceRoles*, which represent different types of roles that can be associated with a particular *DeviceInterface*. For example, one way to coordinate the actions of multiple *DeviceInterfaces* is to identify them by the role they all play. Thus, all *DeviceInterfaces*, which are "edge" roles and which peer with Company X, are different than either *DeviceInterfaces* that are not edge roles or do not peer with Company X. Put another way, a *Logical-DeviceRole* affects the entire device, whereas a *DeviceInterfaceRole* affects traffic on a particular interface of the device.

This relationship is shown in Figure 9-26. Note that there are again the same three roles (CE, PE, and P) as there were before. However, these three roles are *LogicalDeviceRoles* and focus on defining the required logical characteristics of a particular type of router. For example, the Provider may use a different protocol for connecting the CPE router to the PE router versus connecting other types of routers playing different roles.

The fact that a router can have a *CPEPhysicalDeviceRole* and a *CPELogical DeviceRole* simply means that roles are being used to abstract the physical and

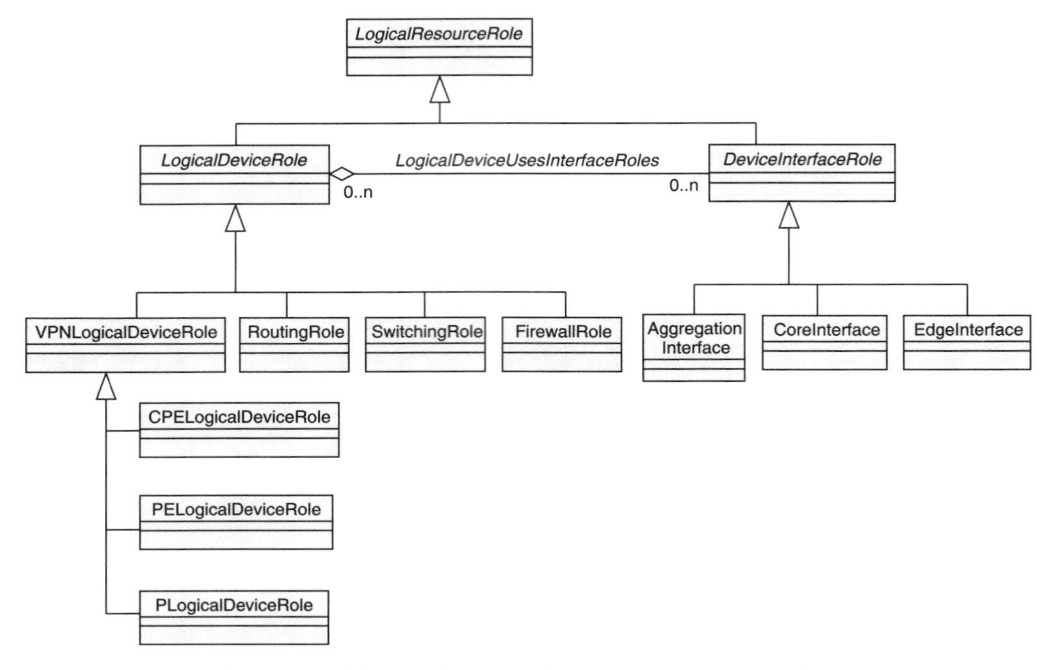

Figure 9-26 More of the DEN-ng *LogicalRole* hierarchy.

logical requirements of that device. This characteristic is unique in the DEN-ng model.

9.4.4 Customer

A *Customer* is represented by a particular kind of *PartyRole*. As such, a *Customer* is modeled as an *Individual* or an *Organization* (which are two of the subclasses of Party) that buys *Products* and *Services* from the enterprise or Service Provider (or receives free *Products* or *Services*). This is modeled as a *Party* having a *PartyRole* of *Customer*. This provides the flexibility to associate other types of roles with the *Party* having a *PartyRole* of *Customer*.

Customers can also be other *ServiceProviders* who resell *Products*, other *ServiceProviders* who lease the enterprise's resources for use, and so forth (Figure 9-27).

Figure 9-27 uses a *PartyRole* to represent different types of people or organizations that purchase *Products*. Note that the SID model can also be used to model the actual purchasing process. This would use the *BusinessInteraction* classes.[17] However, this model is sufficient for our current purposes.

Figure 9-27 depicts a *Customer* as a type of *PartyRole*. Different *PartyRoles* can purchase *Products*, as defined using the *PurchasedBy* association between *PartyRole* and *Product*. *PartyRoles* are aggregated into *Party* through the *hasPartyRoles* aggregations. This enables either Individuals or Organizations to assume the role of *Customer* in order to purchase *Products*.

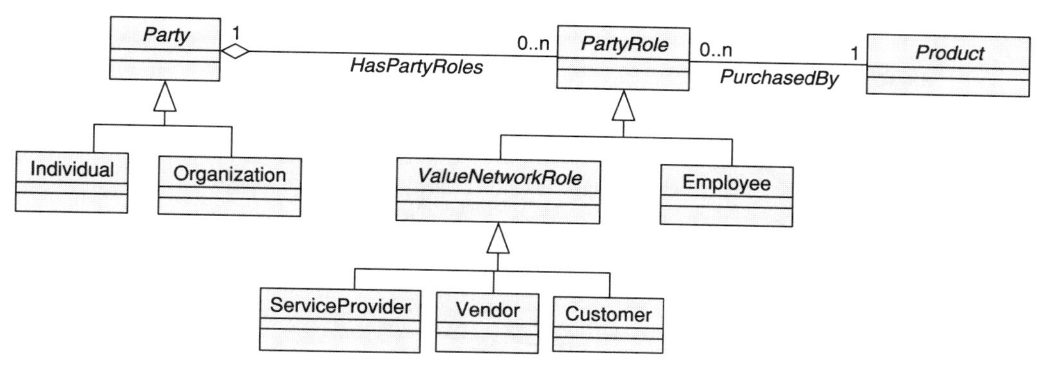

Figure 9-27 Modeling a customer in DEN-ng.

9.5 The System View

The system view is primarily used to address the needs of the system analyst and designer, architects, and, to a lesser degree, implementers. The system view is concerned with adding detail to the business view to set the stage for defining implementations. This is done as follows:

- Start with understanding why the business entities were designed the way they were
- Capture system-level information and decide how best to model that information

It is extremely important to start the system design by reusing the existing entities defined in the business view. This helps cement and formalize the relationship between the business and system views.

In general, there are two obvious ways to do this:

- Extend existing entities defined in the business view through the addition of attributes, methods, relationships, and constraints, and/or
- Define new system entities that are related to the business entities

This is shown in Figure 9-28.

Figure 9-28 illustrates a system activity that is bounded by two system processes. This interaction is in the form of a number of system entities. Every system entity is either derived from a business entity or related to a business entity in some way. This is pictorially represented by the "embedded" business entity shown in one of the system entities. The color coding represents that three business entities gave rise to five system entities.

The interface specification describes how different system entities exchange and reuse information. This is shown in more detail in the bottom of Figure 9-28, which defines two system entities, *Customer* and *Product*, that each expose an interface. This interface defines how Customer information is related to Product information. The SID and DEN-ng models separate the definition of the interface from its implementation.

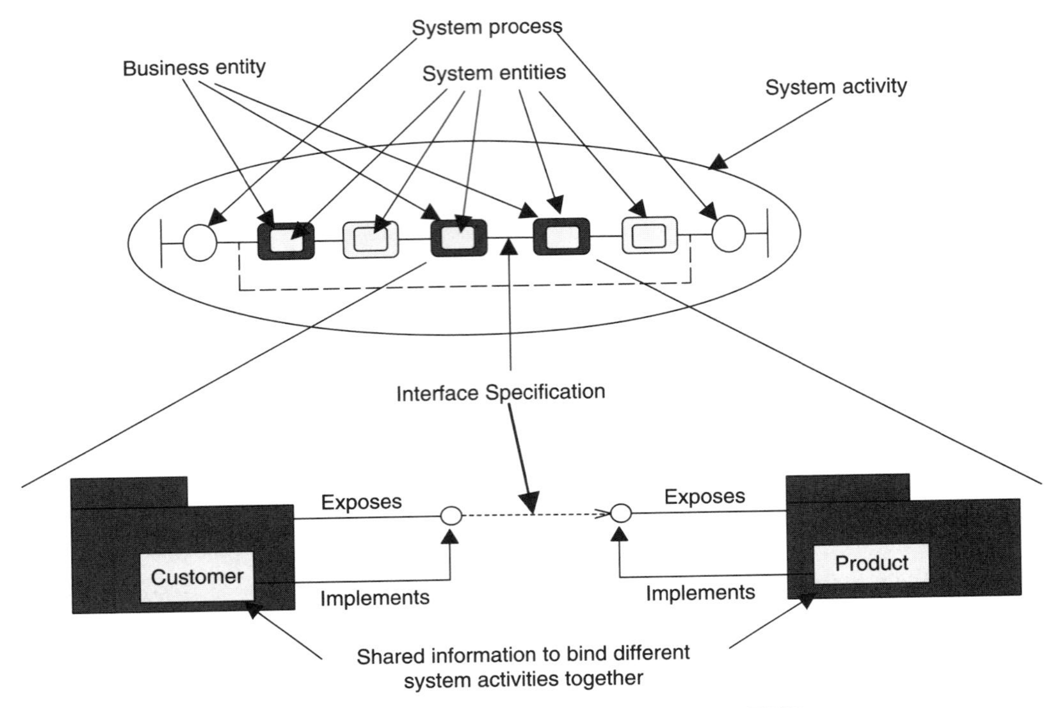

Figure 9-28 The system view of a customer in DEN-ng.

The following sections will demonstrate examples of both of these methods for different types of system entities.

9.5.1 System-Level Product Extensions

At first glance, one might think that a Product is purely a business concept. Instead, the SID and the DEN-ng models take the approach that a Product has system and implementation artifacts. Therefore, these artifacts *must* be implemented in the system and implementation views. This enables business needs, rules, and requirements to be more easily implemented in the system design and implementation.

Figure 9-17 defined the business view of how *ProductSpecification*, *ProductOffering*, and *Product* relate to each other. It also defined VPN-specific concepts, modeled as subclasses, of these three base classes. In the system view, the structure of the relationships between these objects is not altered. However, each of these classes has attributes, methods, relationships, and constraints added to its business view definitions. This in effect "grows" the business definitions into system definitions, as illustrated in Figures 9-29a and 9-29b.

Figure 9-29b shows three examples of how a *ProductSpecification* changes in the system view when compared with the business view. The simplest of these is the addition of a new attribute (*operationalStatus*). This attribute is an enumerated integer that defines the operational state of the *Product* and is used to coordinate and link a *Product* to its constituent services and resources. (This is

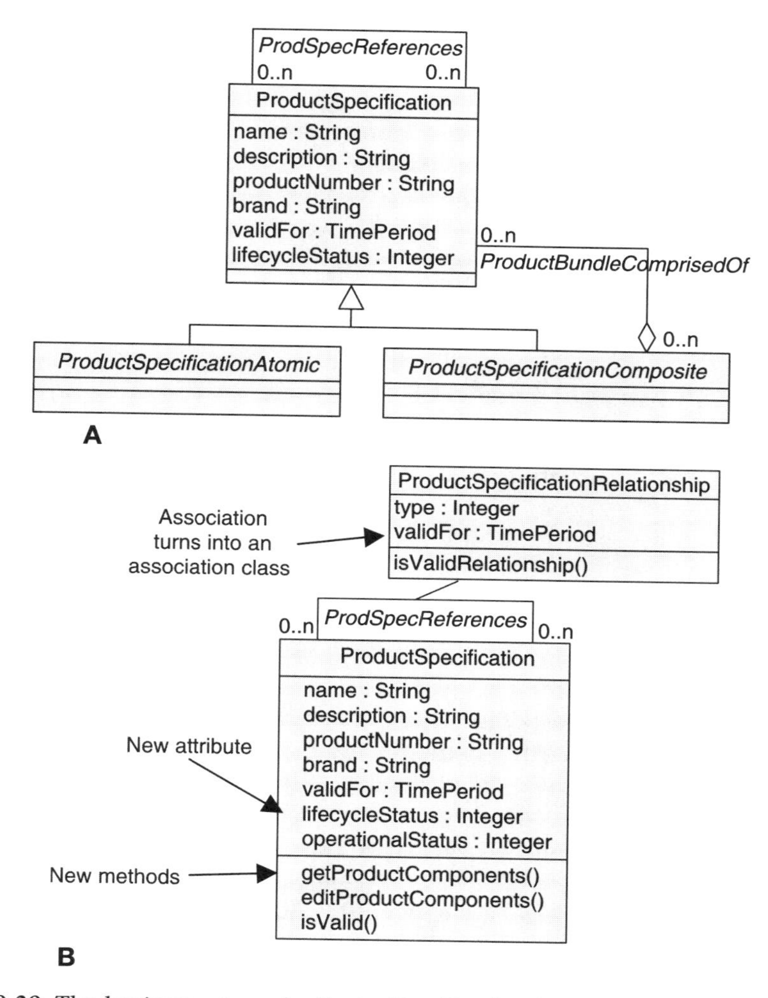

Figure 9-29 The business view of a *ProductSpecification* (A). Transforming a *ProductSpecification's* business view into a system view (B).

only one exemplary attribute—other attributes are also added in the full model; they have not been shown for the sake of simplicity.)

Just as there is value in hiding lower-level system details from a higher-level business view, there may also be value in hiding higher-level business concepts from a lower-level system view. For example, the branding of a *Product* is probably not of concern to system designers or implementers. This has been largely not spoken of until now to keep the discussion simple and focused. DEN-ng supports this requirement by identifying concepts (attributes, relationships, methods, and constraints) that are appropriate only for one or more particular views.

The second example of how a system view differs from a business view is shown at the bottom of Figure 9-29b and is the addition of methods. Methods are almost never shown in the business view, since methods imply an implementation, and other ways can be used to signify the presence of actions to be taken in the business view.

The final example is the "transformation" of the *references* association from a simple association to an association class. Again, association classes are usually not appropriate at the business view, because they contain information that is fundamentally not business related.

Similar changes to *ProductOffering* and *Product*, as well as to the subclasses of these three classes, are needed. They each use the principles explained above to transform business views into system views.

9.5.2 System-Level Service Extensions

Service is fundamentally more complicated than *Product*. This is because at the business level, *Service* has a set of limited functions that relate to business goals. However, when we start to consider the design and implementation of a *Service*, additional concepts come into play. This takes the form of additional classes that need to be created and related to the existing classes defined in the business view (as well as considering the addition of attributes, methods, relationships, and constraints as appropriate).

As an example of this, consider the concept of a *Service*. Figure 9-29 shows the system view of the *Service* class. To keep the concepts simple, we have shown only some of the additions of the system view.

Again, we see the addition of attributes and methods in the *Service* class itself. It still has its two primary subclasses, *CustomerFacingService* and *Resource FacingService*. For this example, we are going to leave the *CustomerFacingService* fragment the same as it was shown in Figure 9-19. However, the *Resource FacingService* fragment has changed.

Two of the subclasses of *ResourceFacingService* are shown in Figure 9-30. *QoSService* represents a generic specification for defining the different types of sub-services that are required to implement a specific type of QoS. Three of the subclasses of *QoSService* are shown in Figure 9-30. These subclasses define specific technologies that can be used to implement a *QoSService*. From a system point-of-view, a *QoSService* abstracts the ability for different types of technologies to be programmed to provide a better-than-best-effort service.

The programming of QoS is represented as a type of *NetworkForwardingService*, because when traffic is forwarded, it can be differentiated from other traffic. During the forwarding process, packets can be queued or dropped; these are two examples of different types of QoS. Thus, there are two fundamental types of *NetworkForwardingServices*. A *TrafficIdentificationService* is used to classify and mark packets to separate one traffic flow from another traffic flow. Once this is done, different means of defining how packets are forwarded can be defined. These are all different types of *TrafficConditioningServices*.

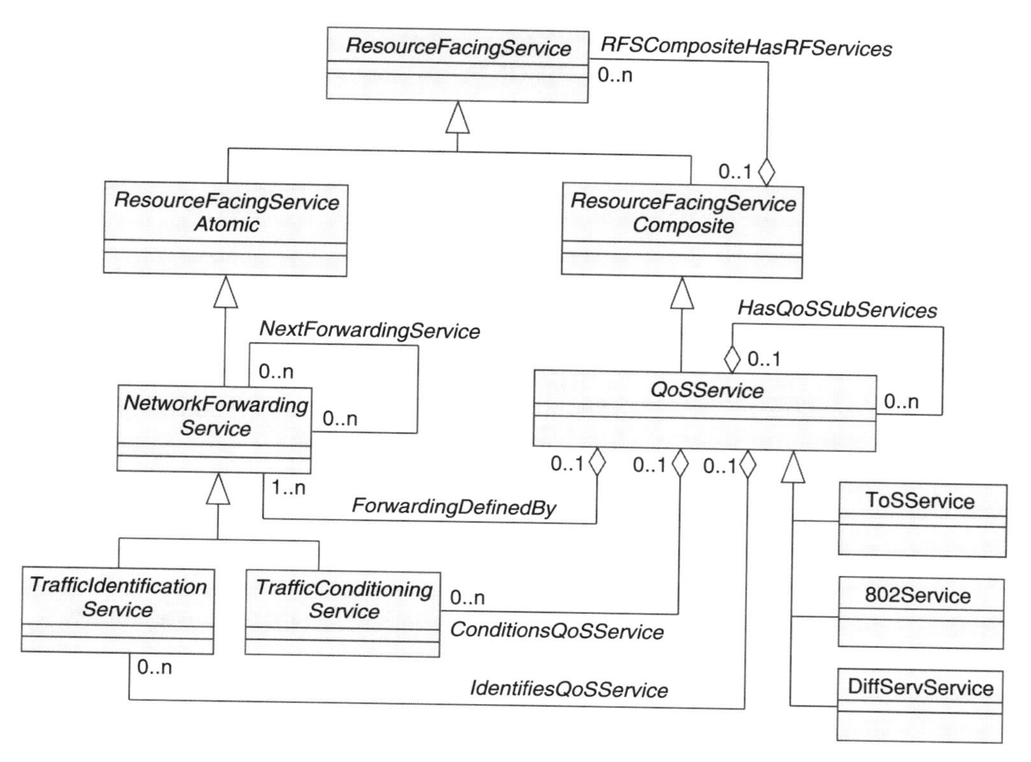

Figure 9-30 The system view of a *ServiceBundle*.

Thus, we see that a given *QoSService* technology can be programmed using a particular set of *NetworkForwardingServices*. A set of *QoSServices* can be used to implement a particular *CustomerFacingService*. These two relationships are represented by the *ForwardingDefinedBy* and *CFServiceRequiresRFService* relationships, (Figure 9-30).

9.5.3 System-Level Device Extensions

Whereas the business view defined devices in terms of high-level functionality (such as the role they play), the system view takes those concepts and translates them into a form that can be programmed. In effect, the system view serves as a bridge between the high-level business definition of a device and how that device is programmed.

When the leap is made from a business model to how something is programmed, the model changes significantly because additional factors and semantics that were not present in the business view need to be appropriately represented. This transformation will be illustrated by taking a simple example: *PhysicalPort*.

Figure 9-31 shows a simple business model for *PhysicalPort*, using the DEN-ng model. *Hardware* is an abstract base class representing any type of hardware

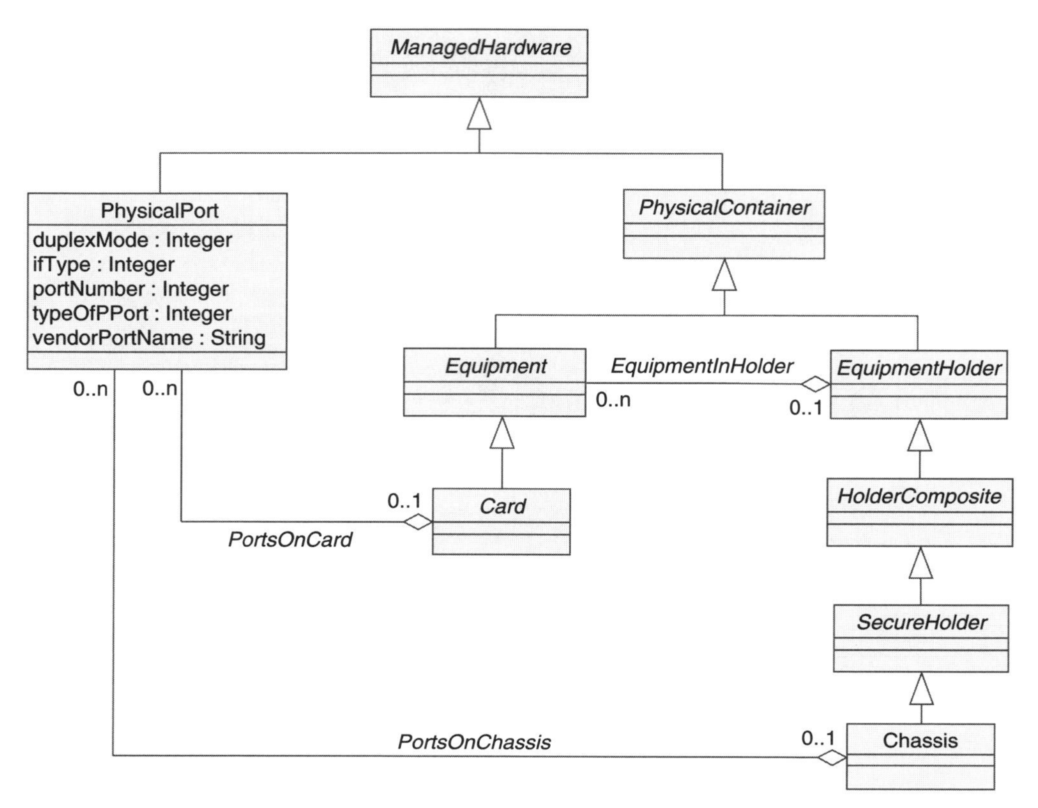

Figure 9-31 The business view of a *PhysicalPort*.

unit that exists as an atomic unit that is not a *PhysicalLink* or a *Physical Connector*. *Hardware* is defined as any component that has a distinct physical identity and can be a component of a *ManagedDevice*. An object has a physical identity if it has a physical manifestation that enables it to be held and have a label attached to it. Thus, software, files, protocols, and policies are not physical objects.

This class hierarchy models a *PhysicalDevice*, such as Routers, Switches and Firewalls, as a managed object that performs one or more end-user functions. A *PhysicalDevice* consists of a set of *Hardware*. *Hardware* has three principal subclasses. Equipment represents hardware components, such as *LineCards* and *Chips*, that fit into the Device and are managed components that are directly needed by the Device to perform its primary end-user function(s). For example, a line card that performs routing and forwarding functions is a subclass of *Equipment*. *EquipmentHolder* is an abstract base class that represents physical objects that are both manageable and able to host, hold, or contain other physical objects. Examples of physical objects that can be represented by instances of this object class are *Racks*, *Chassis*, *Shelfs*, *Cards*, and *Slots*. The difference between subclasses of *EquipmentHolder* and other subclasses of *Hardware* is that all *EquipmentHolder* subclasses are *dedicated* to holding, or containing, other *Hardware*. Finally, *AuxiliaryComponents*, such as *PowerSupplies* and *Fans* are required for proper operation of the Device but have a function that is different than the primary end-user function(s) of the Device.

EquipmentHolder and *AuxiliaryComponents* do not form a complete function on their own – rather, their purpose is to support the function(s) performed by their containing Device. However, the functionality of *DeviceComponents* is directly related to the primary purpose of the device, whereas the functionality of the *AuxiliaryComponent* is not. For example, a *LineCard* is a type of *DeviceComponent* because it supplies routing and forwarding functions, which are related to why one uses a router and not some other type of device. A power supply is an *AuxiliaryComponent* because, while power is clearly required by a router, power is not one of the salient characteristics that differentiate a router from other types of Devices.

One of the principal purposes of the *Hardware* class is to serve as a convenient aggregation point where relationships to different types of *Hardware* can be placed. Figure 9-31 shows three subclasses of *Hardware: EquipmentHolder*, *Equipment*, and *PhysicalPort*. Relationships that involve any or all of these subclasses can be defined using *Hardware* as a participant. This obviates the need for building (in this case) three separate sets of relationships (one to each of the Hardware subclasses) to represent the same thing. In Figure 9-31, there are two separate relationships to *PhysicalPort*, because different types of devices build *PhysicalPorts* on either *Cards* or directly on a *Chassis*.

Before we show the complete transformation to a system view of a *PhysicalPort* in DEN-ng, we shall first examine some system-oriented particulars of the model in Figure 9-31. Two examples of this will be provided.

To emphasize this point, Figure 9-32 compares the DEN-ng system models of the *Hardware* and *ManagedHardware* classes. Several important points can be made:

- The purpose of the system view of the DEN-ng *Hardware* class is to represent how a hardware object physically appears. Thus, it defines width, depth, height, and weight attributes. These are common to all types of hardware objects.
- The purpose of the system view of the DEN-ng *ManagedHardware* class is to define specific attributes that represent how hardware is managed. Thus, a clean separation between a physical hardware object (which is represented by the *Hardware* class) and a *manageable* physical hardware object (which is represented by the *ManagedHardware* class) can easily be achieved.
- *PhysicalConnectors* are not managed in the sense of having operational characteristics or alarm information; therefore, they are a sibling of the *ManagedHardware* class.
- Continuing the example, *PhysicalConnectors* are associated with *Hardware*, not *ManagedHardware*, because there may be some *Physical Connectors* that are used to connect Hardware that is not inherently manageable.

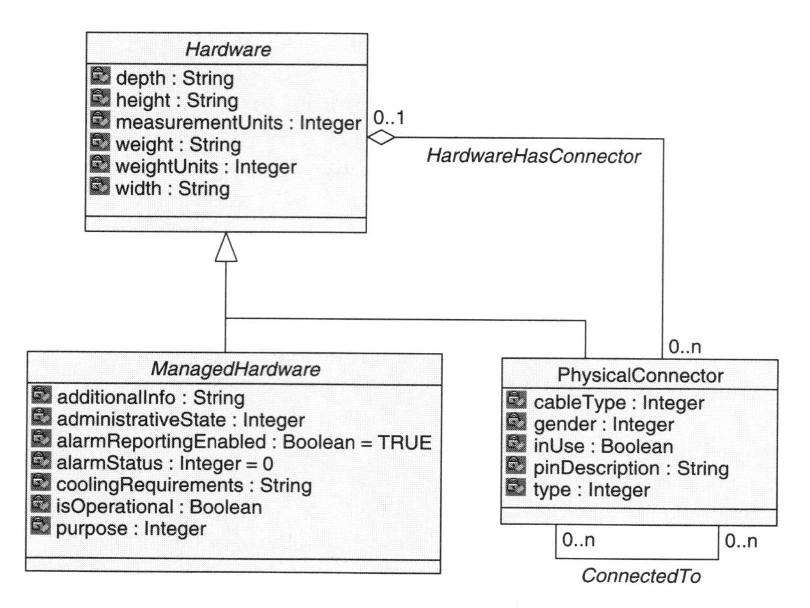

Figure 9-32 Comparing the DEN-ng *Hardware* and *ManagedHardware* classes.

The second change will be to insert a new class called PhysicalContainer into the hierarchy. PhysicalContainer will be defined as an abstract base class that is used to add additional semantics to the ManagedHardware class. Its three attributes define whether a ManagedHardware object can be removed and/or replaced, and whether this action requires power to be removed when the action is performed (Figure 9-33).

In addition to the three attributes, one method and one relationship is defined. The *containerHasHardware* aggregation is actually an aggregation class, not just a simple aggregation. The *ContainerHasHardwareDetails* aggregation class represents the semantics of the *containerHasHardware* aggregation. It defines the particular physical location that this *Hardware* occupies in its containing *PhysicalContainer*, defines if it is the only object contained by this *PhysicalContainer*, and provides methods to tell how many other objects are contained within this *PhysicalContainer*.

The *isCompatible* method is used to verify whether the referenced *Hardware* component may be contained by or inserted into this *PhysicalContainer*. This is a method, because the semantics that determine whether a particular *Hardware* component can be properly inserted into a given *PhysicalContainer* vary by a number of factors. Besides obvious size and weight considerations, a set of other considerations must be evaluated. These include electrical compatibility, determining if there is enough available power and cooling to support the new container, and more esoteric functions. For example, many devices commonly have different types of slots. For example, a *Chassis* may have dedicated *Slots* for holding *SystemCards*, *LineCards*, *MemoryCards*, and so forth. Thus, in this example, a check must be made to ensure that a particular *Card* is being inserted into a *Slot* meant to hold that type of *Card*.

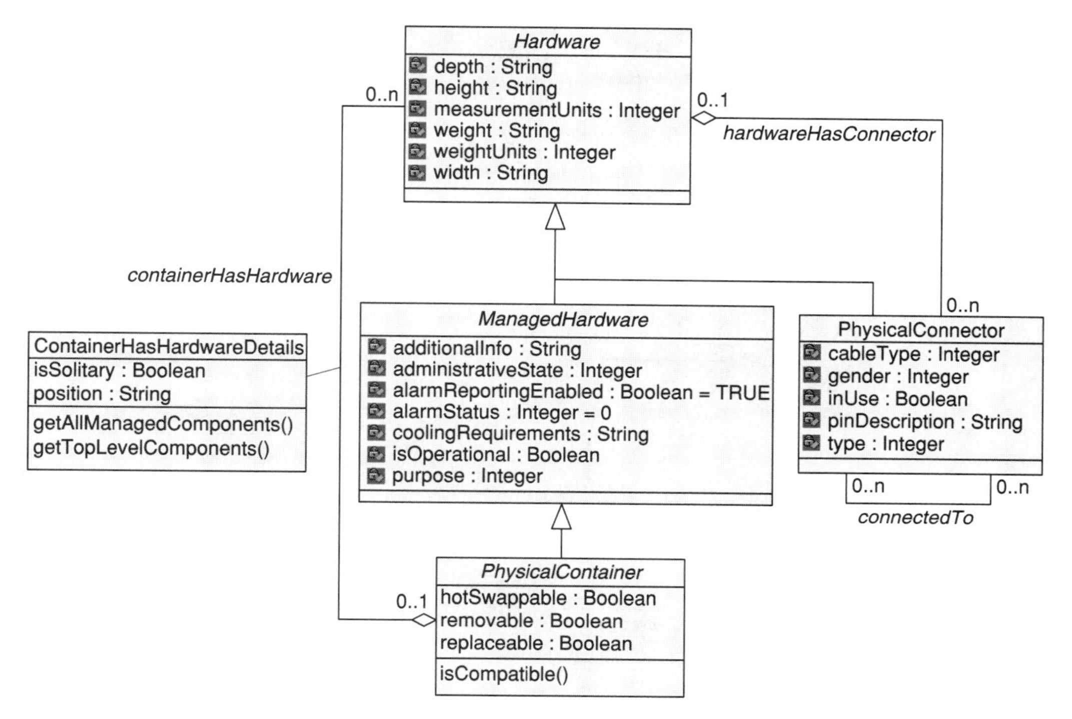

Figure 9-33 The DEN-ng *PhysicalContainer* class.

This is a very good example of the level of detail that is necessary to consider when building a system model. If the model is intended to represent the current state of managed objects, such considerations must be taken into account. This will enable the model to be used to manage the *life-cycle* of the managed objects, including when physical or logical changes are performed.

Figure 9-34 shows part of the DEN-ng system model for *PhysicalPort*. In this model, the attributes and methods for all classes except *PhysicalPort* have been left off for simplicity.

Depending on the scope of the business view, one could argue that all, some, or none of the attributes shown are pertinent. The SID defines all five attributes in reference 13 (Addendum 5PR to GB922), which is the business view of *PhysicalResources*. For the purposes of this example, we will examine one attribute in particular: *appPortName*.

The purpose of *appPortName* is to provide a means to define an application-specific name of this port. This is different from the *commonName* attribute, which represents a system-wide naming structure for all *ManagedEntities*. For example, a system may decide that a common naming attributes for all *Physical Ports* for all of its devices should follow the following rule:

Name = deviceName + "Port" + x

where deviceName is the name of the device (which is either a PhysicalDevice or a subclass of PhysicalDevice), "port" is the string "port", "x" is a monotonically

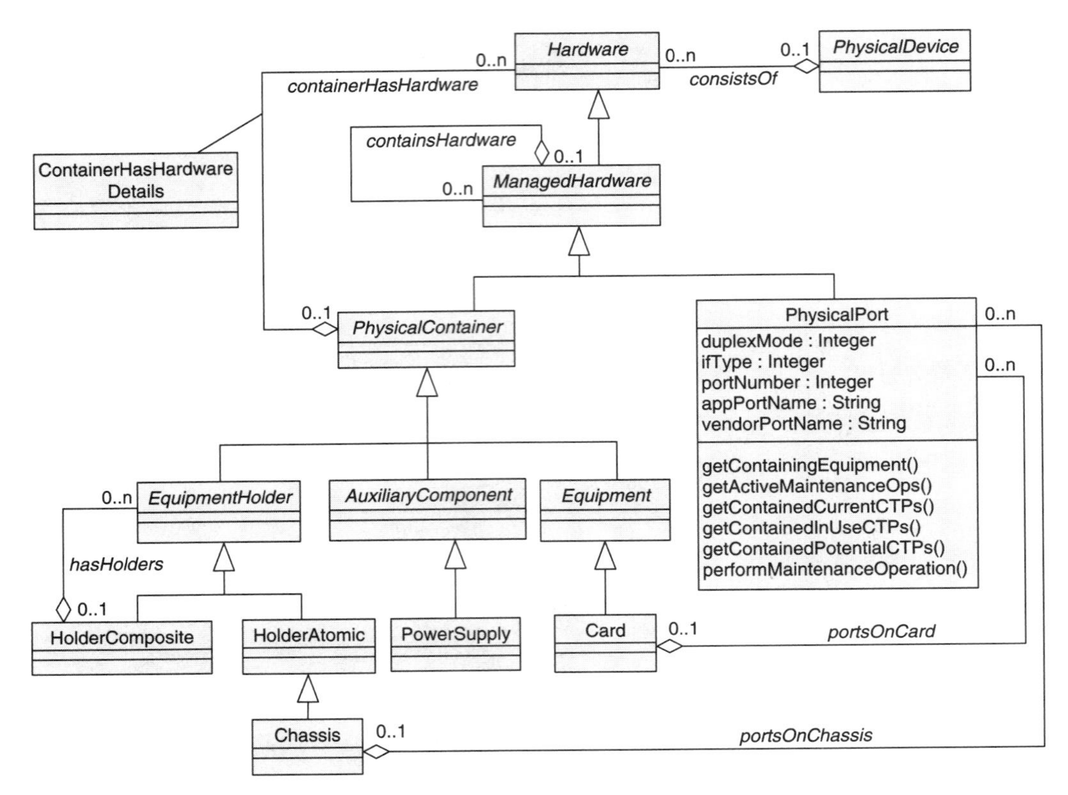

Figure 9-34 The system view of a *PhysicalPort*.

increasing unique integer, and "+" means concatenation. Thus, we could end up with names such as "internetGatewayPort4." This conveniently defines a common naming convention on each device's ports, making it easier to build network topology diagrams and to maintain physical inventory. However, for programming the device using vendor-specific languages (e.g., CLI), we need to use the name of the port that the particular vendor uses. Hence, we can have two names for our port—one used by the management system (i.e., *appPortName*) and one used to map to the vendor-specific name that is assigned to that port (i.e., *vendorPortName*).

In either case, the business view certainly will not contain the set of methods shown. The purpose of these methods is to relate this particular *PhysicalPort* to different types of management concepts. Specifically:

- *getContainingEquipment* is used to retrieve a specific *Equipment* or *EquipmentHolder* that contains this *PhysicalPort*.
- *getActiveMaintenanceOps* is used to retrieve all persistent maintenance commands that are currently enabled on this *PhysicalPort*.
- *getContainedCurrentCTPs* is used to retrieve the "current" CTPs that are contained within a specific *PhysicalPort* or CTP, at specific layer rates. A CTP is a *ConnectionTerminationPoint*.[14] A "current" CTP is defined

as a CTP that is either cross-connectible or cross-connected, in the current mapping configuration.

- *getContainedInUseCTPs* is used to retrieve the "in use" or "actual" CTPs that are contained within a specific *PhysicalPort* or CTP, at specific layer rates. An "in use" CTP is defined as a CTP used by a network in any state (including pending) or a CTP that is terminated and mapped.
- *getContainedPotentialCTPs* is used to discover the contained CTP(s) in this *PhysicalPort* and returns the complete object information for the contained CTP(s).
- *performMaintenanceOperation* is used to perform the specified maintenance operation.

The relationship between a *PhysicalPort* and a CTP was not shown in Figure 9-31 for the sake of simplicity. This is yet another example of concepts that need to be added to the simple business view entities in order to support system design and implementation.

9.5.4 System-Level Customer Extensions

The system level extensions to the *Customer* objects take two forms. On the business side, additional detail is provided that can be used to define how a *Customer* is billed (for example) for a particular *Product*. On the system side, depending on what *Product* the *Customer* purchased, additional entities are required to "connect" the *Customer* to the *Services* purchased.

For example, if the *Customer* has purchased a CPE router, someone needs to install that router, configure it correctly, and connect it to the service provider's backbone. This requires two types of objects:

- Objects that are used to identify the *Customer*'s traffic, so that it can receive the proper *Service* for which the *Customer* has contracted, and
- Objects that represent the programming of the *Customer*'s CPE router

Both of these will be covered in more detail in this chapter and in Chapters 9 and 10.

9.6 The Implementation View

The implementation view is focused on adding further detail to the system view to enable these concepts to be programmed. The advantage of having an implementation view is that the model can be used to bind concepts in the business and system views to the actual implementation. This helps ensure that the network can be configured to implement fundamental business and system rules.

For *Product*, the notion of an *MPLSVPNProduct*, *ProductSpecification*, and *ProductOffering* are required. Although the structure of their relationships is not altered, each of these classes has attributes, methods, relationships, and constraints added to its business view definitions. This in effect "grows" the business definitions into system artifacts.

This view dips into vendor-specific proprietary methods, which prevents detailed examples from being given. However, the following high-level example, taken from the Intelliden product, can be used to serve as a general example of the types of entities that are required in this view, and more importantly, how this view is used in an actual product. More detail will be provided in Chapter 11.

For the purposes of this example, consider some of the programming tasks of a PE router. Once the CPE router announces a new customer route to its ingress PE, the PE needs to do five important things:

- PE creates VRF entry
- PE creates MPLS Label
- PE checks its export *policy*
- PE converts IPv4 routes into VPN IPv4 format
- Ingress PE announces VPN IPv4 route to egress PE using iBGP

We will consider the first of these five tasks, the function of the VRF entry, for our example.

A VRF, or Virtual Router and Forwarding, table is a special forwarding table used for our particular type of VPN. Every CE-PE connection is associated with one or more VRFs. A VRF enables a customer site to be connected to the service provider network by one or more ports. The service provider associates each port with a VRF. When an IP packet is received, its destination IP address is looked up in its associated VRF, which determines how to route the packet. If an IP packet arrives that is not associated with any VRF, the packet's destination address is looked up in the router's default forwarding table, and the packet is routed accordingly. In effect, the router's default forwarding table is used to forward "public" routes, such as those from neighboring P or PE routers, and the router's VRFs are used to forward "private" routes.

The underlying information model will therefore model *how* these two different types of forwarding tables are used. There is no need for the information model to represent the actual contents of these forwarding tables, since they change frequently and are, after all, controlled by the routing protocols that dynamically adjust network routes. However, the *decision* as to whether to consider a received IP packet as belonging to a VPN or not can be modeled. The advantage of doing this is that it enables such decisions to be made in a consistent and controllable fashion.

When a PE router receives a packet from a CE device, it must determine the particular connection over which the packet arrived, since this determines the VRF (or set of VRFs) that can be used for forwarding that packet. One of the ways to determine the connection over which a packet arrived is to determine the physical interface over which the packet arrived. Depending on the connection, the router may also look at some of the packet's layer-2 header information (e.g., the DLCI field if the connection is a frame-relay interface).

This is a good example for showing why the physical and logical representations of a router need to be both modeled and related to each other. Physical characteristics of a device can be modeled, and the model can be used as a template for populating inventory systems. In this case, it can be used to provide not just the particular physical interface, but also the physical characteristics of that particular *PhysicalPort*'s parent router. Logical functions provided over that

PhysicalPort are modeled using a combination of protocol models, software models (to represent the functionality of the router's operating system), and more detailed service and policy models.

9.7 Summary

This chapter discussed how the business, system, and implementation views can be coordinated using the DEN-ng information model. The beginning of the chapter explained the difference between the DEN-ng and SID information models. DEN-ng was shown to be focused on modeling network elements and services in as much detail as necessary. It also was designed from the beginning to provide business, system, and information models. In contrast, the TMF SID is proceeding at a slower, but more generic, pace. It is concentrating on building out a comprehensive business view of all of the main managed objects that are needed to manage an environment.

DEN-ng has promoted the use of policy management and process management to holistically work together to control the behavior of a managed environment. DEN-ng does this by using a finite state machine to define the appropriate states in the managed environment for each making up that environment. In this approach, policy management is used to define the particular state of a managed object and whether it should remain in that state or transition to a new state; process management represents the mechanisms that can be used to accomplish the goal set by policy.

VPNs were then briefly described, so that important example-managed entities could be identified. These managed entities are either represented directly by DEN-ng, or by DEN-ng concepts such as device roles.

Important parts of the DEN-ng model were then described briefly but in detail. One of the features of the DEN-ng model is its definition of Service. Whereas other models such as the DMTF CIM represent a Service as simply another type of logical element, DEN-ng (and the SID) represents a Service as existing only in the context of a Product.

To further emphasize this, Services are organized into *CustomerFacing Services* and *ResourceFacingServices*. *CustomerFacingServices* are *Services* that are visible to *Customers*. This is in direct contrast to *ResourceFacingServices*, which support *CustomerFacingServices* but are not seen or purchased directly by the *Customer*. For example, a VPN is an example of a *CustomerFacingService*, whereas the sub-services that perform different types of routing between network devices making up the VPN are examples of *ResourceFacingServices*. The Entity-EntitySpecification pattern is one example of several important patterns that are used repeatedly in DEN-ng and the SID. It is used to capture the common and invariant characteristics and behavior of managed entities.

The DEN-ng Service and Resource models are built to define these domains in great detail as well as to enforce their association with other domains, such as Product. Both the *PhysicalResource* model and the *LogicalResource* model were briefly described. Innovative features, such as the use of roles to define physical characteristics and usage, were described in more detail, because of their ability to help abstract different systems and solutions. The Party model was then explained. Here, roles are used extensively to define the different functions that an individual, group, or organization can have.

The rest of the chapter then talked about the different entities that are present in the business, system, and implementation views. The DEN-ng models were built so that entities defined in one view could be related easily to entities from another view. This was shown by providing different examples of several different types of important business, system, and implementation entities.

Chapter 10 will build on these concepts, and Chapter 11 will provide specific examples of how these principles have been realized in a commercial product.

9.8 Recommended Further Reading and References

The following is a set of references for this chapter.

1. Strassner, J., *The DEN-ng Information and Data Models*, work in progress.
2. Shared Information and Data model specification of the TMF, It Frontpiece: GB922 proper (*Shared Information/Data (SID) Model—Concepts, Principles and Domains*, Version 3.0, July 2003) and the following Addenda (as of this writing, it is always growing in scope) that are not specifically called out in other references:
   Shared Information/Data (SID) Model—Addendum 1A—Common Business Entity Definitions—Agreement, GB922, Version 3.0, May 2003.
   Shared Information/Data (SID) Model—Addendum 1L—Common Business Entity Definitions—Location, GB922, Version 3.0, June, 2003.
   Shared Information/Data (SID) Model—Addendum 1P—Common Business Entity Definitions—Party, GB922, Version 3.0, May, 2003.
   Shared Information/Data (SID) Model—Addendum 2—Customer, GB922, Version 3.0, May 2003.
   Shared Information/Data (SID) Model—Addendum 5RO—Resource Overview, GB922, work in progress.
3. NGOSS Technical Specifications: TMF (*www.tmf.org*):
   NGOSS Architecture Technology Neutral Specification, TMF053, Version 3.0, April 2003.
   NGOSS Architecture Technology Neutral Specification (Contract Specification), TMF053B, Version 3.0, February 2003.
   NGOSS Architecture Technology Neutral Specification (Behavior and Control Specification), TMF053C, Version 1.0, February 2003.
   NGOSS Architecture Technology Neutral Specification (Metamodel Specification), TMF053D, Version 1.0, February 2003.
   NGOSS Architecture Technology Neutral Specification (Distributed Transparency Services Specification), TMF053F, work in progress.
   NGOSS Architecture Technology Neutral Specification (Policy Management Specification), TMF053P, work in progress.
4. TMF, *eTOM, The Business Process Framework*, GB921, Version 3.5, June 2003.

5. This is the Provider Provisioned VPN working group of the IETF. Its home page can be found at http://www.ietf.org/html.charters/ppvpn-charter.html
6. Davie, B., Rekhter, Y., *MPLS: Technology and Applications*, Morgan Kaufman Publishers, 2000, ISBN 1-55860-656-4.
7. Rosen, E., et al. *BGP/MLPS VPNS*, draft-ietf-ppvpn-vfc25476 03.txt, October 2002.
8. TMF, *Shared Information/Data (SID) Model—Addendum 3—Product Business Entity Definitions*, GB922, Addendum 3, version 3.0, May 2003.
9. TMF, *Shared Information/Data Model—Addendum 4S–QoS–Quality of Service Business Entity Definitions*, Version 1.0, June 2003.
10. Rosen, E., Rekhter, Y., Bogovic, T., Brannon, S., Vaidyanathan, R., Carugi, M., Chase, C., Fang, L., Chung, T., De Clercq, J., Dean, E., Hitchen, P., Smith, A., Leelanivas, M., Marshall, D., Martini, L., Morrow, M., Srinivasan, V., Vedrenne, A., *BGP/MPLS VPNs*, draft-ietf-ppvpn-rfc2547bis-03.txt, October 2002.
11. TMF, *Shared Information/Data (SID) Model—Addendum 4SO—Service Overview Business Entity Definitions*, GB922, Addendum 4SO, Version 3.0, June 2003.
12. TMF, *Shared Information/Data Model-Addendum 1-POL-Policy Business Entity Definitions*, Version 1.0, July 2003.
13. TMF, *Shared Information/Data (SID) Model—Addendum 5PR—PhysicalResource Business Entity Definitions*, GB922, Addendum 5PR, Version 3.0, June 2003.
14. ITU-T, *Generic Network Information Model*, recommendation M.3100, July 1995.
15. Baumer, D., Riehle, D., Siberski, W., Wulf, M., *The Role Object Pattern*, Download PDF from: http://www.riehle.org/papers/1997/plop-1997-role-object.html
16. The following URL provides a business-level description of AT&T's Managed VPN Services: http://www.att.com/abs/serviceguide
17. TMF, *Shared Information/Data (SID) Model—Addendum 1BI—Business Interaction Definitions*, GB922, Addendum TMF053, Version 3.0, May 2003.
18. TMF, *Shared Information/Data Model-Addendum 5LR-LogicalResource Business Entity Definitions*, Version 1.0, July 2003.

Application of PBNM Concepts

This chapter will provide several different examples of how policy is used in a system. The Policy Continuum figures prominently in the recommended approach for building PBNM systems because it enables the business, system, and implementation views to be seamlessly integrated. In addition, the multiple uses of policy will be emphasized through examples in this chapter.

10.1 Introduction

The purpose of this chapter is to show how policy is used in different situations in PBNM systems. There are two common ways that policy is commonly described. One group of people thinks of policy as a way to change lines in the configuration file of a device. Another group of people thinks of policy as only appropriate for expressing rules in the business world.

Of course, both are right. Unfortunately, neither of these represents the use of the full potential of how policy can be used. The DEN-ng concept of policy is epitomized through the use of the Policy Continuum. The Policy Continuum is the basis for defining how policy is defined and used for different constituencies, ranging from the business analyst to the system designer to the network administrator. These uses, as well as others that are suited for different users of the Policy Continuum, constitute the potential of PBNM. The holistic use of the Policy Continuum, and its ability to integrate the needs of the business and system worlds, is key to realizing the potential of PBNM solutions.

This chapter is focused on using policy to manage, as well as control, the configuration of devices. The Policy Continuum will be used to express policies at different levels of abstraction and integrate them together to form a cohesive set of policies.

10.2 Policy Approaches

There are two fundamentally different approaches that can be used to develop a PBNM system. They can be categorized as static versus dynamic. In either case, it is important to note that this categorization is based on how the policies are *applied:* In both cases the policies remain *constant*—it is the *environment* that is changing. Thus, the approaches differ in how the environment is to be controlled.

10.2.1 Static Versus Dynamic

Static approaches are also called pre-provisioned approaches. Here, the idea is to pre-provision the network according to some pre-defined scheme and plan into that pre-provisioning the ability for different behaviors to take effect when conditions warrant. The classic example of this is to use Differentiated Services (DiffServ)[1] to define traffic conditioning according to different pre-defined levels (e.g., Gold, Silver, and Bronze) and group traffic into sets whose forwarding behavior (i.e., how the packet is treated within the router between its input and output interfaces) is the same. This is done by applying various rules to classify traffic at the edge and "mark" it (via setting appropriate bit values in the IP header) according to a set of pre-defined "code points." These code points define a particular type of forwarding behavior (e.g., dropping, metering, policing, shaping, queuing and scheduling) that should be applied.

Dropping means the selective dropping of packets, whether randomly or algorithmically. This is used to tell TCP-based traffic to slow down the transmission rate. Metering is the act of limiting the amount of incoming traffic to a pre-defined rate, and is also called "rate limiting." Policing combines metering with dropping and optionally [re-] marking to indicate that further conditioning is necessary to downstream elements. Shaping is similar to policing, except that it is applied on the egress [whereas policing is applied on the ingress] and contains buffering of packets [policing does not provide any buffering]. Queuing is the act delaying the transmission of a packet by storing it and possibly transmitting the packet in a different order than it was received. Scheduling is the act of defining which packet to send from which queue when the interface is free.

The essence of a static approach is to allow changes in the environment to be accommodated by the rules embedded in the network. In other words, policies are used to initially provision the network, and then the network runs on "autopilot" until something significant happens that requires the download of new policies.

This works based on the ability to program complex behaviors into the network that "do the right thing" in response to small or medium environmental changes. For example, DiffServ has the notion of "drop probabilities" within a particular class. As traffic rate increases, the probability of dropping a packet increases. When a packet is dropped, and the application sending the packet is a TCP-based application, then the dropping of a packet causes the transmission rate of the application to be slowed. Thus, we have a simple self-regulating mechanism that is appropriate for many types of traffic. (Note that I said "many types of traffic" and not *all* types of traffic. Clearly, dropping packets in a VoIP application corresponds to dropping parts of a conversation, which is not desirable.)

What if "doing the right thing" is too complicated in order to be pre-provisioned? This is where dynamic approaches are used. Note that DiffServ can also be used in a dynamic approach. For example, suppose that the input interface of a DiffServ router is monitored in order to observe the number of input flows, and a policy exists that will change the traffic conditioning based upon statistics of this interface (such as number of packets dropped). This has the effect of dynamically adjusting a pre-provisioned mechanism. Note that the policy itself is static—it is the changing statistics that cause the policy to be invoked.

However, the classic example of a dynamic mechanism is of course signaling, such as with RSVP. In this approach, a client asks for a new feature from the network, such as additional bandwidth. The client signals its need of this new feature by explicitly asking for new bandwidth. This, of course, requires a decision to be made as to whether this request should be granted or not. Such requests cannot be "pre-provisioned," since they cannot be anticipated beforehand.

10.2.2 A Better Taxonomy: Pro-Active Versus Reactive

Clearly, these approaches can be mixed. One can adjust pre-provisioned features just as easily as one can pre-provision certain additional features that are not invoked until necessary. Thus, a better way to categorize these approaches is by classifying them as pro-active versus reactive.

Pro-active changes generally take the form of changing the configuration of a device by pushing a new set of commands to it. Pro-active changes are usually implemented as an attempt to avoid an undesirable state. For example, there are many different ways to condition traffic. Not only can traffic be rate-limited at the input (often called "policing"), it can also be regulated at the output (often called "shaping"). The network devices can be pre-programmed to perform either of these functions, or the network can be programmed to incorporate these functions at any desired time. Typically, this is done in a closed-loop fashion by observing traffic patterns and comparing them with desired traffic patterns. Note that this is simply another variation of using a Finite State Machine (FSM)—a device dropping too many packets can be thought of as being in an undesirable state. Thus, we can use one or more policies (as described in Chapter 6) to develop configuration changes that alter the traffic conditioning functions of the device in such a way as to stop the dropping of too many packets. This set of configuration changes can be thought of as transitioning the state of the network device back to an allowed state.

Reactive mechanisms enable parts of the network to offer high QoS guarantees without having to over-provision those parts of the network that need to offer additional guarantees. Traditionally, specialized protocols, such as RSVP,[2] were used to convey those additional guarantees.

Advantages and Disadvantages of Pro-Active Mechanisms

One of the important advantages of pro-active mechanisms is that sets of them can be easily configured to define how to change device configurations given an event and/or condition. Once these changes are implemented in the network, the network can then run in a relatively hands-off manner.

Another important advantage of pro-active mechanisms is scalability. As will be seen, signaling and other reactive mechanisms come with a relatively high price—the maintaining of individual state for each device that is participating in a signaling path. Pro-active mechanisms don't require such state because they represent decisions taken to implement a pre-defined set of behaviors.

The main drawback to pro-active mechanisms is that they cannot have any idea of which resources specific applications entering at different points of the network need. This will tend to waste precious shared resources. (i.e., instead of efficiently allocating extra resources exactly where they are needed, pro-active systems will configure entire portions of the network to act a particular way).

Advantages and Disadvantages of Reactive Mechanisms

The obvious advantage of reactive mechanisms is that they provide a way to explicitly signal additional needs on demand. As we will see, this enables a more efficient use of the network. However, signaling mechanisms offer several important additional advantages that shouldn't be discounted.

Signaled information is, by definition, additional information that is to be applied to a particular flow or set of flows. The beauty of signaled information is that it can traverse the same network path as the traffic that it is intended to manage. This additional specificity differentiated it from pro-active mechanisms, since the latter by definition cannot usually be linked to specific paths. Thus, the advantage of signaling mechanisms is that specific QoS needs for specific paths can be easily communicated. Furthermore, if the needs cannot be met for some reason anywhere along the path, an explicit notification can be provided. This enables an application to pinpoint exactly where it cannot get the QoS that it desires, which enables the application to either search for a different path, accept degraded QoS, or postpone its request.

The most important advantage of reactive mechanisms is the ability to embed additional information into the signaling mechanism. That is, in addition to requesting (for example) additional bandwidth, RSVP can be used to carry additional information, such as user and application information. This is described in many different RFCs. Yadav et al., Identity representation[3] can be used to securely identify the owner and the application of the communicating process in RSVP messages in a secure manner. Bernet and Pabbati[4] described policy elements that can be used to identify different application traffic flows. Finally, Herzog[5] described a preemption priority policy element that defines a relative importance, or ranking, within the set of flows competing for preferential treatment.

The importance of signaling here is that without signaling, such information is typically unavailable. For example, user information is not available in standard packet headers. Even if it was, it and other types of information (such as application identification) would typically be hidden by the use of IPsec, which is often used to secure traffic. The ability to signal information can greatly simplify management, because typically services are allocated in terms of users and applications. This provides us important information to be able to map users and applications to terms that are recognizable to network devices, such as IP addresses and ports.

Of course, nothing in life is free. The advantages of reactive mechanisms such as signaling come with several attendant costs. The first is additional set-up and processing time for all devices, hosts, and end-systems that are using the mechanism. Thus, the first rule is to not use such mechanisms for short-lived flows, or the flow is likely to be finished before the network is finished being (dynamically) configured.

A second-order effect from this additional set-up and processing time is an increase in network traffic due to the increased overhead required for signaling to be used. Thus, for this method to be effective, flows using signaling should ideally generate much more traffic than is required to set up their traffic conditioning.

However, the main problem with such mechanisms is the need to carry state information for each device participating in the flow. Otherwise, the device has no idea whether it should grant the use of additional resources, how long to grant them for, and other factors.

The Solution: Use Both

Integration of these two approaches is being discussed in several fora. One good example is the IETF ISSLL working group,[6] which was formed to address (in part) how to integrate these two efforts—Integrated and Differentiated Services. There are several important works that form part of the output of this working group that define how this can be done. Bernet et al.[7] defined how Integrated Services can be supported over a Differentiated Services network. This in effect provides the best of both worlds. For example, different RSVP flows can be aggregated and tunneled over a common DiffServ network.

There are other important variations of this general theme. For example, Baker et al.[8] described the use of a single RSVP reservation to aggregate other RSVP reservations across a common transit routing region.

10.2.3 The Role of Policy

Policies are essential to the use of either pro-active or reactive models. Policies enable configuration changes, such as how many resources, are to be applied for any given use. For this to be done, however, we need to revisit the notion of policy subjects and policy targets. The definitions of these two terms, from Chapter 2, are repeated here for convenience:

A policy subject is a set of entities that is the focus of the policy. The subject can make policy decision and information requests, and it can direct policies to be enforced at a set of policy targets.

A policy target is a set of entities that a set of policies will be applied to. The objective of applying policy is to either maintain the current state of the policy target, or to transition the policy target to a new state.

Using these definitions, we see that policy subjects are the controlling entities in the policy equation, with policy targets being the means by which policies can be used to change the behavior of the system. However, applying policy correctly is more complicated than it appears.

Equating Different Views of the Same Policy

The first problem is that different policy subjects have different views of the network. Thus, unless a common unifying factor (such as FSM) is used, the same policy executed by different policy subjects could have markedly different behavior. The most likely cause of this would be because the different policy subjects would make different decisions because they were monitoring different objects. The advantage of using a FSM in this case would be to enable the different contexts to be mapped to the same state.

Capability Mapping

The next and bigger problem is that different devices have different capabilities. This manifests itself in several different ways. For example, two different devices could have different commands and/or different features. This makes it very difficult to use these devices to condition the same flows in concert.

Consider two devices that have slightly different capabilities. The same policy cannot be applied unaltered to both devices, unless the different capabilities have the exact same command. This is highly unlikely if the devices have different end-user capabilities. But the problem is actually more involved than this. Mapping to the same commands assumes that each device is able to process the same events and respond to the same conditions. This means that for each of the three fundamental building blocks of a policy (event, condition, and action clauses), a mapping must be done to define equivalent functions in each device. Furthermore, it also assumes that the end result of executing the policy has the same side effects. This is in general a bad assumption to make in this case. The presence of different capabilities is often a good indication that the devices internally are implemented quite differently. For example, turning on fancy queuing might have different effects due to different implementations.

As another example, consider the sub-function of classification. Suppose the two devices both need to classify traffic, and the policy uses Differentiated Services Code Points to specify how the device is supposed to classify traffic. If one of the devices is not capable of handling Differentiated Services, then neither device will be able to read the policy and perform the classification portion of the policy (which may be part of the condition or action clauses). Cases like this can be mitigated somewhat using the policy continuum. Deciding that Differentiated Services Code Points should be used is a technology-specific feature. This means that there are two levels above the continuum (the business and system levels, to be exact) that each contain a higher-level specification of this classification function. Thus, a sophisticated PBNM system could refer to the original policy continuum to retrieve a higher-level specification of the classification function. (Note that in this example, it is very important to ensure that the values of the Differentiated Services Code Points map to the (more limited) values of the other mechanism (e.g., ToS). Otherwise, the devices cannot be guaranteed of performing the same actions.)

Accommodating Device Limitations

While different capabilities make it difficult to ensure that different devices are performing the same function, often devices have a set of restrictions or limitations that must be accounted for when a policy is translated. Two common examples are rule depth and condition complexity.

Rule depth manifests itself by limiting the number of rules that can be loaded or executed at a given time. There may be many reasons for this, such as a lack of memory or processing power. The problem lies in the complexity of the environment that the PBNM system component is operating in. Often, complex environments require many different rules to be executing concurrently. The only real solution to this problem is to try and reformulate or combine rules so that the limitations of the device do not adversely affect the deployment of the policy rules.

Condition complexity refers to how many different condition expressions are in a single condition clause, as well as how many different types of operators are used to combine different expressions into a single clause. Both of these factors can restrict the complexity of the condition that can be implemented by a device. Sometimes, conditions can be simplified by converting to a different representation. For example, in Boolean logic, there are some expressions that are very difficult to implement using conjunctive normal form (i.e., a Boolean expression defined as a logical ANDing of [logical] ORed terms), yet fairly simple to implement using disjunctive normal form (i.e., a Boolean expression defined as a logical ORing of [logical] ANDed terms),. Options such as these should be used to see if the condition can be translated into a form that the PBNM system component can handle. Otherwise, the only recourse is for the PBNM system component to inform the rest of the system that it is unable to process that condition or rule.

Accommodating Different Programming Models

Different devices can also have different programming models that make controlling them together very difficult. This takes two very different forms.

The first form is when two different devices use completely different programming models, yet both need to participate in implementing or enforcing a particular policy. The most common example is where one device uses a vendor-specific CLI, while the other device uses SNMP. The obvious problem is that there is no common information model that equates the actions of the vendor-specific CLI commands to the actions of the SNMP commands.

The second form is a variant of the previous problem, in which a device uses one programming model to configure a function and a different programming model to monitor that same function. Without the use of a common information model, it will be very difficult to correlate the same concepts in each programming model.

In either case, the PBNM system must provide a mapping between the commands of these different programming models. It is highly recommended that a single common information model, such as DEN-ng, be used to do this. In fact, DEN-ng has a model wherein CLI, SNMP, TL1, and other management methods can be used as part of the same overall task. These and other problems greatly complicate PBNM system design. This is why policy and information modeling need to be used together to solve these problems.

10.2.4 Abstracting Network Interface Management into Network Service Management

One of the important ideas behind Policy Management is that policies control services. This means that the programming of network device interfaces is of secondary importance—the main objective is to provision a service. This is a fundamentally different approach than currently used in most management systems. Figure 10-1 shows how current network management systems program network

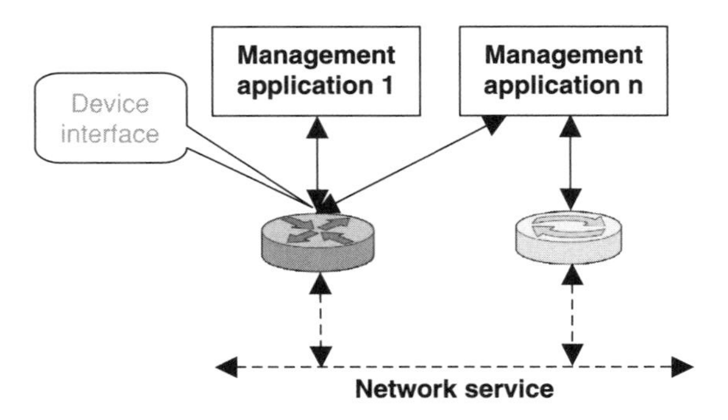

Figure 10-1 Current approaches program device interfaces, not network services.

services. In this approach, a set of n management applications are responsible for programming a set of m device interfaces. Note in particular that a given management application may be responsible for programming multiple device interfaces from multiple devices.

The problem with this approach is that network services are not being programmed. Rather, device interfaces are being programmed. This means that the burden of constructing a network service is placed on the administrator. Specifically, the administrator must know what applications can manage which device interfaces, how to build a part of a service on a device interface, and what role each device interface plays in the construction of a network service.

Figure 10-2 shows how current network management systems program network services. In this approach, the objective is for the PBNM system to program the network service directly. Clearly, device interfaces are still programmed. However, they are no longer the primary focus—the service is instead the primary focus.

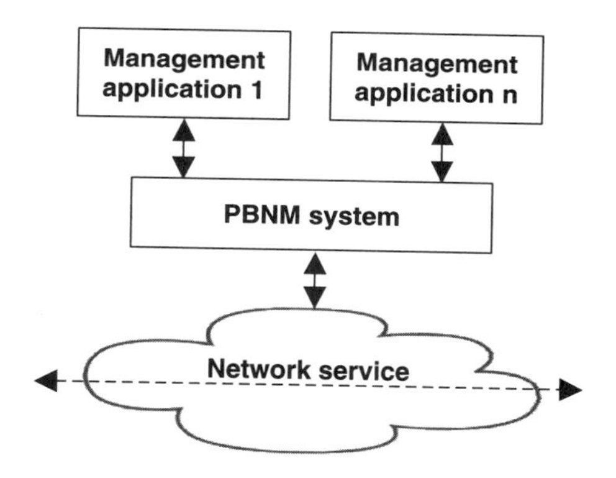

Figure 10-2 The PBNM system to program the network service directly.

In Figure 10-2, the "EML" (Element Management Layer) of traditional TMN[9] systems is still there. The point, however, is that in the PBNM approach, the EML is just another way to instruct the device to do something. This is radically different than the purpose of the EML in traditional TMN systems, which is to control and manage the set of devices. This shift reflects the fact that the PBNM approach views the network as a provider of services, and therefore needs to define the service first and then instruct the device what to do to meet the needs of the service.

In the TMN approach, the EML is the focus because the service is present only in the brains of the designers. There is no "service" when the device is being programmed by CLI through a device interface, because there is no view or definition of the service in the CLI! (This also applies to SNMP, TL1, and other programming models). Rather, there is only a view of what that particular device interface can do.

Thus, in the PBNM approach, the PBNM system as a whole is responsible for programming the network service. It abstracts the task of programming individual device interfaces. Put another way, policy in a PBNM system is used to control a set of Policy Targets that collectively provide a network service. This means that each Policy Target must interpret the Policy (and the Policy Rules that make up the Policy) in the same way.

This enables a different approach to building a network service to be taken compared to the non-policy approach shown in Figure 10-1. Specifically, the PBNM approach can model the semantics of a network service, such as the relationship between the roles played by different objects supplying the network service, directly. This is usually impossible to do using a non-PBNM approach because the methods used (e.g., CLI and SNMP) do not have the concept of a role, or constructs that show directly the dependencies and relationships between different objects that support a service.

10.2.5 Policy Information

As we have seen, there is a large variety of policy information. Each type of policy information may come from a different source (e.g., statistics from an MIB, device capabilities inferred by examining CLI, and so forth). However, unless this information is collected and related to each other, then the policy user cannot be assured of seeing all of the information that may be required to support a given policy rule.

This should not be taken to mean that all such policy information should have to reside in a single information repository. While there are certainly advantages to structuring the PBNM system to operate in that manner, the fact of the matter is that different types of policy information will be generated in different ways, accessible using different protocols from diverse management repositories. Instead, what is needed is to ensure that the PBNM system has a single way of ensuring that different applications can search for and define the set of management information that they need.

One way of accomplishing this goal is to designate a primary repository, such as a directory, to either contain the desired management information directly, or to identify where the management information is located. In the case of a directory, a special datatype, called a labeledURI, is available. This not only defines where to access the data, it defines what protocol to use and optionally some starting arguments.

10.2.6 Administering the PBNM System

There is a dearth of information about administering the PBNM system. Clearly, managing a PBNM system must be simpler than managing the individual components of a PBNM system. Otherwise, there is no real motivation for building the PBNM system in the first place.

However, as with any new system that looks promising, natural human tendency is to try and do too much with the system. Armed with cries such as "scalability" and "consistent enforcement of business rules," people will try and build a PBNM system that does everything to every device for every application. When this occurs, the system will implode and people will blame policy management. Instead, people must be careful to manage their expectations.

Building phased implementations is extremely important for PBNM systems. There are two reasons for this. The first and most obvious one is that a phased approach is always technically more efficient, reusable, and easier to build than the "big bang" approach. The second reason, however, is much more subtle. PBNM represents a different paradigm. For example, it will often prescribe structure and process where there has been none of either. The classic example lies in configuration management. Instead of telling an engineer to go fix a problem and giving that engineer unfettered access to the device, PBNM defines policies and processes that govern both the construction of the configuration change as well as the deployment of the configuration change.

This different way of accomplishing the same task may take a while to be accepted by the engineering community. It serves as a good example of some of the socio-political problems that may be encountered when PBNM systems are introduced.

Paradoxically, one of the greatest benefits in implementing PBNM systems is to realize its potential in implementing business rules to drive the configuration of the network. This is also one of its hardest problems. PBNM systems will be measured on whether they make it easier for the administrator to manage network behavior through a single tool or process, rather than having to use a variety of tools and methods. My own particular worry on this debate is whether PBNM systems are judged fairly or not. Many people like to compare distributed PBNM systems to a single non-PBNM tool or set of tools. This is probably driven by the above problem of people viewing a PBNM system as a panacea. If these are the only elements being compared, then PBNM systems will surely lose. The problem in this comparison is more than the obvious difference in architectures and functionality—the real problem is that the changes in the underlying methodology have been ignored. Most every non-PBNM tool that is currently used today requires communicating with individual network elements multiple times. There is a pronounced lack of integration and coordination that is achievable in this approach. Usually, many of the non-PBNM tools cannot exchange and share data with each other at all. This approach requires the administrator to be knowledgeable not just in how to run different tools, but in the underlying capabilities and functionality present in the network elements themselves.

Worse, this approach requires similar functions to be implemented multiple times in different devices in order to derive or represent what should be common information that is freely shared among all participating applications. This is symptomatic of a larger problem: most every non-PBNM tool currently used today operates in its own stovepipe environment.

The PBNM system can in fact significantly reduce administration costs through its use of the shared information model and its ability to abstract the functionality of the network without requiring detailed knowledge of each element in the network. To best see this, compare the old way of managing the network, as shown in Figure 10-3, to the new way which uses PBNM system, shown in Figure 10-4.

As can be seen in Figure 10-3, without PBNM, each application is responsible for managing in an individual manner the set of network elements and interfaces that it can manage. This places a large amount of responsibility on the administrator, as that administrator must know how each application works and controls its set of network elements and/or interfaces of a network element. Since each device has its own programming language, this approach is analogous to a

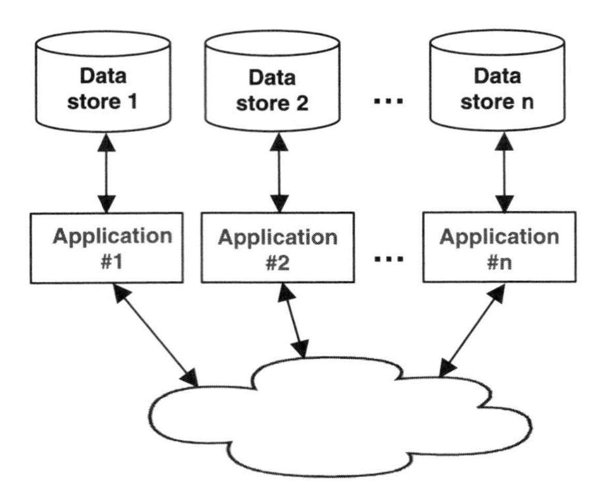

Figure 10-3 Current network management (Without PBNM).

person learning not just different languages, but different dialects of different languages (where the different dialects correspond to different versions of the device operating system).

Also note that this approach encourages the building of application-specific data repositories. Common data is ignored; instead, applications concentrate on building tailored data stores that efficiently store, manage, and retrieve just the data that they need.

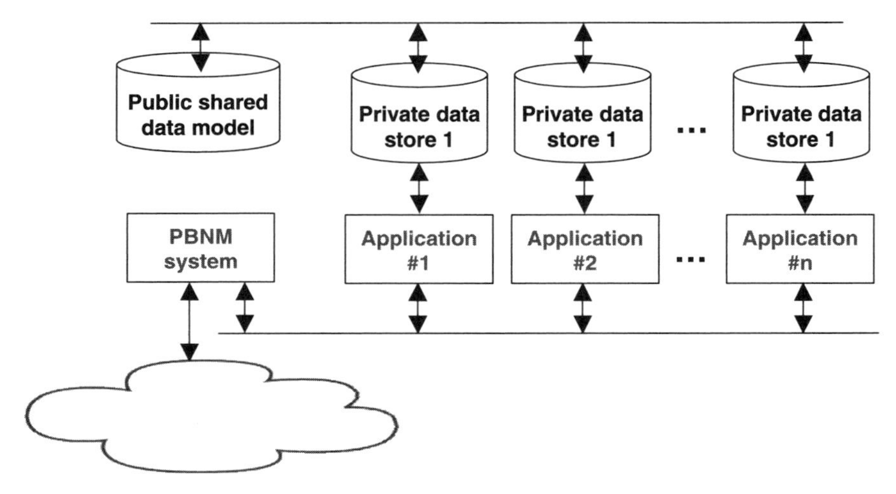

Figure 10-4 PBNM network management.

Figure 10-4 shows the PBNM approach. In this approach, we see two important differences. First, management information has been divided into public and private data. While this in itself is not new, what is new is how the public and private data are used. In a PBNM solution, any data that is exchanged between different components *must* be defined as public data. This ensures that the data can in fact be exchanged, since both components "know" about the data. Furthermore, any such common data is defined to be part of a public shared information model. The second difference is that the PBNM system itself is used to abstract away the differences between the different programming models of the network devices being managed. Now, the administrator only has to learn a single management application instead of multiple ones. This benefit is enabled by the use of a public shared information model.

10.3 QoS Policy Usage Examples

This section will use the results of the previous section to provide three different examples of how PBNM systems can be used. Each of these three examples will be concerned with using policies to configure and enforce a particular type of QoS.

A common thread that emerges from these three application examples is that policy-based management does not solve the entire problem by itself. Just because a PBNM system is being used does not mean that people no longer have to understand how a network works! Rather, the point of policy-based management is to make life easier for the administrator.

For example, suppose that a network is shared by multiple departments. If it happens to be the end of the month, the finance department may need prioritized use of the network to ensure that its large (and very important) monthly reports are delivered on time. Without a PBNM system, the best that can be hoped for is for the network administrator to either ask other network users to limit usage of the network during these times, or for the network administrator to manually prioritize the traffic corresponding to the finance application. The former is very difficult, not just because human nature seeks empowerment and does not like being told what to do, but more importantly because it is very hard for a given user to understand what impact (if any) the application that they are running will have on the finance application. The latter is manually intensive and therefore error-prone. Thus, PBNM is a means for the network administrator to pro-actively manage the network, as opposed to reacting to how users are using the network.

10.3.1 Providing Differentiated Services for Applications Using a Shared Network

The objective is to enable different applications to get different services levels from a common network, according to some prioritized usage defined by policy. We are interested in providing this preferential access because without predictable response for different types of applications, it is impossible to offer value-added services that people will buy. We therefore have at a minimum two distinct uses of policies:

- Policy to define what application gets what type of service
- Policy to define how the traffic belonging to a particular application is conditioned

Conceptually, what we want to do is to classify which applications should get priority treatment. We will define priority treatment as the ability to gain preferred access to shared network resources. Assuming that we have many different applications (say, 20), we decide that we will aggregate traffic into three different classes: Gold, Silver, and Bronze. Our traffic will be assigned to the "Gold" class, meaning that it will always get preferential use of shared resources.

One way of satisfying these goals is to detect the traffic emitted by various applications, and then mark those packets according to a scheme wherein the traffic of each application is placed into a particular class of service. Traffic from each class of service gets appropriately conditioned so as to provide a level of preferred access to shared resources. This is illustrated in simplified form in Figure 10-5.

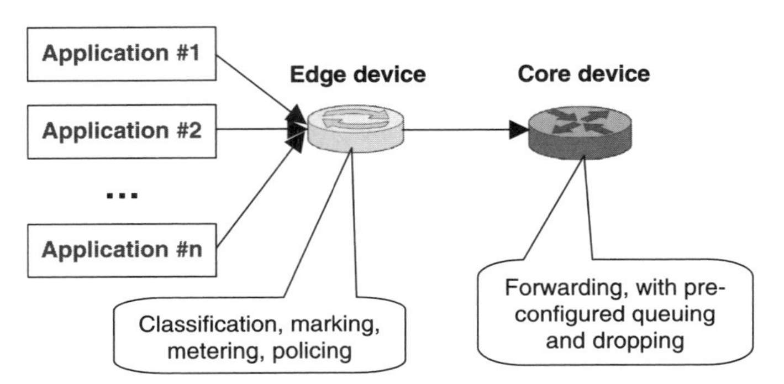

Figure 10-5 Basic traffic conditioning functions.

Note that all of the different traffic conditioning functions are not shown in Figure 10-5 for the sake of simplicity. The point of this basic approach is that at the edge, we apply complex conditioning functions to classify traffic into different groups, or aggregates. We may do some basic policing and other functions to limit the rate and volume of traffic to conform to the capabilities in the core of our network, but this is done so as to make the job of classification more effective. Once the traffic hits the core of the network, the job of the routers is to forward traffic as fast as possible. They may apply queuing and dropping functions, as specified by the packet markings that they receive, to adjust traffic as necessary so as to satisfy class of service parameters. However, their main goal is to forward traffic. Hence, we can use the simplified representation shown in Figure 10-5.

This idyllic situation is complicated by the fact that some applications generate more than one type of traffic. For example, H.323 traffic generates many types of different TCP and UDP traffic flows. Database applications routinely open ports on demand that do not conform to a pre-defined set of port numbers. These and other factors make it very difficult to efficiently classify a particular flow as belonging to a given application.

Clearly, it is wasteful to classify all of those flows with the same marking. However, it is important that different flows receive the conditioning that they need in order for the overall application to have the desired end-user experience. Thus, we need some intelligence in the PBNM system that knows that different flows corresponding to a single application can have different markings, as long as the overall traffic conditioning for that particular application meets the objectives of its class of service.

This approach to providing differentiated services is what the DiffServ working group[1] set out to accomplish. DiffServ describes a set of QoS capabilities that can be delivered from one end of the network to another.

DiffServ Approach

DiffServ is an approach that instructs the network to deliver a particular kind of service based on the QoS marking specified by each packet. This specification is made using the 6-bit DiffServ Code Point (DSCP) setting in IP packets or source and destination addresses. The DSCP is a tag that instructs each device in the network to perform appropriate traffic conditioning functions, such as metering, shaping, dropping and queuing.

DiffServ defines four Per-Hop Behaviors, or PHBs. A PHB is defined as the externally observable forwarding behavior applied at a DiffServ-compliant node to a DiffServ Behavior Aggregate (remember, packets from multiple sources or applications can belong to the same behavior aggregate). The four standard PHBs are:

- Default PHB[10]
- Class-Selector PHB[10]
- Assured Forwarding (AF) PHB[11]
- Expedited Forwarding (EF) PHB[12,13]

The default PHB specifies that a packet marked with this DSCP will receive the traditional best-effort service from a DiffServ-compliant node. Packets arriving at a DS-compliant node whose DSCP value is not mapped to any other PHB will also get mapped to the default PHB.

The Class-Selector PHB is a set of DSCPs intended to preserve backward-compatibility with any IP Precedence scheme currently in use on the network. These Class-Selector PHBs retain most of the forwarding behavior as nodes that implement IP Precedence-based classification and forwarding. Class-Selector PHBs ensure that DS-compliant nodes can coexist with IP Precedence-based nodes. This set of PHBs enable routers that are compliant with DiffServ to be used with routers that are not compliant with DiffServ but are compliant with ToS. (Note that from the DEN-ng point-of-view, this is NOT a complete mapping. While this does identify the basic functionality that needs to be supported by both types of routers, it does not define the complete behavior and semantics of each router. For example, it does not take into account side effects, such as resources consumed, of a router. Often, the functionality of a router is diminished when an advanced type of queuing is being run, due to the increased resources need to support the advanced queuing function. DEN-ng seeks to model these and other features as *capabilities* that are either available or not available; any dependencies, as well as pre- and post-conditions that arise from using that particular capability, are also modelled.)

The Assured Forwarding PHB is nearly equivalent to Controlled Load Service available in the Integrated Services model and is used to define a method by which different behavior aggregates can be given different forwarding assurances. This is usually done by first defining the type of buffering that will be used and then allocating a percentage of the available bandwidth per class. Since different applications have different characteristics, different buffering and queuing strategies are employed. For example, one would not use class-based weighted fair queuing for voice because it is too fair (meaning that it will not properly discriminate voice traffic and attempt to fairly interleave voice packets with other application packets) and will not be able to deliver the latency and jitter guarantees that voice requires.

As stated previously, the purpose of defining a behavior aggregate is to group different applications together that have the same general characteristics. The AF specification defines three drop precedence values that you can use (within each class) to further differentiate between different types of applications. If the applications are all TCP-based, then dropping a packet tells the application to slow down its transmission rate. In effect, this penalizes traffic flows within a particular behavior aggregate that exceed their assigned bandwidth.

Finally, the EF PHB can be used to provide a Guaranteed Bandwidth Service. Applications such as Voice over IP (VoIP), video, and online trading programs require this kind of service. The EF PHB specifies a low loss, low latency, low jitter, and assured bandwidth service.

Figure 10-6 shows a simplified view of the DEN-ng model to represent these concepts.

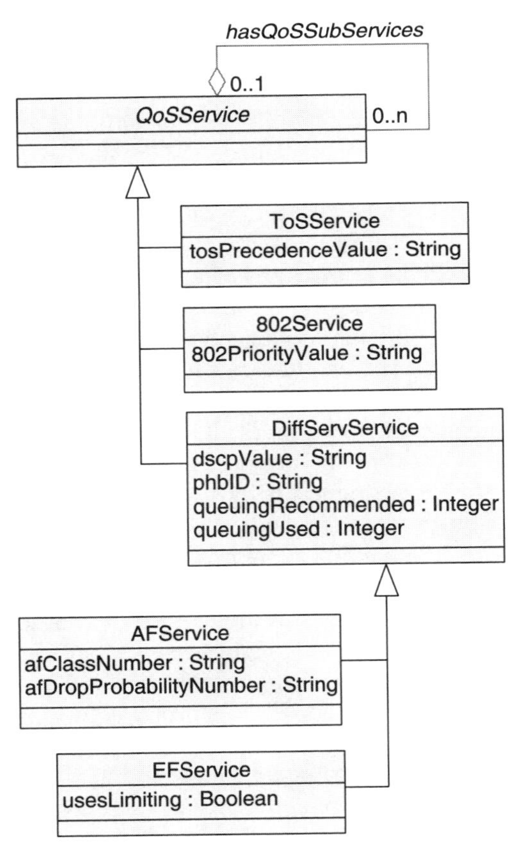

Figure 10-6 Simplified DEN-ng DiffServ model.

Using DEN-ng to Help Implement DiffServ

Although the different types of services are specified by DiffServ, the specific implementation (e.g., queuing and dropping algorithms) for each service is NOT specified. The EF and AF specifications "hint" that class-based weighted fair queuing and priority queuing, respectively, should be used, but neither mandates the use of a particular type of algorithm. This is where the information model can help standardize the implementation, of these types of services. The idea is to define a template that specifies how a particular PHB will be implemented, and attempt to use this template with every instance of that particular service. If a particular device *foo* cannot support the exact features defined by this template, then the DEN-ng concept of *capabilities* can be used to map between the features that *foo* has and the desired features that are specified in the template.

For example, assume that there are different types of routers available in the network. Some are DiffServ-compliant, while others use different means (e.g., the ToS byte in the IP header) to recognize different types of traffic. Furthermore, assume that these different routers have different capabilities.

In this particular situation, DiffServ gets us part way to a solution, but it cannot specify the complete solution. This is because it only specifies the marking to be used to identify a PHB—it doesn't specify *how to implement* the PHB. Furthermore, it cannot be used to specify the implementation of how different device capabilities are mapped to each other. Finally, as pointed out above, it cannot be used to define what capabilities are no longer available when a given capability is used.

Thus, we arrive at the motivation for using an information model, such as DEN-ng, to represent DiffServ and other capabilities of network devices in a common representation. The DEN-ng information model was developed to facilitate normalization of functions (so that mapping between functions could be more easily accomplished) and to define dependencies and inter-relationships between functions using the notion of capabilities.

DEN-ng provides the ability to normalize different functions, and map them to device-specific implementations. For example, the information model can define "AF11" (which is a particular DSCP) as being implemented with class-based weighted fair queuing, along with a particular set of values for weighted random early detection (a type of dropping algorithm). This corresponds to the "template" definition of this DSCP. If a particular device doesn't have class-based weighted fair queuing, but does have other types of queuing mechanisms, then the information model can identify these other types of queuing mechanisms and organize them under a common ontology. The PBNM system could then determine the set of queuing mechanisms that a given device has, and choose the appropriate one to use to implement AF11 for that device. Clearly, this method can be used for any desired function, not just queuing.

Since DEN-ng has the concept of a capability, it can relate any feature or function of a managed element being modeled to a capability of that managed element. This link enables relationships between one feature and other features to be established, which lets us define dependencies between these features. For example, capabilities can be used to specify the particular type of traffic conditioning that would be performed for a given type of traffic. Note that this doesn't replace DiffServ—rather, it enhances the use of DiffServ. A simplified example was shown in Chapter 9 and is repeated here as Figure 10-7 for simplicity.

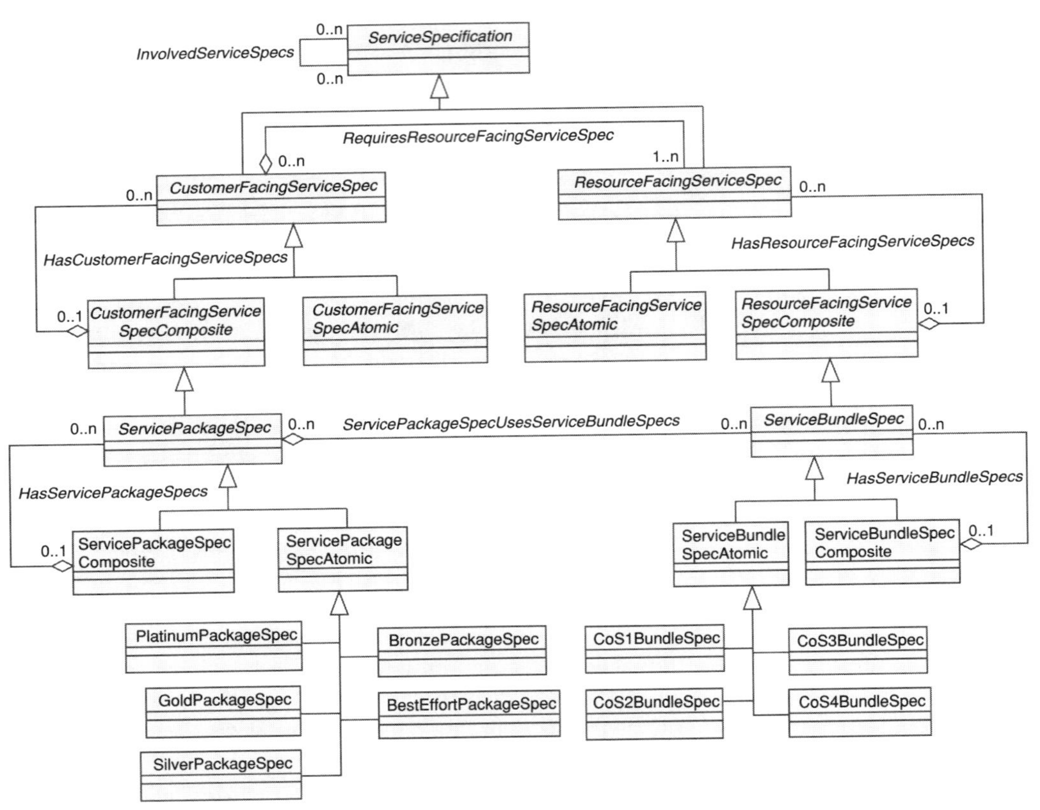

Figure 10-7 Information model for QoS service implementation.

This figure represents a (very) simplified DEN-ng model for representing the relationships between QoS and other *ResourceFacingServices* and how they may be offered as a *ServiceBundle* to a *Customer*. This model works in the following example.

In DEN-ng, there are two basic types of Services. *CustomerFacingServices* are services that customers purchase, whereas *ResourceFacingServices* are services that are not purchased directly by the customer but are still required to support a *CustomerFacingService*. This relationship is shown at the top of the hierarchy in Figure 10-6.

For example, consider an MPLS VPN. Customers buy VPNs, so an MPLS VPN is an example of a *CustomerFacingService*. Further assume that this particular type of MPLS VPN is as defined in Rosen et al.[14] This type of VPN mandates the use of BGP for route advertisement. However, Service Providers do not sell BGP. Thus, we also have the concept of a Service that is related to the *CustomerFacingService* but is itself not sold. DEN-ng defines this as a *ResourceFacingService*. This enables the various supporting services that are needed to model a particular *CustomerFacingService* to be modeled (and, more importantly, to have their relationships established to the *CustomerFacingService*), while keeping the semantics between a *CustomerFacingService* and a *ResourceFacingService* separated.

In this example, the Service Provider offers three different service packages—Gold, Silver, and Bronze—which may represent the grouping of services. This is illustrated in Figure 10-8.

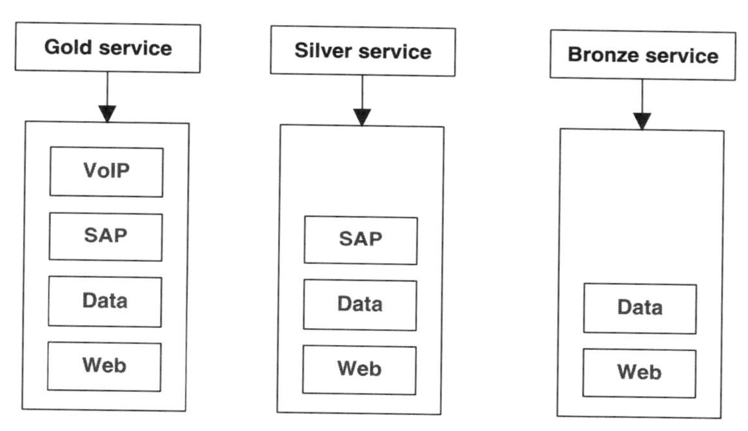

Figure 10-8 Bundling of *CustomerFacingServices*.

For example, by grouping the same services into Bronze, Silver, and Gold, the Service Provider can represent better performance for the Gold Services than the Silver Services. Referring to Figure 10-8, we see that all three services are given Data services. The difference is that the quality of Data Service is best in Gold Service and worst in Bronze Service.

These different packages may also be used to represent access to a Service not available at a lower level. For example, as represented in Figure 10-8, VoIP may be available only if the customer purchases Gold Service.

The concept of a *ServiceBundle* enables the Service Provider to group a set of related Services together as a package. Different *ServiceBundles* enable a Service Provider to model the concept of "upgrading" to a higher performance Service Bundle, in order to access new or better performing Services.

The discussion so far has defined the set of classes necessary to build a class structure to represent different types of *Services*. The next part of the discussion will examine how to represent QoS.

The next step in the process is to realize that each Service in the *ServiceBundle* needs its own set of *QoSServices*. A set of *QoSServices* are needed because different devices with different capabilities are used as a set to provide a Service. In this example, we had two different types of routers—one that used the ToS byte and one that used DSCPs—to indicate how to condition traffic. Therefore, a *Service* in a *ServiceBundle* may need both *DiffServServices* as well as *ToSServices* (where *DiffServService* and *ToSService* represent a set of classification, marking, and traffic conditioning services based on using DSCPs and ToS settings, respectively). The job of the PBNM system is to recognize this and provide a means for all of these different services to be represented.

This process is formally defined in the DEN-ng model as follows. A *QoSService* is a subclass of *ResourceFacingService*, which can be related to a *CustomerFacingService* through the *CFServiceRequiresRFServices* aggregation. This aggregation is used to define the set of *ResourceFacingServices* that are required to support a particular *CustomerFacingService*. Thus, we have the ability to define which set of *QoSServices* are required by any particular *CustomerFacingService*.

This is a powerful concept. For example, it enables a set of *CustomerFacing Services* to all use the same *QoSService*, thereby ensuring that their side effects will be very similar. In this example, the different *CustomerFacingServices* can still be distinguished from each other through the use of classification and marking, and traffic conditioning can similarly be assigned different values and hence different effect. However, the power of the DEN-ng design is that it *masks* these details from the business analyst. Thus, all the business analyst needs to be concerned with is which customer gets which level of QoS, not with how each level of QoS is implemented.

Alternatively, different *CustomerFacingServices* can use different *QoS Services*. The difference in the *CustomerFacingServices* is now explicit. Note, however, that the same fundamental services—classification, marking, and traffic conditioning—are provided in either case. This gives the network designer maximum flexibility in defining the set of network services that a particular Customer receives.

We need some way to map between DSCPs and ToS settings. This can be done by using the information model to first define such a mapping and second, to relate instances of the *DiffServService* class to instances of the *ToSService* class. By ensuring that both *DiffServServices* and *ToSServices* are defined as *QoSServices*, they are identical from an abstract point-of-view. This is important because in each case they use a similar set of services that are keyed off of a marking in the IP header of the packet. The mark of a good information model is the ability to represent concepts that occur in the environment being modeled in a simple, natural means. This abstraction is the first step in doing this.

We can define a mapping between values of a ToS byte and DSCP values by instantiating these two classes and writing simple OCL expressions to relate the eight different ToS settings to eight corresponding DSCP values. This by itself is important, but we need to define what these values mean. The problem that we face is that there is no standard that defines which specific services are required to implement the desired traffic conditioning. Thus, we represent this as two additional types of *ResourceFacingService*—*TrafficConditioningService* and *TrafficIdentificationService* classes.

TrafficIdentificationService and *TrafficConditioningService* abstract the fundamental processes of classifying and marking packets and defining how those packets are to be treated internally by the routers. However, we need a way to describe the different mechanisms that can be used in implementing *DiffServServices* and *ToSServices*. Therefore, DEN-ng defined the notion of a *NetworkForwarding-Service*. This is the base class from which all functions that operate on network traffic in the forwarding path of a network device derive from.

A previous attempt at modeling these features was built by the IETF and later modified by the DMTF. In simplified form, it looks as shown in Figure 10-9:

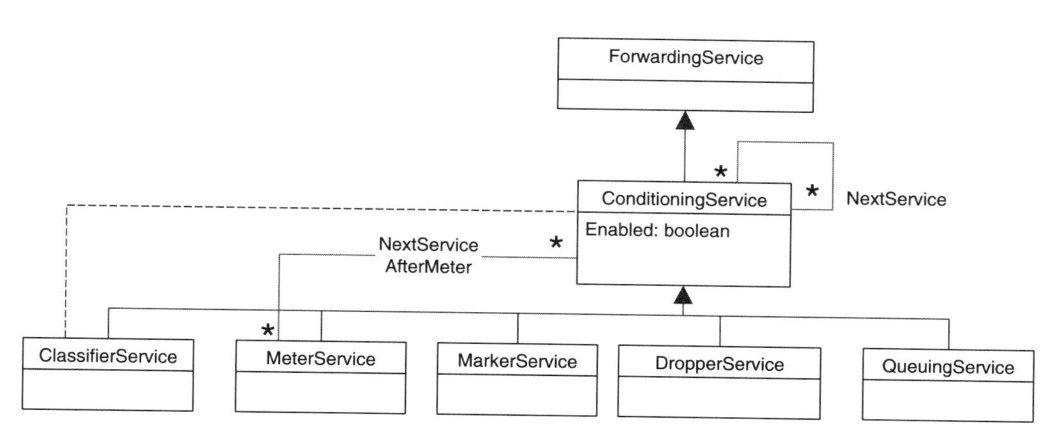

Figure 10-9 Simplified CIM representation of traffic conditioning and classification services.

There are several problems with this model. First, classification and marking are mistakenly categorized as *ConditioningServices*. These two services have nothing to do with traffic conditioning. Furthermore, it does not differentiate between a metering service (also called a policer), which limits the traffic to a specified maximum rate, versus a shaper service, which delays transmission (by buffering the traffic) so that the total resulting traffic is limited to a particular value. The implementation of a policer versus a shaper has different architectural elements and different commands and therefore should be separated in the model.

In the DEN-ng model, neither the *ClassifierService* nor the *MarkerService* are *TrafficConditioningServices*. Rather, they are generalized services that perform the function of further identifying network traffic, through classification and packet marking. This was the motivation for calling them *TrafficIdentification Services* in the DEN-ng model. Note also that the two associations *NextServiceAfterMeter* and *NextServiceAfterClassifierElement* technically go to a subclass of *ClassifierService*, but this has been simplified in the figure.

This enables the *TrafficConditioningService* to focus on representing specific mechanisms inside the router that can be programmed by the PBNM system to supply different types of traffic conditioning. The key to specifying which traffic conditioning mechanisms are to be used is by marking the packets appropriately; this is represented by the *MarkerService* in the DEN-ng model of Figure 10-6. In order to mark the packets, the traffic must first be classified into different flows that may in turn be aggregated together, so that the same traffic conditioning can be applied to them. This is the purpose of the *ClassifierService*. A more complete portion of the DEN-ng QoS model that shows these points is shown below.

By defining a *TrafficConditioningService* class and a *TrafficIdentificationService* class as defined in Figure 10-10, we can relate one or more of each of these two types of *Services* to a *DiffServService* as well as a *ToSService*. To be exact, we can define a particular set of *TrafficConditioningServices* and *TrafficIdentificationServices* to be used to implement a *DiffServService*, and either the same or a different set of *TrafficConditioningServices* and *TrafficIdentificationServices* to be used to implement a *ToSService* (or any other QoSService, for that matter). Furthermore, the types as well as the values of *TrafficConditioningServices* and *TrafficIdentificationServices* can be different for the *DiffServService* compared to the *ToSService* (or any other QoSService, for that matter). For example, the *ToSService* could use a single *MeterService* with two attribute values, whereas the *DiffServService* might need two or more cascaded meters to provide multiple metering levels, each with its own set of attribute values that are different than those used for the *ToSService*.

Figure 10-10 shows a simplified view of the DEN-ng top-level QoS model. Attributes and other detail have been supressed to simplify the figure. There are four important points to make in this figure. First, *CustomerFacingServices* aggregate *ResourceFacingServices* using the *CFServiceRequiresRFServices* aggregation. Second, *ServicePackage* aggregates *ServiceBundle* using the *ServicePackageUsesServiceBundles* aggregation. Third, the (business) attributes for the QoSService subclasses are shown to emphasize their different natures. However, despite these differences, they can each define their own *set of*

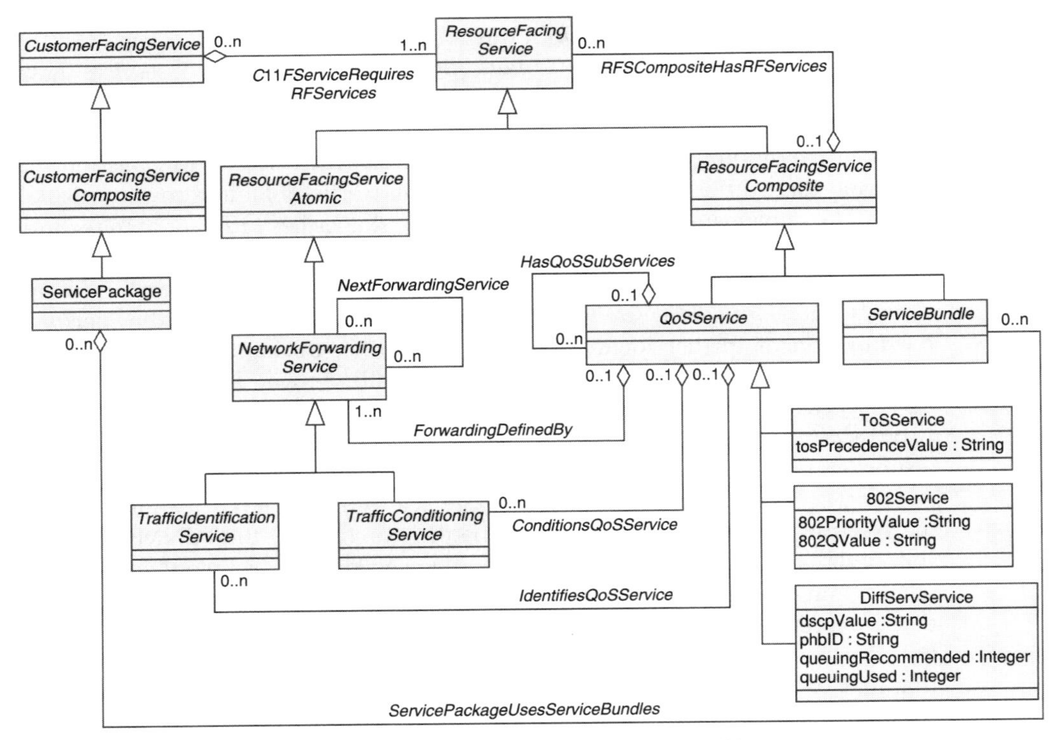

Figure 10-10 DEN-ng representation of QoSServices.

NetworkForwardingServices using the *ForwardingDefinedBy* aggregation. Finally, a *ServiceBundle* aggregates QoSServices using the *RFSCompositeHas-RFServices* aggregation.

In spite of these differences, however, the important point about this design is that the *TosService* and the *DiffServService* are related to each other, since they come from the same definition (*QoSService*). This happens *naturally* through the definition of the DEN-ng model.

The *identifiesQoSService* aggregation is used to define the set of *ClassifierServices* and *MarkerServices* that are needed to identify a particular *QoSService*. Note that this is an aggregation, because sophisticated traffic classification may require more than one *ClassifierService*. For example, consider Figure 10-11.

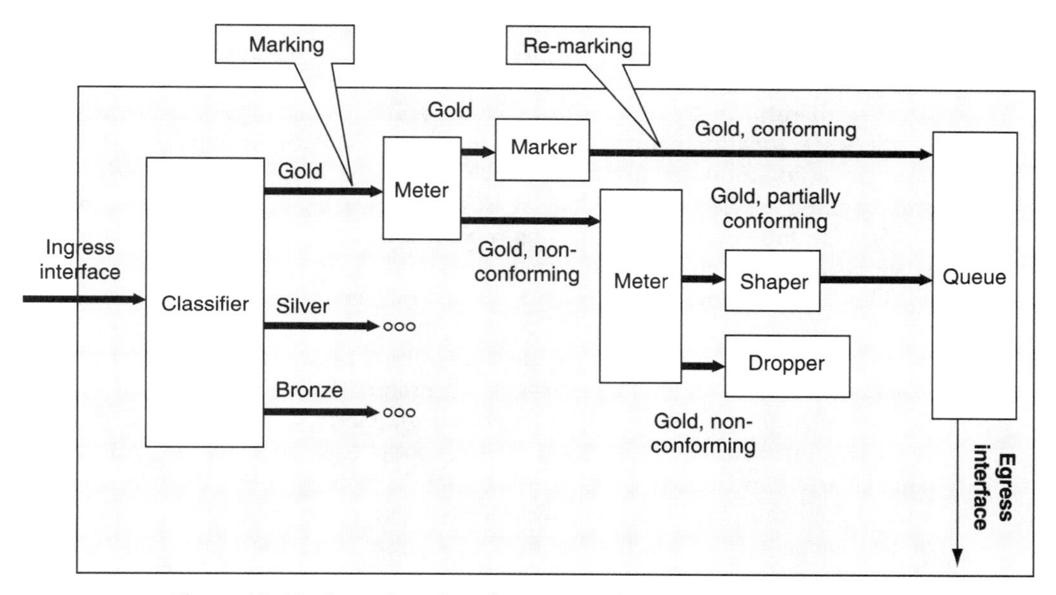

Figure 10-11 Complex classification and marking in the forwarding path.

Figure 10-11 shows Gold traffic being split into three types of sub-traffic flows—flows that are completely, partially, and nonconforming to the basic traffic profile defined for Gold Service. (Remember, this is a conceptual realization of the different functions provided inside the router). This can be implemented in several ways. The simple method in Figure 10-11 shows two cascaded meters being used to separate the traffic into three types of flows. We could also have used two sets of classifiers.

This figure also shows the close bond between *ClassifierServices* and *Marker Services*. Here, we have used two different *MarkerServices*, at different points in the traffic conditioning process, in order to remark packets after some preliminary service (such as metering) has been performed. Note that both are conceptually part of identifying traffic and in so doing often define which particular *TrafficConditioningService* is required.

The *ConditionsQoSService* is used to specify the set of *TrafficConditioning Services* that are required by a particular *QoSService*. This enables different traffic conditioning services to be specified for a particular *QoSService*. Alternatively, as stated above, it is possible that a particular implementation might want the exact same type of *TrafficConditioningServices* for two different *QoSServices*. For example, if the traffic consists of voice and data, then regardless of whether we are using a *DiffServService* or a *ToSService*, we still want to use a form of priority queuing for voice and a form of class-based weighted fair queuing for data (please note that these subclasses exist in the DEN-ng model, but are not shown to make the diagram easier to read). In this particular example, what is important is to specify an appropriate scheduling algorithm, which controls how these different queues are combined and serviced by an output interface of the device. This is present in the DEN-ng QoS model but is not shown for the sake of simplicity.

Note that the model shown in Figure 10-10 enables another form of differentiation. In the last example, specifying advanced traffic conditioning functions using DSCPs is more flexible than it is using a set of ToS markings. This is because there are 64 DSCP values versus 8 ToS values, and also because the DiffServ specification defines dropping as part of the Per-Hop Behavior. This can be easily accommodated in the DEN-ng QoS model by simply relating dropping and queuing services to a DSCP, versus only relating a queuing service to a *ToSService*. However, in the DEN-ng model, this is not done, because ToS-based services will require *DroppingServices* as well.

The DEN-ng model represents these different mechanisms (and others not shown, such as a scheduler) as a set of building blocks that can be combined in different ways, according to the capabilities of the network device. This generic binding of different services is represented by the *NextService* association, which enables any *NetworkForwardingService* to follow any other *NetworkForwardingService*. The DEN-ng model uses OCL to constrain which mechanisms can be first or last in the interface, as well as which can follow which. For example, it would be technically incorrect to make a dropper the first or last element (e.g., the first or last mechanism after the input interface and before the output interface, respectively) in the forwarding path. Similarly, following a dropper by a dropper doesn't make sense, whereas following a meter with another meter does indeed make sense. Note that this level of specificity does not exist in the IETF and CIM models.

Three problems would surface if this and other dependencies were modeled in detail. First, the model would become very complicated and therefore much harder to understand. Second, the more explicit the model becomes at a granular level, the more likely it is that our generic model will conflict with how different vendors are implementing these device-specific mechanisms. Third, if this behavior isn't specified, then the degree of control that can be exerted over the device is less.

If OCL isn't available, then the alternative is to either not specify this at all or to build a set of different associations to model individual behavior. If we want the model to be UML compliant, then we must remember that the different associations are not necessarily subclasses. This is because in UML there is a distinct difference between associations and association classes.[15] This is a problem with the DMTF CIM model in general—it is not UML compliant[16] and represents all associations as association classes. The DEN-ng model can represent these subtleties because it is derived from the UML metamodel and is therefore UML compliant.

In summary, the DEN-ng approach is to model a generic capability in a simple, extensible way, and then constrain it appropriately. Thus, the basic DEN-ng QoS model uses the *NextService* association as shown in Figure 10-10 and offers two ways to constrain it. The preferred way is to provide OCL expressions that restrict how the *NextService* association is applied. The DEN-ng specification also discusses how to subclass this model to explicitly model more detailed, and restrictive, combinations of *NetworkForwardingServices*. This latter may be appropriate if the capabilities of specific vendor devices are being modeled.

Putting It All Together

We have had a brief glimpse of DiffServ- and ToS-based services, and now is the time to see how policy is applied to implement them.

The basic approach is to integrate the policy and QoS models into one single information model. This is done is using the Policy Continuum. Recall that we have throughout this book defined five different levels in the Policy Continuum—business, system, network, device, and instance levels. The (generic) QoS model fits in at the system and network levels, and partially at the device level. The Policy Model, of course, is applicable to all levels of the Policy Continuum. The difference is because the Policy Model is itself a federated model, consisting in reality of five different models. Each of these five models is focused on a particular level of the Policy Continuum. In contrast, the QoS model is concerned with modeling QoS in a generic fashion, and thus concentrates on a subset of the Policy Continuum.

The business view is concerned with ensuring that the applicable business rules and processes are used to *direct* how the PBNM system is to be managed. It does this in terms of business entities, such as *Customer*, *Service*, and *Product*. Thus, the business rule "*John gets Gold Service*" can be expressed in terms of a *Customer* (John) who has bought a *Product* that provides a *Service* (Gold Service) for one or more applications that he is using.

The system view expands on this view, detailing the composition of key objects that are to be managed. For example, if the customer has contracted for VPN-based connectivity, *how* will that connectivity be provided? At the business level, it is a "blob," but at the System level, the "blob" has been transformed into a specific type of VPN (e.g., an MPLS VPN using BGP to advertise routes and OSPF to connect the different Provider Routers together).

Note the importance of the use of a single information model. This enables the same concepts to be shared among the business and system views, even though they are at two different levels of abstraction.

The network view further specifies how to implement the Service by defining the set of technologies that will be used to implement the Service. However, this is done in a device-independent way: it may define the use of RSVP to reserve bandwidth, but this definition is independent of how any particular vendor implements support for RSVP. Thus, the RSVP model will be limited to representing only those features and functions that are present in the appropriate standards or commonly implemented by multiple vendors.

The device view binds this approach to specific device features. This is the level wherein different devices are chosen to support the VPN service, and so differences in support and implementation of standards like RSVP become important. This level also represents a melding of the standards-based DEN-ng model and extensions to this model that represent how specific vendors implement features defined in the standard. For example, DEN-ng defines a QueueService and even defines the basic characteristics of several popular queuing algorithms (e.g., class-based weighted fair queuing). However, it only defines those features that are described in the standards, or are otherwise commonly implemented by multiple vendors. Thus, in order to bind this specification to the specific vendor devices that are implementing class-based weighted fair queuing, it is necessary to extend the DEN-ng model to represent the specific vendor implementation.

(As an aside, note that there are many different implementations of class-based weighted fair queuing, even from the same vendor. This is because in order to perform functions such as class-based weighted fair queuing at line rate, the algorithm must be implemented using ASICs (application-specific integrated circuits.) Different Products use different ASICs, even if they are manufactured by the same vendor. For example, this commonly occurs when one company acquires another company. Hence, even a single vendor can have significantly different algorithm implementations for different products.)

Finally, we need to *program* the device. Each device has its own programming model. If the device view can describe extensions to the DEN-ng framework that represent the capabilities of the device that can be used, then the instance view can in turn translate this model into the appropriate vendor-specific CLI commands that are needed. The instance and device views work hand in hand. The instance view models vendor- and device-specific commands, while the device view translates these vendor-specific features into extensions of a common framework. We therefore can represent vendor-specific features as part of a common set of capabilities, which enables us to map different vendor-specific features to each other. *The effect of this is to be able to view the system in a standard way while being able to program the system using different vendor-specific programming models.*

Thus, we see that networking, customer, and other objects come into and go out of focus, according to what view is being used. However, policies are always present, regardless of what view is being used. In order to see this, let's continue the development of our example.

In the example "*John gets Gold Service,*" several underlying business policies come into play, including:

- Why is John entitled to get Gold Service?
- What does Gold Service provide John?
- How will the PBNM System ensure that John gets Gold Service?
- What will the PBNM System do if John doesn't get Gold Service?

Business rules are used to define what users receive which types of Services, as well as what applications can be accessed within a particular Service Offering.

Different classes of service are often used to provide better than best effort service to either a specific set of users and/or a specific set of applications. This answers the first question.

To answer the second question, first recall Figure 10-7, which illustrated the use of classes of service to provide access to special applications. However, it also can be used to designate differences between traffic from the same application. For example, consider two flows from the same application. One belongs to a Gold Service, and the other belongs to a Bronze Service (perhaps because the two users have different service contracts). Even though this traffic is from the same application, Gold Service will ensure that the first user gets "better" service than the second user. Here, "better" can mean many things, depending on the particular application that is being used. For example, it may mean faster download times, or provide a application with less latency and jitter.

The fourth question is also easy to answer. Clearly, a policy needs to be used to define the action(s) taken if the system violates its contractual obligation to provide Gold Service. If policy is not used, then each violation will need to be addressed on an individual basis, which cannot scale. As discussed in Chapters 1 and 2, the use of policy is preferred to ensure that a consistent and extensible response can be applied to a large number of conditions. (Remember that the basic DEN-ng policy model, as explained in Chapter 6, is a triplet consisting of event, condition, and action clauses. This strategy enables policies to be defined that describe what actions to take when [for example] an SLA is violated). The structure of the event-condition-action policies facilitates associating the "normal" policies with their counterparts that describe "violations."

The third question is related to how the system is being managed. Assume that the PBNM system uses a model to represent the various components of the system that provides Gold Service. One way to ensure that business rules are used to implement which Users get which Services is to translate the *concepts* present in the business view into a set of *objects* that can be managed. Specifically, if DEN-ng is used, then the management of objects will be done by defining Finite State Machine (FSM). The FSM approach defines how to model the behavior of a managed entity using a set of transitions that identify the current and desired states that a managed entity should be in. The model is used to identify the set of statistics, attributes, and other characteristics of a managed object that define the current state, or characteristics, of that object. For example, we may know that a software object may be in the "installed" state but not in the "deployed" state because the measurement of certain statistics, via the reading of particular object values, confirms that this object is installed but not yet deployed. The object values are, of course, defined in the information model.

This means that a *set* of models needs to be used. This is because, in general, different states in the FSM correspond to different attribute values as well as different sets of objects. In addition, different relationships and constraints may be activated, based on the entity changing states. These changes require different instances of the same model to be used to track the transition between states.

Therefore, a given model may be used by one or more of the five different views in the Policy Continuum, as shown in Figure 10-12. The DEN-ng approach to PBNM uses an FSM in this fashion. As shown in Figure 10-12, each

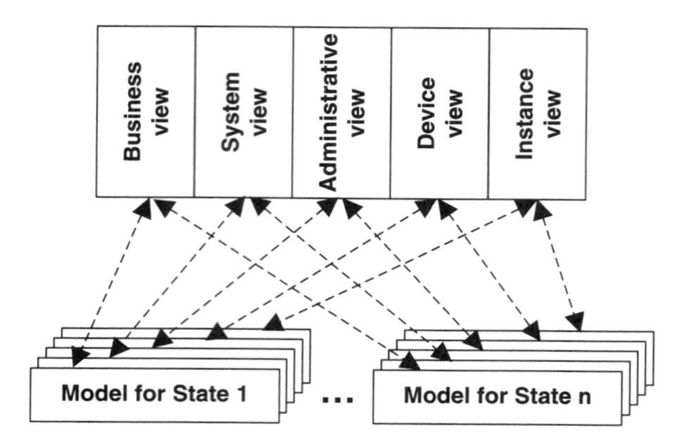

Figure 10-12 The use of individual models to represent FSM states in the policy continuum.

view can make use of one or more models. Each model can represent one or more managed objects. Thus, *as the view of the managed object is changed in the Policy Continuum, the model for that managed object changes appropriately*.

This is a key feature of the DEN-ng model and cannot be over-emphasized. Put another way, the DEN-ng view states that each of the views in the Policy Continuum is equally important. Consequently, each view has its own model. The information model serves to unite each of these different views, and relate the different objects that are present in one view to a potentially different set of objects in another view. Note that this feature is unique to DEN-ng and the SID, and is a by-product of the NGOSS effort.

Some managed objects are not present in a particular view because they are not germane to that view. In fact, the appearance of a particular managed object in a given view is indicative of the role played by a particular managed object in that view. The disappearance of a managed object simply means that it is not relevant or that it has been replaced by another object in that view. This is illustrated in Table 10-1.

The business view focuses on a high-level description of how to manage the environment. At this level, the relationship between Products, Services, and Customers is usually of paramount importance. No details of how the Service is going to be implemented are provided, because they are not important at this level of abstraction. Rather, the business view supplies a consistent view of common business entities from which network services are derived. Policies are used to describe the fundamental business relationships (e.g., why does John get Gold Service, and what happens if this contract is violated) present at this level.

The system view defines the particular approach that will be used and the supporting concepts and objects needed to implement that approach. In this example, it defines the concept of different Service offerings that each use different Classes of Service. It defines the concept of a ResourceFacingService—a service that is not purchased by the Customer, but is nevertheless needed to support the CustomerFacingService. It defines roles and device capabilities and the

Table 10-1 Sample Managed Objects Corresponding to Different Levels of the Policy Continuum

Level in the Policy Continuum	Sample Objects in This Level of the Policy Continuum
BUSINESS VIEW (John gets Gold Service)	Customer (with John as an instance); Product; Service; ServiceOffering; CustomerFacingService; Applications; SLA
SYSTEM VIEW (define multiple Classes of Service)	Business objects plus ResourceFacingService; DeviceRoles; DeviceCapabilities; QoSService
NETWORK VIEW (decide to use DiffServ and ToS; use RSVP for Bandwidth guarantee)	System objects plus define all subclasses of QoSService and NetworkForwardingService needed to build a complete representation of the different QoS services for each CoS
DEVICE VIEW (decide on vendor-specific features; pick specific devices and software releases)	Use models of each vendor-specific device (derived as refinement of above standard model); use Device Capabilities to map between their functional differences
INSTANCE VIEW (write device-specific CLI)	Use above models to generate CLI for each state in the FSM

notions of QoS. All of these objects add additional detail to the model. They also offer new opportunities for policies to be used to control the objects introduced at this level. It may be viewed as a link between pure business policies and device configurations.

At the administrative level, the particular technologies (DiffServ and ToS markings in this example) that will be used are specified. This in turn defines the various objects necessary to manage the implementation of these technologies. It helps define the mapping between these different technologies, although it does not define their configuration. Policies are used to ensure that the technologies are each managed according to the higher-level business policies that they are supporting.

At the device level, it defines the basic parameters of devices and helps choose which OS release and line cards of which devices should be used through the identification of these basic parameters. However, since it does not model device-specific features, it can not be used to specify the configuration of the network devices being used. Policies are used to manage the intricacies of the specific devices that are being used to support the higher-level policies already defined.

The instance level models the CLI that configures each device in each state. Policies are used to manage how device configuration is created and deployed, as well as how different device configuration changes are coordinated with each other.

We can therefore draw a very important conclusion:

Policies ensure that the semantics of different levels of the Policy Continuum are applied to Managed Objects at appropriate levels of the Policy Continuum.

In other words, policies are not just a way to manage a device. Remember that fundamentally, a policy represents goals and a method for achieving those goals.

Therefore, they can also be used to ensure that the semantics at different levels of the Policy Continuum are properly matched up and supported.

More importantly, this approach enables us to define a policy as a way to control the transitioning of a current state to a new state. Here, "control" of a state transition means defining the semantics of the transition—who is in charge of the transition, what the transition is used for, when it occurs, where it occurs, and why it is being done. This uses the DEN-ng concept of archetype—who, what, when, where, and why—to control the state transition. The "what" part of this is typically implemented by using the policy to select one or more processes to implement the state transition. Thus, we have the following important conclusion:

Policies are used to control the state that a Managed Object is in at any given time; the state itself is modeled using an information model such as DEN-ng.

This leads to one last conclusion:

The policy model is used to control which mechanisms to use, when to apply them, and when to change device configurations. Integrating policy and device management is done by integrating the policy model with appropriate models of the device using a Finite State Machine.

For example, the QoS model can be thought of as specifying the changes needed to a current device configuration in order to provide a desired traffic conditioning; given this, the policy model can be thought of as controlling which processes are used to implement the appropriate device configuration changes. The next two sections will examine two very different examples and relate the above conclusions to those examples.

10.3.2 Protected Traffic

This section generalizes the previous example and shows how a set of applications that generate different types of traffic can be protected as a group (compared to the traffic of other applications). Remember from Figure 10-7 that in addition to protecting the traffic as a group, traffic of that particular class of service will be treated better (or worse) than other traffic from that *same application* according to the class of traffic that each flow belongs to.

Assume that our objective is to protect the applications used by a particular group of users. For example, consider the Human Resources (HR) group. They have specific access privileges that restrict other users from accessing their applications and data. Furthermore, depending on the nature of the work, HR operations can take a significant portion of the network's bandwidth. The problem is that HR users are not the only users of the network, so how do we protect HR traffic from other traffic from other users, so that we can be sure that HR users will always get their relative priority of service in the network?

To keep this example simple, we can assume that the HR users are all on their own private subnet. This helps in identifying the HR users. However, these users will most assuredly need to interact with other users in the environment. Thus, we still need to come up with a way to tell the network that when HR traffic is seen, it should be given a relative priority over other traffic types. Assume that this Enterprise has four Classes of Service—Gold, Silver, Bronze, and Best Effort. Gold is reserved for special circumstances, or for special applications like VoIP. We will aggregate different traffic into Silver and Bronze Classes of Service (leaving any other traffic to be aggregated into the "Best Effort" class), as well as sets of applications that all need to be treated collectively better than "normal" traffic. The business policy to express this is very simple:

Rule (1):
Under normal conditions, HR users should get Silver Service.

The System view first translates the above to the following form:

Rule (2):
If
(conditions are normal) AND (traffic is from the HR subnet) OR (traffic going to the HR subnet)
Then
(mark HR Traffic as Silver-Service traffic)
Endif

So far, the translations have been simple rewording of business conditions into a form that can be more easily manipulated. The next step in this process is to translate the business abstractions present in Rule (2) into a form that can be manipulated by network-based PDPs and PEPs. This has the advantage of being able to refer to these common abstractions (e.g., Silver Service and HR Traffic) throughout the system. Note that we are not removing Rule (2)—it is still very useful to business analysts. For example, if another rule involving HR users was later needed, it could be compared to Rule (2) to ensure that it didn't conflict with the requirements of Rule (2). Many times, conflicts at this (high) level are easier to resolve that conflicts at a more detailed level (e.g., the instance view, where CLI must be compared).

Rule (2) is concerned with classifying traffic—nothing more, nothing less. In particular, it is important to realize that Rule (2) does NOT specify the specific traffic conditioning that will be performed on HR traffic—it simply says that HR traffic under normal conditions should get Silver Service, which is defined in terms of a particular QoS level. In fact, note that we haven't even specified yet how Silver Service will be identified to the network.

In order to perform this classification, the environment (not the packet!) is tested to see if conditions are normal or not. If conditions aren't normal, then this rule fails immediately. Presumably, the system is designed so that another rule would test to see if conditions aren't normal, in order to identify what actions to take under abnormal conditions. Alternatively, the fail-action clause of a DEN-ng policy rule could be used. If conditions are normal, then the traffic is tested to determine if it originated from or is destined to go to the HR subnet. If this is true, then the packet will be marked as belonging to Silver Service.

The classification of packets is a networking concept. Thus, it requires the business concepts present in Rule (2) to be translated to a networking form. This is where the network view is used to refine the concepts defined in the system view. In essence, what is needed is to translate terms like "HR subnet" to a form that is meaningful to network device classification mechanisms. In other words, a router has no idea what an "HR subnet" is. However, it can easily recognize an IP address. Thus, if we can translate terms in Rule (2) to an equivalent form that consists of networking terms, our router (or other network device) will be able to evaluate the Policy Rule.

To simplify things, drop the condition for the network to be in a normal state from this point on. The translation of classification and marking is started as shown in Rule (3):

Rule (3):
If
(IPSourceAddress IN 192.213.16.0/24) OR (IPDestinationAddress IN 192.213.16.0/24)
Then
(set DSCP to AF2)
Endif

The abstraction "HR Subnet" has been replaced with a set of addresses in CIDR format[17] that correspond to the set of IP addresses that make up the HR subnet. The marking statement is simplified to just show a mapping to use DiffServ. AF2 stands for Assured Forwarding, class 2.[11] AF defines four classes of service, with lowest being AF1 and highest being AF4. Note that we have not yet specified how the traffic will be conditioned—we have simply specified that it is to be marked AF2. The conditioning specification is discussed later. Note that up to this point, policy has been used strictly to specify what should happen in abstract terms. This will change as we progress to the device and instance views.

(Clearly, Rule (2) must be more complicated to deal with other situations. Rule (2) is being deliberately simplified so that we can concentrate on following the execution path of a particular set of policies for a given type of traffic. Note also that instead of looking for a particular subnet, the test could just as easily have targeted a specific set of applications, machines, or other sources of information.)

At the network level, we have defined the use of DiffServ. This means that we need to choose a particular DSCP. We can also specify different drop precedences for each traffic class. The use of different drop precedences enables different traffic that is aggregated by a single DiffServ class to be further differentiated. Let's assume that our HR subnet will have low drop probability, putting it effectively as the most important traffic in Silver Service. Heinanen et al.[11] recommend a DSCP of 010010, which is AF2 using a low drop probability. Now we know how to mark the packets. In order to do the actual conditioning, we need to understand the capabilities of the different devices that are being used.

Most devices offer a large assortment of different queuing and dropping mechanisms. However, not all dropping mechanisms are compatible with every queuing mechanism. Furthermore, there are often very different restrictions and constraints on the use of these mechanisms. These are some of the reasons why an information model is valuable for representing these dependencies.

To continue the example, let's assume that our candidate device is going to use class-based weighted fair queuing and weighted random early detection (CBWFQ and WRED, respectively) to implement Silver Service. This combination can be defined by the use of a particular DSCP in the network view, but the specification of these mechanisms is implemented in the device view. The device view is also responsible for equating this combination of services to other sets of different services, so that each device can provide the same relative QoS even if it has different capabilities. For example, a second device might use flow-based weighted fair queuing. If the traffic is to use both of these devices, then the particular settings of CBWFQ and its dropping algorithm must be equated to a given set of settings for flow-based weighted fair queuing and its dropping algorithm. Otherwise, the two devices might not be performing the same type of traffic conditioning for the same packet, which could have the effect of violating the SLA.

Since the device view is very different in nature and level of abstraction than the network view, the policies that are used to manage it are similarly different in nature and abstraction than those of the network view. The idea at this level is to define a "traffic class" for traffic conditioning that is independent of the actual policy used to bind that traffic class to the device. From the DEN-ng point-of-view, the traffic class is a named object that can be managed. It will contain expressions using the DEN-ng PolicyStatement class (see Chapter 7) that assert various conditions to see if this packet belongs to this traffic class or not. This is a special use of DEN-ng policy—the PolicyRule is used to determine the functions used in the device to classify and mark traffic. PolicyRules can also defined to specify the type of traffic conditioning (e.g., dropping, queuing, etc.) that is to be done, as well as control when each of these device configurations should be updated.

These examples of using PolicyRules are all examples of specifying policy rules to define the configuration of device mechanisms that should be implemented (i.e., installed) in the device in advance of the device being used to condition packets. In other words, an application other than the network device can use these classes and PolicyRules as inputs and produce device-specific commands as output. When the network device receives these commands, it updates its configuration file (as one example) and is then ready to condition packets.

A simplified view of what happens from this point is as follows. Once a packet is received, it is classified and marked. The marking specifies the appropriate traffic conditioning mechanisms to be used on that packet.

Finally, the instance view would replace the policy definition with device-specific (or another appropriate communication mechanism) CLI code. Here, "device-specific" means that the version of the device's operating system, as well as specific hardware dependencies, must be taken into account. Thus, Cisco devices using the 12.x train of IOS will use Cisco's "modular" CLI to implement this using *class map*, *policy map*, and *service policy* commands. Note that a Cisco device that uses an 11.x-based version of Cisco's CLI, as well as Juniper's JUNOS operating system, doesn't have these commands (though configurations can be made that will do similar things). Thus, at the network level, there is no difference between the two Cisco devices, but there is a difference between both Cisco devices and the Juniper device. At the instance level, there is a very pronounced difference between all three devices.

The use of policy for expressing and translating business rules into device configuration commands has been emphasized. However, there is more to the use of policy than this. Policy provides a useful abstraction to enable different entities having very different capabilities and roles to be managed and controlled as a cohesive unit. For example, Silver Service may require hosts and firewalls to be reconfigured in addition to routers and switches. Most firewalls and hosts have very different programming models than routers and switches. Silver Service in effect normalizes the programming models of these different devices, ensuring their effective coordination.

10.3.3 Using Signaling

Signaling can be used with or without the provisioning approaches previously described. In signaling, the idea is that important information is valid at a particular time, or because a particular set of conditions are active. This is fundamentally different than provisioning approaches, which seek to embed rules within a device so that the device knows how to respond when it detects different types of traffic. Nevertheless, there are similarities between signaling and provisioning policies. This section will provide examples to clarify its use.

Requesting High QoS

Both provisioning and signaling can be used to request QoS. However, the way in which each requests QoS is fundamentally different. Provisioning is limited by the extent in which network traffic capabilities can be understood. This understanding takes the form of building into each device's configuration commands that the ability to recognize and appropriately condition each type of traffic flow that is received. In other words, network traffic patterns are *anticipated*, and devices are programmed to react to a particular flow, match it against their installed rules, and take action. Clearly, the problem is that if the current traffic patterns do not match what was pre-provisioned (or do not properly accommodate those patterns), then business rules (at least!) will be violated, and the network will not support the desired behavior.

A problem that is potentially far worse is when traffic guarantees are desired. If the network isn't over-engineered, then significant variances in its load can cause it to be over-subscribed. Over-subscription does not work well for pre-provisioned networks trying to support traffic guarantees because the guarantees themselves get compromised. However, over-engineering is not desirable, because most of the time the resources are wasted. Signaling offers the ability to adjust network resources on a granular basis. In addition, it can provide additional important information on a per-flow basis to supplement the use of pre-provisioned policies, as described in the next section.

Supplying Useful Identification Information

PBNM systems rely on the ability to identify sources and/or users of traffic as part of the classification process. However, there are many difficult problems that are often encountered. For example, applications may dynamically negotiate for port numbers, making it impossible to pre-define a classification rule to detect the application based on its port number. Worse, if IPsec is used, ports are encrypted and therefore cannot be used at all in the classification process. Finally, if there are community machines (e.g., a "guest" machine, or one for general usage), how can the PBNM system tell which user is using a particular machine?

Fortunately, most signaling applications have the capability of generating messages that can be used to describe the traffic and/or the users or applications sending or receiving the traffic. For example, the RSVP protocol[18] has the ability to send additional information describing users[19] and applications.[20] This can be used to help identify users and applications so that further processing may be applied. In the case of IPsec traffic, hosts provide an SPI[21] that can be used as classification criteria instead of ports.

Thus, we see that signaling applications can be used to supply critical information in addition to asking for specific requests such as bandwidth. The power of signaling applications is that resources can be requested (as well as information supplied) from particular devices along a path. This enables the network service to be fine-tuned to suit the current needs.

Signaling Example

RSVP is arguably the most popular protocol used for requesting QoS resources from the network. Policy can be used to control whether to admit or reject an RSVP request based on the request's attributes and the specified policy.

A signaling policy can be used for several things. Three examples are:

- to control the admission priority of resources
- to provide preemption support
- to provide mapping of services signaled by RSVP (or another suitable protocol, such as COPS) to differentiated services in a core network

The first two functions are used to control whether to accept or deny a request, and what to do if there are many different flows competing for the same resource. Admission priority controls what applications can have their resource requests granted, while preemption priority defines a relative ranking among the set of admitted flows that are competing for shared resources. The third enables different technologies, used by different control approaches (signaling vs. provisioning) to be mapped together.

An admission request decision can be based on comparing an RSVP TSPEC (specification of the traffic flow) or FLOWSPEC (the amount of QoS resources requested) against a meter. Metering is the function of monitoring the arrival times of packets of a traffic stream, and determining the level of conformance of each packet with respect to a pre-established traffic profile. This allows basing an admission decision both on the properties of the reservation request itself as well as on the current temporal resource allocation.

For example, consider the following policy:

Allow resource assignment via RSVP for flows coming from the HR subnet up to a total aggregated rate of 256Kb/sec.

The meter is used to track the current state of resource allocated to the HR subnet, and compares any new request for resources against a 256Kb/sec traffic profile. In this policy, individual resources will be admitted, so long as their cumulative rate doesn't exceed 256Kb/sec.

Policy can be used to control and/or modify RSVP messages. Sample actions include:

- Replace/add DCLASS object in RSVP message
- Replace/add Preemption priority object in RSVP message
- Trigger an error/warning RSVP message
- Instruct the RSVP node to proxy RSVP message as if sent by the RSVP end nodes

The first two examples enable explicit control to be exerted over the behavior of the nodes involved in the signaling decision. The third example—triggering warnings and errors—enables end-nodes to be notified that their resource reservation is about to be adversely impacted (e.g., about to expire). The final example is an optimization. If the device is (for example) part of the boundary of a DiffServ core network, it may be more efficient to simply map the RSVP request to a specific PHB, rather than forwarding the RSVP Path message.

10.4 Accommodating Changes

The preceding section has described how policy can be applied to existing, or planned, circumstances. This section will briefly address how policy can be applied to ease the management of new changes to the environment.

10.4.1 New Users

The simplest and most common example of accommodating changes is ensuring that the system can accommodate additional users that are using existing services. The simplest way to handle this is to use an abstraction to represent a user, and bind this abstraction to the appropriate class of service. In DEN-ng, this abstraction is, of course, a *PartyRole*. This enables the *Service* Provider to define a set of *Services* in terms of *PartyRoles*, instead of having to explicitly identify a particular user or group of users. The advantage of this approach is that existing policies don't have to change as long as they use *PartyRoles* (instead of individual user names).

10.4.2 New Services

Similarly, it is important to be able to abstract service offerings. This can be done in several ways. Two of the more powerful ways are using the DEN-ng *Device Roles* and *ServiceBundle* classes.

DEN-ng abstracts both the logical as well as the physical characteristics of devices in terms of roles. For example, an MPLS VPN needs devices playing the Customer Premise, Provider Edge, and Provider Core roles. These roles specify different logical and physical characteristics that devices playing those roles must have.

A *ServiceBundle* is a class that represents a set of related services. The idea is that the user can choose one *Service* (say, Silver Service) that is part of a *ServiceBundle*. Later, if the user wants to upgrade, the task of upgrading the user's Service is easier, since the *ServiceBundle* defines how Silver Service is related to other *Services* in the *ServiceBundle*.

10.4.3 New Application Traffic

It is much harder to accommodate new types of application traffic in an existing network. Policies can be used to characterize the needs of this application, and help assign it to one of the existing classes of service. This abstraction often extends into the modeling of *Products*. DEN-ng defines a *Service* as part of a *Product*. After all, *Services* do not exist in the ether! They require a delivery vehicle. This delivery vehicle can be modeled as a *Product*.

DEN-ng models a *Product* as an externally facing representation of a service and/or resource procured by customers. *Services* are modeled as an intangible realization of a *Product* (or something provided in support of a *Product*). A *Resource* is part of the infrastructure utilized by a *Service* or *Product*. This is shown in Figure 10-13.

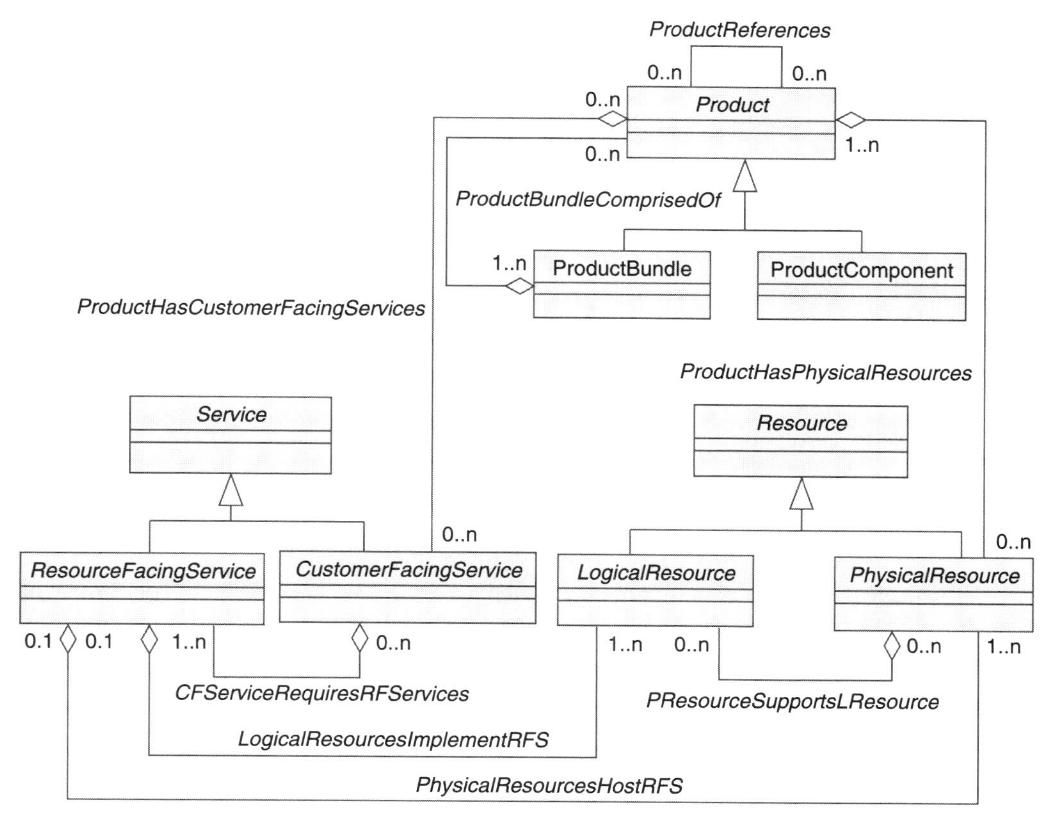

Figure 10-13 Relationship between *Product, Service,* and *Resource* in the DEN-ng model.

Figure 10-13 shows that DEN-ng models a *Product* as containing zero or more *Services* (through the *ProductHasCustomerFacing* aggregation) and/or zero or more *Resources* (through the *ProductHasPhysicalResources* aggregation). Note in particular that the cardinality of each of these aggregations is "1" on the *Product* side, meaning that a *Product* can optimally contain *Services* and/or *Resources* that can be used by a *Customer*. *Services* have invariant portions, defined by *ServiceSpecification* and its subclasses, as well as changeable portions. Changeable service characteristics and behavior are characterized as either *CustomerFacingServices* or *ResourceFacingServices*. (Remember that *Customer FacingServices* are *Services* that are bought by a *Customer*, meaning that *CustomerFacingServices* are directly visible as *Products* or *ProductComponents* to the *Customer*; *ResourceFacingServices* are services that are not purchased directly by the customer but are still required to support a *CustomerFacing Service*).

Service can also be linked directly to *PhysicalResources*, as well as *LogicalResources* (e.g., a router and a specific protocol required to run that Service). This enables the physical and logical dependencies of Services on network and system resources to be specified. Examples of these dependencies are the *LogicalResourcesForRFS*, *PhysicalResourceHostsRFS*, and *PResource SupportsLResource* aggregations in Figure 10-13. Thus, *instead of managing Services outside of the network, they can be managed as a part of the network*.

This strategy works well when the needs and characteristics of the *Service* are already well-known. Even more important, however, is the knowledge of how the *Service* is being *used*. Otherwise, the *Resources* required to run this *Service*, as well as the usage patterns of users that are using this *Service*, are not known. This could have an adverse affect on the provisioning of the *Service*. For example, consider a news application that enables users to have news and current events displayed when their systems are idle. Of course, nothing in life is free, even news service. There are many such examples of these applications. The early ones caused severe performance impacts on many large corporate networks, due to all of the requests from within the corporate Intranet for data from the application servers on the Internet. New applications have a certain novelty associated with them, and can easily spread throughout a corporation as colleagues see a new "harmless" gizmo and want it too. This becomes the worst of all worlds—a user community desiring a service that the corporate IT department does not want deployed. The problem is that when early versions of such applications were deployed, there wasn't PBNM in place, and so this application received much better service than the corporate IT departments wanted to give it. This is a common problem when PBNM is not deployed—there is often no easy way to define what service a given application should receive.

If PBNM was installed, the IT staff could have easily regulated its use by either providing its own special class of service or filtering it completely if the network is congested. This is, of course, because it is not a business-critical application.

10.4.4 Network Failures

An often overlooked capability of PBNM is dealing with network failures. The basic approach is to define a set of policies that are used only during failure conditions. Such policies will not conflict with other policies (assuming that the DEN-ng approach is used) because they are differentiated by the event that triggers them (remember that DEN-ng policies have an event, a condition, and an action clause). In other words, two policies could cause very different configuration commands to be downloaded to a device and not conflict, because one is intended for normal operating conditions and the other is intended for failure conditions.

Different types of network failures have different effects on services delivered to end-users. For example, link failures will cause traffic to be re-routed. The danger in re-routing, of course, is that the characteristics of the new path may be different than the characteristics of the original path. For example, if the original path is traffic engineered, a new path may not have the same traffic conditioning characteristics as the original path.

The DEN-ng model has three concepts that will help us; each can be used as part of an approach to fix such problems. First, the concept of policy domains is defined. This defines sets of managed objects that are controlled by sets of management applications. The idea is to define specific policy domains that have a set of policies active to provide appropriate traffic conditioning. Defining a common policy domain enables complementary success and failure policies to be associated with that policy domain, and therefore with the managed objects that are in the policy domain.

The second DEN-ng concept is that of a policy target. This explicitly defines objects that are targets of a given policy. A policy target is a powerful abstraction that removes the requirement to explicitly identify what managed objects the policy should affect.

The third concept is to trap explicitly on failure state. This requires the use of a set of information models that represent the different states of both managed objects as well as systems.

The recommended approach is for explicit policy rules that deal with failure conditions to be constructed. Policy domains and policy targets can be used to ensure that the scope of these failure policies is identical to similar non-failure policies. The event-condition-action triplet of DEN-ng further enables failure and non-failure policies to be correlated with each other. This also enables the power of the FSM to be used—failures have their own states, and their own transition mechanisms to return to non-failure states. The main disadvantage to this approach is that there will now be many more policies that are active in the system. However, this is not considered a difficult problem, because the events used to trigger failure versus non-failure policies will be very different. There is an increased amount of storage that is necessary, and care must be taken to ensure that searches are optimized to deal with larger numbers of policies, but it is felt that the advantage of using policy to control behavior during failure conditions far outweighs the disadvantages of increased storage and more sophisticated searching and maintenance operations.

The Fine Grain NGOSS Catalyst projects[22] have proven that a system can be made tolerant of failure. In fact, as part of the Fine Grain phase 2 Catalyst, it was shown that certain customers would have their services protected by smart re-routing of their traffic. This was done by defining a *policy* that governed which Customers were granted this "extra" protection. The use case for doing this was to show how a Service Provider could protect (or at least minimize) the cost of violating an SLA for a premium Customer.

10.5 Summary

This chapter discussed how policy is used to control signaling and provisioning applications. First, policy approaches were categorized as being either pro-active or reactive. Changes that are pro-active typically take the form of changing the configuration of a device, whereas reactive approaches concentrate on dynamically adjusting devices. This dynamic adjustment is made through the use of signaling applications. It was stressed that these two approaches are in fact complementary and can be used together.

Policy is more than changing commands in a configuration file. Policies can in fact be used for many other things. This chapter provided four different examples of how policy, in conjunction with information modeling, is used in PBNM systems. The first example was to equate different views that different policy subjects have of the network. Unless this is done, the same policy executed by different policy subjects could have markedly different effects on their policy targets. The second example was to ensure that different capabilities of different devices are equated, so that these different devices can provide the same relative service. The third example was similar—this time, models represented a set of limitations and/or restrictions on functionality that occur when a particular device function is used, and policies are used to equate their application. The final example was to accommodate different programming models—again, a mapping is provided between the different programming models in order to equate the different programming models, and policy is used to control when different functions are used.

These approaches are used in building a new type of network management. Traditional approaches center on configuring the interface of a device. This makes it difficult to provide end-to-end management for two reasons. First, a device interface doesn't know about a service, because it only has a view of that particular interface. A service spans multiple devices. Second, most network vendors build dedicated EMSs to manage a portion of the capabilities of a device. This means that multiple applications, with multiple programming models, are needed to manage the capabilities of a single device. This is exacerbated when multiple vendor devices are used. PBNM is a different paradigm, in which the PBNM system as a whole is responsible for programming network services. PBNM concentrates on understanding not just the syntax, but the semantics of commands used to build network services, so that multiple services can peacefully coexist with each other.

There are several advantages in administering PBNM systems. First, since the PBNM components form an integrated application, the number of different applications that an administrator needs to implement a service is dramatically reduced. Second, PBNM systems rely on separating application data into public and private data. Public data is designed to be shared among participating application components.

Three usage examples were provided. Perhaps the most important one was that policy-based management doesn't solve the entire problem by itself. Rather, the point of policy-based management is to make life easier for the administrator. This includes making policies easier to apply as well as making their application more consistent.

The DiffServ example first concentrated on describing the system capabilities using parts of the DEN-ng models. The example concentrated on showing how the information and policy models were used together to represent the translation of different types of policies using the Policy Continuum.

Specifically, DEN-ng was used to model device capabilities (such as different types of traffic conditioning algorithms) in a device-independent way. These capabilities were associated with high-level business policies that controlled when they were applied. Device-specific commands were accommodated by treating the DEN-ng model as an extensible framework, and building vendor-specific subclasses to represent vendor-specific functionality, which could be added to the DEN-ng QoS and policy models on an application-specific basis. Thus, we have achieved two important goals:

- Modeling vendor-specific functions using an approach based on standards
- Associated high-level business rules with business concepts (such as Product and Service) and linking device configuration commands to both of these

The DiffServ example expanded on the second point and briefly explained how each level of the Policy Continuum would be used to implement these services.

A second example discussed protecting particular traffic, and ensuring that the service provided to the protected traffic was not adversely affected by the changing environment. This example again used the Policy Continuum, but concentrated on how policy expressions at one level of the Policy Continuum were translated to different levels of abstraction. The concept of mapping between different capabilities was also discussed.

The final example discussed the use of signaling. It was emphasized that signaling is not just for requesting additional bandwidth. In fact, a very important use of signaling is to provide decision-making applications with more information in order to perform their functions. RSVP was used as an example to show how user and application identification and IPsec information could be used to add critical information needed to make decisions.

The power of signaling applications is that both additional resources that are being requested, as well as additional information provided via the signaling protocol, traverse a specific path or paths in the network. This is often a welcome alternative to over-engineering, since it enables a set of resources on a specific path to be fine-tuned without having to waste resources that aren't used.

The final subject covered in this chapter was to show how policy could be used to accommodate changes to the managed environment. New users, new services, and new application traffic all lend themselves to being described using a powerful set of abstractions that policy can manipulate. Generic examples (in terms of roles) as well as specific examples (using some of the concepts present in the DEN-ng model) were provided. In this context, network failures were viewed as a change to the environment that could be identified using dedicated events. Policy domains and targets were recommended to be used to identify devices that could be affected and associate with them separate policies for normal operation and operation under a failure condition.

10.6 Recommended Further Reading and References

The following is a set of references for this chapter.

1. The home page of the Differentiated Services working group of the IETF is: *www.ietf.org/html.charters/diffserv-charter.html*
2. The home page of the Integrated Services working group of the IETF is: *www.ietf.org/html.charters/intserv-charter.html*
3. Yadav, S., Yavatkar, R., Pabbati, R., Ford, P., Moore, T., Herzog, S., *Identity Representation for RSVP*, RFC2752, January 2000.
4. Bernet, Y., Pabbati, R., *Application and Sub Application Identity Policy Element for Use with RSVP*, RFC2872, June 2000.
5. Herzog, S., *Signaled Preemption Priority Policy Element*, RFC3181, October 2001.
6. The home page of the ISSLL working group of the IETF is: *www.ietf.org/html.charters/issll-charter.html*.
7. Bernet, Y., Ford, P., Yavatkar, R., Baker, F., Zhang, L., Speer, M., Braden, R., Davie, B., Wroclawski, J., Felstaine, E., *A Framework for Integrated Services Operation over Diffserv Networks*, RFC2998, November 2000.
8. Baker, F., Iturralde, C., Le Faucheur, F., Davie, B., *Aggregation of RSVP for IPv4 and IPv6 Reservations*, RFC3175, September 2001.
9. ITU-T, *Principles for a Telecommunications management network*, Recommendation M.3010, May 1996.
10. Nichols, K., Blake, S., Baker, F., Black, D., *Definition of the Differentiated Services Field (DS Field) in the IPv4 and IPv6 Headers*, RFC2474, December 1998.
11. Heinanen, J., Baker, F., Weiss, W., Wroclawski, J., *Assured Forwarding PHB Group*, RFC2597, June 1999.
12. Davie, B., Charny, A., Bennett, J.C.R., Benson, K., Le Boudec, J.Y., Courtney, W., Davari, S., Firoiu, V., Stiliadis, D., *An Expedited Forwarding PHB (Per-Hop Behavior)*, RFC3246, March 2002.
13. Charny, A., Bennett, J.C.R., Benson, K., Le Boudec, J.Y., Chiu, A., Courtney, W., Davari, S., Firoiu, V., Kalmanek, C., Ramakrishnan, K.K., *Supplemental*

Information for the New Definition of the EF PHB (Expedited Forwarding Per-Hop Behavior), RFC3247, March 2002.

14. Rosen, E., et al.: *BGP/MPLS VPNs*, draft-ietf-ppvpn-rfc2547bis-03.txt, July 2002.
15. OMG Unified Modeling Language Specification, Version 1.4, September 2001, section 3.41–3.49.
16. Strassner, J., ed.: *Mining Information from the DMTF CIM into the TMF SID*, July 2002.
17. Fuller, V., Li, T., Yu, J., Varadhan, K., *Classless Inter-Domain Routing (CIDR): an Address Assignment and Aggregation Strategy*, RFC1510, September 1993.
18. Braden, R., (Ed.), Zhang, L., Berson, S., Herzog, S., Jamin, S., *Resource ReSerVation Protocol (RSVP)—Version 1 Functional Specification*, RFC2205, September 1997.
19. Berger, L., O'Malley, T., *RSVP Extensions for IPSEC Data Flow*, RFC2207, September 1997.
20. Berger, L., O'Malley, T., *RSVP Extensions for IPSEC Data Flow*, RFC2207, September 1997.
21. This is a TMF member's only document that describes the Interface Implementation Specification of the Fine Grain NGOSS Catalyst Project. It is specified in the following document: TMF839v1.5. and in the following (members-only) web site: *www.tmforum.org/sdata/documents/TMFC1379%20TMFC 1000%20TMF839v1[1].5.pdf*
22. Blake, S., Black, D., Carlson, M., Davies, E., Wang, Z., Weiss, W., *An Architecture for Differentiated Services*, RFC2475, December 1998.

New Directions in Policy-Based Management

This chapter will explore new directions in PBNM. First, the current state-of-the-art will be explored, in order to see how PBNM solutions have been used up to now. Then, different products will be examined to see how policy has been implemented commercially. Finally, an example of a relatively new product—Intelliden—is examined as a precursor to future developments in PBNM solutions.

11.1 Introduction

The purpose of this chapter is to examine the future of Policy-Based Network Management (PBNM) applications. We will view this from five different angles: (1) a sampling of various types of magazine, conference, and journal articles, (2) analyst opinions, (3) educational and research work, (4) standards bodies and fora work, and (5) products that various companies are currently manufacturing.

This is certainly not an exhaustive study. As of November 2002, a search for "policy-based network management" in *www.google.com* yields an astounding 43,200 hits. Rather, this section concentrates on providing a sample of research from a small set of companies that have either been following PBNM for over a couple of years, or which provide a very different point-of-view.

11.2 A Sampling of Press and Magazine Articles

This section will provide a small sampling of different types of articles that have occurred in leading magazines and journals. The purpose of this section is to deduce the research areas that are getting the most interest and attention. We can then link these to work in various fora and manufactured products.

11.2.1 Magazines, Journals, and Conferences

First, it is important to realize that research in PBNM is global phenomenon—it is *not* limited to, or even emphasized in, "just" the United States. For example, ZDNet in France[1] offers a variety of Webcasts and white papers on PBNM. These include samplings from companies, magazine articles, and white papers written by consultants. The United Kingdom, Germany, Japan, and many other countries in addition to the United States have active research in various areas of policy management.

Magazines and Journals

One of the first articles summarizing PBNM appeared in Conover.[2] Also of interest are several articles from Network Computing.[3,4]

Another good source of information is the Distributed Systems Engineering Journal,[5] which has a set of interesting articles on policy-based management.

A recent article[6] provides an excellent summarization of the motivation for PBNM solutions. The thesis of this article is that increased concerns about network asset utilization and security are encouraging organizations to look seriously at how PBNM solutions can help enterprise IT shops. This article emphasizes some obvious and not so obvious benefits of PBNM solutions. First, the obvious ones:

- Provide QoS to sensitive, prioritized traffic
- Stop over-provisioning the network (which has cost and management benefits)

- Manage the combination of more users, more applications, and more sophisticated requirements of those applications
- Manage how different applications with different needs are prioritized according to business rules
- Satisfy different user expectations (e.g., the president gets platinum service, the vice-presidents get gold, and the engineers get silver)

More subtle benefits include:

- Ability to reveal true network efficiency and resource utilization
- Enforce application-level security (i.e., ensure that users can only access information and run applications that they are authorized to use)
- Enable business rules to drive the configuration of network services

This article makes the key point that PBNM is not just about network management. Rather, PBNM also concerns application management, security management, and identity management.

While some articles are negative, and some complain about the lack of standards progress for PBNM, the majority of magazine articles talk about the promise of PBNM solutions. An increasingly popular theme is the relationship between how an organization runs its business and how PBNM can be used to accomplish those goals. Most of these articles do not focus on low-level applications, and serve as a good introduction to the subject.

A good example of technical journal articles is the *Journal of Network and Systems Management JNSM*. A special issue of the will be devoted to policy-based management of networks and services and will be published in September 2003.[7]

Verma[8] is an example of a technical journal article. This paper discusses architectural implications of PBNM solutions, and shows how administration can be simplified as a result of implementing PBNM solutions. Flegkas et al.[9] is another example of a technical journal article, in which PBNM solutions are used to extend the functionality of management systems. This paper again starts with defining a policy architecture for managing QoS in DiffServ networks. It then discusses the use of policies to control IP DiffServ networks (both positive and negative effects), and describes its implementation in high-level terms.

Technical journal articles tend to concentrate either on theoretical advancements or applications. Rarely do they provide the whole solution, but they do provide insight into what benefits such applications must provide.

Conferences

The Workshop on Policies for Distributed Systems and Networks[10] is specifically devoted to all aspects of policy management. The fourth conference in this series was held in Lake Como, Italy, on June 4-6, 2003. This is arguably the premier conference on policy management and policy applications, with a very good mix of academia, consultants, and vendors attending each year. This provides a good combination of current results, practical applications, research papers, and theoretical studies.

There were two conferences that focused on policy sponsored by Upperside. These were the IP Policing conferences[11] that were held in 2001 and 2000.

A recent conference[12] also focused on policy. The conference commenced with an all-day tutorial on policy, given by the author. It then had three days of papers focused on policy management and applications.

Networld + Interop, a network-oriented conference presented by Key3 Media, offered a special tutorial on Policy, QoS, and DEN, from 2000–2002.[13]

Please note that the ACM Role Based Access Control conferences[14] are closely related to PBNM. Recall that chapter 5 used the notions of roles to abstract the functionality provided to a client. The RBAC and SACMAT conferences are important for researchers and practitioners alike, as they cover theoretical and practical aspects of roles and access control, and how they are related to PBNM.

As expected, most conference papers are technical in nature. The overwhelming majority of articles assume that PBMN is good and real, and they seek to extend its functionality. Technical articles provide a fairly balanced combination of theoretical considerations and practical applications. In one interesting paper,[15] PBNM was used to enable the network to respond to the business needs of an organization. These papers are much more technical than magazine articles, and tend to be on a par with technical journal articles. Again, these articles do not provide a specification as to how to build such systems. Rather, they provide a summary of what the systems must do, and sometimes additional insight into key functionality provided by the system and/or a summary description of the architecture employed.

As we will see, in order to get additional information on how PBNM solutions are built, the best source of information is from educational organizations and, sometimes, white papers from vendors.

11.3 Analyst Opinions

There are several analyst firms that have been tracking PBNM. There are many more analyst firms that are covering PBNM than just Gartner and The Burton Group. However, these two firms are representative of two different opinions that characterize analyst opinions. This section provides brief excerpts and references to their studies.

11.3.1 The Gartner Group

Gartner has mixed opinions on Policy. Gassman[16] recently wrote a short research note stating that the cost of PBNM technology, along with associated costs of managing policies and policy-based applications, often is more than alternative solutions. This note advocates simplicity, and is worried that most end-to-end PBNM applications are too complex. This is because the probability that heterogeneous devices are used in an end-to-end application is very high, which in turn means that different functionality must be controlled. Rather, this article recommends PBNM solutions to be limited to edge applications.

It should be noted that Gartner does cover vendors that manufacture policy products. However, these reviews are based on using these products to accomplish a single specific task. Gartner seems to support the use of policy for individual devices, but not for end-to-end configuration.

While simplicity is certainly desirable, I do not agree with the conclusions of this article. Nothing comes for free, so if it this application is too complicated for PBNM solutions, what other solution can be used? Of the proposed alternatives in this article:

- buying more and/or faster data circuits is often not an option due to greatly decreased spending targets
- the other five alternatives all do not deal with network problems—they instead offer alternatives for servers, applications, and other non-network devices.

The benefits of PBNM solutions, as described in the preceding chapters of this book, are not matched by any other approach. While they may entail more complexity, they also offer more benefits in terms of ease of administration, greater ability to program the network to realize business objectives, and more efficient management of network resources.

11.3.2 The Burton Group

In contrast, the Burton Group has been supportive of PBNM since it was first introduced in the IETF and made a "mainstream" product. Kosiur[17] states that while PBNM can give IT managers greater control over their system and network resources, the technologies that comprise PBNM are still young and relatively unproven.

Kosiur[18] is more encouraging. While still acknowledging that the technologies which make up PBNM systems are relatively young and evolving, many networking vendors have produced first-generation PBNM products that can be used to control a variety of applications, such as QoS. This is despite the absence of mature standards. Kosiur[19] addresses the needs of Service Providers. Paradoxically, while PBNM promises to lower the number of personnel that are required to manage the network, its deployment will require additional personnel and training. This report hypothesizes that Service Providers will most likely outsource their policy needs to a new kind of Service Provider that specializes in using PBNM solutions to provide provisioning services.

Later reports concentrate on further architectural definition and implementation, with a heavy emphasis on directories and, lately, XML technologies.

In contrast to technical journal and magazine articles, analysts have varying opinions. Analysts will take a while to examine products and interview key personnel. Once an analyst firm comes out with an opinion, it will generally stay with that opinion, watching vendor developments closely, unless clear evidence to the contrary is found. Analysts will try and emphasize trends and innovation. If a particular topic is covered by multiple analysts, it is something to be watched closely.

11.4 Educational and Research Work

There are several prominent universities that are focusing on PBNM efforts. It is beyond the scope of this book to detail all of the various efforts in the educational and academic communities. Therefore, this section will concentrate on arguably one of the best known universities that is working on PBNM—the Department of Computing Science at the Imperial College of Science, Technology and Medicine, University of London (DSE).[20]

The group is headed by Professor Morris Sloman. Dr. Emil Lupu plays a prominent role in overall research and direction of the group. Other members of the group can be found in the DSE home page.[21] The publications page[22] has links to the publications of four of the leading researchers of this group. Links to current and past projects of the group can also be found.[23]

There are many active avenues of research in the Distributed Software Engineering Group of DSE. One of the more interesting areas of focus is distributed systems management and security policies. One of the founding principles for this research is the premise that management is itself a distributed activity. Domains are used to partition responsibility. Similar to the definition of domains given in Chapter 2, domains in this context are collections of managed objects which have been explicitly grouped together, so that a common set of management policies can be applied to them. Work on domains and domain management is continuing.

Much of the DSE research centers on the use of the Ponder language. In Ponder, they have defined four types of policies, as summarized in Chapter 5. Ponder takes these four types of policies and applies them within a role-based framework. This framework can be used to specify policies as well as analyze them for conflicts. Related work has produced tools and methods for configuring and managing distributed services and applications. Specific work is centered on configuration and mobile applications.

In addition, active work on security management and trust issues, as well as distributed systems monitoring, is ongoing.

Drs. Sloman and Lupu were instrumental in starting the Policy conferences, and continue to play an active role in international policy activities.

11.5 Examples of Standards-Driven Applications—The TMF

The three fora that are most well known for pursuing PBNM solutions are the IETF, the DMTF, and the TMF. Of these, the IETF's Policy Framework working group[24] was the first to seriously start work on formally modeling policy. Its work was based on the original DEN policy model. The DMTF then used this model and added CIM naming classes, plus some minor modifications.[25] The TMF instead has adopted the new DEN-ng policy model. This is because it is more than just an information model (See Chapter 6).[26] It also has architectural implications.[27-29] As mentioned in Chapter 6, this is one of the things that differentiates the DEN-ng policy model from previous policy models. The remainder of this section describes the Fine Grain NGOSS Catalyst program, which used an early version of the DEN-ng policy model.

11.5.1 TMF Catalyst Programs

One of the unique features of the TMF is its emphasis and support of Catalyst programs. TMF Catalyst programs are affectionately called the "living lab" of the TMF. The purpose of a Catalyst program is to demonstrate real situations encountered in building an OSS, and to show how problems arising in these situations can be effectively solved using NGOSS principles. Thus, Catalyst programs demonstrate to members and non-members alike that NGOSS can be used to solve problems that cannot be solved by current technologies and products.

The purpose of a Catalyst Project is to share implementation experiences needed to drive implementation and products in the global marketplace. Catalyst Programs are sponsored by one or more Service Providers and/or System Integrators. They involve different vendors of COTS (Common Off The Shelf) software and hardware. In principle, this means that all of the participants of a given Catalyst program will share and use common software. In the case of the Fine Grain NGOSS Catalyst, one of the common things that was shared was the DEN-ng (later to become the SID) information model.

11.5.2 The Fine Grain NGOSS Catalyst Program

The Fine Grain NGOSS Catalyst project[30,31] defines a new type of highly distributed, policy-based architecture for building a next generation OSS. Its primary objective is to create a componentized architecture capable changing systems, software and structure in an automated way. Here, the word "componentized" means an atomic unit of deployment whose functionality is manageable using an external application (such as a PBNM system).

The "Fine Grain" name is derived from the granular approach taken by the team in building business processes. This approach mandated the use of many small, service-based components, which can be reassembled to perform more powerful business functions. The functions of the OSS are then built by combining simple services to develop more complex services.

Fine Grain NGOSS is a Java-specific implementation that exploits Sun Microsystems' Jini Network Technology's spontaneous networking philosophy. The objectives of Fine Grain Phase II were comprehensive and included demonstrating:

- Adherence to all NGOSS principals (core and extended)[32]
- Jini / J2EE integration[33,34]
- Support of legacy network element(s) through the implementation of a Jini Service for management and provisioning of network services
- Support of "legacy" application(s) (example: billing) through the implementation of a Jini Service for intercommunication between applications and the OSS
- Support of wireless devices (e.g., a phone and/or PDA) through the implementation of a Jini Service for intercommunication between applications and the OSS
- Enhancement to Phase I architectural elements (e.g., dynamic policies and business processes, an enhanced information model, and contracts)

The Fine Grain demonstration (done at TeleManagement World in May, 2002) emulates a Network Service Provider (NSP) providing the access and transport network facilities to an IP Service Provider (ISP) through a Wholesale offer (e.g., through a bandwidth offer and/or connectivity offer). The ISP provides an MPLS VPN service to the End-Customer through a retail offer. (All of these concepts are modeled in DEN-ng as well as in the SID; DEN-ng currently adds QoS models that are not yet part of the SID.) The roles that each participating vendor played in the overall approach are shown in Figure 11-1 below.

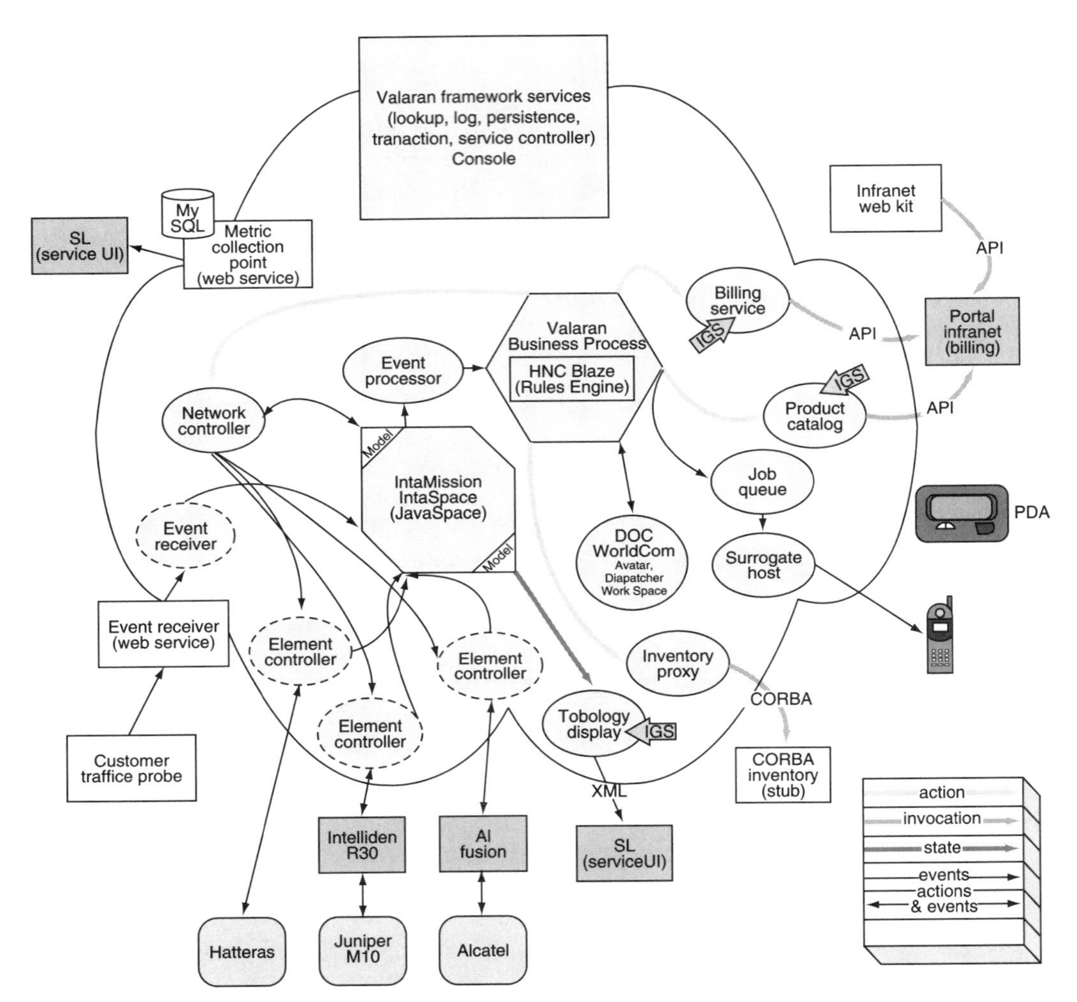

Figure 11-1 Roles played by different vendors in fine grain, Phase II.

The Fine Grain demonstration assumes that a 100 Mbps Ethernet tributary service exists for a given corporation. This is aggregated into Gigabit Ethernet access via a Hatteras Networks switch, and passed to an (emulated) ISP via Juniper M10 routers. The ISP network runs over an Alcatel SONET transmission network, organized as a separate business unit. MPLS private-line services isolate customers and network traffic management is provided with trunk side MPLS routing. In each case, the VPNs terminate on hosted network servers

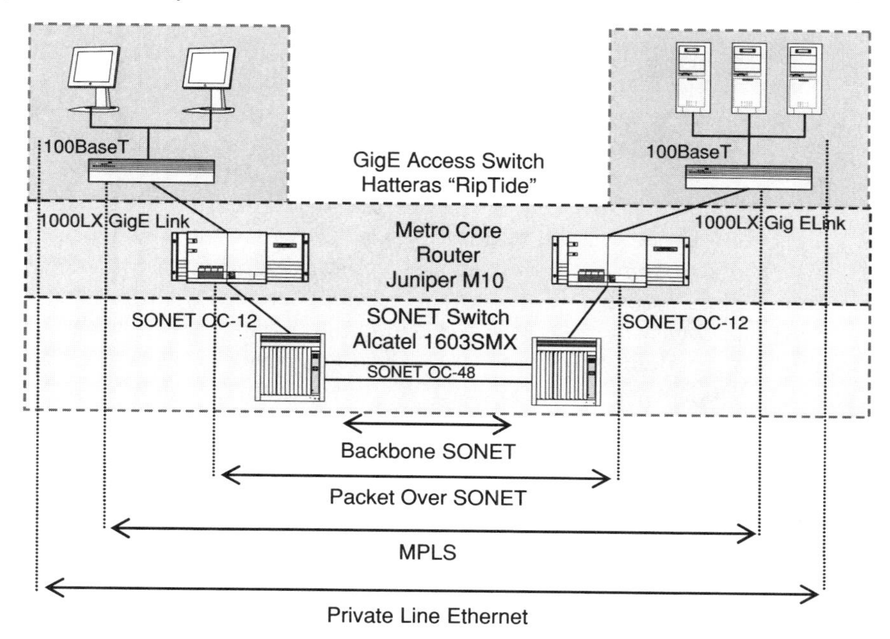

Figure 11-2 Simplified network connectivity in the fine grain, Phase II demonstration.

which provide an Intranet product to the customers. This demonstrates the ability to integrate legacy network elements (SONET devices) with modern devices (Juniper routers) with ultra-new devices (Hatteras metro switches). A simplified view of the network setup is shown in Figure 11-2.

One of the innovative points demonstrated in Fine Grain Phase II was the ability to discover new cards being plugged into devices. Sometimes, different parts of an organization can't communicate as efficiently as desired. Time lost between installing a card in a network device and configuring that card translates to lost revenue. Fine Grain solves this problem by automatically detecting when cards are plugged in. Fine Grain generated events that corresponded to a card being plugged into a device. Software ran that discovered the resources made available, and provisioned services in response to this event according to policy. Specifically, policies are used that decide how to configure a device (or set of devices) according to which customer is using what service.

Another innovative use of this Catalyst project is to protect critical business traffic. This is done through the trapping of failure and performance events. While all failures are important to spot and correct, failures associated with business-critical applications (as well as premium customers) have larger penalties associated with them. Associating network failures with customers enables the Service Provider to adjust resources to fix problems according to business priority. Alternatively, if backup connections are available, then the Service Provider may choose to reroute business-critical traffic to avoid violating critical SLAs. This ability for the network to automatically reconfigure itself under stress to match SLA contracts is very innovative, and likely to be part of future offerings.

It is important to note that this Catalyst project could not work without a common information model. In this case, a combination of the SID and DEN-ng models were used (DEN-ng covered areas that had not yet been worked on by the SID team). Specifically, both the ISP as well as the NSP utilize a service-centric design and a shared information model that promotes sharing and reusing of management data. The use of a shared information model enables two important benefits. First, multiple Service Providers can now work together and share information through a set of common definitions provided by the shared information model. This enables the different Service Providers to coordinate different product offerings, services, and functionality through the use of common data definitions. Second, it lets Service Providers define and offer different services independent of the particular transport architecture that each uses.

The use of policy enables service activation to be reliable, secure, automated and controlled in real-time. The NSP is now aware of new services and functionality as soon as they are physically (and/or logically) available. This helps reduce loss of revenue, and improve cost control. Policy also enables consistent operation to be provided for different activation scenarios. Finally, it enables business rules to drive the configuration and operation of the network.

This Catalyst project emphasized building an NGOSS-compliant architecture. The analyst firm RHK covered the TeleManagement World show,[35] which is a twice-a-year event sponsored by the TMF to showcase Catalyst programs and its work to the industry.[36]

11.6 Product Directions

There are a wide number of vendors offering a large variety of "policy-based" solutions. They include network equipment vendors, the makers of packet shaping, bandwidth management and other point products, and the various management platform and middleware suppliers. In testament to the marketing draw of PBNM solutions, most vendors will describe their products as policy-based.

Recall that in Chapter 2, we defined *policy-based management* as the usage of policy rules to manage the configuration and behavior of one or more entities. In this book, the use of an object model that expresses the policy rules as instances of a class hierarchy, along with the use of a management method (i.e., a finite state machine), is implied in this definition. This last point is critical. Most models are "open-loop" systems, in that they represent just the state of a set of managed entities. DEN and DEN-ng, as well as the emerging policy work in the TMF, all use a finite state machine to model the various states that a managed entity can have in its life-cycle, and the operations required to transition a managed entity to a new state (or to keep it in a current state). This enables models to be built to represent the entire life-cycle of the managed system.

This section provides a brief listing of different types of PBNM solutions, ending with more detailed descriptions of several modern PBNM products.

11.6.1 Point Product Examples

This section will provide two examples of different types of point products. These are products that are designed to use PBNM techniques to manage bandwidth, QoS, and other networking components. A good reference article that discusses the advantages and disadvantages of managed bandwidth (vs. over-provisioning) is Passmore.[37] These devices will in general lack much of the full QoS functionality in a network vendor's device. However, they have been designed to do a particular job, and they usually do that job just as good (if not better) than other, more general devices or systems that try and manage "policy" in the more general sense. This is because both of these point products have specific, low-level knowledge describing what mechanisms they need to control for each network vendor's product. This type of low-level information is difficult to obtain and use for different functions (e.g., QoS and security). Hence, point products *usually* can offer more features that are under "policy control" than other, more generic, solutions.

Allot Communications

Allot is a 5-year old company.[38] Allot makes three main products—NetEnforcer, NetPure, and NetPolicy—that each associate various types of network actions to business policies. These three products provide a set of hardware and software tools to enable policy to help improve overall network performance and control how the network is used (e.g., by which users and applications).

Allot's NetEnforcer product line is a set of LAN appliances that offer bandwidth management and traffic shaping, primarily aimed at QoS and SLA enforcement. They are designed to be installed at the LAN-WAN boundary to manage bandwidth through traffic conditioning. Allot makes two different models of NetEnforcer, one for Service Providers and one for Enterprises. The Service Provider model enables each user to be assigned to a particular SLA, which has associated bandwidth limits. NetEnforcer then builds different types of QoS policies to enforce each SLA.

The Enterprise model first auto-discovers applications that are run in a network, which enables an administrator to manually determine which protocols and traffic flows are most affecting the performance of the network and therefore should be managed. Once this is done, NetEnforcer provides a form to prioritize different QoS attributes to control bandwidth and implement them. In theory, this enables NetEnforcer to adjust the bandwidth used by different users and applications, so that business objectives are always met despite having different users and loads on the network.

The two versions of NetEnforcer are thus very similar. Both use policies to control and allocate bandwidth to users and applications. The difference is in how the bandwidth is defined per user or per application. Each NetEnforcer application performs traffic conditioning prioritization, conditioning, policing and shaping. NetEfnforcer also includes using signaling techniques, such as RSVP, to signal bandwidth reservations through the network.

Three add-on products can supplement the functionality of NetEnforcer. NetAccountant is a policy-based accounting appliance that provides usage-based accounting, billing and reporting of selected transactions. NetBalancer is a policy-based appliance that helps balance traffic to a set of servers, as well as assigns different priorities to applications and users that use a particular server. This enables the functionality of the server to be tailored to suit the needs of different clients. Finally, CacheEnforcer is a hardware appliance that defines policies to control different types of caches (e.g., web server caches, video server caches, and FTP server caches). This enables copies of data to be stored geographically closer to requesters of data, which in turn reduces the load on the network for retrieving content and reduces bandwidth congestion of costly Internet links.

NetPure is a policy-based "content filter," which uses policies to manage HTTP traffic. The theory is that business policies should be used to control how Internet connections (and other types of web traffic) are used. It does this by inspecting URLs that are visited, and ensures that the content in each web page that is retrieved matches accepted usage policies.

NetPolicy is a policy-based management suite of software products that provide real-time usage information describing the applications and users that are under policy control. It enables bandwidth usage to be viewed and various types of traffic usage reports to be generated.

The Allot devices all aim at managing bandwidth. They are primarily edge devices, and are designed to sit at the edge of a policy domain and perform their tasks before traffic enters the network. Thus, they are termed a "point device" because they are not designed to perform an end-to-end PBNM function. Unfortunately, they don't have an external API that can be used. This means that they cannot be embedded within another OSS component, and instead must be used in a stand-alone manner.

Packeteer

Packeteer[39] makes a set of products that manage bandwidth and control application traffic using policies.

The PacketShaper is a hardware appliance that manages bandwidth for a set of applications. PacketShaper utilizes "layer 7" classification capabilities to automatically identify the different applications that are being used in the network. It provides a wide variety of identifying applications, such as by source and destination address and port, MAC address, IP precedence bits, MPLS label, DSCP, and others. It enforces policies that manage bandwidth and priorities allocated to different applications. It also monitors traffic utilizations and round-trip response time.

AppVantage is a hardware appliance that is installed on a LAN segment that connects to a WAN device. This appliance is aimed at identifying managed application services. It serves as a physical point of separation between an ASP (Application Service Provider) and its customers and partners. It monitors each application service that passes through it and tracks application and network performance on both sides of it. It provides detailed reports on response time, throughput, discarded packets, and other metrics. While PacketShaper is designed to be used at the edge of a network at branch offices, AppVantage is designed to be used at the edge of a network of a subscriber's location. Though AppVantage is not strictly a "PBNM solution," it does provide valuable metrics that can be used by PacketShaper to provide an overall PBNM solution.

PacketWise is the core technology used in Packeteer and AppVantage. It consists of traffic identification and classification, traffic conditioning, monitoring network and application behavior, measuring and controlling network performance, and reporting on the results. It can be OEMd to partners, enabling them to embed some or all of the above functionality into the products of the partner. PacketWise is interesting because of its combination of traffic identification, conditioning, and monitoring functions. However, it is unclear how standards-compliant it is, or whether its object models are available to be used by external applications.

Packeteer also makes two products for centralized management. Policy Center is a software package that simplifies the task of deploying and using multiple PacketWise-based applications throughout the organization. It is specifically focused on reducing the number of administrative tasks that must be done when multiple instances of PacketWise-based applications are deployed and require coordination. For example, it can manage software upgrades to Packet Wise components, summarize and report on statistics generated by all PacketWise-based applications, define and distribute new traffic classes, configure groups of PacketWise-based appliances, and so forth.

ReportCenter aggregates metrics from different PacketWise-based applications and creates appropriate reports. It is aimed at identifying and analyzing various trends that are detected at different points in the network. Thus, it can serve as an effective tool to provide the detailed statistics required to aid in optimizing the network.

Thus, Packeteer is another example of a point product. It performs traffic identification, classification and conditioning for a large number of different types of applications, and then manages the priority and bandwidth allocated to each managed application. It differentiates between subscriber-based managed application services (using AppVantage) and provider-based managed bandwidth (other products). It, like Allot, does a very good job on specific features, but does not purport to do end-to-end policy-based management.

11.6.2 Network Vendor Examples

This section will provide three examples of different types of network vendor policy products. These are Element Management Systems that are designed to manage one or more types of vendor devices using policies.

Nortel Examples

The Optivity Policy Services is arguably Nortel's best-known set of policy-based management applications.[40] Optivity is a suite of software applications that are used for managing data and telephony networks. Each application uses policy to define business rules that are transformed to control specific network features, and each uses a directory running LDAP (lightweight directory access protocol) to define, store and manage policies.

Optivity Policy Services is one of the products of the Optivity product line. It is focused on defining, enforcing, and reporting on various QoS features that can be used to manage bandwidth, control which users can access what resources and run which applications, and other performance-oriented features.

The Optivity Policy Services software includes three important management applications: Optivity Switch Manager, Optivity Network Management System, and Optivity Telephony Manager.

Traffic identification is based on examining the source or destination IP address or port, protocol, inbound DSCP, VLAN ID or Tag, and 802.1p priority tag. These are then matched against the appropriate policy rule to determine the type of traffic conditioning that should be applied. Metering and policing, as well as provisioning based on DiffServ, are all supported. For edge routers that are running BayRS software, URL content filtering, as well as other forms of deep-packet filtering, can be performed to provide added functionality.

Optivity can manage both CLI-based Nortel devices as well as Nortel devices that use COPS-PR. However, the methods for creating policies and configuring the devices are very different. For example, in a CLI-based device (such as a Bay switch or a Passport device), Optivity "learns" the different interfaces from a device. These interfaces are then configured through the Optivity management console. Roles are also defined and assigned through the Optivity management console. Policy management is done on device interfaces that have first been created in the management console. Roles are then created and assigned to the interfaces of the modeled device, which are then assigned to an interface manager. This is because the policies are distributed using COPS-PR, but need to be translated to CLI commands. This translation is done by the interface manager. Policy management is done by associating a policy with the device's reported role.

In contrast, in a COPS-PR device (such as a Business Policy Switch or a Business Communications Manager), interfaces and roles are both configured through the device management console and reported to the policy server. Policy management is done by associating policies with one or more roles of the device.

In order to add a CLI device to the Optivity policy server, its IP address and SNMP community string (for reporting on policy filters) must be known. Besides having SNMP reporting enabled on the network devices that are to be managed by Optivity (in order to check the policy filters installed on them), the EMANATE SNMP master agent (available as an installation option) must also be installed with the policy server. This dependence on SNMP in general and outside tools in particular limits the effectiveness of Optivity being a complete PBNM solution.

In either case, policies are defined as consisting of four components—traffic conditions, actions, shapers, and schedules. Optivity enables any number of conditions to be trapped, but only allows for a single action (optionally with a single shaper) to be part of any single policy. This means that complex policies need to be represented as a set of simpler policies, each having only one action. It should be noted that certain policies cannot be accurately represented using this approach. This is why nested policy groups and policy rules were put in the IETF policy model,[41] as well as in the DEN-ng policy model (as described in Chapter 6).

Policies can be applied immediately or scheduled for installation at a future time. Multiple schedules can be assigned to a single policy, in which case they are logically ORed. This has the effect of acting as a series of enablers—if the installation time overlaps with any one (or more) of the schedules, then the policy is activated.

Roles are treated in Optivity as they are defined in the IETF approach (i.e., a role is an attribute that can be used to group together a set of device interfaces). A role acts as an identifier that associates a number of devices or device interfaces with a similar function. Roles are used as a common identifier between policies and devices.

Optivity can associate a given group of devices having a common role with a particular policy. A policy is applied to a device only if both the device and the policy are associated with the same role or group of roles. To enforce this in the UI, Optivity creates a policy as a part of a role object.

Unfortunately, the interface to roles is very different, depending on whether the device is a COPS-PR device or a CLI device. A COPS-PR device, such as a Business Policy Switch, must have its Roles created for its interfaces through its own management console. When a COPS-PR subsequently connects to the policy server, it reports its roles to the policy server and the roles appear as objects in the Optivity management console. When the role objects are associated with policy objects through the Optivity management console, the policy server can then begin to send policy configuration information to the COPS-PR devices.

For CLI devices (for example, a Passport 8600 or BayRS router), roles must be created through the Optivity management console. When a role is created, one or more policies are associated with it using the console. When the interfaces for a CLI device are configured, one or more roles must be associated with each device interface. When the policy server connects with a CLI device (through the interface manager), it uses the roles to determine which policies to send to each interface. The interface manager translates the policy information into CLI commands.

One problem with Optivity is its restrictions on how roles are used. Multiple roles acting together are called a role combination. While this is potentially powerful, the two largest restrictions are policy restrictions and the relative difficulty to identify conflicts between different policies. For example, a role combination cannot have associations with policies that act on both layer 2 and layer 3 traffic. Optivity can prevent users from creating a single policy that violates this and other conditions; however, the application has no way of anticipating which combinations of policies will be sent to a device as a result of association under a common role.

To Optivity's credit, it does provide an ability to test role combinations for such conflicts using its Role Combination Preview feature. This can also be used to test for policy priority conflicts. However, this still places the burden of *manually* determining if a problem exists on the administrator.

Another problem is the lack of functionality in QoS configuration that Optivity offers, along with ensuring that enforcing one policy won't hurt other deployed policies.

Finally, Optivity does not have an external API that can be used. This means that it cannot be embedded within another OSS component, and instead must be used in a stand-alone manner.

Avaya Examples

Avaya's Policy Manager (formerly called CajunRules) is interesting, in that it also manages some Cisco devices (currently the 25xx, 72xx, and 75xx) as well as Marconi's ESR5000 and ESR6000 series of switches in addition to Avaya devices (the P series of Cajun devices). Most networking vendors only manage their own products.

The Avaya Policy Manager uses a Directory Server to store information and communicate (via LDAP) with some of the devices it manages. The Avaya Policy Manager also uses the Enterprise Directory Gateway (EDG 1.1.1) to communicate with all DEFINITY telephony switches in order to prioritize VoIP traffic for both IP hard and soft phones. The Avaya Policy Manager can also be used with Lucent's QIP (which is a DNS and DHCP policy manager). This enables policies to be used to define how addresses are allocated and managed, and for those addresses to in turn be used by the Avaya Policy Manager. Unfortunately, the GUIs for these products are fairly different.

The Avaya Policy Manager consists of a Server and a Console user interface application. The Avaya Policy Manager enables the network to be managed more efficiently by installing policies onto the devices that make up the network. These policies consist of various rules that define how traffic will travel across the network. They define various actions, such as permitting, denying, or prioritizing traffic based on QoS and Access Control parameters. In addition, the Avaya Policy Manager enables Voice over IP (VoIP) traffic coming from a DEFINITY switch to be managed.

Avaya uses LDAP to communicate with its directory server. Devices are managed and configured using SNMP (except that Cisco IOS CLI is used for the Cisco devices that are managed). CIM and DEN are mentioned, but it appears that their schemata are not used in this implementation. Quite possibly, they are referring to the generic use of the directory when referring to these terms.

The Avaya Policy Manager takes a simple approach to policy management. First, business objectives must be defined. This is done by prioritizing applications and users, access controls, and schedules. Second, these objectives are mapped onto network policies. Finally the policies are administered using the Avaya Policy Manager.

The Avaya Policy Manager defines domains as a collection of targets (devices and/or interfaces). This means that a domain is restricted to only containing enforcement targets. Devices must be added to a domain in order to apply a policy to them. A device/interface can be added to a single domain only. However, different interfaces of the same device can be put into different domains if they need to get different policies.

Domains are used to distribute policies to their members. All members of the policy domain share the same abstract policy (see below). This policy can be customized for individual members or sub-groups of the domain. All members of the policy domain share the same Time Zone. If the target is a Cisco device, then it can have policies applied to individual interfaces on the same device.

Policies are created independently of domains. Each policy can be associated with one or more domains. A policy can be customized for individual targets, validated, and enforced. If it passes validation on that domain, then the policy can be deployed to members of the domain.

The Avaya Policy Manager has two levels of policy presentation. The first level is referred to as an abstract policy. Abstract policy describes network-wide high-level policy rules defined for a particular action. For each policy, a name and a default action (Deny All or Allow Any) must be specified. The second type of policy is referred to as a custom policy. These are polices that can be customized to fit individual targets.

The Avaya Policy Manager has an interesting notion of full, partial, or no enforcement. This notion occurs because the devices that it manages have varying degrees of functionality, and it allows the same policy rule to be applied to multiple devices of differing functionality. Thus, depending on the rule, a device may be able to implement and enforce all of the actions of the policy rule, some of the actions of the policy rule, or none of the actions of the policy rule. These three conditions give rise to policy enforcement being full, partial, or none.

The Avaya Policy Manager provides a means to affect partial enforcement choices by configuring a rule as "mandatory". This makes the rule preferred over other rules and guarantees that the rule gets enforced, or if not possible, the whole policy is rejected as not valid.

While this has more functionality than previous examples, it places the burden of policy enforcement on the administrator. This approach should be compared to both Cisco and Intelliden, which are covered later in this chapter.

The Avaya Policy Manager works by first defining the network infrastructure that policies operate in. A network consists of one or more networks, subnets, hosts, host groups, and/or applications. An Application can be a specification of IP protocol and/or port parameters. Applications have one or more members, which represent specific servers where the application is located or subnets that the originating application traffic is restricted to.

Cisco devices and interfaces are handled differently than Avaya devices and interfaces. If device interfaces are added or removed, these changes must be input manually to the Avaya Policy Manager.

Cisco Example

Cisco makes a dizzying array of network and element management products to manage its devices. There are a total of 2 optical managers, 39 CiscoWorks applications, and 84 other product entries.[42] This large number of network management solutions is because of the complexity and number of different Cisco products, the wide diversity in functionality offered as a function of operating system (type and version) and hardware, and the different management foci of these products. Therefore, this section will discuss one of these tools—the QPM (QoS Policy Manager) add-on to the CiscoWorks network management software.

This tool was chosen as an example because it covers a very large number of different types of devices and QoS functions. For example, the following different types of QoS mechanisms are available (note that not all of these are available on every device—this is a function of the type of operating system software (e.g., IOS versus CatOS) that is being run, the version of the operating system, the available memory and type of processor, and other factors):

- Supported Scheduling Methods
- Supported Modular CLI (MQC) Features
- Supported Marking, Policing, and Shaping Methods
- Supported Link Efficiency Features
- Resource Reservation Protocol Support
- Supported Frame Relay Traffic Shaping Features
- Access Control Policy Support
- QoS Analysis (Monitoring) Capabilities
- Supported Queuing, Shaping, and Policing Features for Layer 3 Devices
- Supported QoS Features for Switches Running IOS

Note the use of the word "supported." This means that not all products support all of the features for this particular QoS feature. For example, there are a large number of different scheduling methods. Some of them are listed below:

- First In First Out Queuing
- Class-Based Queuing
- Custom Queuing
- Priority Queuing
- Fair Queuing (Flow-based Distributed WFQ)
- Weighted Fair Queuing
- Distributed WFQ and QoS group DWFQ

(There are other scheduling variations, but these are hardware dependent and only run on a few models, so are not listed in the set.)

QPM is built to integrate with CiscoWorks. This provides a measure of lifecycle management of policies that isn't present in other products examined so far. QPM 3.0 is made up of multiple tools: Analysis, Policy Configuration, Deployment Control, Device Management, Reporting and General Administration.

The simplest way to use QPM is to first produce a baseline profile of network traffic that you want to use policies to improve. This includes the different applications that are running on the network, and the traffic distribution of those applications. If some subset of those applications are identified as more important (e.g., more deserving of shared network resources when the network gets congested), then these are targets for QoS conditioning. In order to properly apply QoS, however, the administrator must understand the nature of the applications and how they are used. This will help the administrator understand the scope and nature of the policies.

Creating this baseline is an important part of any QPM usage. The monitoring data can be collected on a real-time, or on a periodic (historical) basis. In this way, a set of QoS policy configurations can be analyzed to determine if they are working as expected. If not, then the network must be further analyzed (e.g., the capacity of the network devices, in terms of memory available, processor speed, etc., along with the capacity of the network links (link speeds, overhead, and so on) that they apply to) to see where traffic conditioning can help. QPM has some ability to help in this analysis, but only for specific applications as opposed to general network design.

Once the traffic baseline is created, QoS policies can be created and deployed. Devices can be added to the system manually, using a CSV (comma separated value) file, or through Resource Manager Essentials, the inventory management component of CiscoWorks. QoS policies define the QoS actions that will be applied to specific data packets. These policies are managed within policy groups, which are applied to a specified set of network elements through deployment groups.

QPM can be used to configure QoS policies throughout the network, whether the device is an edge, aggregation, or core device. QPM can be used to manage the following QoS mechanisms:

- Packet Marking (ToS, DSCP, layer 2 CoS value, MPLS experimental CoS value, Frame Relay DE bit, and the trust state of a port)
- Traffic Policing for Limiting Bandwidth and Marking Traffic (transmit, drop, mark and transmit, or markdown); policing can be defined on a microflow, aggregate flow of the same interface, or aggregate flows across multiple interfaces using CAR, Modular QoS CLI, or a simple two token-bucket policing scheme
- Traffic Shaping for Controlling Bandwidth (traffic shaping attempts to smooth the traffic flow to meet your rate requirements by buffering the packets); shaping algorithms include generic, frame relay (which adaptively adjusts the buffers according to bursting parameters), distributed (among processors) or modular (which supports adaptive shaping for non Frame Relay applications)
- Queuing Techniques for Congestion Management for Outbound Traffic (these include class-based QoS queuing, where multiple actions can be applied to different traffic classes, fair queuing, weighted fair queuing, distributed weighted fair queuing (for special hardware interfaces), priority queuing, custom queuing (to ensure minimum bandwidth for one or more traffic flows), and several methods for managing switch congestion (weighted round robin, as well as queues with thresholds)
- Queuing Techniques for Congestion Avoidance on Outbound Traffic (these include RED and WRED)
- Management of Voice and Other Real-Time Traffic (these include Low Latency Queuing, a combination of class-based QoS and strict priority queuing), IP RTP (strict priority queuing for RTP traffic), link fragmentation and interleaving (to reduce delay and jitter on lower speed links), CRTP (compression of RTP header information to reduce delay), and Frame Relay Fragmentation (to prevent delay on Frame Relay links)
- Managing Traffic Through Access Control—access control policies permit or deny traffic that matches the filter definition in the specified direction
- Signaling Techniques (marking and RSVP)

These are explained in more detail in various web sites.[43-48]

These policies must then be monitored to ensure that their actions are producing the intended results on the network traffic. QoS monitoring is a combination of defining/executing policy tasks, collecting data from the IOS class-based QoS MIB or CAR MIB on Cisco routers, and viewing charts and graphs with traffic and QoS statistics. QPM allows the user to build real-time and historical monitoring tasks specific to devices, interfaces and QoS policies. Data collected is stored in a database and a file can be exported for additional analysis.

QPM does this through a web-based GUI. QPM's GUI takes care of translating policies defined using its GUI into the device's command line interface (CLI) commands. This is an important task, since the combination of hardware and operating system version determine the type of QoS features that are provided by a particular device.

The QPM Policy Configuration application is used to define end-to-end QoS policies for network devices. Groups of QoS policies can be defined for specific sets of devices, interface types, and interface properties, such as VLANs. Policies can be defined and then applied to existing device interfaces, or policies can be built before any interfaces are actually defined, and used to control the building of interfaces. Either is simplified using QPM's global libraries of policy building blocks (which includes groups of IP addresses and host names, as well as protocol and port definitions for various applications). Policy templates can be created in order to share common policies across different device groups. A nice feature of QPM is its ability to upload existing QoS configuration commands into QPM. QPM incorporates these commands into its own GUI for you to work on later.

QPM organizes policies into one or more deployment groups, so that the administrator can control when the policies are deployed, and to which devices. The policies themselves can either be deployed directly through QPM, or indirectly using a configuration file. QPM facilitates correct deployment of policies by letting the administrator examine the device log. It also saves the previous configuration, so that a rollback can be performed if an error occurs.

QPM differentiates between policy groups for voice applications and policy groups for other applications. QPM provides a set of voice templates that correspond to the guidelines given in Cisco Qos for IP Telephony.[46]

In conclusions, QPM provides a large number of detailed QoS mechanisms that span most of the popular routers and switches made by Cisco. This is actually a very large number of combinations, since each device can have hardware that provides QoS specific functionality, as well different versions of IOS (or CatOS) that provide different commands. QPM does a good job of building QoS configurations and installing them on devices. However, it is limited to "just" QoS configuration, which means that other tools must be used in addition to QPM to manage the entire functionality of the device.

11.6.3 Management Platform Example

There are a large number of Network Management Systems that are in use today. A brief description of Hewlett-Packard's PolicyXpert management system is provided here, as it is one of the few examples of a policy management tool plugging into a network management system (in this case, HP OpenView). HP has discontinued selling this product directly, but is actively supporting it through an OEM arrangement with Hitachi.[49]

The goals of the PolicyXpert product were to ensure that QoS for mission-critical traffic, as well as differentiated QoS for other traffic, could be configured and managed through policies. Unlike the policy products built by most network vendors, it was far-reaching in both services (e.g., it aimed to provide secure configuration of firewalls, multicasting, VPNs, and others) as well as for heterogeneous devices from different vendors (e.g., Cisco routers, Lucent routing switches, HP servers, Intel Network Interface Cards, and Packeteer traffic shapers). Through OpenView integration, PolicyXpert should also be able to provide information for accounting and billing support, and run on multiple platforms.

The approach was to define policies that were based on business requirements (such as SLAs), and distribute these policies into the environment managed by HP OpenView. Agents would be used to enforce policies by translating them into appropriate configuration commands. HP OpenView could also monitor the status of each policy.

PolicyXpert supports three main types of QoS:

- Prioritized Class of Services through different queuing, scheduling, and dropping algorithms
- Assured bandwidth through provisioning a committed access rate with acceptable bursting parameters
- Reserved bandwidth through RSVP signaling to dynamically reserve path-specific resources for QoS

PolicyXpert depends on use of the COPS protocol[50] to communicate its policies. If the target device does not support COPS, then a configuration proxy agent must be used to translate the COPS policies into a form that the device can understand, as is shown below:

PolicyXpert implements a variety of IETF standards. Control over DiffServ (classification, conditioning, queuing, and dropping algorithms) and RSVP are two of its main features.

HP also provides an SDK, so that external developers could customize the functionality of PolicyXpert to suit their own application-specific needs. Specifically, PolicyXpert uses a C++-based API to manage all aspects of communication between the policy configuration and/or proxy agent and the policy server. Classes are provided for all main features of PolicyXpert, and can be exercised using the appropriate class methods. The SDK even enables new proxy agents and new devices to be added. However, for this to be done, the developer needs to be familiar with how the device communicates and functions, so that it can be represented using extensions to the standard PolicyXpert classes.

HP has licensed some of the technology used in PolicyXpert from Intel's Communication Architecture Labs.

The PolicyXpert product is an interesting product. It covers the largest number of different types of products of any solutions reviewed so far. However, it is still focused mostly on QoS. It is an SDK, and so its real power can best be realized by a developer. Its one big limitation is that its only link to business rules is through SLAs. This lack of formal modeling of business rules makes it harder to use PolicyXpert to integrate the business and system worlds of an organization.

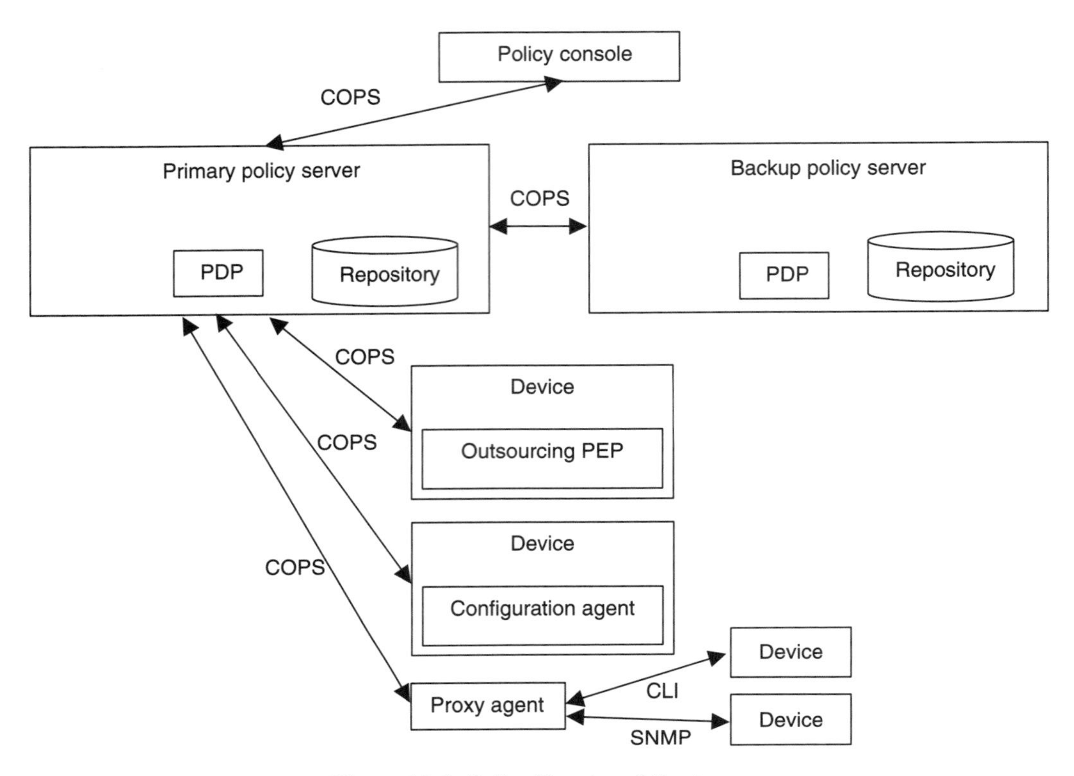

Figure 11-3 PolicyXpert architecture.

11.6.4 Configuration Management and Service Activation—Intelliden Example

Intelliden[51] has taken a new and different approach to policy management. Intelliden's approach is based on first providing a firm foundation of configuration management (via its 2.x products), and then building various types of service activation features on top of this foundation.

Configuration Management versus Provisioning

There is a world of difference between configuration management and provisioning. Provisioning includes all actions that are necessary to order, activate and manage a service. Since the focus is service activation, provisioning tools change parts of the configuration of a device to enable the activation of the desired service. These changes represent only a small portion of the total device configuration, and often ignore the other services currently being supported by the device. ***This is a key theme that applies to all of the previously surveyed projects.***

In contrast, configuration management systems are built to understand all of the functions, and hence the entire configuration, of a device. They track who changed a configuration when, where and why, and archive changes to each configuration to enable a previous working version to be installed if a problem is encountered. While there are many products that *claim* to be configuration management products, very few actually do. This is because they don't separate the process of constructing a configuration change from the process of deploying a configuration change. Intelliden uses policies to control each of these activities (and their sub-activities) and workflow-driven processes to implement each change.

This difference is fundamental and powerful. Provisioning systems first assume that the network is already working the way you want it to, and then attempt to activate services on top of the network. Provisioning systems that are focused on activating one type of service do not know about other deployed services—and they do not care, because it is not their job. In contrast, configuration management systems can detect and deploy the smallest of changes in the configuration of a device. These configuration changes can be specific to a single service, or applicable to a wide variety of services. This enables network engineers to know what services the device is supporting at any given time, and hence, the state of that device. Engineers can thus manage the network, not just isolated parts of it, and ensure that all supported services are managed efficiently and effectively.

More importantly, configuration management systems enable daily configuration tasks to be performed using the business processes of your organization. After all, no one *should* just telnet into a router and starts changing its configuration! If configuration management is implemented, then business policies can be written that formalize rules such as these, so that the organization can have better control over who makes configuration changes when and how. This is important, because there are defined processes that govern how a configuration file is built and deployed, even for the smallest changes. Configuration management systems ensure that the applicable set of business processes, on a per-business-unit basis, is enforced for any given change. Good configuration management systems provide audited records of who changed what when, where and why.

The Intelliden R-Series is unique in that it controls the construction of a configuration change as well as manages the process of deploying that change. This involves invoking the appropriate approval process, followed by scheduling and deploying it, for each configuration task. The Intelliden R-Series then validates that the new configuration installed on the device is correct, making the configuration process "bullet-proof" and eliminating costly configuration errors.

Configuration management has two other important benefits. Services can be protected through engineers archiving configurations that correspond to a working service. They can then revert to those configurations if subsequent changes misfire. And since the Intelliden R-Series knows exactly what is being used on a device, it can tell you if valuable functionality is used or not.

Motivation for Configuration Management

There are three big issues in front of us today concerning network devices and network management. The *lack of a consistent product model*, despite all the standards that abound, results in different vendors implementing traffic conditioning using different algorithms that have different side-effects. Worse, in our current world of Mergers, Acquisitions and Divestitures, completely different devices that happen to have the same logo and vendor name on them are bought that act in completely different manners. Thus, *there is no predictable behavior when you mix different products from the same vendor, let alone different devices from different vendors*.

Arguably worst of all, *there is no standard for sharing and reusing data*. This is important for two reasons. First, without such a standard, the different components of a system can't talk to each other and exchange and/or reuse management information. Second, there's no way to enable different constituencies, like the business community and the network community, to tell each other what they need. **This in effect divorces the network from how the rest of the business is run.**

Consider a network that has some 802.1p switches feeding a set of routers. Some of the routers are DiffServ-compliant, while others aren't. We now have an immediate set of design choices to face, because these different technologies (802.1p, ToS, and DiffServ) each have their own definition of QoS. There is no RFC that defines a mapping between these technologies (note that the DiffServ RFCs do define a *recommended* mapping to ToS values, but this is just a recommendation and doesn't map to 802.1p at all). From the system and implementation points-of-view, it's even worse, because DiffServ doesn't define HOW to implement traffic conditioning—it simply specifies that these 6 bits in an IP packet header mean that it should be treated in a certain manner relative to other markings.

Clearly, if we add non-IP technologies, such as Frame Relay and ATM, the problem is exacerbated, as there is no RFC (or other standards-based documents) that tries to map each technology's QoS definitions (e.g., UBR, CBR, ABR, etc. of ATM and CIR of Frame Relay) to each other or to IP technologies.

Until DEN and DEN-ng, there was, in fact, no mapping. DEN-ng is a layered object-oriented information model, which enables different management users to utilize different concepts in thinking about, defining, programming, and managing the system. It defines abstract concepts, such as traffic conditioning, and then provides a framework in which different vendor implementations can be plugged in. This enables different functionality from different vendors to interoperate.

DEN-ng is more than "just" a standard representation of data—it also defines semantics and behavior of, and interaction between, managed entities. It does this in an extensible way through strict adherence to the UML metamodel. It is organized as a federated, layered model—a framework of frameworks—so that different models, standards, and extensions can all be plugged into a common framework. By representing information in a common framework, new data can be added in an interoperable fashion, since it is based on existing concepts that management applications know about.

Architectural Approach

Figure 11-4 illustrates how most services are provisioned and activated in today's networks. This figure shows two different *CustomerFacingServices*—an IPsec VPN and an MPLS VPN, and two different *ResourceFacingServices*—

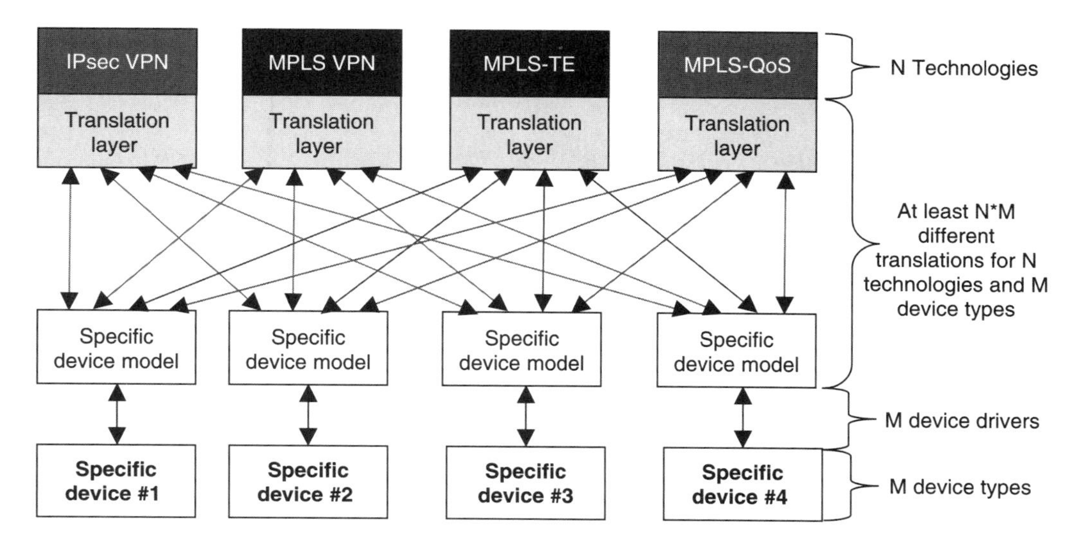

Figure 11-4 Common service-activation architecture.

MPLS Traffic Engineering and QoS, implemented in an MPLS networks. Each of the *CustomerFacingServices* requires a translation layer to integrate its application-specific needs into a common framework. For N technologies and M devices, this approach results in *at least* N*M translations (note that it could be more, since functionality is dependent on hardware and operating system version, as well as other factors). This is why vendors often build element management systems to provision a small number of services for a small number of device types.

Unfortunately, this makes it very difficult to manage multiple services, because now entire element management systems must know and understand multiple services, and how they relate to each other. Alternatively, the organization is locked into the devices and tools that each vendor decides to write.

In contrast, Intelliden has built a layered model, conceptualized as follows:

Each of the different CustomerFacingServices and ResourceFacingServices can plug into the DEN-ng information model. This is possible because the purpose of CustomerFacingServices and ResourceFacingServices is to abstract the service into a set of concepts that can be implemented in a technology-neutral fashion. Once this has been accomplished, an XSD (XML Schema Definition) is built for each combination of vendor, type of device, model of device, and operating system version (the VTMOS that appears in the third layer in Figure 11-5). Thus, DEN-ng serves as a common technology-neutral normalization layer, which is extended by the {vendor, type of device, model of device, operating system version} four-tuple technology-specific XSD. This four-tuple works because each of the four parameters in the XSD helps determine the end functionality that is being modeled.

Device configurations are built from stored knowledge. Unlike other products, the user can't apply an incorrect command to a device because the product only presents legal options for a specific device command. This

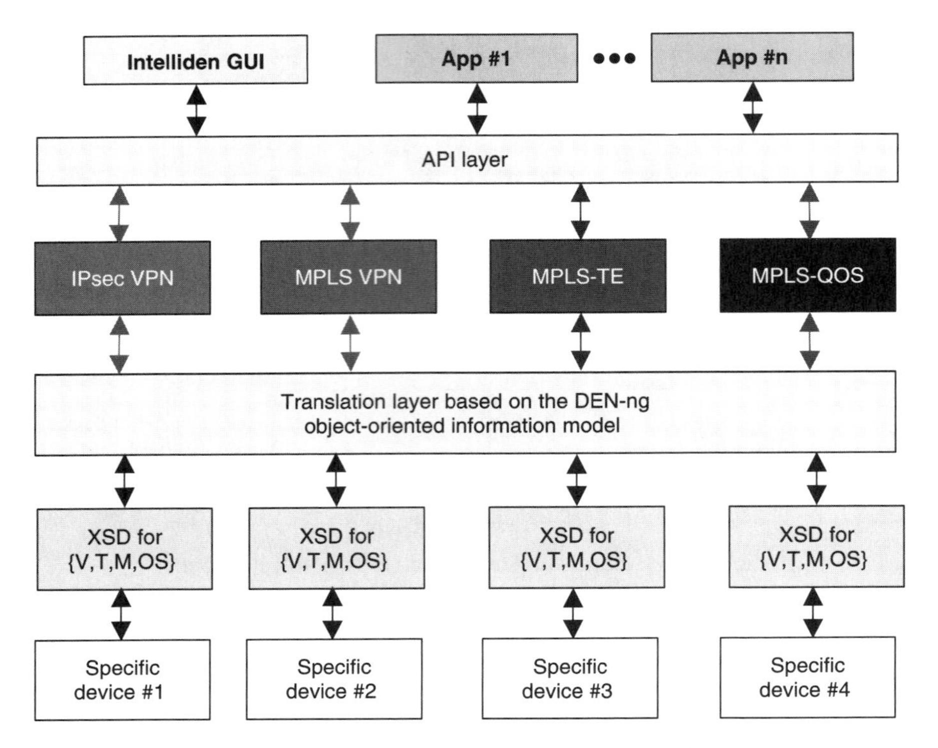

Figure 11-5 Intelliden's new service-oriented architecture.

low-level characterization of device functionality enables different commands from different devices to be mapped to each other, since models of each command can be built that represent how that device behaves, and what resources it requires.

Thus, vendor-specific features have been abstracted, so that the developer can concentrate on functionality and let Intelliden do the required translation to a particular {V,T,M,OS}. This modeling represents an enormous amount of work. Intelliden has built automated software that constructs these models, both to minimize errors and to increase scalability. The real benefit, however, is the reduction on upper layers of the management system—they are now presented a single interface, instead of having to deal with the intricacies of multiple different programming models.

The key point is that DEN-ng itself is being used to model device functionality. This enables both business rules as well as system and implementation concepts to be represented using the same information model. This helps ensure that business rules can be used to drive device configuration.

Business Driven Device Management

Business Driven Device Management (BDDM) is defined as using business rules to manage the construction and deployment of commands and configuration files for a device or set of devices. In this way, BDDM enables you to operate your network as another profit center of your organization instead of an alien entity that people are scared to touch.

BDDM addresses the two fundamental principles of device configuration: individuality and process. Individuality means that different types of configuration changes require different processes. For example, there is a huge difference between changing the SMTP server address of a device vs. changing how routes are distributed on that same device. The former is simple and straightforward, whereas the latter is very complex, requiring both more skilled personnel as well as several different organizations to be involved in its approval process.

BDDM views the network as a provider of intelligent services. BDDM is predicated on using holistic solutions such as PBNM, because services require a higher-level view of the network than what can be provided by telnetting to a device and typing CLI! Furthermore, businesses don't operate or sell interfaces! Businesses operate and manage services according to the priority and contractual obligations that the business enters into. This mandates intelligent processes that can manage the rich functionality of your network, and ensure that changes to your network devices follow approved processes. Thus, there is a desire to link PBNM solutions to what an organization sells as well as to use PBNM solutions to drive the configuration of the network.

The Intelliden R-Series was built to enforce the different business rules that govern how each type of configuration change is constructed, approved, installed, and verified. This enables the different groups within an organization to use their own processes to reflect their own individual needs for constructing and deploying configuration changes. If one tries to mandate a "one process fits all" approach to configuring devices, the different users of the network will not use the system, because it won't reflect their management needs.

A unique strength of the Intelliden R-Series is that it enables different business rules to be defined that manage the different configuration processes needed as a function of who is implementing the change and the type of change that is being managed. This enables a single system to manage the different configuration processes of a business, and helps turn a network into an extension of how a business operates using BDDM.

The R-Series uses an advanced form of role-based access control to determine which users can perform what commands and functions against which devices. The R-Series assigns privileges to groups. All users must belong to at least one group, though a user can belong to more than one group. The set of privileges that a user receives is a function of the roles that the user is currently playing. Similarly, devices are grouped into domains (called "realms") and permissions are assigned to realms. The permissions are determined by the set of roles that a device assumes at any given time. Thus, the functionality that a given user has is the intersection of the group roles that the user has with the device roles that the user is trying to operate on.

This reflects the real world of network management, where different parts of the network are administered by different administrators. In most other products, an administrator (if the product supports roles) can do anything to any device. In the R-Series, an administrator may have different sets of functionality that apply to different realms.

Business-Driven Service Activation

The next release (3.0) of the Intelliden R-Series will start to deliver business-driven services. These are services whose characteristics and behavior are driven by business demands. They rely on the robust business-driven configuration management functionality that is part of the shipping 2.x version of the R-Series. The models of the functionality of each {vendor, type of device, model of device, operating system version} four-tuple enable Intelliden to understand what can be configured, and how, for each managed device. In this way, a service can be directly related to the configuration of a device. *The philosophy of the R-Series is that managing the device configuration is the first step in managing the services that the device performs*. This is shown conceptually in Figure 11-6.

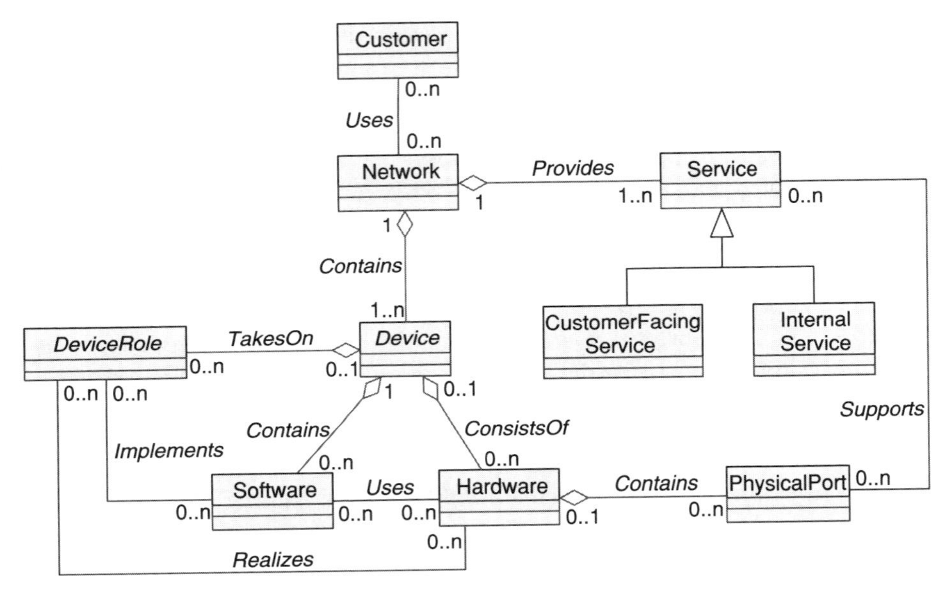

Figure 11-6 Relating software to devices and services.

Remember that DEN-ng also relates other business entities, such as a Product, to both Resources and Services. Therefore, this model enables changes in Products and other business entities can be related directly to configuration commands. For example, if a Customer is upgraded from Silver Service to Gold Service, this usually means that the Customer will be able to run new Services, or that existing Services will have (in this case) better performance. Thus, the configurations of the devices that are carrying the Customer's traffic must be changed. Intelliden is able to relate changes in business entities to changes in configuration files, something that other products cannot do.

Intelliden has, since its first release, provided both an innovative GUI as well as an API. The latter is important, since it enables developers to embed Intelliden functionality in their own branded products. However, there is more to an API than just a set of function calls. In its 3.0 release, Intelliden is making its functionality accessible through web services.

A web service is a piece of self-contained, self-describing, business logic, located somewhere on the Internet, that is accessible through one or more standard-based Internet protocols. It can be published, located and invoked across the web, using standard-based protocols such as HTTP.

Web services are easy to invoke. Simply search for a service in which you are interested (using, for example, UDDI—the Universal Description, Discovery and Integration protocol, which is a "meta-service" for locating web services,[52] retrieves the description (for example, using WSDL—the Web Service Definition Language, which is an XML format for describing network services[52] for the service), and generates service proxy code from the WSDL description. Intelliden 3.0 will provide a web services interface for both its configuration management functionality as well as service activation.

Use of the Policy Continuum

Another distinguishing feature of the Intelliden example is its use of the Policy Continuum. This enables the business needs of an organization to be expressed in a form that will simplify the resulting system design and implementation. Currently, the policy continuum exists as internal processes that drive how Intelliden presents its features to the developer. Future versions of Intelliden will make this a more generic interface.

Strengths of the Intelliden Product

The Intelliden product is one of the new wave of PBNM solutions. It contains several important examples of functionality that are not available in most other current products. This is because of several key architectural points that differentiate Intelliden from other products:

- The use of a standards-based information model (DEN-ng) to represent
  - Business entities, like Customers and Products
  - System entities, such as networks and applications
  - Runtime entities, such as the Intelliden implementation
  - The ability to relate all three of the above entities to each other, enabling changes in one to affect the others
- The expression and coordination of these entities as part of the Policy Continuum
- The realization that fundamentally, configuration changes control service activation, and that different services are related to each other and/or to shared resources of a device or network
- The use of policy management and process management mechanisms to ensure that each configuration change has its own appropriate set of construction and deployment policies and processes
- The use of role-based access control to dynamically determine what actions a particular user can perform on a given set of devices

One final point—services are controlled by device configurations. A device configuration is not just a set of commands—it is a collection of commands that control a set of services. Thus, it is of paramount importance to understand the *semantics* of the services that a particular device is supporting, so that changes in one service can be checked to ensure that they don't adversely affect other deployed services. This is the fundamental weak link of almost all current provisioning products. However, with the combination of policy management and DEN-ng, a solution is at hand.

11.7 Recommended Further Reading and References

The following is a set of references for this chapter.

1. Please see: http://fr.itpapers.zdnet.com/search/searchNew.asp
2. Conover, J., *Policy-Based Network Management*, Network Computing, November 29, 1999.
3. Boardman, B., Saperia, J., *Are You a Control Freak?* Network Computing, pp 35-38, January 21, 2002.
4. Boardman, B., Saperia, J., *No Standards, No Policy, No Management*, Network Computing, pp 51-53, January 21, 2002.
5. Please see the following URL in the IoP Electronic Journals: *The Distributed Systems Engineering Journal*, for articles on policy-based management: http://www.iop.org/EJ/S/UNREG/Q..oNsTxrOEfV7J.9jG1GA/search/0967-1846/1
6. Young, M., *Policy-Based Network Management: Finally?* Business Communications Review, pp 48-51, August 2002.
7. *Journal of Network and Systems Management*, Special Issue on Policy-Based Management of Networks and Services, to be published September 2003.
8. Verma, D., *Simplifying Network Administration using Policy based Management*, IEEE Network Magazine, March 2002.
9. Flegkas P., Trimintzios P., Pavlou G., *A Policy-Based Quality of Service Management System for IP DiffServ Networks*, IEEE Network, special issue on Policy-Based Networking, Vol. 16, No. 2, pp. 50-56, IEEE, March/April 2002.
10. The Workshop on Policies for Distributed Systems and Networks will have its fourth meeting in June 4-6 of 2003. Information on these conferences is: Policy 2003: http://www.labs.agilent.com/policy2003/; Policy 2002: http://www.policy-workshop.org/2002/; Policy 2001: http://www-dse.doc.ic.ac.uk/events/policy-2001/; Policy 1999: http://www-dse.doc.ic.ac.uk/events/policy-99/
11. The IP Policing Conference had a special focus and emphasis on policy. Please see: http://www.upperside.fr/ippol2001/ippol2001pro.htm; http://www.upperside.fr/ ippol2000/ippol.htm
12. The NetCon conference, or Network Control and Engineering for QoS, Security and Mobility, with Focus on Policy-Based Networking, can be found at: http://net-con.utt.fr/netcon/arret_marc_7_octobre_2002/index.html
13. Strassner, J., Wheeler, J., *Policy, QoS, and DEN—Tutorial T248*, Network + Interop.

14. ACM has a series of conferences on Role-Based Access Control. Please see: *http://www.list.gmu.edu/confrnc/rbac/rbac99.html*; *http://www.list.gmu.edu/confrnc/rbac/rbac98.html*; *http://www.list.gmu.edu/confrnc/rbac/rbac97.html*; the RBAC conferences have been merged into SACMAT (the Symposium on Access Control Models and Methodologies after the 1999 RBAC conference. Please see: http://www1.acm.org/turing/sigs/sigsac/sacmat/
15. Strassner, J., *A New Paradigm for Network Management: Business Driven Network Management*, SSGRR summer conference, L'Aquila, Italy, July 2002.
16. Gassman, B., *Consider the Costs of Policy-Based network Management*, Research Note, 5 June 2002.
17. Kosiur, D., *Policy-Based Network Management: What Is It and Who Needs It*? The Burton Group, 16 Nov 1999.
18. Kosiur, D., *Policy-Based Network Management: Architectures and Issues?* The Burton Group, 25 Feb 2000.
19. Kosiur, D., *The Future of Policy-Based Network Management on the Internet*, The Burton Group 18 May 2000.
20. Please see: for general information: http://www.doc.ic.ac.uk/; for the Department of Computing Science: http://www-dse.doc.ic.ac.uk/; for the policy management page: http://www-dse.doc.ic.ac.uk/Research/policies/index.shtml
21. Please see the following home page, which has links to each member of the DSE group: http://www-dse.doc.ic.ac.uk/Research/policies/group.shtml
22. Please see: http://www-dse.doc.ic.ac.uk/Research/policies/publications.shtml
23. Please see: http://www-dse.doc.ic.ac.uk/Research/policies/projects.shtml
24. The home page of the IETF's Policy Framework working group is: http://www.ietf.org/html.charters/policy-charter.html
25. The home page of the DMTF's Policy working group is a members-only home page. Please see: http://www.dmtf.org/download/about/workgroups/sla-WGCharter.pdf
26. *Shared Information/Data (SID) Model—Addendum 1POL—Common Business Entity Definitions—Policy*, GB922, Version 1.0, TeleManagement Forum, July 2003.
27. *NGOSS Architecture Technology Neutral Specification*, TMF053 Version 3.0, TeleManagement Forum, April 2003.
28. *NGOSS Architecture Technology Neutral Specification (Behavior and Control Specification)*, TMF053C Version 1.0, TeleManagement Forum, February 2003.
29. *NGOSS Architecture Technology Neutral Specification (Policy Management Specification)*, TMF053P Version 1.0, TeleManagement Forum, work in progress.
30. A web site was built to describe the Fine Grain NGOSS project in more detail. Please see: www.finegrain.org
31. *NGOSS for Enabling Fine Grain Services Phase II—Interface Implementation Specification*, TMF 842, v 1.0, June 24, 2002.
32. *TeleManagement World 'NGOSS Powered' Catalyst Evaluation Process*, TeleManagement Forum, TR126, v 0.5, July 2001.

33. The Jini specification is available from: http://wwws.sun.com/software/jini/; in addition, the following URL is for a book on the second edition of the Jini specification: http://java.sun.com/docs/books/jini/spec/
34. Please see: http://java.sun.com/j2ee/
35. *Fanning the NGOSS Flame at TMW*, RHK News Alert, November 12, 2002.
36. This is the home page of the TeleManagement World show. Please see: http://www.telemanagementworld.com
37. Passmore, D., *The Big vs. Managed Bandwidth Debate*, Business Communications Review, pp 20-26, May 1998.
38. Please see: http://www.allot.com
39. Please see: http://www.packeteer.com
40. Please see: http://www.nortelnetworks.com/products/family/optivity.html
41. Moore, B. (Ed.), *Policy Core Information Model Extensions*, <draft-ietf-policy-pcim-ext-08.txt>, May 2002.
42. The following three URLs shows all network management products made by Cisco as of December 1, 2002: for optical platforms: http://www.cisco.com/en/US/products/sw/opticsw/index.html; for CiscoWorks: http://www.cisco.com/en/US/products/sw/cscowork/index.html; for all others: http://www.cisco.com/en/US/products/sw/netmgtsw/index.html
43. General links to Cisco QoS resources can be found at: http://www.cisco.com/warp/customer/732/Tech/qos/
44. A good overview of Cisco QoS, with links to white papers on all major QoS technologies, can be found at: http://www.cisco.com/warp/customer/732/net_enabled/qos.html
45. A detailed overview of Cisco QoS capabilities can be found at: http://www.cisco.com/univercd/cc/td/doc/cisintwk/ito_doc/qos.htm
46. A Design Guide for Cisco QoS for IP Telephony can be found at: http://www.cisco.com/univercd/cc/td/doc/product/voice/ip_tele/avvidqos/index.htm
47. General information for Cisco QoS for VoIP can be found at: http://www.cisco.com/univercd/cc/td/doc/cisintwk/intsolns/qossol/qosvoip.htm
48. A Cisco QoS configuration guide for IOS 12.2 can be found at: http://www.cisco.com/univercd/cc/td/doc/product/software/ios122/122cgcr/fqos_c/index.htm
49. For more information on HP's PolicyXpert, please see: http://www.openview.hp.com/products/policyexpert/index.asp
50. COPS is defined in the Resource Allocation Protocol working group of the IETF. Please see: http://www.ietf.org/html.charters/rap-charter.html
51. Please see: http://www.intelliden.com
52. Please see the home page for the organization that oversees all UDDI work, which is: http://www.uddi.org/
53. Please see the home page for the organization that oversees all WSDL work, which is: http://www.w3.org/TR/wsdl

Appendix 1

Guide to UML

This Appendix will provide a brief guide to the most important terminology and symbology of UML. This is not meant to be a replacement for any of the excellent UML references provided—it is simply a quick, convenient reference of the most basic and important UML terminology and representation used in this book. Readers that are not conversant with UML are strongly urged to refer to one of the UML references provided.

A1.1 Introduction

The simplest answer to the question "What is UML" is the following:

"The UML is the standard language for specifying, visualizing, constructing, and documenting all the artifacts of a software system."

UML has many types of diagrams. However, in this book, class diagrams are the only ones used. Class diagrams are used to show how different classes are derived and relate to each other.

The objective of this Appendix is to briefly define the semantics used to build a class diagram.

Classes

A class is a collection of objects, each of which having a common structure. This structure is defined by having a common set of attributes, methods, relationships, and semantics (e.g., constraints as expressed using OCL). This in turn means that classes have the same behavior when viewed by other classes.

Classes are represented in the UML as a rectangle with three compartments, as shown in Figure A1-1 below:

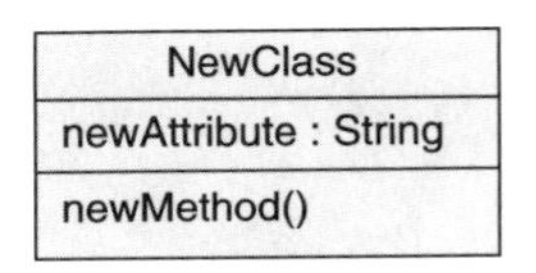

Figure A1-1 The UML representation of a class.

The top compartment is used to define the name of the class; the middle is for listing its attributes; the bottom is for listing any methods. In the middle compartment, we see that the attribute named new Attribute is assigned the datatype of String. Note, however, that UML editors (e.g., tools that are used to draw UML diagrams) usually have the ability to suppress various features of the drawing. In this book, for example, most of the UML diagrams have their attributes and methods hidden to simplify their presentation.

A class represents a concept or an entity within the system being modeled. Classes have structure (their attributes), behavior (their methods) and can interact with other classes (through relationships to those other classes and the application of constraints). Relationships and constraints are talked about in the next two sections.

Relationships

There are two fundamental types of relationships in UML. The first is inheritance. Inheritance is a mechanism by which more specific elements incorporate structure and behavior of more general elements. In other words, a subclass inherits all of the attributes, methods, relationships, and constraints defined by its superclass(es) and then *extends* these characteristics and behavior by adding its own new attributes, methods, relationships, and/or constraints. This is also called *refinement*.

Single inheritance means that a subclass only inherits from a single superclass. Multiple inheritance means that a subclass can inherit from multiple superclasses. This is most often done to endow the subclass with different parent characteristics and behavior. Note that the UML metamodel uses multiple inheritance quite frequently. On the other hand, DEN-ng does *not* use multiple inheritance. This choice was made so that implementations would be easier to build, since not all languages support multiple inheritance.

The process of building one or more subclasses from a common superclass (or set of superclasses) is called *specialization*. The process of gathering common attributes, methods, relationships, and/or constraints from a set of classes and building a common superclass that contains these common elements is called *generalization*.

Single and multiple inheritance are shown in Figure A1-2.

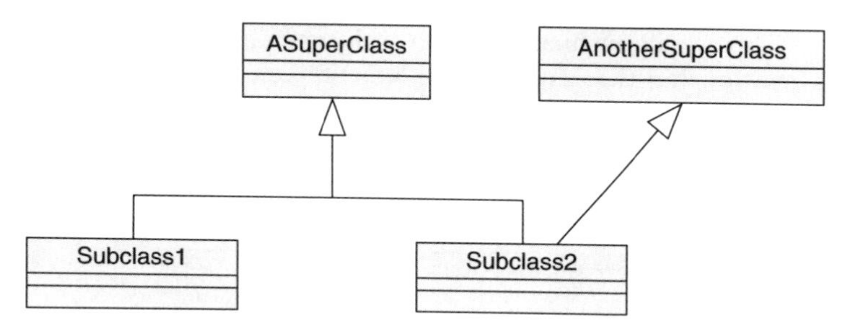

Figure A1-2 Single and multiple inheritance.

In this Figure, Subclass1 is a subclass of the class ASuperClass. Subclass2 is a subclass of both ASuperClass and AnotherSuperClass. Subclass1 uses single inheritance, whereas Subclass2 uses multiple inheritance.

The other three types of UML relationships—Association, Aggregation, and Dependency—all represent a form of connection or communication among or between classes. Specifically:

- An Association is a bi-directional connection between classes. This type of relationship can represent a variety of semantics, such as "object A uses object B," "object A has an object B," and so forth.
- An Aggregation is a stronger form of an association. An aggregation is used to define a relationship between an aggregate element (called the "whole") and its constituent components (called the "parts"), which is why it is also referred to as a "whole-part" relationship. An aggregation signifies that there is a strong coupling between the aggregate object class and its constituent component object classes.
- A Composition is an even stronger form of an aggregation. A composition is used to denote that the composite (or "whole") component owns its constituent components. This means that actions that happen to the composite element are propagated to each of the composite's constituent components.

Before these relationships are drawn, multiplicity needs to be discussed.

Multiplicity

Multiplicity defines how many objects participate in a relationship. It is the number of instances of one class related to one or more instances of the other class. For each association and aggregation, there are two multiplicity decisions to make: one for each end of the relationship.

For example, suppose that the relationship between employees and their employer is being represented. From the point-of-view of the employee, each employee *isEmployedBy* their employer. However, from the point-of-view of the employer, each employer *CanEmploy* an employee.

Associations are defaulted to be bi-directional. If they are uni-directional, then an arrowhead is used to show this restriction. While UML allows the possibility to define higher-order relationships, there aren't any in DEN-ng, so this will not be covered in this Appendix. (This was a conscious decision in DEN-ng, in order to simplify implementation).

This is a convenient point to show the difference between a relationship and a relationship that must be implemented as a class. For example, suppose that an employee can work for multiple employers, and that each employer quite naturally can hire one or more employees. This can be represented in Figure A1-3.

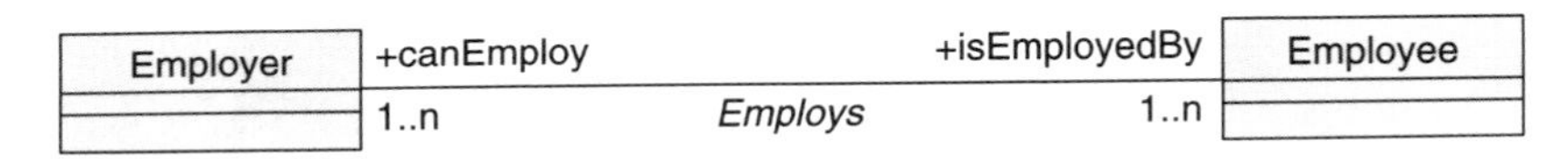

Figure A1-3 A simple model of employer-employee relationships.

Here, the two roles *canEmploy* and *isEmployedBy* are shown (though they are not shown in any of the DEN-ng diagrams) to reinforce the fact that this is a bi-directional association and has slightly different semantics depending on which end of the association it is being read from. Each string "1..n" underneath each role is the individual cardinality of that role. The string "Employs" is the name of the association. This association is read:

An Employer canEmploy one or more Employees. Each Employee isEmployedBy one or more Employers.

Suppose that it is desired to model a salary attribute. Where does it go? It can't go in the Employer class, because that would mean that each Employee would get the same salary. Similarly, it cannot go in the Employee class because different Employers would want to pay different Employees different salaries. A better solution is to make the salary attribute a property of the *Employs* association, as shown in Figure A1-4.

The advantage of Figure A1-4 is that now salary can be determined as a function of the particular *employs* association that is being defined.

Multiplicity is also a very useful tool to help capture and implement the business rules of an organization. For example, a business rule might state that a manager should govern between 3 and 15 employees. This business rule can be expressed via constraints, and when the system is implemented, once can check to ensure that a given manager has between 3 and 15 employees.

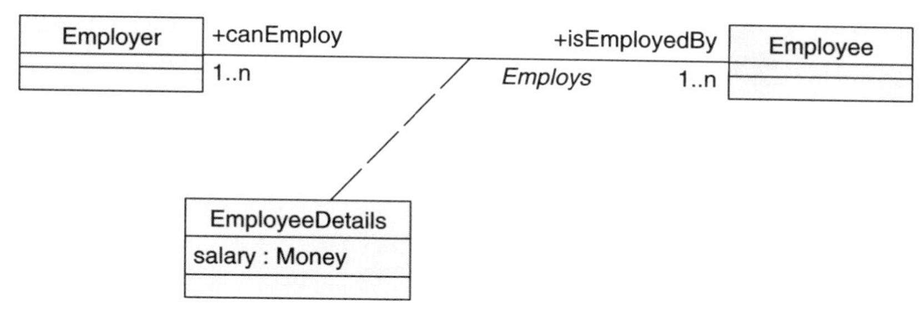

Figure A1-4 A better model of employer-employee relationships.

Class Diagram

A class diagram represents relationships between different classes. Class Diagrams are used to show the existence of classes and their relationships in a static, logical view of a system. It is depicted as a graph of classes that are connected by their various static relationships.

In a class diagram, the power of object modeling lies more in the relationships connecting classes than in the definition of the classes and relationships themselves.

Comments on Non-Essential Policy Terminology from RFC3198

This appendix will contain commentary on important terms from RFC3198 that are for advanced practitioners only.

A2.1 Essential Terminology from RFC3198

This Appendix will summarize the non-essential terms from RFC3198 that are relevant for the advanced practitioner. Only those terms that are relevant to generic PBM system design will be summarized here. Note that RFC3198 is focused on policy-based network management, and in particular is strongly influenced by current IETF work in the areas of differentiated services [Diffserv], policy representation [Policy] and security [IPSP]. As such, it is not generic to PBM, but rather is specific to policy-based network management.

Some of these definitions will be changed, based on implementation experience of the author. In other cases, the definitions of RFC3198 will be modified to make these terms more generic to PBM systems. This is done when a single definition is overloaded with conflicting meanings resulting from different uses that are forced into a single definition. In every case, the RFC definition will first be analyzed to explain why it needs modification, and then a new definition will be provided.

These non-essential definitions from RFC3198 now follow in alphabetical order, except when terms are dependent on each other. Each definition is given in its own section, and comments (where appropriate) are provided immediately below each definition. The new definition, if needed, follows in *italics* and is indented.

Common Information Model (CIM)

RFC3198 defines the CIM as follows: An object-oriented information model published by the DMTF. It consists of a Specification detailing the abstract modeling constructs and principles of the Information Model, and a textual language definition to represent the Model. CIM's schemas are defined as a set of files, written in the language of the Specification, with graphical renderings using UML [UML].

It is this last sentence that is incorrect—CIM is *not* UML compliant. Furthermore, UML is more than "just" a graphical rendering.

Policy Error

RFC3198 defines a policy error as: "Policy errors occur when attempts to enforce policy actions fail, whether due to temporary state or permanent mismatch between the policy actions and the device enforcement capabilities."

This definition needs a small amount of fine-tuning. This is because the definition mentions the concept of *capabilities* without mentioning *which capabilities are applicable at a given moment in time*. This is due both to the lack of capabilities in the policy model of the IETF (and the DMTF), but also because a formal management paradigm has not been identified.

A policy error can be identified by comparing the state of a *ManagedEntity* after a *PolicyAction* has been executed to the expected state of the *ManagedEntity*. If one or more attributes don't match up, or if the state of another *ManagedEntity* isn't correct, then we can assume that a policy error exists.

The definition of a policy error is refined as follows:

A policy error is the result of failing to successfully execute one or more policy actions to either maintain the state of a set of managed objects, or transition the state of a set of managed objects to a new state.

Thus, the easiest way to determine if a policy error exists is by checking the current state of all *ManagedEntities* to see if each of them are in the correct state after a *PolicyAction* has executed.

In DEN-ng, a policy error can occur when either the event clause, the condition clause, and/or the action clause of a policy rule is either malformed or cannot be evaluated. A policy error can also occur if the semantics defining a policy rule (e.g., how it relates to other policy rules in a policy group) are malformed.

The mismatch between the actions of a policy rule and the capabilities of the device is more fundamental that what RFC3198 says. This is a mismatch of the functionality of the policy rule to the capabilities of the device, and thus the policy rule should not be applied to the device in the first place.

Policy Goal

RFC3198 defines a policy goal as: ". . . the business objectives or desired state intended to be maintained by a policy system. As the highest level of abstraction of policy, these goals are most directly described in business rather than technical terms."

This definition has two problems. In the first sentence, the emphasis is made on *business* aspects of a policy. While the business aspects of policy are very important, and arguably drive other aspects of policy, it is incorrect to define a policy goal as always having a business objective. Sometimes, policies are activated to ensure an orderly and consistent state transition, and have nothing to do with an underlying business objective.

The second problem is with the next sentence, where policy "goals" are asserted to be "the highest abstraction of policy". A "goal" is defined in Webster's unabridged dictionary as "the final purpose or aim; the end to which a design tends, or which a person aims to reach or attain." By this definition, a goal can not be the "highest level of abstraction", as the goal presumes a purpose, and that purpose can itself be represented by a policy. Business policies are great examples of pure *rules* that are independent of achieving any particular purpose – rather, they simply embody how an organization (for example) is supposed to operate.

Thus, the following is a better working definition of a policy goal:

"A policy goal represents a purpose that can be achieved using a set of policy rules."

Note that this is often driven by a business objective, but doesn't have to be.

Policy Negotiation

RFC3198 defines policy negotiation as: "Exposing the desired or appropriate part of a policy to another domain. This is necessary to support partial interconnection between domains, which are operating with different sets of policies."

This definition is wrong. While one reason to negotiate policies is certainly to arrive at an agreed course of action between managed objects in multiple policy domains, multiple policies can certainly be negotiated within the same policy domain. For example, two administrators in the same domain could have conflicting goals (represented by conflicting actions). Furthermore, the essential factor of a negotiation (the modification of one or more of the objects being negotiated) isn't captured by the above definition. Hence, we will use the following definition:

A policy negotiation results in the modification of one or more of the Policies that are being negotiated by one or more of the entities performing the negotiation. This may result in the modification of the metadata, data, event clause, condition clause, and/or action clause of one or more of the Policies.

Policy Request

RFC3198 defines a policy request as: "A message requesting a policy-related service. This may refer to a request to retrieve a specific set of policy rules, to determine the actions to enforce, or other policy requests. When sent by a PEP to a PDP, it is more accurately qualified as a "policy decision request" [RFC2753]."

Again, we have the overloading of different meanings that serve different purposes. For this book, it is important to distinguish between requesting a *decision* and requesting other information. The solution, of course, is to specifically qualify the term *policy decision request*, as follows:

A policy decision request is the explicit demand for guidance in how to respond to an event that triggers the evaluation of the PolicyCondition clause of a PolicyRule. A policy decision defines what set of actions, if any, should be executed in response to the request.

This is differentiated from a *policy information request*, as follows:

A policy information request is the explicit request for generic policy information that does not take the form of a policy decision request. That is, a policy information request does not involve evaluating the PolicyCondition clause of a PolicyRule. A policy information request enables different policy components, such as metadata, PolicyConditions, or PolicyActions, to be retrieved.

Policy Translation (also called Policy Mapping)

RFC3198 defines policy translation as: "The transformation of a policy from a representation and/or level of abstraction, to another representation or level of abstraction."

This definition just needs the trivial adding of "and/or" to the second half of the first sentence, yielding:

A policy translation (or mapping) is the transformation of a policy from one representation and/or level of abstraction to another representation and/or level of abstraction.

Note that in PBM, a mapping of policies is one part of performing model mapping.

Index

C

D